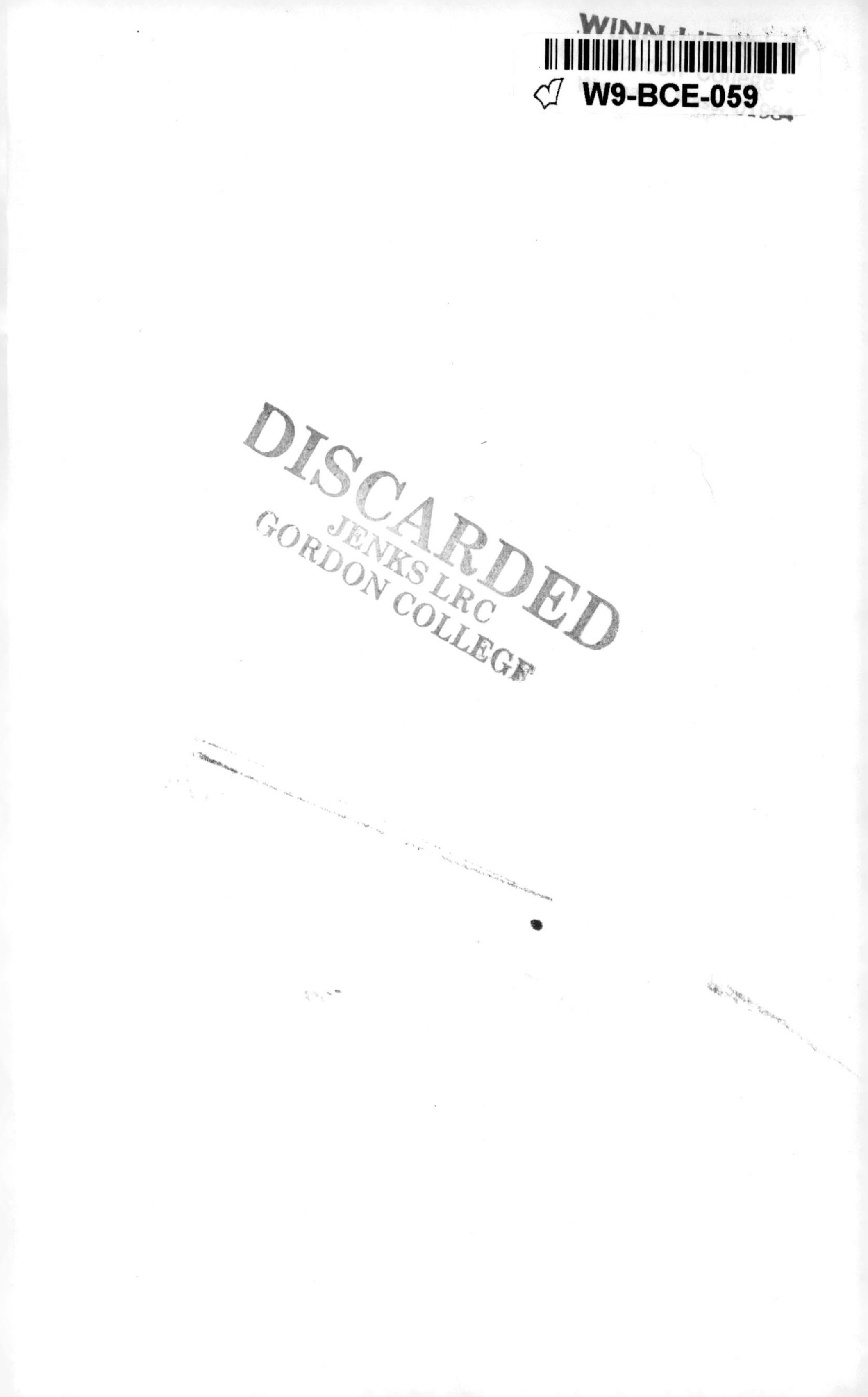

Energy and the
Developing Nations

Pergamon Titles of Related Interest

Bach et al RENEWABLE ENERGY PROSPECTS
Goodman/Love BIOMASS ENERGY PROJECTS
Goodman/Love GEOTHERMAL ENERGY PROJECTS
Goodman/Love SMALL HYDROELECTRIC PROJECTS FOR RURAL DEVELOPMENT
King et al BIORESOURCES FOR DEVELOPMENT
Mossaver-Rahmani ENERGY POLICY FOR IRAN
Stewart TRANSITIONAL ENERGY POLICY 1980-2030

Related Journals*

CONSERVATION & RECYCLING
ENERGY
ENERGY CONVERSION & MANAGEMENT
OZONE: Science and Engineering
SOLAR ENERGY
SUNWORLD
WATER SUPPLY & MANAGEMENT
WORLD DEVELOPMENT

*Free specimen copies available upon request.

Energy and the Developing Nations

Proceedings of an Electric Power Research Institute Workshop, Hoover Institution, Stanford University, March 18-20, 1980

Edited by
Peter Auer

Published in Cooperation with
the Electric Power Research Institute

Pergamon Press

NEW YORK • OXFORD • TORONTO • SYDNEY • PARIS • FRANKFURT

Pergamon Press Offices:

U.S.A.	Pergamon Press Inc., Maxwell House, Fairview Park, Elmsford, New York 10523, U.S.A.
U.K.	Pergamon Press Ltd., Headington Hill Hall, Oxford OX3 0BW, England
CANADA	Pergamon Press Canada Ltd., Suite 104, 150 Consumers Road, Willowdale, Ontario M2J 1P9, Canada
AUSTRALIA	Pergamon Press (Aust.) Pty. Ltd., P.O. Box 544, Potts Point, NSW 2011, Australia
FRANCE	Pergamon Press SARL, 24 rue des Ecoles, 75240 Paris, Cedex 05, France
FEDERAL REPUBLIC OF GERMANY	Pergamon Press GmbH, Hammerweg 6, Postfach 1305, 6242 Kronberg/Taunus, Federal Republic of Germany

Library of Congress Cataloging in Publication Data

Electric Power Research Institute Workshop on Energy
and the Developing Nations. Hoover Institution, 1980.
Energy and the developing nations.

(Pergamon policy studies on energy)
Includes index.
1. Underdeveloped areas--Power resources--Congresses.
I. Auer, Peter. II. Title. III. Series.
TJ163.2.E36 333.79'09172'4 80-29586
ISBN 0-08-027527-3

Printed in the United States of America

Contents

Acknowledgments

The successful planning and execution of this workshop is indebted to a number of individuals at EPRI and Stanford University. Foremost among these were Alan Manne, who helped conceive the format of the workshop and provided continuing advice during its organization, Susan Missner, who acted as the principal liaison between EPRI and Stanford, and Sally Nishiyama, who managed most of the administrative tasks. Further thanks are due to Shahid Iqbal, who recorded and transcribed the workshop discussions, and to Martha Coulton for her help in the preparation of these proceedings.

Preface

The perception of the energy problem may differ as one samples various segments of opinion in the United States. There is good reason for this, for the situation is complex and has many facets. One of these, the international aspect of the energy problem, tends to receive relatively little attention from the general public, yet is of major concern.

Many will grant that oil is somehow at the bottom of our present day problem; and some may even grant that OPEC's control of oil and our lack of control over OPEC add up to the fact that there is a problem today. Even at this level of perception, however, there is a tendency to view the problem in a narrow sense—it's us versus OPEC, those foreign oil-producing countries.

There are many other players, though, in this affair. Oil is big business in world trade. Each day nearly one billion dollars flow into OPEC accounts, at today's oil prices. And this is repeated 365 days of the years. A prodigious transfer of money and wealth on this scale could hardly be sustained for any length of time unless counterbalancing mechanisms existed for returning most of this money to the importing nations. If and when these mechanisms fail, some or all of the oil-importing countries face economic ruin.

The United States is a dominant factor in these transactions, for we have been purchasing about one-fourth of OPEC's exports. Our allies and trading partners among the industrialized nations have been purchasing most, but not all, of the remainder. Some one hundred nations, representing nearly three-fourths of the world's population, are counted among the developing nations or less developed countries (LDCs). Collectively, those LDCs that import oil have been purchasing about one-sixth of OPEC's total exports.

A slightly longish view of events may not be unreasonable, even when considering something as chaotic as the world oil market. The developing nations have demonstrated a remarkable record, for well over a decade, of significantly higher economic growth rates than the industrialized nations. Along with economic growth has come increased industrialization and urbanization and an appetite for commercial forms of energy—for the present—mainly in the form of oil. One need not extrapolate these trends too far to conclude that within this century the LDCs could surpass the United States in the amount of oil they will have to import.

One also need not view a finite world with finite resources as a ground for competition, but one often does. As long as that is the binding perception, world instability and tension will rule, and the future can't be very promising. While many elements may contribute to world turmoil, the availability of energy at affordable prices will surely play a principal role for some time to come. What role the developing nations may have in these future events is something to which we need to pay greater attention. Just as our policies affect them, so their needs, their economic state and political developments will affect us.

This workshop, then, was organized in order to learn more about the energy problems of the developing nations. The designation *developing nations* is, of course, no more than a convenience. The divergences that exist among them are as great as, if not greater than, the divergences between them and the industrialized nations. They are tied together by the simple fact that if one draws a dividing line on the economic ladder, the LDCs occupy the lower rungs. Other than that, it would be difficult to draw many meaningful generalizations.

It seems remarkable, nevertheless, that in an aggregate sense the LDCs weathered the precipitous 1974–75 rise of oil prices in far better shape than the industrialized nations. A dominant question during the workshop was, Can this pattern be continued? Some degree of optimism could be found in the views expressed. Mexico's future appears bright, but there are a few other countries whose natural endowment may propel them into the oil-exporting club. A larger proportion, perhaps, may find that they can become nearly self-sufficient in energy if an intensive search for oil and gas in their country meets with some degree of success. In many developing nations, particularly in Latin America, a very large potential for hydropower development remains. There was considerable sentiment for supporting an accelerated search for conventional energy forms in the developing nations.

That is not to say that alternative forms of energy don't have near-term prospects among the developing nations. The very large scale experiment in Brazil for converting agricultural products to alcohol fuel was held up as something each nation should consider carefully to see whether or not it may be feasible to emulate. Unfortunately, not many nations appear to be as rich in land as Brazil. Other forms of solar energy—renewable energy systems—may find a role in supplying the developing countries with their energy needs. But the somewhat romantic notion that systems found to be uneconomical in the industrialized regions of the north should be shipped to the developing nations of the south met with little enthusiasm from the experts. We may eventually learn that small is indeed beautiful, but for the time being this idea is not only not selling well in Peoria, it's not selling well in China either. Customer acceptance and other factors of market penetration are as important in the rural villages as in the cosmopolitan city centers.

Much more has to be known about the underlying factors of energy demand and its relation to development planning. Progress appears to be being made in this direction. Data is spotty, particularly among the less industrialized nations, but the need for it is becoming widely recognized. Whether price and income elasticities should be the principal means of demand forecasting in developing nations is still an open question, but there was agreement that much work remains to be done in this area.

The near-term outlook for world energy supplies, basically oil, does not appear very promising. The immediate post-Iranian revolution era seems to have left most observers with rather pessimistic views. OPEC's oil output is more likely to shrink than expand, and there may be more violence and disruptions in the Middle East. Energy prices will continue to rise relative to other goods. How, then, will the LDCs continue on their ambitious course of development?

While there was general agreement that the road ahead would be even more difficult than what has been traversed, there seemed to be little agreement on what would be the wisest policy to pursue for the future. Should there be some preferential terms of trade in favor of LDCs, dual pricing of oil perhaps? or should one aim at free trade and avoid protectionism? Differing opinions were heard.

Eventually, we in the United States and our partners in the Organization for Economic Cooperation and Development (OECD) must join together with the many developing nations to solve what is obviously a problem of mutual interest. Sufficient energy for us and sufficient energy for the rest of the world are but two sides of the same coin. There is much that we in the United States can do to help others learn more about energy planning and thereby help ourselves. The final session of the workshop was devoted to this theme and its messages should be given close attention.

Our workshop profited greatly from the presence of many knowledgeable people who were gracious enough to share their wisdom with us. One can only hope that these proceedings will enable the reader to share some of this experience.

Peter Auer

Chapter 1
Energy and Development: Crisis and Transition

Lincoln Gordon

The central theses of this overview are that the developing countries constitute an increasingly important component of the global energy scene, that their problems of energy management overlap those of the industrial world but are even more difficult, and that there is a shared interest in mitigating the short-term crises now confronting them and in securing their full participation in the longer-term transition in energy regimes on which the world is now embarked.

ENERGY AND DEVELOPMENT

Before summarizing today's developing-country energy situation in quantitative terms, it may be useful to reflect briefly on the relationship between energy use and economic development. That relationship is close, but more complex than is often assumed. From the viewpoint of intellectual history, it poses a paradox. Scholars of human social evolution and of technological change virtually equate the progress of mankind with the use of animal and mechanical power to replace hard human labor. The Prometheus myth and anthropological evidence alike find in the controlled use of fire one of the great forward steps in human evolution. Yet the classical economic theories

This paper is based mainly on work done at Resources for the Future, in collaboration with Joy Dunkerley, William Ramsay, and Elizabeth Cecelski, on a project on energy and developing countries financed by the Ford Foundation. The author wishes to express appreciation for this collaboration and support, but takes full responsibility for the presentation here.

of growth and development, associated with such names as Adam Smith, David Ricardo, John Stuart Mill, and Karl Marx, paid almost no attention to energy. For them, the key factors of production were land, labor, and capital, with natural resources somewhat fuzzily assimilated into the concept of land.

This paradox probably results from two causes. On the intellectual plane, it was only during the mid-nineteenth century that there developed a rigorous conceptual understanding of energy. The laws of thermodynamics and electromagnetism were not discovered until well after the basic formulations of classical economics. On the practical plane, energy—in contrast to land, labor, and capital—remained unchanged until very recently. Periodic flurries of concern about fuel shortages were always relieved by improved technology or by the introduction of new fuels, with smooth transitions from wood to coal to oil and gas as the principal sources of commercial energy.

Both historical and cross-sectional intercountry studies have indicated a close long-term correlation between energy consumption and economic growth as measured by national accounts. The ratio of energy growth to growth in gross domestic product (GDP) in the now-industrial countries was apparently higher than unity during the period of intensive heavy industry development, and then fell to unity or below as economies matured, with the well-known shift toward services and more complex manufactured products with high added value in transformation.

The cross-sectional comparative studies appeared to confirm the general relationship, but were complicated by major ambiguities in definition of both numerator and denominator. The widely used figures on per capita GDP, based on national accounts data with currencies converted at current exchange rates and regularly published in the *World Bank Atlas*, overstate the disparities in economic levels by a wide margin. The studies of comparative purchasing power led by Irving Kravis suggest that the average per capita GDP ratio between developing and industrial countries is only about 1:6.4 instead of 1:13, i.e., the relative position of the developing countries is only half as bad as suggested by the *World Bank Atlas*. On the numerator side, the major problem in defining energy consumption is the very heavy reliance of developing countries on traditional fuels—wood and charcoal, dried animal dung, and agricultural wastes—most of which do not pass through organized markets. Until the last several years, few countries even attempted to measure these forms of energy use, but recent sample studies suggest that the traditional fuels account for as much as 35 to 40 percent of total energy consumption in developing countries. They are used mainly for cooking or heating in open fires, with only a small fraction of the contained energy actually delivered in a useful form; it is therefore likely that the overall average fuel conversion efficiency is substantially lower in developing than industrial countries.

Making allowance for all these corrections, a general comparison of energy consumption and energy intensities in the industrial and developing countries is presented in table 1.1. It will be noted that levels of per capita use in the industrial countries are almost nine times those in the developing countries, while the intensity of energy use (measured by useful energy consumption per million dollars of real GDP) is about twice. There is, of course, enormous diversity among individual countries within these two groups, especially in the developing country group. At one end of the spectrum are countries with very limited economic activity beyond subsistence agriculture and livestock tending; at the other end are countries such as Korea, Brazil, and Mexico, with such large and dynamic industrial sectors that they are close to graduation into the industrial category. In between, large amounts of energy consumption are accounted for by relatively poor but populous countries, such as China, India, and Pakistan, where a substantial modern industrial sector is embedded in a primarily agricultural and low-productivity economy.

Table 1.1. Energy Consumption and Gross Domestic Product, 1973: Industrial Countries and Developing Countries

	Unit	Industrial Countries	Developing Countries	Ratio of Industrial to Developing
1. Per capita total energy consumption (including rough estimates for traditional fuels and counting hydropower at equivalent primary fossil fuel rates)	Metric tons of oil equivalent (T.O.E.)	4.22	0.48	8.8
2. Real GDP per capita (Kravis calculations)	U.S.$	3,343	520	6.4
3. Energy consumption per million dollars of real GDP	T.O.E.	1,263	916	1.4
4. Assumed average fuel conversion efficiency	percent	50	35	1.4
5. Useful energy consumption per million dollars real GDP	T.O.E.	631	321	2.0
6. Population	millions	1,119	2,717	0.4

Source: Energy consumption data based on J. Parikh, "Energy and Development," World Bank PUN 43 (1977). GDP data based on I. Kravis et al., "Real GDP Per Capita for More than 100 Countries," *Economic Journal* 88 (June 1978).

Energy Costs as a Constraint on Growth

Much has been written about industrial country growth in this century being based on cheap energy, with the implication that higher energy costs would have entailed much lower growth rates. Cheap energy is of course a magnet for certain highly energy-intensive industries, but the overall share of energy costs in economic output, at least until very recently, has been too small for them to constitute a major engine of growth. The oil price shock of 1973 raised primary energy costs in the United States from 2.4 to 5.2 percent of gross national product (GNP). A rough macroeconomic calculation shows that, had this higher level prevailed over the 25 years from 1948 to 1973, average growth rates would have been reduced from 3.6 to 3.1 percent a year, and the overall quarter-century growth reduced from 140 percent to 115 percent of the starting level. During that economic golden age, European growth rates were consistently higher than American despite unit energy costs double the U.S. levels.

Nevertheless, energy costs and availabilities are potential constraints on growth, and in larger measure than might be suggested merely by shares of GNP. It is always useful to bear in mind the three special characteristics of energy that distinguish it from other raw materials: (a) its pervasiveness, which makes it a part of all production of goods and services as well as an important item of final consumption; (b) its inability to be recycled (a direct consequence of the second law of thermodynamics); and (c) its low elasticity of substitution taken collectively (as distinct from individual fuels). For most developing countries, which still face rapid population growth for several decades ahead, which seek high growth rates in incomes per capita, and whose structural transformations imply a rise in energy intensities, it is little wonder that energy costs and security of supply have become issues of cardinal concern.

A standard analysis of the effect of doubling energy costs on industrial country growth rates takes as an outside measure the share in GNP, assuming that the cost increase is gradual and that energy-related problems are well managed. Thus, if energy now accounts for 5 percent of GNP, the extra resources needed to pay a doubled cost—whether additional exports to finance higher priced oil imports or diversion of resources to finance higher cost domestic supplies—would amount at most to another 5 percent. With "normal" economic growth at 3 or 4 percent a year, that would simply mean a 15- to 20-month delay in achieving any particular level of living; if spread over 10 years, it would imply annual growth rates of 3.1 instead of 3.5 percent. Conservation and substitution effects would, in practice, reduce the impact below this outside limit.

In the case of developing countries, the impact of an energy cost doubling may be much more severe. The critical constraints on their overall growth are likely to be either the levels of domestic savings and capital formation or the capacity to import capital goods and complementary raw materials and components not producible domestically. With average income levels already low, resource diversion is more likely to come out of savings than consumption; savings typically run between 15 and 20 percent of GNP, so that five percentage points would mean reducing investment by one-quarter or one-third. On the balance-of-payments side, it simply cannot be assumed that exports are automatically expansible, so that doubled oil import bills imply either extra borrowing abroad or a compression of nonenergy imports—again by much more than 5 percent. On the other hand, it can be argued that developing countries can shift more readily to lower energy intensities because they do not have so large a capital stock committed to energy-intensive technologies. They may also have greater flexibility in substituting labor for commercial energy.

The Developing Countries and the World Energy Economy

With three-quarters of the world's population, the 127 developing countries, including China, account for about 37 percent of the world's real economic output (22 percent on the nominal *World Bank Atlas* basis). They are believed to hold over two-thirds of the total probable resources of oil and gas, although only one-fifth of the coal. They dominate the supply side of the world's oil trade: Out of a total production of 60 million barrels per day, 34 million are exported, of which 80 percent come from OPEC member countries, 6 percent from Mexico and other non-OPEC developing countries, and only 14 percent from industrial countries.

On the demand side, developing countries now consume over 20 percent of total commercial energy as well as the bulk of the traditional fuels. Figure 1.1 shows graphically how rapidly their commercial energy consumption has increased over the last quarter century; it also indicates the increasing share of liquid fuels in the total. During that period, commercial energy consumption was growing at 7.9 percent a year in the developing countries and 3.8 percent in the industrial countries. On the conservative assumptions that global energy growth will slow down but that developing country consumption will remain at least 2½ percentage points higher than that of the industrial countries—a minimal assumption if there is to be any significant improvement in developing country living standards—the developing countries will account for 25 percent of the total by 1990 and 30 percent by the end of the century.

Those shares are sufficiently large to have a major impact on market conditions for commercial fuels. Traditional fuels used in developing countries will also impinge on the international commercial energy markets, since at the margin, kerosene and bottled gas are often direct substitutes for

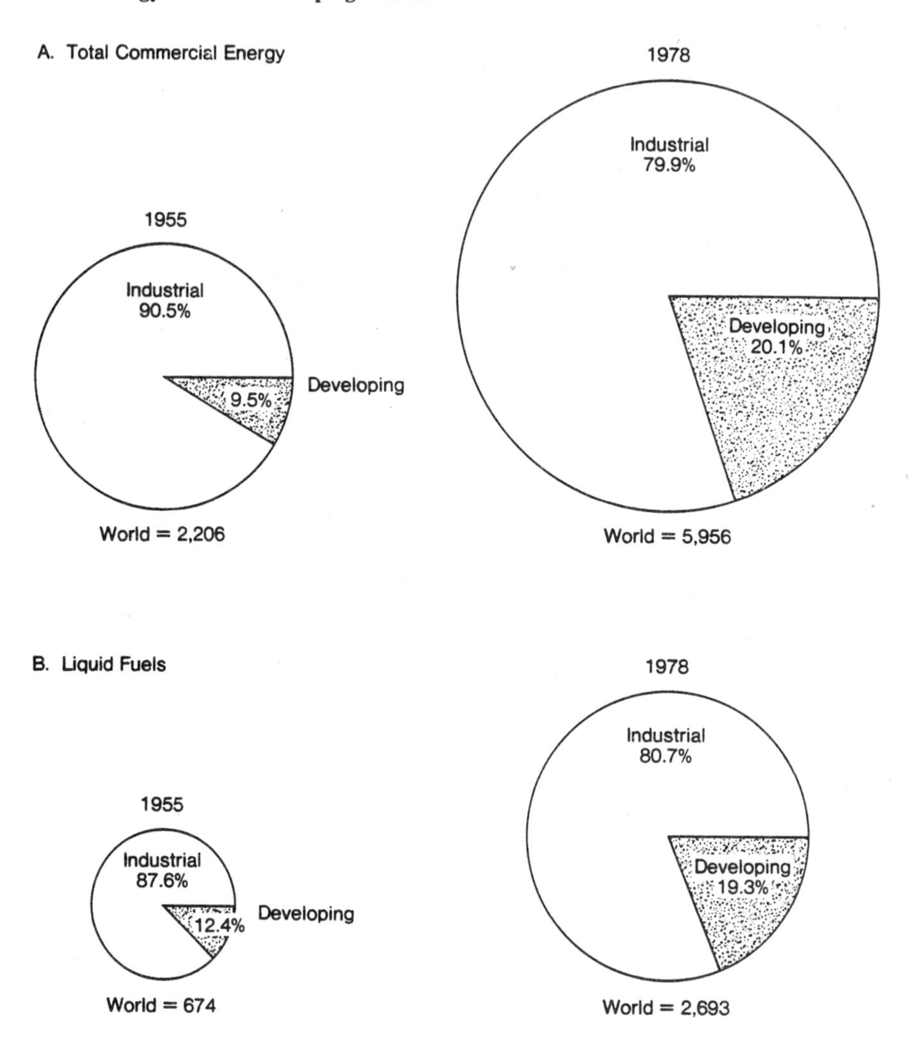

Fig. 1.1. World Consumption of Commercial Energy and of Liquid Fuels, 1955 and 1978 (millions of metric tons of oil equivalent)

Source: Data from United Nations, *World Energy Supplies, 1950–74* and *1973–78*, Series J, Nos. 19 and 22 (New York, 1975 and 1979).

fuelwood, charcoal, vegetable wastes, and biogas. It follows that effective international energy strategies for the coming decades cannot be formulated or implemented without taking the developing countries into account.

CRISIS AND TRANSITION

From the perspective of oil-importing developing countries, the world energy situation combines the elements of short-term crisis and long-term transition. In the short term, the problem is how to cope with sharply higher oil import bills through some combination of adjustments in other imports and exports, oil conservation and substitution, and international private or public financing of the difference—all while minimizing the adverse impact on continuing development. For the longer term, the problem is how to accomplish an orderly transition to an altered regime of energy supply and use involving higher relative costs, different resource demands, and possibly significant modifications in development strategies.

The consequences of mismanagement in either the short- or long-term aspects may be calamitous for economic growth prospects and for social and political stability. Nor can the long-term problems be deferred until the short-term problems are fully resolved. The lead times inherent in basic alterations in the energy regime impose high penalties on failing to get started as rapidly as possible. Comprehensive planning of the energy sector, therefore, has now become an imperative for developing-country governments—not as an isolated exercise, but as an integral part of their general economic management. For most of them, like much of the industrial world, this is uncharted territory. Where problems and potential solutions are shared with industrial countries, they can secure guidance from the industrial world. But many of the issues are site-specific or modified by the physical and human resource endowment of the particular countries. For these, organized international cooperation may be of assistance, but the central responsibility must rest with national authorities.

Short-Term Adjustment

Such expressions as "crisis" and "catastrophe" were widely used in 1974 to describe the situation of the oil-importing developing countries on the heels of the quadrupling of world oil prices during the previous year. There were gloomy forecasts of complete stagnation in development, widespread defaults on debt service, and shrinkage of markets for industrial country exports. In fact, as brought out in table 1.2, the developing countries weathered the storm better than the industrial countries, with only a small reduction in economic growth rates and continued expansion in energy consumption. Rather than becoming a drag on the world economy, their continued economic robustness helped to limit the depth and duration of the recession in the industrial countries.

This unexpectedly smooth adjustment was not uniform among developing countries. In general, the poorer oil importers suffered more than the newly

Table 1.2. Economic Growth and Energy Consumption, 1973–78

	Economic Growth Rates (% per year)		Commercial Energy Consumption Growth (% per year)		Change in Energy Intensity (% per year)	
	1970–73	1973–78	1970–73	1973–78	1970–73	1973–78
World	5.6	3.4	4.8	2.6	−0.8	−0.8
All industrial countries[a]	5.4	2.9	4.0	1.6	−1.3	−1.3
All developing countries[b]	6.7	5.3	8.4	6.4	1.7	1.1

[a] Eastern and Western Europe (excluding Greece, Spain, Portugal, Yugoslavia), Australia, Canada, the United States, Japan, and New Zealand.
[b] All others (including oil producers and the People's Republic of China).
Sources: World Economic and Social Indicators, World Bank, October 1979. *World Energy Supplies*, United Nations, Series J, 1973–1978.

industrialized group. For the oil importers as a whole, the main keys to adjustment were the rapid increase in OPEC imports, which quickly reduced the petrodollar surpluses over the period 1974–78, and the recycling of residual financial surpluses through the commercial banking system, especially its Eurocurrency accounts, which became a major source of credit for developing countries. The volume of international lending was so large that it fully covered the current account deficits of the oil-importing developing countries and in addition made possible a very large buildup of their foreign exchange reserves—from $29 billion to $69 billion in nominal value. Until 1978, moreover, despite the huge increase in levels of international indebtedness, the ratio of their debt service payments to current export earnings had not significantly increased.

Against the record of the late 1970s, is it correct to speak of a new crisis for the oil-importing developing countries in the early 1980s, or can one expect another cycle of smooth accommodation to the 1979–80 doubling of world oil prices? One well-informed school of thought today holds the more optimistic position, arguing that the scale of needed adjustment is relatively smaller than in 1974 and that the petrodollar surpluses are bound to find some channel to the deficit countries which once again will finance their continuing imports of oil and other requisites for development.

This author, however, leans to a more pessimistic view, noting four differences from the past decade:

1. Starting from the plateau of high import levels, the OPEC-surplus countries are unlikely to increase them further in a way offering developing

countries new export markets and new emigrant worker jobs as a means of paying currently for part of their higher oil bills.

2. The supply-demand balance for oil is much tighter than in the mid-1970s, and the OPEC countries seem intent to keep it tight through unilateral production restrictions involving some form of tacit coordination, even if not formal production quotas; they are thus unlikely to permit the real value of oil to be eroded by general price inflation as it was in the 1970s.
3. The burden of indebtedness is therefore likely to increase sharply—both because the annual petrodollar surplus gap will remain large and because very high interest rates raise the cost of debt servicing even when the level of debt is unchanged.
4. Although it is axiomatic that petrodollar surpluses must be recycled in some form or other, there is no automatic mechanism to apportion those OPEC capital exports to the deficit countries most in need. The commercial banking system performed most of that function in the 1970s, but the larger banks now appear to be seriously worried about the creditworthiness of many of their developing country borrowers, and they are also under pressure from monetary authorities to slow down the expansion of international liquidity.

It is also axiomatic that a balance-of-payments deficit cannot exist unless it is financed—either by drawing down reserves or by a combination of public and private borrowing from abroad. The potential crisis of the early 1980s is not only of multiple defaults or near defaults, emergency debt rescheduling, and tightening of credit lines; it also consists of frustrated deficits—of imports of essential capital goods and raw materials required for economic development which must now be sacrificed or replaced by higher cost and less efficient substitutes. An overt financial crisis, involving a chain reaction of banking failures, can probably be warded off through cooperation among central banks and international financial institutions. Whether international cooperation can also be mustered to ward off a hidden development crisis is now an issue of high urgency.

Over and beyond the added resource and financing problems involved in the surging price of oil, many developing countries face a second category of acute problems in the pressures on sources of traditional fuels—deforestation of fuelwood reserves and diversion of animal and vegetable wastes from their fertilizing and soil-conditioning roles into cooking fuels. If sold at world prices, kerosene is no longer a viable substitute for these fuels for low-income consumers. Measures are needed to improve the efficiency of their production and use or to find other substitutes that can be widely adopted in rural areas. In this field, the kinds of energy policies applied by industrial countries to their own needs are unlikely to be of much help to developing countries. This is another area which appears to call for specially organized international cooperation.

Longer-Term Transition

The longer-term transition must evidently aim at an energy regime less dependent on imported oil and, for most countries, less dependent on oil and gas regardless of source. Since the long-run sources are certain to be more costly than oil in the 1960s, although not necessarily more costly than marginal supplies of imported oil in 1980, the ultimate energy regime should seek to economize energy—to reduce the energy input needed to achieve a desired level and pattern of goods and services and to secure that energy in forms which are most efficient in terms of overall resource use—a criterion overlapping but not identical to technical energy efficiency.

The transition is enormously complicated by the fact that the ultimate destination—the energy supply mix of the distant future—is not yet clearly defined. It is like a voyage of Columbus—aimed generally at the Indies but traversing uncharted seas and perhaps discovering in midcourse the West Indies instead of the East Indies. Transitional strategies must therefore incorporate a large component of research and development on new energy sources; they should keep options open as long as possible, and they should maintain maximum resilience to respond to unexpected opportunities or challenges. All these elements of flexibility involve some costs. In addition, the urgency of the crisis requires forward movement on replacement of imported oil and on improvements in energy efficiency, accepting some risks of second-best solutions. A crucial aspect of the crisis, in fact, results from the absence of coherent policies for assuring a gradual and nondisruptive longer-term transition.

These characteristics of the energy transition apply as much to industrial as to developing countries. The developing countries must also deal with a number of special problems of transition management. But it is important to avoid faddish prescriptions or generalizations about the Third World as if it were a homogeneous grouping. In particular, industrial country spokesmen should avoid the temptation to advocate low technology and renewable energy sources as universal solutions to developing-country energy problems. Such preachments are understandably interpreted as advice to abandon the goals of economic modernization or as disguised efforts to keep the world's dwindling hydrocarbon supplies as a preserve of the already industrialized countries.

In general, energy policy should be subordinated to broader development strategies. Since, however, energy costs and availabilities are now a major potential constraint on development, comprehensive energy planning should be an integral part of development policy-making. Energy policy is likely to be especially relevant to strategies of balanced agricultural and industrial development, of economic decentralization, of integrated rural development, and of interregional balance.

There are no easy panaceas for the energy dilemmas of the developing countries, either on the side of supply or in opportunities for conservation. Solutions will be composites of a wide variety of supply and demand measures. Reduction of oil imports is one central objective, but it would be an error to pursue energy self-sufficiency at all costs. In some cases, its achievement would be at the expense of even higher energy costs and added pressures on the balance of payments for imported food and raw materials.

On the supply side, it seems clear that fossil fuels and conventional hydropower will continue to be the main sources of commercial energy for at least the next two decades. That fact, together with the potential economic viability of resources previously unable to compete with cheap oil, gives a high priority to developing country efforts to discover and develop these kinds of resources. But tropical conditions may also afford special opportunities for exploiting direct and indirect forms of solar energy. The nuclear option is relevant only to a relatively small number of highly industrialized developing countries, but that group accounts for a substantial fraction of total developing-country consumption.

On the side of conservation, their generally low levels of per capita energy consumption and their limited use of energy for space heating make it impossible for the developing countries to secure conservation gains in magnitudes comparable with those achieved by the industrial countries. There are, however, important opportunities for more efficient use of both traditional and commercial fuels—in cooking, transportation, and industrial applications.

Convergence of Interest or Conflict

To analyze the elements of convergence and conflict of interest in the energy sector among and within the various groups of countries would require a separate paper at least as long as this one. It can be persuasively argued, however, that in the energy relationship among exporters, industrial-country importers, and developing-country importers, the elements of convergence outweigh those of conflict. They do so because of special features of the global energy market and because of shared interests in a smooth and violence-free transition to a future energy regime.

This shared concern with energy is more tangible and immediate than the general interest of richer countries in economic progress and the reduction of poverty in the developing countries—an interest which can be defended on a combination of broad economic, security, humanitarian, and moral grounds. In the case of energy, all forms and sources of supply, demand, and conservation ultimately affect the balance of the international trade in oil, and oil has a unique impact on the system of international finance, the pressures of inflation, and the resources available for investment and development. A mismanaged energy transition is a potential bottleneck—even a source of potential strangulation—for continued economic growth in all parts of the world. But many aspects of oil production and use are not effectively governed by competitive market forces, and the speed and shape of the transition to a new energy regime will be enormously influenced by government policies.

A smooth transition is obviously in the interests of both industrial and developing-country oil importers. It should also be a major concern to exporters, whose own markets and sources of supply require at least a minimum of world stability and who must consider their own ultimate transition to economic diversification and substitute energy sources. The alternative of competitive scrambles, bilateral special deals, and the threat or use of military force to secure supplies is extremely hazardous, even to those who suppose that they might secure a temporary gain.

Whatever the merits of its specific proposals, therefore, the recent call of the Brandt Commission for an internationally negotiated and agreed upon energy strategy deserves a positive response. Elements of that strategy of special interest to oil-importing developing countries include short-term assistance in financing oil imports, the possibility of price concessions and minimum supply guarantees for at least the poorer countries in the group, collaboration in fossil fuel discovery and development, cooperative applied research in site-specific technologies, and accelerated transfer of proven new technologies for competitive energy production and more efficient use. They would also be partners in any broader arrangements affecting medium-term prices, production levels, and stockpiling.

The urgent need for international cooperation in the energy field should take priority over other global issues and receive separate treatment. For reasons outlined in this presentation, it requires the full representation of developing countries on both the energy-exporting and the energy-importing sides.

Chapter 2
Oil and Gas Prospects in Non-OPEC Less Developed Countries

Everett M. Ehrlich

Oil production from non-OPEC less developed countries (LDCs) outside the Middle East is projected to reach 9.3 million barrels per day in 1985 and 12.1 million barrels per day in 1990. The bulk of this production will come from several producers, notably Mexico, followed by Egypt, Australia, Argentina, India, and Malaysia. Gas production will increase significantly in Thailand, Malaysia, Australia, and Pakistan, and liquefied natural gas (LNG) deliveries from LDCs may total 2.7 billion cubic feet per day by the mid-1980s.

THE SIGNIFICANCE OF LDC SUPPLIES

These supply increases are of interest and importance for two strategic reasons. First, supplies from LDCs outside of OPEC are an important source of uncertainty as to the state of the world oil market and the future price and availability of oil. Second, supplies from LDCs will influence strongly the prospects for economic growth in the Third World, and in turn, in the developed nations, for whom LDCs have been a trading partner of growing importance.

LDC Supplies and the Oil Market

Some analysts of world supplies point out that oil is everywhere, and that its apparent concentration in the Middle East is the product of political and logistical conditions rather than an efficacious worldwide search for oil and gas.[1] Oil prices are therefore limited in the long run by these supplies. The implicit assumption in such an argument is that geology is random, and that the modern industry's concentration on the Middle East may be indicative of future discoveries. Certainly, one-tailed uncertainty exists concerning the potential of such places as the Burwood Bank, the Gulf of Thailand, and the Chad Basin. But the chances of dramatic changes in the oil market in this decade due to new reserves in this group is slim.

In fact, the supply response of LDCs is evidence of an efficacious search. New LDC fields are generally smaller, further from markets, and in more inhospitable environments than those of their OPEC counterparts. LDC oil supplies will approximately double between 1979 and 1985. Yet these increases will not be sufficient to create enough supply elasticity to stabilize the oil market nor to make non-OPEC LDCs net exporters as a group. Supplies of 9.3 and 12.1 million barrels per day in 1985 and 1990, respectively, will leave LDCs with net imports of less than a million barrels per day, according to a study by the Congressional Budget Office (CBO).[2] The same study sees worldwide excess demand of approximately 2 million barrels per day in 1985 and twice that amount in 1990, unless prices rise above the level of $30 per barrel in 1979 increasing at 2 percent in real terms annually. LDC supplies would have to be 50 percent larger than projected here to close this gap.

But the numerical equality of supply and demand does not imply dynamic stability in the oil market. The price increases of the past year, driven largely by consumers, have taken producer nation revenues to levels that transcend their infrastructure, factor markets, product markets, and organization. While development plans cannot be extrapolated to accommodate these revenues, Western capital markets do not provide an asset that producers can confidently use as a long-term store of value. Currency movements, inflation, and political interference all pose a hazard. The prospects for limited Organization for Economic Cooperation and Development (OECD) growth make industrial assets questionable. Producing nations may justifiably fear that once they sell their oil, all they will have left is money. Under these circumstances, oil in the ground becomes a rational economic and political strategy. Price increases may, therefore, generate proportionate supply cutbacks by producing nations, yielding a backward bending supply curve. Such a turning point in the supply response probably occurred in 1979. With prices at $12.70 per barrel, nations such as Iran, Nigeria, and Venezuela ran trade deficits. At $30 per barrel, deficits are unlikely, allowing conservationist sentiment.

Against this background, increased supplies from LDCs seem unlikely to provide real relief. To a limited extent, the interests of the surplus OPEC nations are served directly by new LDC supplies. Such supplies would reduce the necessity of producing at levels that pose economic and political costs to surplus countries in order to stabilize the market. Moreover, LDC producers rarely carry excess capacity that could have an impact on short-term price fluctuations.

LDC Supplies and Self-Sufficiency

Often overlooked in discussion of LDC supplies is how they affect the LDCs themselves. *OPEC Review* defines 100 nations as being less developed countries outside of OPEC.[3] In 1973, 11 of these nations exported oil, 14 produced oil at levels less than their domestic consumption, and 75 imported all their oil.

In 1979, 15 of these nations were exporting, 14 produced less than their domestic consumption, and 71 produced no oil. By 1985, three more nations will be exporting (or at minimum be self-sufficient) and three fewer will have no production.[4] Thus, LDC production has not provided general relief from the economic burden of oil import payments. This is particularly true for the newly industrialized countries. In nations such as Brazil, South Korea, and Taiwan, economic miracles may die under their own weight in the face of debt burdens worsened by oil. These nations, until now growing trading partners, may be forced to reduce their demand for OECD goods and services.

For nations of the Fourth World, higher oil prices can devastate the prospects for development. In a sense, these nations are bystanders in a conflict over the terms of trade between producing and industrial nations, whipsawed between them. Ideally, OPEC and OECD might seek to mitigate the situation by either providing price relief or extending oil and gas development aid where appropriate.

LDC PRODUCERS

The following section provides notes on recent oil and gas activities in selected nations. Table 2.1 summarizes this data. Oil production estimates are provided for 1985. An appendix discusses oil development in the People's Republic of China.

Latin America

Argentina. Argentinian production is estimated at 500,000 barrels per day in 1985. The prospect of rising oil imports has led Argentina to revise drastically its relationship with private capital. One product of the new relationship was a 30 percent projected increase in drilling in 1980 over 1979. Immediate increases in Argentinian production should occur over the next several years as secondary recovery projects are implemented at Camodoro, Argentina's major field, and other fields in the inland San Jorge and Neuquén basins. These jointly could produce 350,000 barrels of oil by 1985. Exploration in this area is proceeding successfully, with 11 new oil and gas formations found in Neuquén in 1978. Inland discoveries were also made in the western San Jorge basin.

Table 2.1. Crude Oil Production, 1979 and 1985: Thousands of Barrels per Day

	1979[a]	1985[b]
Latin America		
Argentina	464	500
Bolivia	33	30
Brazil	159	300
Chile	21	30
Colombia	124	150
Mexico	1,399	4,500
Peru	186	200
Trinidad/Tobago	214	175
Others	5	20
Total	2,605	5,905
Africa		
Angola	143	300
Benin	0	30
Cameroun	32	100
Congo	57	85
Egypt	500	900
Ghana	4	20
Ivory Coast	0	10
Morocco	0	60
Tunisia	105	130
Zaire	21	25
Total	862	1,660
Asia/Oceania		
Australia	440	500
Brunei	225	300
Burma	30	30
India	240	420
Malaysia	270	350
Pakistan	11	30
Philippines	15	100
Other	21	25
Total	1,282	1,755
Total, LDCs	4,746	9,320

[a] Estimated production for first half, 1979. *Source: Oil and Gas Journal*, December 31, 1979.
[b] Projection by author.

If new discoveries are not made in the three large interior basins—San Jorge, Neuquén, and Cuyana (now in decline), then Argentinian self-sufficiency will be sought offshore, in the Tierra del Fuego region. Offshore exploration contracts were signed in 1978, and in 1979 Shell and a CFP-Deminex consortium made additional commitments to Tierra del Fuego and the nearby Río Gallegos areas. Future contracts may extend into the Malvinas basin, the subject of much speculation.

Gas production reached 385 billion cubic feet in 1978. A gas infrastructure is being constructed to serve the area from Bahía Blanca to Buenos Aires. A line from Tierra del Fuego to Buenos Aires has opened and will reach 210 million cubic feet per day capacity. A line from Neuquén to the east coast carrying 350 billion cubic feet per day is planned for 1982. Gas reserves are 6.7 trillion cubic feet, but production increases will depend on the availability of pipelines and the development of end-users according to an industrial plan. Argentina is reported to flare as much gas as it imports.

Bolivia. Bolivian oil production is estimated at 30,000 barrels per day in 1985. Bolivia's major fields, Camiri and the Santa Cruz area fields (notably Río Grande), are in decline, and foreign operators are leaving Bolivia, finding no extensions. A series of late 1970s discoveries were made, suggested by Nehring,[5] as doubling Bolivia's reserves, that include new gas fields at La Vertiente and Vuelta Grande. In late 1979, successful oil and gas tests were made at La Vertiente and Porvenir, and production facilities are being installed. Gas reserves are now 6 trillion cubic feet, but Japanese sources believe the true figure is twice that, and LNG operations may begin via Chile. Gas production will reach approximately 450 million cubic feet per day in 1985, perhaps more if the LNG venture begins.

Brazil. Brazilian production is estimated at 300,000 barrels per day in 1985. Brazil's ambitious offshore program hit bottom in 1979, when only 9 of 42 tracts offered for exploration received bids, and 6 of these were in the interior. Originally conceived to bring about self-sufficiency, offshore development is hoped to raise production to less than half of domestic consumption by the mid-1980s.

The only truly successful area offshore is the Campos basin, and even there, production goals have been lowered from 500,000 to 210,000 barrels per day. While the Garoupa field has failed to live up to expectation, the Enchova field is being developed rapidly, and the Cherne field may reach a level of 90,000 barrels of oil with 26 million cubic feet of gas per day. Three of Brazil's four significant continental shelf 1978 discoveries were in this area, one a possible extension of Enchova.

The Espírito Santo basin just north of Campos is a growing area, now that the Cacao field has been brought on-line. Similarly, prospects are fair in the Rio Grande do Norte area, with a new discovery at nearby Ceará. Further north, in the Amapá area at the mouth of the Amazon, tracts were awarded in 1979, but results have been disappointing and one tract was abandoned at year's end.

In response to this growing pattern of exploratory failure, Brazil has recently liberalized private sector leasing terms, so that operators who conduct seismic tests are no longer obligated to drill, and if they do, they can participate in development and production. Moreover, the focus of exploration seems to be shifting to the inland areas and southern Paraná province. In 1978, the first commercial occurrence in the interior Amazon was recorded at Jurma, about 500 miles inland. A second find was made in early 1979, and a third in early 1980. While this area poses immense logistical difficulty (volcanic activity makes seismic testing impossible), its potential is considered good. A second area of emphasis is offshore Paraná, where traces of hydrocarbons were discovered in late 1979. In 1980, 17 risk areas were put up for bidding there.

Chile. Chilean oil production is projected to reach 30,000 barrels per day in 1985, accompanied by 700 million cubic feet of gas daily. Over a third of this gas will be LNG from the Tierra del Fuego area, where Chilean production will be centered. The recent discoveries there of Ostion and Spiteful have dramatically increased Chilean production. Ostion, brought on in 1979, was hoped to reach 19,000 barrels per day at the end of that year. Spiteful is due to begin production in 1980 at 5,000 barrels per day. Chile's older fields are in decline.

A U.S.–Chilean consortium is planning on LNG deliveries in 1984. Gas reserves are 3.2 trillion cubic feet, and LNG production of 250 million cubic feet per day is seen.

Colombia. Colombian production is anticipated to reach 150,000 barrels per day in 1985, slightly above current levels. Colombia's older southern fields are now being maintained with secondary recovery. Finds are being reported near Venezuela at Mérida, the Llanos basin, and the Meta River areas. The Cartagena area is thought to have potential. All of these may be commercial. Gas is also being discovered in the La Guajira peninsula and further south in the Upper Malagena basin. Gas production of 500 million cubic feet per day is anticipated, a 25 percent increase from existing levels.

Colombia is committed to an ambitious program to more than double current reserves in the next ten years. Private capital is active, and World Bank assistance has been discussed.

Guatemala. Guatemalan production is projected to range around 15,000 barrels per day in 1985. This will come from its Tortugas, Rubelsanto, and Chinajá fields, the latter a recent discovery. Exploratory activity will also take place near Mexico, although it is unlikely that Guatemala shares Reforma geology.

Mexico. Should an oil and trade arrangement be reached between Mexico and the United States, Mexican oil production could reach 4.5 million barrels per day in 1985, and 6.5 in 1990, or more. Mexico's development as an oil and gas producer has changed as a result of the absence of a sizable gas deal with the United States. In the absence of a large foreign market, Mexico has opted for domestic use of gas as a fuel for industrial development and is in the process of tying major industrial centers to the trunk line that originally was to carry gas to the United States. In addition, this failure of both sides to reach a gas deal led to the development of the Campeche oil fields, which, originally, were to be developed after the Reforma fields. Located offshore, northeast of the Reforma, Campeche was found to have a lower gas-to-oil ratio. Its production was therefore substituted for the Reforma's, in order to minimize gas flaring. Thus, while Mexico was once depicted as being unable to accommodate the levels of gas production associated with its oil targets, flaring is now declining and is lower in Mexico than in the North Sea. This parallel development of the Campeche and Reforma fields will give Mexico a sizable production capability in the near future. When port facilities are completed in the early to mid-1980s, exports of over 2.0 million barrels per day could be accommodated.

Reforma production now exceeds 1 million barrels per day, and could double if a market for the ensuing associated gas were to develop. In early 1980, a new field in the Reforma (named Iris-Giraldis) was credited with 1.5 billion in reserves. The Campeche area, which began producing only last year, is expected to reach 500,000 barrels per day soon. To date, wells in this area have averaged 32,000 barrels per day, an exceptional figure. It is unclear what effect the Ixtoc blowout will have on development.

The Chicontepec region is credited with 110 billion barrels of oil in place, but its fractured geology and low porosity suggest recovery of only 7 percent of this resource as opposed to the conventional 30 percent. It will probably require 16,000 wells for development, and therefore may be postponed. In the northeast, large gas fields have been found in the Sabinas region. These are significant because of their proximity to the industrial center of Monterey, and because they are the first major fields found outside the coastal plain or offshore areas.

Thus a commitment to raise its 1982 production goal to 4.0 million barrels per day (from an earlier target of 2.25 million barrels per day) could be supported by Mexico's reserve base. If it is, it will represent a decision to speed the rate of economic development, and will be an apparent shift of thinking. Since severe hyperinflation caused by rapid growth forced a series of draconian economic measures in 1976, the attitude of Mexican planners towards rapid development has been negative. But a reassessment may have been made in the face of pressing needs. Mexican unemployment is estimated at 50 percent. A poor year in agriculture has forced Mexico to use currency to increase food imports. Mexico's transportation infrastructure is poor; rail and port investments are required.

More important, Mexican planners are apparently reaching a consensus of how Mexico will develop. Export markets are seen as superseding domestic ones in the long run, and to act on this premise, industry is being encouraged to shift from its present upland locations of Mexico City and Monterey to a series of towns on the Gulf Coast, with cheap energy, transportation systems, and ports for exportation.

The Mexican economy has had a historical tendency to underproduce capital goods. Imports of foreign capital goods, long a problem, will now be financed by oil revenues. Mexico's conscious policy of seeking exchanges of its oil for foreign financing, technology, and direct investments is promoting an active competition among industrialized nations. Japan, Canada, Sweden, France, Britain, Spain, Israel, and Brazil have all either completed or actively sought these reciprocal solutions.

The new higher production would therefore suggest a higher rate of development for Mexico and a larger Mexican import market. In several years Mexico should become the second largest consumer of U.S. exports. But questions will be raised by these developments. As mentioned, other governments have offered trade and technological concessions—ranging from landing rights at airports to reactor technology—in exchange for Mexican oil. Mexico's decision to diversify its economic trade is already being implemented. The United States' share of Mexican exports could fall from its current 85 percent to below 40 percent without progress in creating a new long-term economic relationship between the two nations.

Mexican planners are said to be considering U.S. asset acquisition as part of their development strategy. Mexico may decide that the most expeditious way to acquire modern technology in its developing sectors is to acquire interests in large U.S. firms involved in them. Thus, Pemex is said to be considering the purchase of a major U.S. refinery and a drilling equipment manufacturer. Mexican firms interested in exporting are also reported to be seeking U.S. firms to assist in their marketing. Thus Mexico may use the revenues created by their expanded production to buy interests in multinationals.

The future U.S.–Mexican relationship, including whether or not production goals are raised, may thus be based on resolutions to outstanding economic issues:

- the types of acquisitions of U.S. business interests by Mexico that the United States will allow
- U.S. restrictions on Mexican agricultural exports to the United States
- the development of border industries or export platforms that assemble U.S.-made component parts in Mexico for re-export to the United States
- the influx of Mexican workers into the United States

Peru. Peruvian production is projected at 200,000 barrels per day in 1985. Until the 1970s production had come from the Talara and Lima areas, where the La Brea fields are among the oldest on earth. Inland Andean fields were developed in the early 1970s by Occidental, where light and heavy oils are mixed for delivery via a new pipeline to the coast. This pipeline has raised current output to 200,000 barrels per day. Discoveries continue to be made in this region (the San Dorissa/San Jacinta area near Ecuador). The Barta field is now producing, and Vivian and Dorissa fields may be economic. Offshore gas is being discovered at Piedra Redonda on Peru's northern coast.

Trinidad/Tobago. Trinidad's oil production should approximate 175,000 barrels per day in 1985. Its large Soldado and Forest Reserve fields are in decline. Teak, Samaan, and Poui, its recent eastern fields, are anticipated to decline by almost 60,000 barrels daily by 1983, although a late 1979 wildcat success may extend Samaan. Successes have expanded the reserves at Galeota, but no report of a production platform has occurred.

Trinidad's gas reserves are established at 6 trillion cubic feet by De Goyler and McNaughton. They will produce for a 500 million cubic feet per day LNG facility.

Others. Barbados produced less than 1,000 barrels per day in 1979, and small discoveries have been made recently. Cuba, which produces less than 3,000 barrels per day, and Jamaica may both find oil in the Pedro Bank formation. Belize, Costa Rica, the Dominican Republic, French Guiana, Guyana, Haiti, Honduras, the Netherlands Antilles, Nicaragua, Panama, Paraguay, Puerto Rico, El Salvador, and Surinam have all had either some preliminary drilling, seismic exploration, or discussion of such, but none can be projected as a producer in this decade.

Africa

Angola. Angolan production is seen as 300,000 barrels per day in 1985. Angolan production and exploration have been disrupted by the recent civil war that brought the MPLA to power. A decree of July 1978 granted a 51 percent share of oil and gas operations to the state company, Sonangol, a condition imposed on Gulf. Angolan production is centered in three coastal areas—the northern province of Cabinda, in the Kungolo area near the Zaire border, and near Luanda, further south. New discoveries remain unproduced in the Kungolo area, and Cabinda is thought to have additional potential. However, political events have brought development and exploration to a halt.

In late 1979, however, new contracts were signed with Texaco and Gulf, and in 1980, a 400,000 barrel per day target was announced for 1985. Given the current production level of 180,000 barrels per day (restoring the highest level recorded before the civil war), such a goal represents a strong commitment to new exploration.

Benin. Norway has extended a loan and assistance program to Benin for development of the Seme field, discovered in 1968 and originally deemed uncommercial. Production in 1985 is estimated at 30,000 barrels per day, equal to the field's planned 1982 initial production.

Cameroun. Production of 100,000 barrels per day may be a minimum estimate for 1985. The outlook for the Cameroun has recently turned very optimistic. Production levels have risen to over 30,000 barrels per day in the three years since the Rio del Rey–Kole area began producing in 1977. Several small fields have been discovered in this area since 1972. Eight new fields were discovered out of 19 exploratory tracts in this area during 1978, and seismic work continues there.

In 1979, significant offshore discoveries were made. The most important was the confirmation of reserves at Sanga, described by one participant as "half as large as Frigg" (North Sea). This source described reserves of one to two trillion cubic feet of gas and 7.5 to 15 billion barrels of oil. Although other participants refrained from making estimates, the find is apparently significant.

Chad. Two areas are now being explored in Chad. In the southern region, one success was recorded out of five wells on a tract held by a Conoco group. Chad is said to be ready and is developing its other field, near Lake Chad. It is unclear if either field will reach production, given their remote location and the low level of Chad's internal use (1,300 barrels per day in 1977).

Congo. Congolese production has risen steadily in the past several years to a projected 75,000 barrels per day in this year. Production of 85,000 barrels per day is estimated for 1985. Development is occurring at the Likoula fields, adjoining Loango and Emeraude (described by Nehring as possible giants). Two offshore discoveries were made in late 1979 at Pointe Noire, flowing at a combined 3,000 barrels per day. In early 1980, the state-owned company, Hydro-Congo, announced its participation in seismic testing at a large coastal tract, possibly in this area.

Egypt. Egypt's production should reach 900,000 barrels per day by 1985. Egypt's largest producing area is the Gulf of Suez and its banks. The offshore discoveries of El Morgan, July, and Ramaelan fields by the mid-1970s have led to a series of new fields. These three fields, now in decline, produce 350,000 barrels per day. Pressure maintenance is underway. GOS–382, 380, and 195, discovered in 1977, may produce 150,000 barrels per day or more. GS–195 alone has been termed larger than the July field. In 1979, EE–85 was discovered, and was termed the "most promising [find] ever made in Egypt."

By the mid-1980s, the Gulf of Suez should produce in excess of 800,000 barrels per day.

Oil and gas are also being found in Egypt's closest areas west of the Gulf of Suez. Two successful wildcats were drilled at Umm Al-Yusr in 1978. In addition, the Nile Delta gas field Abū Qīr, opened in 1979, was reported to be experiencing a major extension in early 1980. The Qattara Depression was also the scene of successful wildcats.

In the Sinai peninsula, the Alma and Abu Rudeis fields have been returned to Egypt and are being reworked. Exploration tracts for the Sinai will be awarded in 1980.

Ghana. By 1985, Ghana probably will have two commercial fields in production at 20,000 barrels per day. Saltpond, discovered in 1977, produced 3,500 barrels per day in 1978, and was slated to reach 5,000 barrels per day late last year. Facilities at the field can handle twice that amount. In 1978, Phillips discovered a new field at South Thano, near the Ivory Coast, and reached an agreement to continue exploration in the area.

Ivory Coast. Production from the Ivory Coast will begin in 1980 when Esso connects a pipeline to Berlier field, due to open this month. Oil and gas successes occurred in the Ivory Coast in 1978 from what appears to be a separate small field near Berlier. Production may reach 10,000 barrels per day in 1985.

Morocco. Morocco's current production comes from its Sidi Fili field, now in advanced decline. Production includes 8 million cubic feet of gas daily. Exploratory efforts continue offshore, but no successes have been recorded.

Occidental Oil has reached an agreement with Morocco to develop its oil shale resources, using its in situ process. In early 1980, the Moroccan government was reported to have given 60,000 barrels per day as a 1985 production target. In 1979, the government spoke of one gigawatt of generating capacity from shale by 1990. In fact, Morocco may have such capacity before the United States.

Sudan. The Sudan has promising prospects both onshore and offshore. In 1975–1979, two wells tested 1,200 barrels per day of oil. In 1978, an onshore tract reported an oil show in the south-central region, where Chevron is sole operator. In 1979, two inland tests proved successful. While the state of geologic knowledge of the area is low, Chevron intends to spend $100 million in the area. No estimate is given of Sudanese production, but some may exist by 1985.

Tunisia. Tunisia's projected 1985 production of 130,000 barrels of oil per day may be accompanied by 12 billion cubic feet of gas. Tunisian oil and gas development is now jeopardized by a border conflict with Libya. Tunisia's large Isis oil field originally was to be developed by Compagnie Francaise de Pétroles (CFP) to 40,000 barrels per day; however, military conflict between Tunisia and Libya has raised the issue of borders on the continental shelf, and Isis is in the zone of contention. An additional disappointment for Tunisian production has been the failure of the Miskar gas field to prove out. Miskar was originally expected to contribute 2 million cubic feet per day of gas, an additional 7 percent of production.

Most of Tunisia's oil and gas production comes from the Ashtart field in the Gulf of Gabès, and El Borma, in the southwest, near Algeria. These older fields require pressure maintenance and face decline. In 1979, an inland discovery was made at Sabria, and awards were made for more secure offshore areas, most notably near the Kerkennah Islands. In the absence of the development of Isis, Tunisian oil production will rise only modestly.

Zaire. Production from the Mibale field and its likely extensions should bring output to 25,000 barrels per day in 1985. Gulf's 1973 discovery of Mibale accounts for virtually all of Zaire's current production of 19,000 barrels per day. One exploratory success occurred in 1978, out of five wells drilled. No other exploratory efforts have been noted.

Others. The Central African Empire, Gambia, Guinea, Kenya, the Malagasy Republic, Mali, Niger, Seychilles, Tanzania, and Togo have all been hosts to some form of exploration, but none can be considered prospective producers.

Asia and Oceania

Afghanistan. Afghan gas production reached 290 million cubic feet per day, and signs are active in the northern gas-producing areas. Reserves, set at 3 trillion cubic feet, could be higher. An oil find was made in 1977, but there has been no development. Gas discoveries have also been made and new fields brought on line. Given the political uncertainties surrounding Afghanistan, no projection can be made.

Australia. Australia will produce 500,000 barrels of oil per day through the middle and late 1980s. Australia's current production largely comes from three offshore fields in the Bass Strait—Barracuda, Halibut, and Kingfish, which produce about 400,000 barrels per day. In 1977, the Lobia field added 150 million barrels in reserves, and in 1978 Fortescue (first thought to be an extension of Halibut) and Seahorse added over 300 million barrels. Mackerel and Tuna are approaching peak output, which Barracuda and Halibut have attained. The Snapper gas field will be in production in a few years. Thus this area will continue to produce through the decade.

The Cooper basin in South Australia was recently the scene of Australia's largest recorded onshore oil test flow at Strzelecki, which has not yet gone to production. Queensland shares this basin, and includes the Bowen–Surat basin, where small discoveries are being made. The Perth area has gas potential.

Australia has significant gas reserves in the northwest, put at 13 trillion cubic feet, centered in North Ranking. Production awaited purchase agreements for a LNG train. Such an agreement was apparently reached in 1980. Deliveries are due to start in July 1986, to average 918 million cubic feet per day.

Bangladesh. While a goal of ten exploratory wells was announced for 1980, Bangladesh gas reserves are now believed to be marginal. Several test wells proved dry in 1977 and 1978, and permits were relinquished.

Brunei. Production is estimated at 300,000 barrels per day in 1985, along with 750 million cubic feet of gas via LNG train to Japan. Brunei has an active exploration program centered around its existing large fields, predominantly offshore. Champion and Magpie fields are being developed shortly and account for 115,000 barrels per day at present. The recent Seria field will reach 50,000 barrels per day at its peak. Discoveries were made near Champion and Seria in 1978.

The Ampa field produces 80,000 barrels per day, and gas for an LNG facility at Lumut of the size given above.

Burma. Burmese production should stay at its current level of 30,000 barrels per day through 1985. Burmese exploration is at a standstill, although in 1980 the government claimed to have undeveloped discoveries and sought Japanese assistance. Burma's Mann field will account for most of domestic production. Like other Burmese fields, Mann is slightly underproduced due to absent infrastructure.

India. Indian production is estimated at 420,000 barrels per day in 1985. Over half of this will come from the Bombay High and Bassein fields in the Gulf of Bombay. In 1978, Bombay had been reported as having water intrusion problems and was in a holding pattern. In 1979, an oil discovery south of the existing fields was made (R–12), and several months later, a goal of 240,000 barrels per day was announced, suggesting the discovery's size. Discoveries are also being made in onshore Gujarat state, near the Bombay fields.

Despite R–12, finds in the Bombay High area have been predominately gas, such as the 1978 Tapti find. Exploration continues on both the west and east continental shelf, in the Bay of Bengal, and in the Assam basin near Burma.

Malaysia. Malaysian production is projected to reach 350,000 barrels per day in 1985. Exploration in all of its forms is proceeding rapidly in all Malaysian areas. Offshore peninsular Malaysia's production has risen from 20,000 to 95,000 barrels per day from 1978 to 1979. Also in this area, a long-standing Thai-Malaysian border dispute has been resolved, allowing development of the Pilong field. In eastern Malaysia, the Sarawak outer continental shelf is becoming a major gas and oil region. Sarawak is thought to have 9.7 trillion cubic feet of gas, and all of Malaysia, 23.4. An LNG facility is to open in a few years at a level of 185 million cubic feet per day. Significant gas discoveries were made east of peninsular Malaysia in 1979.

Oil production is being developed off Sarawak, notably in the Tenana and Betty fields. Oil exploration success has been recorded in all regions.

Pakistan. Pakistan oil production is projected as 30,000 barrels per day in 1985. Most of Pakistan's oil production (now at 10,000 barrels per day) comes from the Potwar region, near Islamabad, where a new oil discovery has been made. World Bank and International Development Association assistance has been offered to develop fields in this area. A government goal of 50,000 barrels per day in 1982 has been announced. A 1976 discovery, Dhodak, confirms the first oil to be found outside the Potwar areas, as well as additional gas, and may make such a goal feasible. Pakistan produced 575 million cubic feet per day in 1978, and output is rising.

Philippines. Philippine production is estimated to be approximately 100,000 barrels per day in 1985. This production will occur off Palowan Island, where the South Nido field was discovered in 1977. Two new fields, Cadlao and Matinloc, were added in this region in recent tests. In late 1979, an extension of Nido was reported. Nido and Matinloc were asserted to share 750 million barrels of reserves. Thus, the Western Philippines may become a major province.

Thailand. Thai oil production is uncertain at present, but some fields may be developed onshore after leasing in 1979. Some oil exists offshore, but border disputes with Kampuchea restrict development.

The Gulf of Thailand is apparently a major gas province. Reserves of 7 trillion cubic feet have been found in one field that is continually being extended through drilling. Seven consecutive successful tests were made in 1978. Deliveries may start in 1981 following a World Bank loan for a delivery system. Initial production will be 150–200 million cubic feet per day.

Vietnam. Before the fall of South Vietnam in 1975, two discoveries in the offshore Bow Valley—Dua and Bach Ho—had occurred. Drilling in the past year has failed to confirm these. Soviet assistance is reported in the Reel River delta onshore, where some finds are rumored. Vietnam's major potential is probably in the Gulf of Tonkin, but border disputes with China will restrict exploration.

Others. New Zealand's gas reserves of 11.2 trillion cubic feet will bring about production of 450 million cubic feet per day in 1985. Production is limited by the absence of a market and the decision not to develop LNG. Production of 10,000 barrels per day is associated. Taiwan is the scene of several small discoveries. Concessions have been awarded in Sri Lanka.

NOTES

1. See Bernardo Grossling, *Window on Oil* (London: Financial Times, 1976).
2. CBO, *Oil Imports in the 1980s: An Analysis of the Risks,* (April 1980).
3. *OPEC Review* 3, no. 3, (Autumn 1979).
4. The reader should keep in mind that LDCs that produce oil usually do import refined products, lacking this capability.
5. Richard Nehring, *Giant Oil Fields and World Oil Resources*, RAND/R-2284-CIA, (1978).

APPENDIX: PROSPECTS FOR OIL PRODUCTION IN THE PEOPLE'S REPUBLIC OF CHINA

This appendix discusses projected oil production levels in the People's Republic of China (PRC). These estimates are based on accounts of Chinese reserves, exploration, and development from Western industry publications, the news media, PRC announcements, and personal communications with informed sources.

Overview

Chinese oil production is estimated at 3.8 million barrels per day in 1985, with a feasible range of 2.4–4.8 million barrels per day. Endpoints, however, represent a combination of either all pessimistic or optimistic assumptions. Chinese reserves may ultimately reach 70–90 billion barrels, but much of this potential is in uninhabited areas or far from commercial centers.

The major uncertainties associated with future Chinese production are the rate of new discoveries and the pace at which field development can occur. The available evidence suggests that the PRC will conduct extensive exploration and development of both its offshore and inland reserves, and will do so with the assistance of Western enterprises. China's modernization program is predicated on export earnings, the most important of which must be hydrocarbons. The failure of onshore reserves in the north and northeast to produce enough for export sales will require the PRC to develop both of these hitherto undeveloped regions. Moreover, the political climate, in bringing China closer to the West, is transforming Chinese oil production into part of its relationship with its new apparent allies.

The likely estimates presented here take a middle ground with respect to both the rates of discovery and development. Modest successes are presumed in new areas (offshore) and in areas where preliminary assessments have been made (inland). No super-giant reserve is expected to be found. On the other hand, the possibility of no new discoveries is used only for minimum cases. With respect to western development, it is assumed that by 1985, Tsaidam basin will be linked to eastern and southern economic centers, but that Tarim and Dzungarian basin will not.

1985 Production at the Basin Level

The CIA's 1977 report, *China Oil Production Prospects*, lists 16 major basins in mainland China and its continental shelf. Each of these is discussed below. Production data are given in Table 2.2.

Table 2.2. Projections of Chinese Crude Oil Production, 1985; in Millions of Barrels per Day

Region	Minimum	Likely	Maximum
Sung-liao Basin	1.10	1.50	1.60
Taching	0.90	1.20	1.20
Fuyü	0.05	0.10	0.10
Other	0.15	0.20	0.30
North China Basin	0.50	0.70	0.90
Sheng-li	0.30	0.40	0.45
Ta-kang	0.10	0.10	0.15
Other	0.10	0.20	0.30
Offshore	0.60	1.10	1.45
Pohai Bay	0.40	0.60	0.70
Kiangsu Basin	0.10	0.20	0.25
Taiwan Basin	0.00	0.05	0.15
Liu-Chou Basin[a]	0.10	0.25	0.35
Western Provinces	0.09	0.32	0.55
Tarim Basin	0.02	0.07	0.15
Dzungarian Basin	0.02	0.05	0.10
Tsaidam Basin	0.05	0.20	0.30
Other	0.02	0.05	0.10
Synthetics	0.05	0.10	0.25
TOTAL	2.36	3.77	4.75

[a] Includes onshore Gu-Dao field.

Sung-liao. Production from this basin is estimated at 1.5 million barrels per day in 1985. Sung-liao contains China's largest field, Ta-ching. In 1978 production from Ta-ching was 900,000 barrels per day, and is believed to be close to peaking. In 1979, however, a Ta-ching extension equal to 25 percent of its existing reserves was reported. Current reports for Ta-ching cite improved drilling speeds and stable well pressures as evidence of a potential "sharp upsurge in production." Ta-ching will also benefit from anticipated improved deep drilling. Ta-ching is expected to produce 1.2 million barrels per day in 1985.

Other fields exist in the basin. In 1979, a discovery further north of Ta-ching was reported. This may prove to be a further extension of Ta-ching. In addition, the exploitation of Fu-yü field was reported to have begun, although it has been reported as disappointing. These other fields are projected to produce 300,000 barrels per day in 1985, with one-third coming from Fu-yü.

North China. Production from the onshore portion of the North China basin is estimated to be 700,000 barrels per day in 1985. Over half of this total comes from the Sheng-li field, currently producing 360,000 barrels per day. Evidence of its possible underdevelopment exists, as in the CIA's 1977 statement that Sheng-li's "reserves are large enough so output should grow rapidly for some years to come." It is projected to produce 400,000 barrels per day in 1985. The Ta-kang field, second largest in the basin, produced 100,000 barrels per day in 1977, and is projected to remain at that level.

The North China basin extends under Pohai Bay to northeast China, near Korea. The Panshan field there was described by the CIA as having a maximum output of 81,000 barrels per day in 1985. In 1980, the PRC announced that the Liahoe field had reached a production level of 100,000 barrels per day, and some believe that Liahoe may include Panshan. The *Peking Review* described Liahoe's reserves as "abundant."

A discovery near the southwestern rim of the North China basin at Nanyang was reported in 1979. Reserves are given as 4 million tons per square kilometer, but since a tract of 4,600 square kilometers was reported, the magnitude of these figures invites speculation. Production in the entire North China basin outside of Sheng-li and Ta-kang is estimated at 200,000 barrels per day in 1985, bringing estimated total production from the North China basin to 700,000 barrels per day in that year.

Chinese Continental Shelf. The Chinese continental shelf begins in Pohai Bay, which shares the North China basin with onshore fields. Progressing south, the shelf contains the Kiangsu, Taiwan, and Liu-Chou basins.

Offshore Pohai. The offshore component of the North China basin (Pohai Bay) is projected to produce 600,000 barrels per day in 1985. Twenty-two successful wells have been reported in Pohai, of which only four produce due to pipeline and storage constraints. A major portion of this production will come from the field, to be developed by the Japanese National Oil Company (JNOC). A production goal of 300–400,000 barrels per day in 1982–1983 was recently announced for this field. JNOC, involved in much of Pohai's development under the umbrella of a proposed $2 billion development loan to China, has set a 1985 target of 600,000 barrels per day. Japanese terms of participation involve a 42.5 percent share of production for 15 years and cost-sharing in the event of no discoveries, which is extremely unlikely.

Kiangsu. Production from the Kiangsu basin, which includes the Yellow Sea, is estimated at 200,000 barrels per day in 1985. British Petroleum (BP) was the first to conduct seismic surveys in the Yellow Sea, but it has been joined by France's ELF–CFP, JNOC, Shell, and Petro-Canada. A great number of firms are participating in the seismic and reconnaissance investigation of the

Chinese shelf because such participation is a PRC precondition for bidding on development tracts. Seismic interpretations of the entire shelf were reported as due in 1980, with lease sales to follow in 1981 and early 1982. In 1980, however, it was reported that exploration and development contracts for the Yellow Sea would be signed in the second half of 1980. Production should be available by 1984, if either schedule is followed, and if reserves are discovered in proportion to the strong consensus concerning this region's potential.

Taiwan. The Taiwan basin, occupying the Taiwan Straits, is projected to reach a production level of 50,000 barrels per day in 1985. No exploration beyond preliminary surveys in the Straits has been conducted. Yet the region is thought to have good potential and is sometimes depicted as an extension of the productive southern Liu-Chou basin. The 50,000 barrel per day estimate presumes the leasing for this area is conducted by 1982, and that initial production commences by 1985. It is possible that no production will be seen by that year.

Liu-Chou. Production from the Liu-Chou basin, which extends from Hong Kong to waters in the Gulf of Tonkin, is projected to rise to 250,000 barrels per day in 1985. About one-third of this total will come from the onshore Gu-Dao field, part of the basin. Reserves for Gu-Dao are reported to be 700 million barrels, which would allow a moderate expansion of output from the current level of 70,000 barrels per day. Discoveries have also been reported off Hainan Island, in the Gulf of Tonkin, and in the Pearl River estuary. These discoveries have been lighter and more sulfur-free than most Chinese oil. Their production may be inhibited by border disputes between China and Vietnam in the Gulf of Tonkin. In 1974, the two nations resolved to create a zone in which neither would conduct oil operations until a border resolution was achieved. The Vietnamese claim that China has violated this agreement, and a Vietnamese attack of an exploration boat has been reported. It may not be coincidental if U.S. firms are invited into this area before their Japanese or European counterparts. JNOC, Mobil, Exxon, Socal-Texaco, Phillips, Amoco, Agenzia Generale Italiana Petroli, BP, Cities Service, Pennzoil, and Union have all been reported to be involved in South Chinese Sea activities.

Major Western Basins. Development of significant inland resources will be the second source of expansion of Chinese oil production. While some geologists see this region's potential as prolific, with more shallow formations and lighter oil than any of China's other regions, the western basins are 1,500 to 2,000 miles from China's eastern economic core. No major infrastructure exists in this region, and oil is currently transported by road and railcar.

Previous plans to build roads, housing, and other improvements have been scuttled. However, a consortium of over a dozen Western firms is now negotiating with the PRC to build this infrastructure, including 2,000 kilometers of highway, 5,000 kilometers of large pipe, and 100,000 housing units. These types of projects will be required before western output can be expanded significantly. On the other hand, Western participation in recovery-related projects could increase production in the short term. Some western China development must be considered likely. The three major basins in the west are Tarim, Tsaidam, and Dzungarian.

Tarim. Production from the Tarim basin in northwest Sinkiang province is estimated at 70,000 barrels per day, reflecting initial production levels. Tarim is the subject of great enthusiasm, due to its many promising formations. Japanese interests are now studying the feasibility of producing Tarim fields, and have set a goal of 160,000 barrels per day in the mid-1980s. This level will require a more concentrated effort at western development than can be safely assumed.

Dzungarian. Dzungarian production is projected as 50,000 barrels per day. This basin contains one major producing field, Ko-la-ma-i, which produced 20,000 barrels per day in 1975, the last year for which data are available. A discovery was reported near the Soviet border in 1978. Dzungarian may be the scene of several discoveries, but, as the furthest basin from the eastern coast, it will probably lack a transportation network in 1985.

Tsaidam. Tsaidam production is projected as 200,000 barrels per day in 1985. The most recent production level observed was 12,000 barrels per day in 1976, but this figure is virtually unrelated to reserves, which Meyerhoff estimated at 1.76 billion barrels in 1970. Soviet analysts have reported many large potential fields that make such an estimate plausible. Meyerhoff also reports 18 known fields in Tsaidam. Tsaidam is also the closest of the major western basins and will be the first focal point of westward expansion.

Other Areas. The Nan Shan basin has one known field, Yümen, which produced 16,000 barrels per day in 1976. An equal level is forecast for Szechwan, although Meyerhoff credited it with over a billion barrels of reserves. Together, they are seen as producing 50,000 barrels per day in 1985. No production has been forecast for the Turfan, Chao-shui, Min-ho, Ordos, Hu-lun-Ch'ih, and Kwangsi–Kweichow basins.

Chienchiang. Chienchiang is a field associated with none of the basins listed by the CIA. Chienchiang's production rose steadily through the early 1970s to a 1975 level of 82,000 barrels per day. A level of 100,000 barrels per day is forecast for 1985.

Lead Times for Field Development

An issue related to Chinese oil development is the speed with which new fields can be brought on-line. While the rate of development cannot be anticipated with certainty, analogy to other fields provides a basis for a range of estimates.

The Chinese offshore basins can be compared to two recent offshore discoveries—India's Bombay High and Brazil's Campos basin fields. Wildcatting in the Bombay High field was reported to have started in November 1972. A first discovery was reported in August 1974, and initial production began by January 1976. Peak output (approximately 200,000 barrels per day) was projected for 1980, while half this rate was achieved in 1978.

Brazil's Campos basin fields were first identified in late 1972, and its largest field, Namarado, was first tested in late 1974. In 1977, Petrobras, the state oil company, commenced a development program slated to bring the basin to a peak level of 230,000 barrels per day in 1980. Part of the delay in setting up a production program for the Campos fields is explained by its depth and distance from shore, analogous to the Chinese continental shelf with the exception of Pohai Bay. Experienced Western firms, however, seem more interested in China's prospects than they were in Brazil's.

The current state of development of the Chinese continental shelf can be compared to these fields. The Pohai Bay, with a few producing wells and knowledge of other structures, is at a stage comparable to Bombay High's in 1974. The agreement with the Japanese to develop the Chengbei field with a 1982–1983 production goal, presumed to begin this year, suggests a level of development comparable to the Campos basin in 1977, or Bombay High in 1976. Given Meyerhoff's estimate of onshore reserves in the Sung-liao basin, in which Pohai Bay is found (8.5 billion barrels), there is a good chance that Pohai Bay will be a larger producing region than Campos or Bombay High, with larger initial production levels forthcoming. The experience of Brazil and India indicates that large production levels may be forthcoming in three to four years, suggesting 1984–1985 as a reasonable target date for large scale production from Pohai. Liu-chow, Kiangsu, and Taiwan basin development will probably follow, in that order.

The western basins are more difficult to analogize, given their large distance from population centers. The most comparable project is the development of the Alaskan North Slope and the TAPS pipeline. TAPS, 800 miles long, took three years to complete—including the time consumed in construction of an access road—over terrain as inhospitable as that of western China. A pipeline from Tsaidam to the Peking area, 1500 miles in length, could be available by the mid-1980s if development plans are made this year.

Chapter 3
OPEC Versus Developing Nations' Energy Needs

Fereidun Fesharaki

INTRODUCTION

The 1970s were dominated by events in the international petroleum market. After decades of cheap and abundant oil, the change in the oil power balance in favor of the producing nations totally altered the world petroleum picture. Not only did oil prices rise from $1.80 per barrel in 1970 to an average of $30 per barrel in 1980, but also the problem of future petroleum supply availability came to the fore.

The increase in the price of petroleum did not bring about the expected massive surge into alternative energy sources in the 1970s. In the industrial countries, there was more talk than action. Environmental objections against coal and nuclear power, suspicion of the oil companies, excessive energy regulations, and short-term political considerations effectively blocked all chances of a serious start-up in the area of alternative energy sources. At the same time, the doomsday predictions of world economic and monetary system collapse due to rising oil prices and surplus revenues of the oil exporters proved to be unfounded.

The real price of oil fell for four straight years, 1974–78, after the quadrupling of oil prices in the last quarter of 1973. A recent study examining the impact of dollar devaluation and inflation on the cost of crude shows that for Japan and West Germany the real cost of crude had declined by around 20 percent in the four-year period. Only by the third quarter of 1979 had the oil price increases fully compensated for the declines in the preceding years. At a real cost of $7–11 per barrel in the third quarter of 1979, the price of oil reached the 1974 level in real terms.[1] In addition, the large OPEC trade surplus of around $60 billion in 1974, had by the end of 1978 been reduced to only $1 billion.[2] The international banking system proved to be far more flexible than many had expected, by successfully channeling the surplus capital to both developed and developing nations.

In retrospect both the industrial and developing nations managed to escape major economic problems in the face of rapid changes in the prices of oil. The industrial economies grew at an average annual rate of 3.4 percent during 1970–78, while the rate of economic growth for the developing world was 5.7 percent for 1970–76 and 4.8 percent for 1977.[3]

In general the declining real price of oil, diminishing OPEC surplus, and relatively successful world economic performance had by the end of 1978 created an environment in which the earlier sense of urgency surrounding petroleum supplies and prices had given way to a sense of complacency reminiscent of the earlier decades of cheap and abundant oil.

The Iranian revolution of February 1979 amounted to a rude awakening of those who had thought the problem of oil supplies and prices would go away by itself. Once again the vulnerability of the world economy to political upheavals in the Middle East began to manifest itself. In just one year, January 1979 to January 1980, oil prices increased by more than 120 percent to an average price of $30 per barrel.[4] The OPEC surplus of $1 billion in 1978 rose to around $50 billion in 1979 and is expected to reach $90–100 billion in 1980.[5]

To assume that the oil price/supply situation in the 1980s can be handled with the same degree of relative ease as in the 1970s may prove to be a grave mistake. The circumstances surrounding the world petroleum markets are structurally different from those of the 1970s. The problem of supply shortages, as we shall see later, is going to be a major problem. These are due not only to physical shortages, but also to policy decisions by the major oil exporters to cut back their production. Oil exporters are going to ensure that oil prices will not fall in real terms again; indeed, production cutbacks are likely to lead to further real price increases. The prospect of unified petroleum prices and an organized petroleum market is dim. This time the situation is far more complicated and far less predictable.

The relationship between OPEC and the rest of the developing world has always been a matter of grave concern to the oil producers. Initially, the non-OPEC LDCs were pleased to see a group of developing countries that succeeded in altering the terms of trade in their own favor. They saw the OPEC success as a glimmer of light towards which they themselves could proceed. OPEC's generous aid, which had by 1976 covered most of their additional oil import bills, was very much appreciated. During the negotiation of the North-South dialogue Conference on International Economic Cooperation (CIEC), the developing nations refused to break ranks with the OPEC nations and maintained their Third World/Fourth World solidarity.

As another round of oil price increases started again in 1978, a number of the non-OPEC LDCs became uneasy about future developments in the oil market, feeling abandoned by their OPEC brethren. Public opposition to OPEC policies was voiced during 1979 at the (UNCTAD) meeting in Manila and the Conference of Non-Aligned Nations in Havana. For the OPEC nations, which have been known to champion the cause of the poor nations, particularly Algeria, Libya, and Venezuela, the current developments are unpleasant. Political and, to a lesser extent, economic support of the developing nations is of great importance to OPEC, not only in the international arena but also in their own domestic politics. To continue to get this support, OPEC will have to look toward new and more flexible policies to help the developing nations overcome their energy problems.

In the following discussion, three major issues will be considered: first, OPEC's petroleum supply prospects; second, the energy/oil needs of the developing nations; and third, how OPEC can help alleviate the LDCs' energy problems.

OPEC SUPPLY POLICY

Table 3.1 lists a number of reputable projections for energy supply/demand of the non-Communist world. All projections indicate that a relatively high level of OPEC production must be maintained in order to fill the gap between supply and demand for energy. The estimates range from 35.5 million barrels per day (Frankel) to 44.6 million b/d (Petroven/Petrocanada) of OPEC oil production in 1985. However, to put this into a realistic perspective, it is necessary to examine the prospects of OPEC supplies through analysis of individual country policy plans or policy trends.

As the OPEC nations began to consolidate their control over their oil industries during the 1970s, they became increasingly concerned with the inevitable exhaustibility of their oil resources. Although a formal production prorationing program was never adopted by OPEC, by the turn of the decade, many had become convinced that conservation of their resources through production cutbacks was the most sensible politicoeconomic choice facing them. A number of factors contributed to this attitude.

1. The governments and not the oil companies now have the responsibility to decide production levels. Thus the general rules of supply and demand applicable to the case of private oil companies are not relevant.

2. Oil is an exhaustible resource whose depletion may throw the OPEC nations back to the dark ages of poverty and domination by industrial powers. The production/reserve ratio is falling, and little possibility of additional large-scale oil discovery exists.

3. The original hopes of relatively prompt modernization and industrialization through the use of oil income have now faded away. It has become

clear to the OPEC nations that the injection of oil money will not by itself bring about rapid development. Self-sustained growth is an evolutionary process, and for that reason the life span of the oil revenues must be extended to ensure a continuous flow of foreign exchange.

4. Countries with surplus funds invested abroad have faced inflation and the declining value of the dollar, while making themselves hostage to the holders of their assets. The freeze of Iranian assets by the United States has had a major negative psychological impact on the oil exporters. The idea of storing their wealth under the ground has an attraction far greater than many other alternatives.

5. Political pressures in many of these countries are mounting in favor of conservation. At the same time, they fear that excessive investments, as in the case of Iran, may disrupt the very fabric of the social order in their countries. Even without revolutions or changes of governments, many of the conservative regimes are likely to cut production to appease the opposition.

6. The OPEC nations see little in the way of conservation in the industrial world. They feel the price has not gone far enough to force a serious conservation policy. Thus they find it unacceptable to deplete their dwindling assets to appease the appetites of industrial nations.

7. Rising domestic demand for oil products is also a major consideration in reducing oil exports. OPEC domestic demand rose from 770,000 b/d in 1970 to around 2.2 million b/d by the end of the 1970s. It is expected to reach 3.9 million b/d in 1985, 6.3 million barrels per day in 1990, and 16.7 million barrels per day by the turn of the century.

8. The oil exporters realize that there are no major substitutes for their oil in the next 10 to 20 years. They also know that supply reduction is the key to price increases. Reduced supplies, at the same time, give them enormous political power, which they can exercise for their own benefit, as well as obtaining some concessions for other LDCs.

9. Oil exporters can obtain added value through refining and petrochemicals.

A number of OPEC countries have now publicly declared their intention to cut back production. Many others are likely to do so in the near future. A number of countries, particularly Saudi Arabia, have exceeded their production ceilings to cushion the impact of supply interruptions and rising prices on the industrial world. Their position is likely to change in the future because of domestic political pressures and their inability to influence the oil market in a significant way.

Table 3.2 shows the position of oil exports in 1980. There is no country in OPEC today which does not have some kind of restriction on production. The five large Persian Gulf exporters which account for 70 percent of OPEC capacity, 68 percent of OPEC production, and 62 percent of the world petroleum trade, are all producing below capacity for policy reasons. In 1980, OPEC's production is expected to decline by 1.65 million b/d and its

Table 3.1. World Energy Demand and Supply in 1985—Various Forecasts[a]

	Lichtblau[b] Case B/C 1978	Frankel[b] 1978	SRI[b] 1978	Exxon 1978	Petrocan/ Petroven 1978	WAES/MIT[b] Case D&C 1977	OECD Demand and Supply Only CIA 1978	OECD 1977
Demand								
Million b/d oil eq.	122.4	120.3	120.5	123.9	131.2	114.1–123.2	98.0	101.9
Quadrillion BTUs	(259.1)	(254.0)	(255.0)	(262.3)	(277.8)	(241.5)–(260.8)	(207.5)	(216)
Annual GNP								
growth rate	4.0–4.5		4.2	4.0	4.4	3.4 5.2	4.2	3.9
E/GNP multiplier	0.9–0.8		0.81	0.88		1.0–1.0	—	0.84
Energy demand	3.7	3.0	3.4	3.5		3.4–5.2	—	3.3
Supply	83.9	84.3	120.5	82.7	85.2	78.7–87.7	59.4	65.9
Oil and HGL								
(non-OPEC)	24.4	24.5	58.9 (*incl. OPEC*)	24.8	25.5	22.4–23.5	17.0	17.7
Natural Gas:								
Million b/d oil eq.	21.1	21.4	22.7	18.9	21.7	19.7–22.2	14.1	15.5
Trillion cubic ft.	(43.5)	(44.2)	(47.0)	(39.0)	(44.9)	(40.7) (45.9)	(29.1)	(32.0)
Coal:								
Million b/d oil eq.	22.6	21.0	23.4	23.4	21.7	18.8–21.6	15.4	17.3
Million short tone	(2175)	(2020)	(2250)	(2225)	(2038)	(1809)–(2078)	(1482)	(1662)
Nuclear:								
Million b/d oil ol.	8.1	7.5	6.2	7.5	·8.0	10.1–12.0	6.3	9.3
Gigawatts capacity	(270)	(250)	(206)	(250)	(300)	(336) (400)	(210)	(310)
Hydro/Geo/Other:								
Million b/d oil eq.	7.7	8.9	8.1	8.1	8.3	7.7–8.4	6.6	6.1
Quadrillion BTUs	(16.3)	(18.8)	(17.2)	(17.2)	(17.6)	(16.3)–(17.8)	(14.0)	(12.9)

Table 3.1. (Continued)

	Lichtblau[b] Case B/C 1978	Frankel[b] 1978	SRI[b] 1978	Exxon 1978	Petrocan/ Petroven 1978	WAES/MIT[b] Case D&C 1977	OECD Demand and Supply Only CIA 1978		OECD 1977
Net nonoil imports				1.1			3.1		3.4
Natural Gas:									
Million b/d oil eq.							1.8		2.2
Trillion cubic ft.							(3.7)		(4.6)
Coal:									
Million b/d oil eq.							1.3		1.2
Million short tons							(2.8)		(116)
CPE[c] net oil imports	−0.5	−0.5	−1	−1.1	−0.1		2.5		−1
Residual demand for OPEC oil (includes OPEC's own requirements)	38.0	35.5	N.A.	40.1	44.6	36.0–39.0	34.9	OECD	35.0
							2.3	Non-ODEC LDC imp.	−0.8
Oil as a % of energy demand	51	47	49	52	53	49–91	1.3	Other D.Cs	1.2
							4.0	OPEC	3.5
							45.0	Total demand for OPEC oil	37.9

Note: Figures do not add due to rounding. Some studies project two or more scenarios. In that case only the base case scenarios have been presented.

[a] Non-Communist World.

[b] Lichtblau: Petroleum Industry Research Foundation, Frankel: Petroleum Economics Ltd., SRI: Stanford Research Institute, WAES: Workshop on Alternative Energy Source.

[c] Centrally Planned Economics.

Source: U.S. Department of Energy.

Table 3.2. 1980 Anticipated OPEC Production Level—Million B/D

	Sustainable Capacity 1979[a]	Expected 1980 Production[b]	Decline from 1979[c]	Type of Restriction
Algeria	1.1–1.2	1.1	—	Resource Constrained
Ecuador	0.25	0.215	—	Resource Constrained
Gabon	0.25	0.21	—	Resource Constrained
Venezuela	2.4	2.15	0.15	Resource/Policy Constrained
Indonesia	1.65	1.5	0.1	Resource Constrained
Nigeria	2.4	2.15	0.21	Resource/Policy Constrained
Libya	2.1	1.7	0.4	Resource/Policy Constrained
Iran[d]	4.5	3.0 [i]	—	Policy Constrained
Iraq[e]	4.0	3.5	—	Policy Constrained
Kuwait[f]	2.5	1.75 [i]	0.52	Policy Constrained
Saudi Arabia[g]	9.5–10.5	9.0 [i]	0.2	Policy Constrained
United Arab Emirates	2.48	1.75	0.07	Policy Constrained
Qatar	0.65	0.51	—	Resource Constrained
Neutral Zone[h]	0.6	0.57	—	Resource Constrained
Total OPEC[j]	34.38–35.48	29.11	1.65	
Domestic Consumption	—	2.43	—	
Export Availability	—	26.68	1.60	

Table 3.2 (Continued)

[a] Capacity which can be maintained for several months. Does not necessarily reflect the maximum production rate without damage to the fields. Generally 90 to 95 percent of installed capacity.

[b] & [c] Based on official and unofficial declaration of production levels compared to the first nine months of 1979. For countries where no declaration is made, the 1979 production level is assumed.

[d] Iran's capacity has declined from 7 to 4.5 million b/d after the Revolution, due to fall in pressure and lack of necessary maintenance. Iran's stated target for 1979 was 4 million b/d. In the second half of 1979 it produced between 3.5–4 million b/d. Although no official announcement is made regarding 1980 production level, political and technical factors strongly suggest 1 million b/d cutback from 1979 target.

[e] Iraq has hinted a few times that it will reduce production by 600,000 b/d in 1980. Still the government is noncommittal, saying any cutback will depend on market developments. It is assumed here that in fact production will not be reduced in 1980.

[f] Kuwait has had a production ceiling of 2 million b/d for several years. It has declared its intention to reduce production to 1.5 million b/d. Since Kuwait is to some extent bound to exceed its desired limit in solidarity with Saudi Arabia, it is assumed here that production will be kept at 2 million b/d for the first half of 1980 and reduced to 1.5 million in the second half.

[g] Saudi Arabia's previously anticipated expansion of reserves to 14–20 million b/d has not materialized due to the Saudi's lack of desire to invest heavily in such expansion as well as some technical difficulties. The sustainable capacity is estimated by CIA to be 9.5 and Petroleum Intelligence Weekly to be 10.5 million b/d. However, more recent testing of capacity suggests a figure of 10.2 million b/d. The Saudi oil production ceiling of 8.5 million b/d was lifted for part of 1979 to cushion the impact of disruptions from Iran and reduce the rise in prices. It has been announced that production ceiling will not be operational for the first quarter of 1980. It is assumed here that production will be maintained in the first half of 1980 at 9.5 and the 8.5 million b/d ceiling will be imposed in the second half of 1980.

[h] Equally shared between Saudi Arabia and Kuwait.

[i] Kuwait and Saudi Arabia together produce 750,000 b/d above their desired level of production and may in the case of any political flare-up choose to go back to their desired ceiling. Iran's production is also subject to political and technical interruptions.

[j] Based on estimates by the OPEC Secretariat: S.A.R. Kadhim and A. Al-Janabi "Domestic Energy Requirements of OPEC Member Countries." Paper presented to the OPEC Seminar, *OPEC and Future Energy Markets*, October 1979.

exports by around 1.60 million b/d. Many industry experts, as well as Saudi Arabia, have publicly forecast a glut for 1980. They expect the large stockpiles in industrial countries to be drawn down and recessionary forces to reduce demand significantly. While it is likely that demand for oil will fall in 1980, to make a glut, it must fall by more than the 1.65 million b/d which OPEC is expected to take out, or major additions must come from Mexico and the North Sea. Even if a glut develops, it will be short-lived: Production will fall to adjust supply to demand. Because Iranian oil is overpriced compared to similar crudes in the Gulf, Iran's oil will be the first to suffer a decline in offtake. Yet a glut, no matter how temporary, may once again serve to distract attention from the major problem, which is long-term supply.

Table 3.3 considers the prospects of exports from OPEC in the next ten years. It shows that except for Iraq, the major Persian Gulf producers are expected to reduce production. Saudi Arabia and, to a lesser extent, Kuwait face a floor in their production schedule below which their costly gas production systems remain underutilized (see notes to table 3.3). It also shows the large increase in domestic consumption within OPEC which severely limits exportable oil. OPEC exports are expected to decline from 28.28 million b/d in 1979 to 22 million in 1985 and to 17.29 million b/d by 1990. This indicates a nearly 40 percent cutback in available OPEC supplies. Considering that OPEC accounted for 83 percent of the world trade oil in 1979, it can be seen that the impact on the world market will be colossal. Still, this forecast is relatively optimistic, since it assumes increasing Iraqi production, relatively steady Iranian production, and large Saudi production. Any political upsets like the Iranian crisis could again disrupt production and further reduce supplies.

In the aftermath of the Iranian revolution, the balance of power within OPEC has drastically changed. Unlike 1976, when a Saudi threat to flood the oil market was sufficient to stop excessive price increases, today Saudi Arabian power as the swing producer has been severely curtailed. By stretching the Saudi production nearly to the limit, the other members have partially taken the initiative away from the world's largest exporter. When Saudi Arabia deliberately priced its oil $4/b below similar crudes at the June 1979 OPEC Ministerial Conference in the hope of checking the price spiral, this was seen as a test of power within OPEC. As it happened, prices rose even further and spot prices reached $45/b. The greater portion of the price difference went to increase the profits of the oil companies which make up Aramco, the principal purchaser of Saudi oil. Non-OPEC exporters, notably the British, Norwegians, Soviets, Mexicans, and Chinese, sold their oil at the higher tier price. Consumers had to pay the higher tier price. The current underpricing of Saudi crude by $2/b is going to prove to be yet another futile exercise. The Saudis appreciate that they can no longer affect the developments in the market as before, but they may hope that this repeat exercise can finally persuade their allies, particularly the United States, not to expect them to perform miracles. Once this fact is accepted by the Western world, then the Saudis may have little reason to keep production at such high levels.

In the current oil market situation, any small OPEC exporter can influence the course of developments in the oil market by even temporarily reducing production by 0.5 to 1 million b/d at the right psychological time. Panic buying and large stockpiling will start again, pushing up spot prices and a price spiral like that of 1979 will emerge. Today the majority of OPEC producers believe that their production policy should correspond to their own economic development planning and not to the international demand for petroleum. This means extending their production life by reducing exports. This group has for the first time been placed in a position to translate its thoughts into action. OPEC as an institution may well remain intact throughout the 1980s, but its functions are likely to be different from those of the past. The price-fixing function through the well-publicized biannual conferences may be the first to go. OPEC may become a consultative body devoted to research, as well as an institution which coordinates general trade and economic relations with both the industrial and developing nations. Certainly, OPEC's weakened position as an organization does not mean that its member countries are weakened. Indeed, it underscores even greater influence and power which its members can exercise without having to take collective decisions under an OPEC umbrella.

To put the international petroleum supply situation in perspective, a brief discussion of export potential of the major non-OPEC exporters may be useful. As can be seen from table 3.4, the non-OPEC petroleum export potential is no more encouraging than the OPEC potential. The current exports which account for only 6 percent of the total world trade are likely to decline at least by 50 percent.

Future Oil Prices

The analysis in the preceding pages indicated that the supply of petroleum from OPEC and non-OPEC sources is likely to decline drastically during the course of the late 1980s. While the projections in table 3.1 called for OPEC production of 35–45 million b/d by 1985, the likely production level is expected to be around 26 million b/d. With OPEC domestic demand for petroleum rising to 4 million b/d in 1985 and 6.3 million b/d in 1990, available crude for export will be even further reduced. Indeed, OPEC exports are expected to decline from 28.8 million b/d in 1979 to 17.3 million b/d in 1990—a drop of nearly 40 percent.

The current conventional wisdom for facing the supply shortages is to reduce demand for petroleum through conservation and a set of fiscal and

Table 3.3. Current Reserves and Production and Likely Exports in the 1980s—Millions b/d

	Proven Reserves[a] (billion barrels)		Production	Domestic Consumption	Exports	Production	Domestic Consumption	Exports	Production	Domestic Consumption	Exports
	1973	1979	1979[b]			1985[c]			1990[c]		
Saudi Arabia[d]	138	165.7	9.20	0.25	8.95	6.3	0.46	5.84	6.3	0.84	5.46
Iran[e]	65	59.	2.98	0.64	2.34	3.0	1.04	1.96	2.7	1.66	1.04
Iraq[f]	29	32.1	3.38	0.20	3.18	4.0	0.34	3.66	4.5	0.53	3.97
Kuwait[g]	64.9	66.2	2.25	0.04	2.21	1.5	0.06	1.44	1.0	0.09	0.91
Qatar	7.0	4.0	0.50	0.01	0.49	0.5	0.01	0.49	0.5	0.02	0.43
United Arab Emirates[h]	22.8	31.3	1.82	0.05	1.77	1.8	0.07	1.73	1.0	0.11	0.59
Neutral Zone	16.0	6.5	0.56	—	0.56	0.5	—	0.5	0.5	—	0.5
Total Persian Gulf	342.7	364.8	20.69	1.19	19.5	17.6	1.98	15.6	16.5	3.25	33.25
Venezuela	13.7	18.0	2.34	0.29	2.05	1.9	0.5	1.4	1.3	0.79	0.51
Nigeria[i]	15.0	18.2	2.33	0.16	2.17	1.9	0.27	1.63	1.8	0.42	1.38
Libya	30.4	24.3	2.06	0.08	1.98	1.6	0.17	1.43	1.2	0.31	0.89
Indonesia[i]	10.0	10.2	1.60	0.36	1.24	1.6	0.59	1.01	1.5	0.92	0.53
Algeria	47.0	6.3	1.14	0.14	1.0	0.9	0.27	0.63	0.9	0.45	0.45
Gabon	1.1	2.0	0.21	0.02	0.19	0.2	0.03	0.17	0.2	0.05	0.15
Ecuador	5.7	1.2	0.21	0.06	0.15	0.2	0.09	0.11	0.2	0.12	0.03
Total OPEC	465.6	445	30.58	2.30	28.28	25.9	3.9	22.0	23.6	6.31	17.29
OPEC as % of World	70	69	50	—	83	—	—	—	—	—	—

[a] Proven reserves denote reserves which can be recovered with current prices and technologies. Excludes secondary and tertiary recoveries. Secondary recovery alone can boost reserves in the Middle East by at least 50 percent. The proven reserve figures widely quoted in the literature are questionable to say the least. Often fields of 50–500 million barrels which have a cost of production of over $1/barrel are considered too small, too expensive and are not included in reserve figures. The figures for Iraq and Kuwait are generally believed to be wrong. Iraqi reserves are believed to be 4 to 5 times larger than the quoted figure—second only to Saudi Arabia—while Kuwaiti figures are thought to be exaggerated. Most OPEC nations, as a matter of policy, do not declare their reserve figures, and the information is generally provided by the oil companies.

[b] Reflects January–October 1979 production.

Table 3.3 (Continued)

[c] Production estimates are based on the author's expectations of production cutbacks due to political/economic and technical reasons. These estimates exclude the possibility of major political upheavals. Domestic consumption forecasts are quoted from the OPEC Secretariat.

[d] Many powerful voices in the Saudi Royal family prefer a cutback in production to 4–4.5 million b/d. Nevertheless, barring major political changes, one can assume that the Saudi production preference in the 1980s will be at a level with which its massive gas gathering project for associated gas can operate at full capacity. The gas-gathering project, which is estimated to cost between $14–21 billion has a collection target of 3.3 billion cubic feet a day and will be ready by 1982. To feed the new system, oil production levels of 5.7 million b/d from Ghawar and Abqaiq fields together with 550,000 b/d from Berri field are sufficient. Therefore, a production level of 6.3 million b/d is assumed here to be likely production for the 1980s. It is highly unlikely that the Saudis will invest heavily in expansion of capacity, as a larger capacity will place them under international pressure to increase production.

[e] Iran's production is expected to remain relatively steady during the 1980s, although an increasing share will go to domestic consumption. It is assumed that by 1990, Iran will try to export one million b/d to pay for its import needs. Technical problems by this time will seriously curtail additional production.

[f] The poor relationship between the postrevolutionary governments of Iraq and the oil companies in the 1960s was the reason behind ignoring the development of Iraqi oil reserves and capacity. Indeed, the expansion in production of a number of other Middle East producers came at the expense of Iraq. The situation was reversed in the 1970s when Iraq, nationalizing its oil, heavily expanded capacity and production. In 1975, when OPEC production as a whole fell by 11 percent due to declining demand, Iraqi exports rose by 15 percent. Despite temporary cutbacks occasionally, it is clear the the Ba'athist government of Iraq, in its quest for influence and power in the region, will push production and capacity higher in the 1980s. This means that Iraq will probably be the only country in OPEC expanding production. Iraqi capacity in 1990 is expected to be around 5–6 million b/d.

[g] Kuwait has expressed its intention to cut back production to 1.5 million b/d in the short term and maintain a 100:1 reserve/production ratio in the long term. Like Saudi Arabia, Kuwait's flexibility is restricted to some extent by its associated gas facilities. The difference with Saudi Arabia is that all the gas is used domestically: 33 percent for desalination; 25 percent for oil company use; and 9 percent for reinjection. Together with liquid petroleum gas (LPG) plans, ideally 3 million b/d production are required for full capacity utilization, a production level long abandoned by Kuwait. Still, production is expected to decline, but only gradually. Like Saudi Arabia, the United Arab Emirates (UAE) and Libya, Kuwait will accumulate surplus investments abroad that will bring in large interest income in the 1980s.

[h] Increasing tensions between emirates that produce oil and those that do not, together with uneven distribution of oil income within the seven sheikdoms which form UAE, will not allow a sharp cutback in production in the next five years. If the federation is to continue in one piece, oil income will have to be channeled to poorer sheikdoms in larger quantities. Still, by 1990 the lower production level of 1 million b/d is expected to generate sufficient income for UAE's needs.

[i] As the two most populous nations of OPEC, these countries' need for ever-increasing oil income is expected to remain high. Indonesia, the poorest OPEC country, is expected to maintain production at full capacity, while Nigeria is expected to reduce production by only 215,000 b/d compared to its declared target of 2.15 million b/d for 1980. Recent reports indicate an expansion in the production capacity of both nations is likely in the 1980s. According to *The Oil Daily* (November 1, 1979), Indonesia's exploration and development efforts have increased substantially, and a production capacity of 2 million b/d by 1985 is likely. During 1990–2000 a capacity of 2.5 million b/d is not impossible. For Nigeria, a recent report by the U.S. Department of Energy and U.S. Geological Survey *Report on the Petroleum Resources of the Federal Republic of Nigeria* (October 1979) indicates that with increased drilling, prospects for capacity expansion to 3.5 million b/d is there. The report, however, anticipates a production range of 2–2.5 million b/d could be maintained during the latter part of the 1980s through the year 2000. In the case of both countries, the available production capability will not necessarily mean that they will produce at those levels.

Table 3.4. Present and Future Export Potential of Major Non-OPEC Oil Exporters (in million Barrels per Day)

	1979[a]			1990			
	Production	Consumption	Exports	Production	Consumption	Exports	Type of Restriction
Norway[b]	0.4	0.2	0.2	0.7	0.3	0.4	Policy Constraint
United Kingdom[c,d]	1.5	1.9	−0.4	2.0	1.9	0.1	Policy/Resource Constraint (expected)
Canada[c]	1.6	1.8	−0.2	1.7	1.8	−0.1	Resource Constraint
Mexico[e]	1.4	0.9	0.5	2–2.5	1.6	0.4–0.9	Policy Constraint (expected)
Egypt[f]	0.5	0.2	0.3	1–1.5	0.6	0.4–0.9	Policy/Resource Constraint
Malaysia	0.2	0.1	0.1	0.5	0.2	0.3	Resource Constraint
Oman	0.3	Negligible	0.3	0.3	Negligible	0.3	Resource Constraint
China	1.9	1.7	0.2	3.5	3.0	0.5	Resource Constraint
Soviet Bloc	12.1	10.9	1.2	10.5	12.8	−2.3	Resource Constraint
USSR[g]	11.7	8.4	3.3	10	10	0	Resource Constraint
East Europe	0.4	2.3	−1.9	0.5	2.5	−2.0	Resource Constraint
Other Communist Countries	—	0.2	−0.2	—	0.3	−0.3	Resource Constraint
TOTAL	19.9	17.7	2.2	22.2–23.2	22.2	.0–1.0	
% of world	32	—	6	—	—	—	

[a] 1979 figures are for the first half of the year.

[b] Norway has imposed a production ceiling for reasons of "social policy." The ceiling is expected to be maintained during the 1980s.

[c] UK and Canada are assumed to maintain domestic demand at current levels.

[d] UK output of oil is expected to peak by 1982 at 2.8 million b/d before it begins to decline, unless unexpected new fields are found. There are strong indications that the government will impose production ceiling on output in the early 1980s in order to increase the life span of its exports and avoid becoming a net importer of oil in less than a decade. Another consideration is flaring of the associated gas, which the government wishes to avoid. Already, the government has ordered reduction of output from 185,000 to 100,000 b/d for Esso/Shell's Brent oilfield to avoid flaring. Also, a now radicalized Labor Party victory at the next elections is sure to result in some kind of production ceiling. It is assumed here that production will be steadily maintained at 2 million b/d for 1980s.

[e] Mexico is expected to impose production ceiling in the early 1980s. The limit is likely to be in a 2–2.5 million b/d range. The government has often hinted that its production policy will be based on "domestic economic needs with a view to spreading the life span of reserves"—a borrowed OPEC catch phrase.

[f] Egypt's production is expected to reach 1 million b/d by 1982. It may wish to follow other producers by imposing a ceiling or stretch itself to the limit by producing 1.5 million b/d. No official or unofficial hints on future policy.

[g] The Soviet Union is expected to be able to produce 10 million b/d, at most, between 1985 and 1990. If its domestic demand grows at over 5% as in 1973–1978, domestic demand will reach nearly 14 million b/d making the USSR a major oil importer. It is assumed optimistically here that the Soviet Union will restrict the growth of demand to 2 percent annually. The same goes for Eastern Europe: a traditional demand growth of 5.5 percent would imply imports of nearly 4 million b/d. It is assumed here that demand growth is held down to 2 percent year.

Source: Based on data from *BP Annual Statistical Review of the World Oil Industry*; *Petroleum Intelligence Weekly*; *CIA World Petroleum Market in the Years Ahead (1979)* and *CIA International Energy Statistical Review*; World Bank: *Petroleum and Gas in Non-OPEC Developing Countries (1978)*.

administrative procedures in order to lessen the impact during the transition period into alternative energy sources. However, demand may not be a determining factor for future price levels. Since the OPEC nations' propensity to reduce production is only limited by the prospects of a catastrophic impact on the world economy and/or a threat of military action, a cutback in demand for oil will be more than matched by a production cutback. This is not to undermine the importance of conservation and demand reduction; rather, this is a realistic assessment of the current mood within OPEC. If the world economy takes no measures to reduce demand, the supply cutbacks are likely to lead to major price increases. However, if demand for oil is gradually reduced, the price increases would be small but regular. The projections in tables 3.2 and 3.3 are based on an expected set of policy measures which are likely to be adopted by the OPEC exporters; however, external demand considerations may serve to speed up or reduce the pace of such policy implementations. Indeed, reductions in world demand for petroleum serve as guidelines for OPEC nations to adjust their export level downwards. The 1980s will probably witness large price increases and a permanent tight market for petroleum. But how fast the prices are increased will depend in part on how energy demand is managed by the world economy.

It would be a mistake to assume that high petroleum prices will lead to a glut which could weaken prices. Indeed, the excess of supply over demand does not necessarily mean a weakening of prices. In the first two months of 1980, supplies exceeded demand by one million b/d, while oil stocks were at their peak of 700 million tons. Yet official prices have been increasing and spot prices are still $2–5 per barrel above official prices. Any glut in petroleum supply will be temporary and will be eliminated by production cutbacks with a time lag of three to six months. In fact, petroleum prices themselves are expected to determine production levels in some instances: Prices will be increased, and any demand decline will result in reduced liftings from producing areas. Any reduced lifting may well serve as a new

Table 3.5. Expected Oil Prices in the 1980s (U.S.$/barrel)

	1980	1985[a]			1990[a]		
	Actual	Low	Medium	High	Low	Medium	High
Average Price (Constant 1980 prices)	30	30	38	48	30	49	78
Average price[b] (Current prices with 7% average annual inflation)	30	42	53	66	59	93	144

[a] Low: no real price increase; medium: 5% annual real price increase; high: 10% annual real price increase.
[b] 7% inflation rate is based on the World Bank's expected annual inflation rate in the 1980s.

ceiling for production. The possibility of a decline in the real price of petroleum, as was the case during 1973–78, is extremely unlikely. The rate of real price increases is likely to remain in the range of 5–10 percent annually. Again, it should be emphasized that demand considerations are likely to have far less influence on OPEC's pricing policy, compared to their assessment of the economic and political impacts on the world economy and/or OPEC's own security problems.

An argument often used by the OPEC ministers to justify oil price increases is the cost of alternative energy sources. Without going into the pros and cons of the validity of this argument, one may simply note that the average cost of alternatives has gone up from $7 per barrel of oil equivalent in 1973/74 to between $35 and $60 per barrel.[6] In fact, the estimated increase in the cost of alternatives has overtaken crude price increases. However, it is the feeling of the author that costs of alternative energy sources have not and will not be a major factor in the actual OPEC decision-making policy on the price issue.

LDC's ENERGY/OIL NEEDS

Energy demand in the net oil-importing developing countries (NOIDCs) grew at a high rate during the 1960s and 1970s. In fact, the oil price shock of 1973/74 did not significantly affect the growth in demand. Energy consumption grew at a rate of 6.7 percent annually during 1960–76 (compared to 4.1 percent for the industrialized countries and 4.5 percent for the world) and is expected to grow by 6 percent annually through 1985—from 12.6 million barrels per day oil equivalent in 1976 to 20.4 and 28.5 million b/d oil equivalent by 1985 and 1990.[7]

Treating the non-OPEC LDCs (both oil importers and oil exporters) as one unit, as many studies do, masks the large variations between different LDCs. While total non-OPEC LDC oil consumption rose from 5.7 million b/d in 1973 to 7.5 million b/d in 1978, six countries—Mexico, Egypt, India, Brazil (oil producers), and Taiwan and South Korea (non-oil producing but fast growing)—accounted for 75 percent of the rise in oil consumption during 1974–78. Most of the remaining increase of about 400,000 b/d was spread among Malaysia, Thailand, the Philippines, and Pakistan. The majority of poor NOIDCs increased their oil consumption only marginally.[8]

The larger and/or fast growing LDCs' shift to energy-intensive industrialization has prevented a reduction in energy/GNP ratios despite rising oil prices. Domestic economic policies have been a major factor in determining energy use. Brazil has passed over higher oil prices to the domestic consumers, while Taiwan until December 1979 kept prices artificially low to encourage its export-oriented economic growth. In almost all non-OPEC

LDCs, some kind of administrative policy measures have been used to shield consumers from higher prices. Tax rebates, subsidies, and limited price pass-alongs for refiners and distributors have often been employed—accompanied by major additions to LDCs' cumulative debt burden. Kerosene prices have often been kept low because of their impact on the rural poor, and electricity and transport costs have been controlled because of their impact on food prices and the urban poor.

Prospects for the 1980s

For the 1980s, growth of demand in the large high income LDCs is expected to take the lion's share of additional energy requirements. Between 1978 and 1982, total non-OPEC LDCs' oil consumption is expected to increase from 7.4 to 9.3 million b/d—a growth rate of roughly 6 percent. This growth rate implies an oil consumption growth rate of 7 percent for the 16 major LDCs[9] and 2 percent for the remaining LDCs.

The rate of growth of supply as compared to demand is subject to differences in the studies undertaken. The World Bank feels that during the 1980s, it may be possible to tap the oil resources of 38 oil importing LDCs with no current oil production,[10] while the CIA, with its shorter timeframe of up to 1982, is generally pessimistic about such immediate prospects. Table 3.6 shows the CIA analysis of net oil import requirements of LDCs in 1982. It

Table 3.6. Projected Net Oil Import Requirement of Non-OPEC LDCs (in millions of barrels per day)

	1978	1980	1982	1979–82 Implicit Annual Rate of GNP Growth
Production	4.6	5.6	6.7	—
Consumption	7.4	8.3	9.3	—
net imports	2.8	2.7	2.6	—
of which:				
Argentina	0.1	Negligible	Negligible	4.0
Brazil	0.8	1.0	1.0	6.0
Mexico	−0.3	−0.7	−1.1	8.0
Peru	Negligible	−0.1	−0.1	5.2
Egypt	−0.3	−0.4	−0.7	7.0
India	0.4	0.3	0.4	4.0
Philippines	0.2	0.2	0.2	6.0
South Korea	0.5	0.5	0.7	9.0
Taiwan	0.3	0.4	0.6	10.0
Malaysia	−0.1	−0.1	−0.1	8.1

Source: The World Oil Market in the Years Ahead, pp. 4 and 36.

projects the growth in production of oil to be nearly offset by growth in demand for oil, particularly by the oil producers and other high-income LDCs. Total net oil import requirements of LDCs is projected to fall from 2.8 million b/d to 2.6 million b/d from 1978 to 1982.

To show the projections of non-OPEC LDCs' energy supply and demand, tables 3.7 and 3.8 are provided. These two tables compare seven studies—most of which were completed in 1978—and their projections for 1985. The following conclusions may be drawn from these projections:

1. Energy/GNP multiplier is unlikely to fall below 1.1.
2. Indigenous energy production is likely to more than double from 6.8 million b/d oil equivalent to 14.6–16.7 million b/d oil equivalent.
3. Residual demand for imported oil is likely to be in the range of 0.2–3.6 million b/d.
4. Indigenous oil production was expected to rise from 4.1 million b/d in 1977 to between 7.7 and 9 million b/d by 1985.
5. Most of the increase in indigenous production is likely to come from Mexico, Egypt, and Malaysia (oil exporters) and India, Argentina, Brazil, Peru, and Bolivia (net oil importers).

The total picture of non-OPEC LDCs' oil demand and supply does not seem promising in the short term or medium term. Any combination of supply and demand shows at best a balance—which implies that a great number of oil importing LDCs are going to remain in a difficult position while a small number of LDCs will earn substantial sums through their oil exports.

Currently 48 of the 74 oil-importing developing nations depend on oil for at least 90 percent of their commercial energy requirements; only a few (such as India, Korea, Pakistan, and Zambia) are less than 50 percent dependent on oil owing to their extensive use of coal, natural gas, and hydroelectric power.[11] Most of these 74 countries are likely to remain oil importers in the 1980s, facing supply shortages and prices substantially above the current rates. What makes the position of poor oil importers even more precarious is that under conditions of supply squeeze, they will face fierce competition for securing oil not only from the industrial countries but also from the other high-income LDCs. Also, their access to international finance will be even more limited, while their lack of effective political or economic power internationally provides little incentive for the oil exporters or the oil companies to give them any kind of preference.

Commercial versus Noncommercial Energy

All projections cited in table 3.7, as well as the figures quoted, relate to commercial energy only. However, noncommercial energy (sometimes re-

Table 3.7. Energy Balance of Non-OPEC LDCs in 1985

	Actual 1976	Frankel 1978	IEA 1978	Petrocan Petroven 1978	WAES/MIT D and C Cases 1977	
Demand						
Million b/d oil eq.	8.9	17.2	17.3	16.8	14.9	18.2
Quadrillion Btu	(18.8)	(36.4)	(36.6)		(31.5)	(38.5)
GNP growth rate	5.3	N.A.	5.6	5.4	4.1	6.1
E/GNP multiplier	1.39	N.A.	1.24	1.1	1.2	1.2
Energy demand growth	7.3	N.A.	6.9	5.9	4.9	7.3
Indigenous supplies	6.8	16.7	15.6	15.0	11.6	14.6
Oil and NGL	3.4	8.5	8.9	7.7	5.1	6.7
Natural Gas: Million b/d oil eg.	0.8	1.9	1.5	2.3	2.1	2.4
Trillion cu. ft.	(1.7)	(3.9)	(3.1)	(4.7)	(4.3)	(4.9)
Coal: Million b/d oil eq.	1.6	3.4	2.7	2.5	2.2	2.7
Million short tons	(154)	(327)	(276)	(243)	(217)	(260)
Nuclear: Million b/d oil eq.	engl.	0.2	0.4	0.6	0.4	1.1
In gigawatts capacity		(6.6)	(13)	(17)	(13)	(37)
Hydro/geo/other:						
Million b/d oil eq.	1.0	2.7	2.1	1.9	1.8	1.7
Quadrillion Btu	(2.1)	(5.7)	(4.4)	(4.0)	(3.8)	(3.6)
Net non-oil imports	−0.5	−0.9	−0.2	−0.3	−0.4	−0.2
Natural Gas: Million b/d oil eq.	−0.5	−0.3	−0.2	−0.4	−0.5	−0.5
Trillion cu. ft.	(−1.0)	−(0.6)	(−0.2)	(−0.9)	(−1.0)	(−1.0)
Coal: Million b/d oil eq.	—	−0.6	—	0.1	0.1	0.3
Million short tons		−(58)		(7)		
Residual demand for imported oil	2.6	0.2	2.1	1.7	3.3	3.6

Source: U. S. Department of Energy.

Table 3.8. Non-OPEC LDC Oil Production Possibilities, 1985 (1000 barrels per day)

	Actual 1977	CIA 1978	IEA 1978	World Bank 1978	Petrocan 1978	Texaco 1978
Argentina	440	430	575	548	600	470
Bolivia	35	100	115	110	100	
Brazil	162	505	610	753	600	625
Chile	22	39	43	41	100	
Colombia	140	155	175	137	100	
Guatemala	—	15	—	25	100	
Mexico	990	3,940	3,500	2,329	2,200	2,700
Peru	90	200	220	192	200	
Trinidad & Tobago	230	200	235	180	300	
Subtotal W. Hemisphere	2,109	5,584	5,473	4,315	4,300	4,685[a]
Angola	195	200	235	220	500	
Cameroon		100		27		
Congo	35	70	125	16		
Egypt	450	1,000	1,000	1,233	1,000	476
Tunisia	87	150	145	151	100	
Zaire	24	50	80	96		370
Subtotal Africa	1,791	1,720	1,585	1,743	1,700[a]	1,261
Bahrain	54	20	52	55		
Oman	350	250	250	370	700	606
Syria	200	200	210	219		270
Brunei/Malaysia	397	610	625	521	600	492
Burma		70		41		
India	200	500	575	493	500	427
Pakistan		50		41		
Taiwan		7				
Other Asia				600	100	117
Subtotal	1,291	1,707	1,712	2,340	1,900	2,007[a]
Others not listed	38					
TOTAL	4,139	9,011	8,920	8,398	7,700	7,953

[a] Totals do not add up because production was not identified in a number of countries.
Source: U.S. Department of Energy.

ferred to as traditional or untaxed energy) constitutes a major portion of LDC energy use. Such forms of energy supply about 5 percent of world energy consumption, but they account for about 50 percent of the energy production of oil importing LDCs and more than 85 percent of the energy requirements of rural areas. Some low-income LDCs, such as Nepal, Mali, and Tanzania, rely on noncommercial sources for 90 percent of their energy needs. The demand for such fuels is dominated by household uses, primarily cooking. About half the world's population today cooks with noncommercial energy.[12]

To what extent does the exclusion of noncommercial energy distort the projections of energy supply and demand for LDCs? The answer is not easy, given the lack of reliable data and information on noncommercial energy. It is, however, realistic to assume, first, that the LDC governments are not likely to change their attitudes towards such fuels in the short or medium term and may continue to pay little attention to their development; and second, that since noncommercial energies are excluded both from demand and supply projections, if noncommercial energy supply grows at such a rate as to satisfy increasing demand in rural areas, the exclusion of the noncommercial energy sector may not significantly affect the conclusions reached for the commercial sector. This is not meant to undermine the vast potential for expansion of noncommercial energy supply, but it is a realistic assessment of its likely impact in the next few years.

OPEC ASSISTANCE

Given OPEC's intentions to help the developing world, as a matter of policy, a great deal of assistance was provided in the 1970s. While these policies were generally effective until 1977/78, the developments in the past two years and expected future developments require new policy approaches to the problem. The key problems in the 1980s are expected to be supply shortages and major price increases. In the following, OPEC traditional assistance policies and options for the future will be examined.

What OPEC Has Done So Far (1973–79)

OPEC Aid: Since the 1973/74 oil price increase, the OPEC nations have entered the aid donors league on a large scale. Except for Ecuador, Gabon, and Indonesia, the members of OPEC have actively participated in providing aid to LDCs. Aid was provided through five channels: the OPEC Fund, Specialized Banks, national development funds, international organizations such as the International Monetary Fund, the International Board for Reconstruction and Development, and the United Nations. (IMF, IBRD, UN), and bilateral aid (see tables 3.9 and 3.10). Basically, aid was provided both for balance of payment difficulties and on a project basis. During 1974–76, OPEC aid (excluding multilateral contributions) amounted to $14.6 billion, while the LDCs' net oil import bill rose by $28 billion in the same period. However, if the net oil import bill of the five major LDCs (Taiwan, South Korea, India, Brazil, and Argentina) is excluded, then it can be seen that OPEC aid more than compensated for the increase in the oil import bill of the LDCs.[13]

Table 3.9 shows the multilateral OPEC aid agencies' contribution during 1976–78, while table 3.10 shows total flows from OPEC to LDCs until 1978. Total net OPEC flows peaked in 1975 and then slowly began to fall to $5.3 billion in 1978—corresponding to 3.3 percent of combined GNPS of OPEC member countries. However, concessional Official Development Assistance (ODA) aid has remained relatively steady at $5–6 billion a year, constituting over two-thirds of total flows during the same period. As a percentage of GNP, ODA flows fell from 2.7 percent in 1975 to around 1 percent in 1978, as compared to 0.35 percent from the Development Assistance Committee (DAC) members of OECD.

OPEC aid's ODA element is far larger than OECD aid. The ODA portions of OPEC aid carry a grant element slightly lower than the average grant element of the industrial nations' aid. Also, OPEC aid has the major advantage of not being tied. This increases the flexibility and the real worth of aid significantly. At the same time, OPEC aid, which initially gave preference to Muslim nations and which had political ties, has since 1976 been expanded to include all LDCs without regard to religious and political ties. For instance, the OPEC Fund has so far extended 113 loans to 66 nations, including Communist countries (Vietnam, Mozambique, and Laos), with grant elements of 60–70 percent. However, since 1977 OPEC aid has not kept up with the increased oil import bills of LDCs, primarily due to the declining real price of oil and diminishing surpluses.

Price Discounts: A number of OPEC countries, notably Iran, Iraq, and Saudi Arabia, have in the past supplied some quantities of oil to LDCs at discount prices. In some cases, there was an outright discount, in other cases, part of the price was accepted in local currency or deferred as long-term low-interest loans. Since these discounts were awarded in secrecy, little information is available on the terms. However, it is understood that the quantities involved were small and no such discounts are known to have been given since 1977.

How OPEC Might Help in the Future

Aid: OPEC assistance, which began to decline in the late 1970s, is likely to be substantially increased in the 1980s. With oil sales of over $300 billion and a likely surplus of $100 billion in 1980, prospects for OPEC aid looks encouraging. At the same time, OPEC finance ministers have agreed in principle to turn the OPEC Special Fund into an OPEC development agency with an additional $2.4 billion capital which would raise its funds to $4 billion—less than 15 percent of which has already been disbursed. It is, however, unrealistic to assume that the increase in OPEC aid can cover the additional oil

Table 3.9. Total Net Flows Committed by Arab/OPEC Multilateral Institutions ($ million)

Institution[a]	1974	1975	1976	1977	1978
AFESD	127.2	200.7	336.5	365.3	1.3
AFTAAAC	—	—	0.5	4.1	6.8
ABEDA	—	—	79.6	77.0	72.8
GODE	—	—	250.0	1,495.0 (3)	100.0[c]
IDB	—	—	—	120.2	240.0[c]
ISF	—	—	6.0	8.0 (3)	8.0[c]
OAPEC Special Account	79.0	—	37.1	—	—
OPEC Special Fund	—	—	42.7	243.0	155.0
SAAFA[b]	80.3	71.6	56.5	13.2	—
Others	41.4	70.1	207.2	400.0 (3)	450.0[c]
Total	327.9	342.4	1,016.1	2,725.8	1,033.9[c]

[a] AFESD = Arab Fund for Economic and Social Development; AFTAAAC = Arab Fund for Technical Assistance to Arab and African Countries; ABEDA = Arab Bank for Economic Development in Africa; GODE = Gulf Organization for Development in Egypt; IDB = Islamic Development Bank; ISF = Islamic Solidarity Fund; SAAFA = Special Arab Aid Fund for Africa (merged with ABEDA in 1977).
[b] Arab African Bank; Arab International Bank; Arab Investment Company; Arab Petroleum Investments Corporation (Apicorp).
[c] Estimate by OECD Secretariat.
Source: OECD Development Corporation, 1979.

import bill of the LDCs, which amounts to $18 billion between 1978 and 80.[14] In addition, OPEC has floated the idea of an OPEC bank of $10–20 billion, but seeks contributions from the industrial nations.)

Aid/Energy: Very little OPEC aid has gone into indigenous energy production in LDCs. The OPEC Fund, which has so far played a small role in channelling OPEC aid, had committed $64 million by 1979 in aid for power generation projects. The commitments were for Bangladesh, India, Nepal, Pakistan, Sri Lanka, Morocco, Rawande, Sudan, Tunisia, the Dominican Republic, Honduras, and Jamaica.[15] This represents a minor activity in terms of LDCs' energy needs since (a) the sums involved were insignificant and not even all funds have yet been disbursed, (b) power generation, in most cases, meant electric power from imported oil, and (c) no attention was paid to development of nonoil energy sources. In the 1980s, OPEC may consider linking a larger portion of the now expanded OPEC Fund to projects involving indigenous energy sources.

Two-tier Price System?: The idea of a two-tier price system for oil sales to LDCs was totally rejected by OPEC in mid-1970. It was thought that the oil might be reexported as crude or refined products, it might encourage

Table 3.10. Total Net Flows (Disbursements) from OPEC Members: 1973–78[a]

	($ Million)						(As % of GNP)					
Donor Country	1973	1974	1975	1976	1977	1978p	1973	1974	1975	1976	1977	1978p
Algeria	29.8	51.4	42.2	66.6	73.1	55.2	0.34	0.41	0.29	0.41	0.38	0.23
Iran	4.9	739.4	936.1	807.3	315.5	333.4	0.02	1.59	1.79	1.22	0.38	(0.40)
Iraq	11.1	440.2	254.4	254.7	135.2	211.2	0.21	4.15	1.92	1.59	0.70	0.96
Kuwait	555.7	1187.1	1712.2	1875.7	1864.6	1150.8	9.21	10.90	14.23	13.28	13.04	6.10
Libya	403.8	263.2	362.8	363.2	287.0	548.9	6.25	2.21	3.21	2.45	1.62	3.01
Nigeria	5.7	134.8	347.5	176.8	35.7	7.7	0.05	0.60	1.31	0.52	0.09	0.02
Qatar	93.7	217.9	366.7	240.3	265.9	133.6	15.62	10.90	16.90	9.79	10.65	4.61
Saudi Arabia	334.9	1622.1	2466.7	2817.3	2709.7	1746.7	4.43	7.02	6.67	6.70	4.86	2.78
UAE	288.6	749.4	1206.6	1144.5	1395.2	761.4	15.96	11.12	16.29	11.89	12.11	6.63
Venezuela	17.7	483.4	473.8	392.2	510.5	346.7	0.11	1.66	1.75	1.26	1.43	0.85
Total	1745.9	5888.9	8169.0	8138.6	7592.4	5295.6	1.89	3.35	4.01	3.30	2.54	1.59
Official Development Assistance (ODA) (Concessional)	1307.8	3446.6	5516.9	5594.7	5848.9	3703.6	1.42	1.96	2.71	2.27	1.96	1.11

p = provisional.
[a] Includes all concessional, nonconcessional, bilateral and multilateral flows.
Source: OECD Development Corporation, 1979; World Development Report, p. 157.

oil-intensive industrial activity, and more importantly, it might lead to pressure from the industrial countries to purchase oil at the lower price. The latter objection was perhaps the decisive one in killing the proposal. However, the idea has been revived, with different variations, by a number of countries, particularly Iraq. The Iraqis propose that LDCs with per capita incomes of under $300 and oil imports of less than 10,000 b/d be fully compensated for the increase in their oil import bill, and LDCs with per capita income of $300–1000 and oil imports of under 100,000 b/d be compensated for 25–30 percent of their additional oil bill through long-term interest-free loans.[16] It is not likely that a formal two-tiered price system will be established, but, there is a good chance that OPEC will indirectly compensate the developing world for a portion of their additional oil bill. Another possibility, not seriously considered as yet, is the deferment of a portion of the price—say, above $25–30—as long-term interest-free loans.

Term Contract/Supply Security: During the second half of 1979, when spot prices rose to nearly double the official (term contract) prices, the LDCs, who were not party to a long-term (9–12 months) contract, were placed in the difficult position of purchasing crude at spot prices. Most OPEC nations, uneasy about the discrimination against LDCs, decided to sell oil at official contract prices to LDCs. This helped LDCs escape the volatile spot market and ensured security of supply as well. As long as one or more LDCs could secure direct delivery by getting access to tankers, they were given special preference for term contracts. A number of OPEC nations even reduced their spot sales to accommodate term contracts for LDCs. In a number of cases, when LDCs were already purchasing oil on a term contract and official prices were increased, the purchasing LDCs were partially compensated. Iraq alone provided $170 million in long-term interest-free loans to poor LDCs between June and December 1979 to cover the surcharges on Iraqi contract crude. A number of examples for 1979 and 1980 are cited in table 3.11.

A large number of contracts were entered into by OPEC and LDCs both for 1979 and 1980. Since spot prices in July 1980 are not much above official prices, the benefit to buyers of contract crude is not as great in 1980 as it was in 1979. Nevertheless, such crude contracts assure LDCs of supply security and shield them from violent price fluctuations of the spot market.

During the 1970s, the pattern of crude supply to major international oil companies drastically changed. While in 1973, 92 percent of the world petroleum supply was handled by the oil companies, by 1979, this ratio had fallen to 58 percent. Direct state-to-state deals rose from 1.5 million b/d (5 percent of OPEC trade) in 1973 to 5 million b/d by 1979 (16.5 percent of supplies).[17] The increase in direct sales (state-to-state and commercial) came at the expense of the international oil companies, whose liftings have been drastically

Table 3.11. OPEC Nations' Direct Contract Crude Sales to Developing Nations–Selected Countries

Seller	Buyer	Contract Date	Quantity
Iran	Bangladesh	1979	NA
Iran	Portugal	1979	NA
Iran	Brazil	1979	NA
Iran	Philippines	1979	NA
Iran	India	1980	125,000 b/d
Iran	Spain	1980	30,000 b/d
Iran	North Korea	1980	NA
Iran	Brazil	1980	60,000 b/d
Iran	South Korea	1980	60,000 b/d
Iraq	India	Oct.–Dec. 1979	1,400 b/d (gas oil)
Iraq	India	Oct.–Dec. 1979	572 b/d (kerosene)
Iraq	India	Oct.–Dec. 1979	10,000 b/d
Iraq	Argentina	1980	NA
Iraq	Cyprus	1980	13,800 b/d
Iraq	Zambia	1980	3,000 b/d
Iraq	Yugoslavia	1980	100,000 b/d
Iraq	Greece	1980	44,000 b/d
Iraq	Tanzania	1980	NA
Iraq	Brazil	1980	400,000 b/d
Iraq	India	1980	118,000 b/d + 10,000 b/d (kerosene)
Saudi Arabia	Lebanon	Jan.–Sept. 1979	332,000 b/d
Saudi Arabia	Spain	1980	150,000 b/d
Algeria	Lesotho	1980	NA

Source: Petroleum Intelligence Weekly and *Middle East Economic Survey.* Various issues.

reduced. For 1980, major cutbacks have been announced by Iran, Iraq, Kuwait, and the UAE for sales to the traditional purchasers of their crude—the international oil companies.

In the 1980s the trend is certainly towards increasing state-to-state deals, particularly for the developing nations. The problem with direct sales to LDCs is that their small demand size makes tanker transport and refining uneconomical. LDCs must consider joining together on a regional basis and placing larger orders for OPEC oil, which would make its transport and refining economically viable. OPEC is expected to continue to give preference to the national oil companies of the developing world.

LPG Trade: After oil, natural gas is the most important resource in many of the oil-exporting countries. Around 38 percent of the world's proven gas reserves are located in OPEC countries, while Iran alone has 20 percent of world gas reserves. Currently gas production in OPEC is in the neighborhood of 268 billion cubic meters, 50 percent of which (nearly one billion barrels of crude oil equivalent) is being flared. Despite the important role which gas might play in the world energy picture, the future of gas exports from OPEC is not promising for the following reasons:

1. Gas prices are in most cases linked to fuel oil No. 6, an inferior fuel, while gas competes favorably with all petroleum products.
2. Transport of gas through LNG tankers or pipelines is extremely expensive—5–10 times the transport cost of crude.
3. Liquefaction and regasification terminals mean little flexibility in destinations of exports.
4. Liquefaction and pipelines require major capital investments.
5. The netback on gas sales is at best 50 percent of the netback on crude sales (for equal Btus).

Rather than exporting their natural gas, most OPEC nations are planning to use an increasing volume of it for secondary recovery (gas injection) and domestic consumption.

While the future prospects of natural gas exports are limited, the future of trade in liquid petroleum gas (LPG) is promising. LPG is a mixture of butane and propane found dissolved in crude oil or in associated and nonassociated gas. It can be recovered in special gas separation facilities or as a refinery byproduct. Compared to LNG, LPG requires far less manufacturing and transportation costs. It provides flexibility for both the exporter and the consumer. Saudi Arabia, Kuwait, and the UAE are involved in massive gas gathering programs which will make substantial volumes of LPG available for export. Saudi Arabia's LPG production is expected to reach 15 million tons, Kuwait's, 4–5 million tons, and the UAE, 5 million tons, by 1985. The demand for LPG was around 10 millon tons in 1977/78 and is not expected to grow at rates faster than the petroleum products' demand. This could lead to large surpluses of LPG supply over demand which may result in weakening of LPG prices or serious underutilization of capacity. This provides a strong case for including LPG in any aid package for LDCs.[18]

LPG consumption in developing nations has been growing; however, its growth potential is far greater than the historical rates suggest. LPG competes favorably with the most widely used petroleum product in developing countries—kerosene—both for cooking and heating. In the developing countries production and supply of middle distillates—of which kerosene constitutes the largest portion—has created a major problem for their governments. Refinery yields do not match demand patterns—that is to say, the percentages of middle distillate demand in LDCs is higher than the percentage refinery yields of these products. The mismatch has forced the developing nations either to increase refining capacity unnecessarily, or to install costly isomerization units to increase middle distillate yield, or to import kerosene at high costs. Even OPEC nations themselves face similar problems. Iran and Indonesia, for instance, are forced to import kerosene themselves.[19]

The OPEC nations' increased trade in LPG with the developing nations is, therefore, a possibility which benefits both sides. OPEC can reduce its surplus LPG while helping LDCs in an effective energy aid mechanism. LDCs can benefit from such trade by getting supply security, buying the product at relatively cheaper prices and saving themselves the ever-increasing marginal costs of obtaining additional kerosene.

Development of LDCs' Indigenous Petroleum Supplies: The issue of developing indigenous energy resources of LDCs as a means of reducing their dependence on imported oil has only recently received attention from the international community. Following a decision of the World Bank's executive directors in July 1977, the bank undertook a program for financing petroleum production projects. A number of studies by the bank staff and the French Petroleum Institute, as well as the U.S. Department of Energy, indicated the vast potential for petroleum production of major oil-importing LDCs.[20] The World Bank's initial $500 million fund was looked upon as a catalyst for raising more funds and encouraging the oil companies to enter into exploration and development in LDCs. However, the bank's well-intentioned move has not yet met with much success, for reasons beyond the scope of this paper.

OPEC's interest in indigenous energy resources of LDCs started with the Ministerial Commission on Long-Term Strategy,[21] which held its first meeting in Geneva in June 1978. The commission, chaired by the Saudi oil minister, Sheikh Zaki Yamani, commissioned a number of studies by experts on the future role of OPEC. One of the studies is supposed to explore various means of helping the development of indigenous petroleum reserves of LDCs. Though the final report has not yet been submitted to the member governments,[22] it is reasonable to assume that this area will receive special attention from OPEC members, given the current and likely problems which LDCs are going to face in securing their petroleum needs.

OPEC's help in developing LDCs' petroleum reserves could take different forms, none of which have yet been thoroughly explored. The options include:

1. Cooperation with the World Bank in providing funds for geological surveys, exploratory drilling, and development
2. Providing finances for surveys and drilling through independent consultants and oil companies

3. Joint ventures of OPEC national oil companies with LDC energy authorities as well as possibly international oil companies
4. Providing free advice on contractual and other legal matters by involving the OPEC Secretariat and experts from the OPEC national oil companies
5. Helping LDCs get the best possible deal from the oil companies and guaranteeing smooth marketing of their oil if exports are possible

The potential impact of OPEC participation in the development of the oil resources of LDCs could be great. Even a symbolic OPEC presence would serve to reassure both LDC governments and the international oil companies of their contractual arrangements. Oil companies are worried about expropriation, political instability, and low return on their investments, while LDCs are inexperienced in such dealings and often impose terms which deter oil company involvement. OPEC members lack the expertise required for direct involvement, though countries such as Iran, Iraq, Algeria, Venezuela, and Indonesia have some technical expertise which can be used in development of the oil resources of LDCs. The most likely development is OPEC cofinancing together with the oil companies and LDC governments through joint ventures. So far only one country—Kuwait—has publicly announced its intention to enter into LDC petroleum ventures. Kuwait plans to start exploration in four or five LDCs with a view to commercial development of oil reserves in these countries.[23] Though the scope of Kuwait's plan is not clear, this area certainly offers a great deal of potential for the future.

SUMMARY

The world petroleum market in the 1980s is likely to witness shrinking petroleum supplies and rising real prices. The shortfall in supplies is likely to come about as a result of OPEC production cutbacks due to domestic political pressures and problems of international financial instability.

Both the industrial and developing nations which got through the 1970s with a relative degree of ease are unlikely to be able to repeat the same performance in the 1980s; while a number of LDCs are likely to become large exporters of oil, the majority of the developing countries will remain dependent on oil imports.

The OPEC aid program which was effective through 1978 is not likely to be able to cope with the developments of 1979/80 and the years ahead. Indeed, it is both unfair and unrealistic to expect OPEC to compensate LDCs for their rising oil bills—which have risen from $12 billion in 1978 to a possible $30 billion in 1980. Thus other measures will have to be found to *partly replace aid with energy aid.* These options, some of which have not yet been explored by OPEC, include: financing energy projects, term contracts, price discounts, LPG trade, and the development of the indigenous petroleum reserves of LDCs. The options considered are by no means exhaustive, and there is no certainty whether one or a number of these options will be adopted by OPEC. Nevertheless, since LDC energy problems are likely to affect these countries' relations with OPEC, and since the LDCs' goodwill is of paramount importance to the oil exporters, it is fair to assume that policies along these lines to supplement direct aid may be adopted to help LDCs alleviate their energy problems.

To what extent OPEC assistance will help LDC development problems is a question which cannot be answered. However, it is unrealistic to assume that OPEC assistance by itself can solve either the energy problems or the broader development problems of the developing world without cooperation from the industrial world and without major attempts by the LDCs themselves to reduce their energy dependence.

NOTES

1. *Petroleum Intelligence Weekly* (New York, November 19, 1979).
2. *World Financial Market* (New York: Morgan Guaranty Trust Company, November 1979).
3. *World Development Report* (Washington, D.C.: World Bank, 1979).
4. See F. Fesharaki, "Iran's Energy Picture After the Revolution," *Petroleum Intelligence Weekly*, Supplement, September 24, 1979; and *The Middle East Economic Survey*, Supplement, September 3, 1979. Also see F. Fesharaki, "Revolution and Energy Policy in Iran: International and Domestic Implications," paper presented to APESC III Workshop, Honolulu, HI., 1980 (unpublished), February 1980.
5. *World Financial Market*, p. 4.
6. Bechtel Corporation estimates costs of synfuels at $40–60 per barrel oil equivalent. See *The Economist*, October 6, 1979. OPEC Economic Commission Report indicated a cost of $35–56 per barrel oil equivalent for alternatives.
7. *World Development Report*, p. 35.
8. *The World Oil Markets in the Years Ahead*, p. 31.
9. Argentina, Brazil, Mexico, Egypt, India, South Korea, Taiwan, Singapore, Thailand, Malaysia, Colombia, Peru, Philippines, Pakistan, Chile, Syria.
10. *Energy Options and Policy Issues in Developing Countries*, The World Bank Staff Working Paper No. 350, 1979.
11. *Energy Options and Policy Issues*, p. 9.
12. *World Development Report*, pp. 41–43.
13. "Oil and Arab Cooperation." *OPEC Bulletin*, 4, no. 1 (1978).
14. LDCs oil import bill was approximately $12 billion in 1978, $20 billion in 1979, and is expected to meet $30 billion in 1980.
15. For details see OPEC Special Fund, Third Annual Report (1979).
16. This idea was proposed at the meeting of nonaligned nations in Havana in 1979 and later at the December 1979 OPEC Conference. OPEC aid is made conditional on reciprocity by the developed countries to compensate LDCs for the increase in the cost of their manufactured exports in a $15 billion joint fund.
17. *Petroleum Intelligence Weekly*, February 25, 1980.

18. The idea was originally suggested by Bijan Mossavar-Rahmani in "OPEC and the Indian Ocean in the 1980s," *OPEC Review* 3, no. 3 (Autumn 1979).
19. See F. Fesharaki, *Development of the Iranian Oil Industry* (New York: Praeger, 1976).
20. *Petroleum and Gas in Non-OPEC Developing Countries: 1976–1985*, World Bank Staff Working Paper No. 289 (April 1978); *Energy Options and Policy Issues in Developing Countries*, World Bank Staff Working Paper No. 350 (August 1979); *A Program to Accelerate Petroleum Production in Developing Countries*, World Bank, January 1979; *The 1985 Oil Production of 21 Non-OPEC Oil Producing Nations*, U.S. Department of Energy, March 1979; *Energy Supply and Demand Balance and Financing Requirements in Non-OPEC Developing Nations*, U.S. Department of Energy, February 1979.
21. Members of the Commission are Iran, Iraq, Saudi Arabia, Kuwait, Venezuela, and Algeria.
22. The report is expected to be first submitted to the OPEC Ministerial Conference in Algeria in June 1980 and later to the OPEC Summit Conference in Baghdad during October 1980.
23. Interview with Sheikh Ali Al-Khalija, the Kuwaiti Oil Minister with the Middle East Economic Survey, February 11, 1980.

Chapter 4
Long-Term Value of Hydroelectric Energy for the Developing Nations

John R. Kiely

The energy from falling water is one of mankind's oldest forms of energy, and has been one of the key factors in the economic and social development of most of the countries in the temperate zones of the world. During the last 50 years or so, however, the abundance of low-cost fossil fuels coupled with the technology for making their energy available over long distances has meant that in many parts of the world hydroelectric power has not been developed to anything near its full potential.

The World Energy Conference's long-range energy study, *World Energy Looking Ahead to 2020*, (1978) concluded that only 16 percent of the reasonably available hydroelectric power in the world has been developed. The study also showed that by far the largest number of undeveloped potential hydroelectric sites are located in the developing countries. Ellis Armstrong, former head of the Bureau of Reclamation, prepared a chart for the study that illustrates this point graphically (see figure 4.1).

As we enter the 1980s, a new and entirely different set of energy economic factors has developed which will have a profound effect on future sources of electrical energy. There has been a quantum jump in the cost of oil and a very substantial escalation in the cost of coal. As a result, the economics of the energy from falling water has improved vastly in relation to coal and astronomically in relation to oil. The conclusion in this paper is that in the face of constantly escalating fossil fuel cost, it is extremely important to look ahead and over the next 20 to 30 years to see the long-range benefits of a hydroelectric plant built now, as compared to a fossil-fuel plant built now.

While some previously proposed hydroelectric facilities were ruled out for a variety of unfavorable considerations, a careful reevaluation of these facilities is now necessary in the light of the current situation. Further, the finite

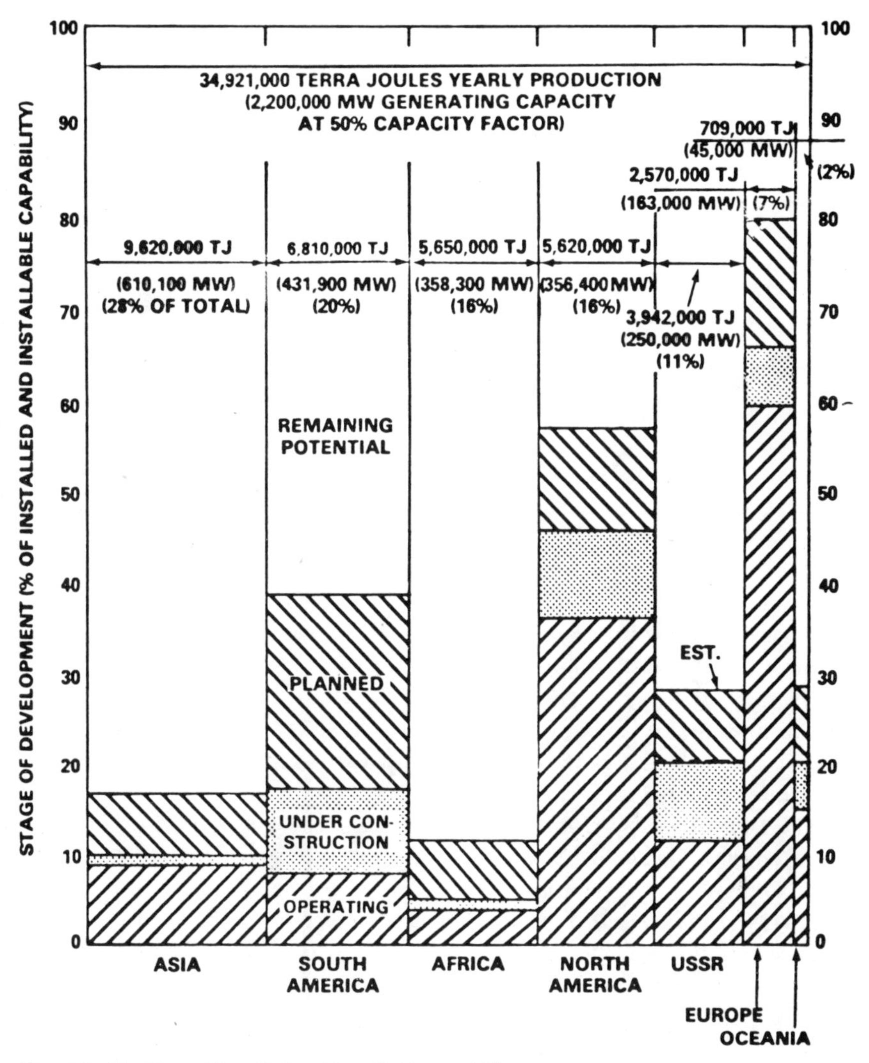

Fig. 4.1. World total installed and installable capability

quantity of fossil fuel and uranium resources used in current light water reactors makes the renewable hydroelectric power resources increasingly important and attractive.

A simplified illustration of the current economic situation is shown in figure 4.2. This graph compares the annual operating costs of a 350 MW 50 percent capacity factor hydro plant costing $2000/kW, with a 250 MW 70 percent capacity factor coal plant costing $1000/kW. Coal is priced at $50 per ton in the first year of plant operations and escalates 8 percent annually.

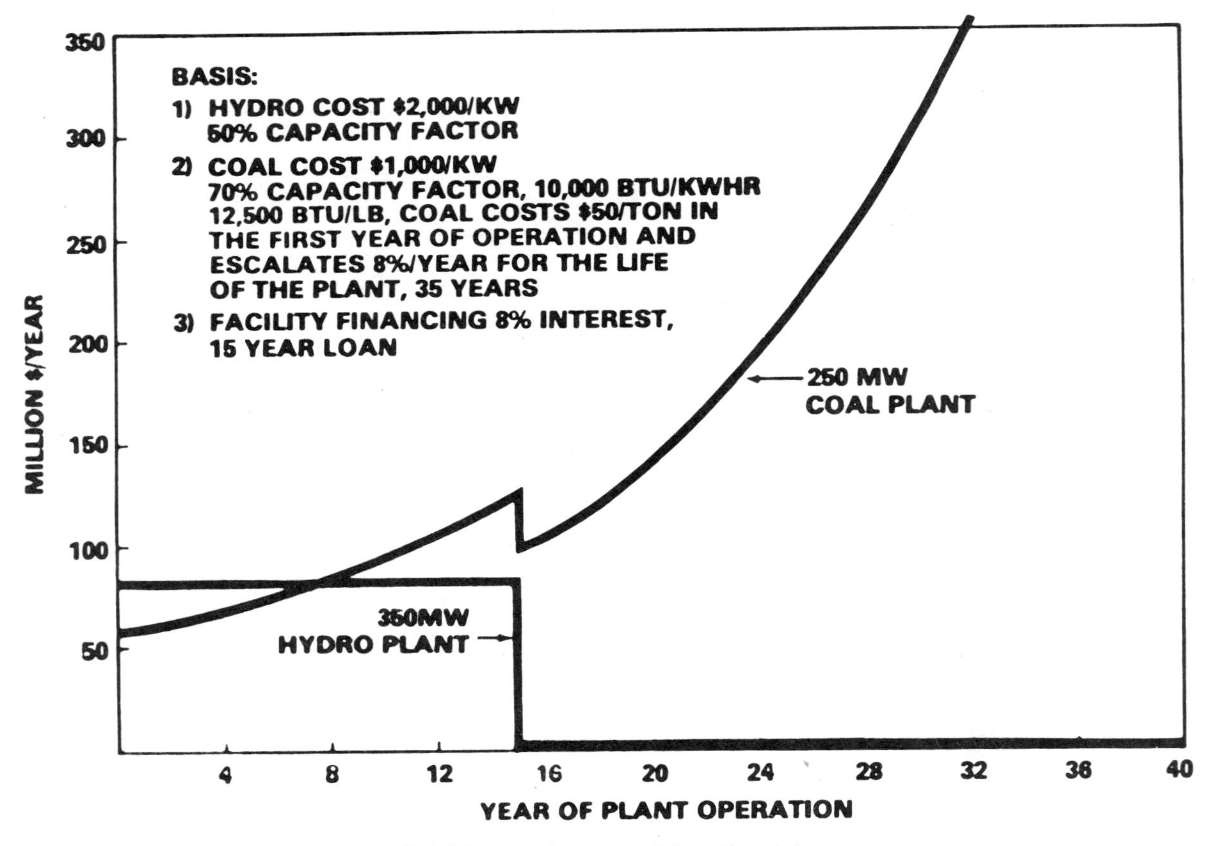

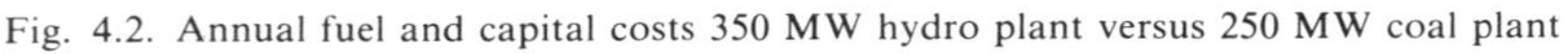
Fig. 4.2. Annual fuel and capital costs 350 MW hydro plant versus 250 MW coal plant

The 50 percent and 70 percent capacity factors are typical of hydro and thermal electric generating facilities respectively. The plant sizes were selected such that both would produce about the same amount of electricity in the course of a year. The financing of the plants was assumed to be the same. The cost of money and the repayment period are typical of World Bank and other international financial institution loans to developing countries for energy facilities. As can be seen, the $2000/kW hydro plant is dramatically less costly over the long term. If a developing nation were able to arrange more moderate financing arrangements, the benefits of hydro would be even greater.

Many developing countries cannot justify moderate or large- size generating facilities. Figure 4.3 compares a 70 MW hydro plant with a 50 MW coal plant. Even in this small size range, hydro has a substantial advantage over coal.

Figures 4.2 and 4.3 are a bit simplified since only capital and fuel costs are included. If operating and maintenance are included, the economic advantage of hydro is enhanced even further. This reflects the larger number of operating personnel and increased maintenance requirements associated with a thermal electric facility.

Figure 4.4 provides an easy way to convert fuel costs from dollars per million Btu, dollars per ton of coal, and dollars per barrel of oil to cost per kWhr on the basis of a plant heat rate of 10,000 Btu per kWhr. Considering that oil currently costs approximately $36 per barrel on the spot market, it is interesting to note, that on an equivalent cost per Btu basis, coal would cost $150 per ton. Coal and oil are, of course, not directly comparable in terms of cost per Btu but certainly this points out the potential for substantial increases in the cost of coal.

Figure 4.5 shows the capital cost/kWhr for typical hydro and coal plants at varying interest rates and levelized yearly payments. This figure is based on a 15-year repayment period which is typical of subsidized financing available to developing nations.

Another way of looking at the comparative costs of coal and hydro is shown in Figure 4.6. The previous figures have all addressed the macro economics situation or cash flow out of a developing country as a result of a major investment in a new generating facility. This does not reflect the typical financing conditions that an operating company within the country would incur over the life of the facility. Hydro plant life is approximately 50 years, and coal-fired plant life is approximately 35 years. Under normal circumstances, the payback period on the loan is the same as the life of the plant, which is the basis for Figure 4.6. The graph compares the justifiable capital cost for a hydroelectric facility with the cost of a $1000/kW coal plant when the cost of coal escalates 8 percent a year over the life of the coal plant. For example, assume 8 percent interest and coal costing $30 per ton in

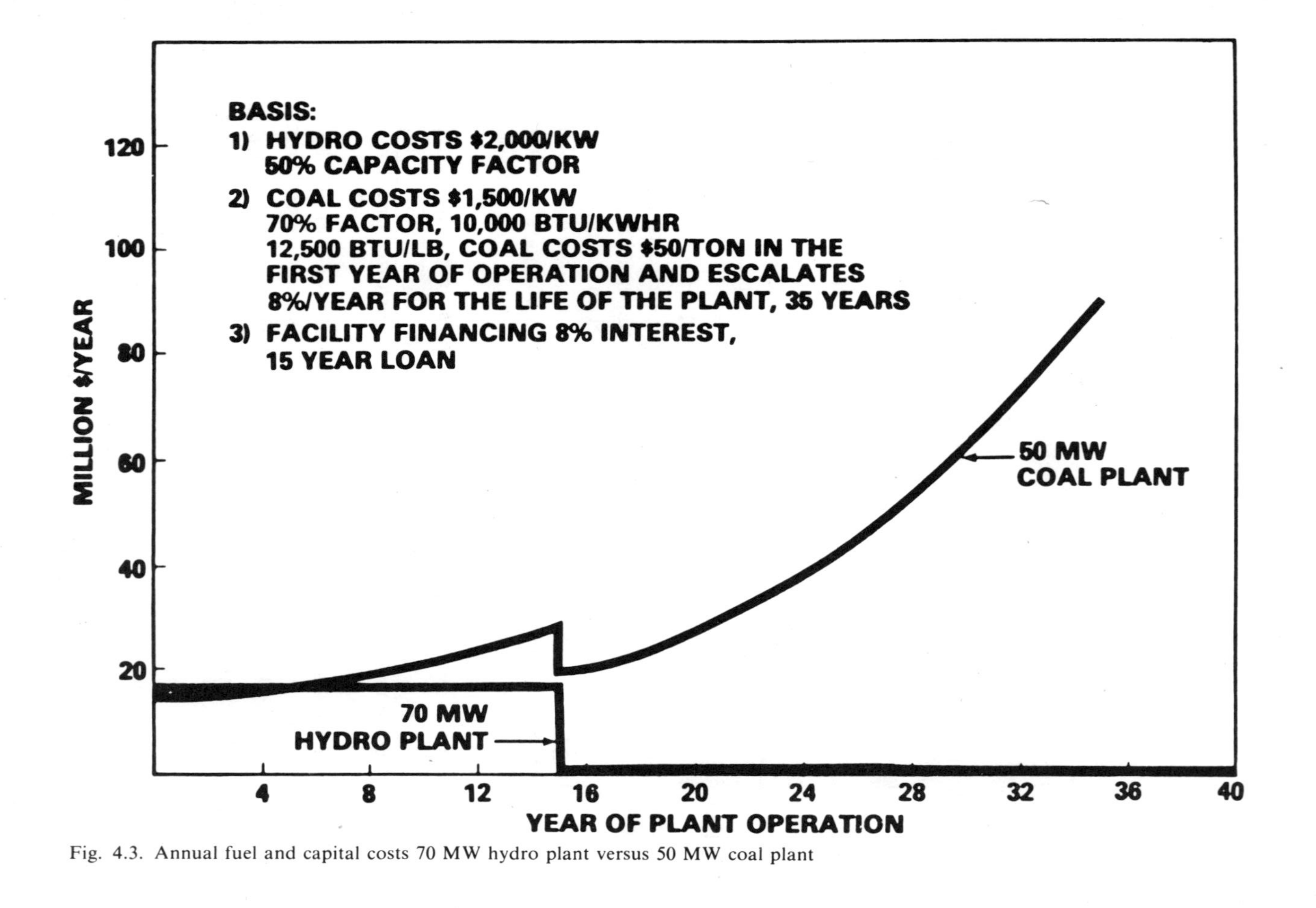

Fig. 4.3. Annual fuel and capital costs 70 MW hydro plant versus 50 MW coal plant

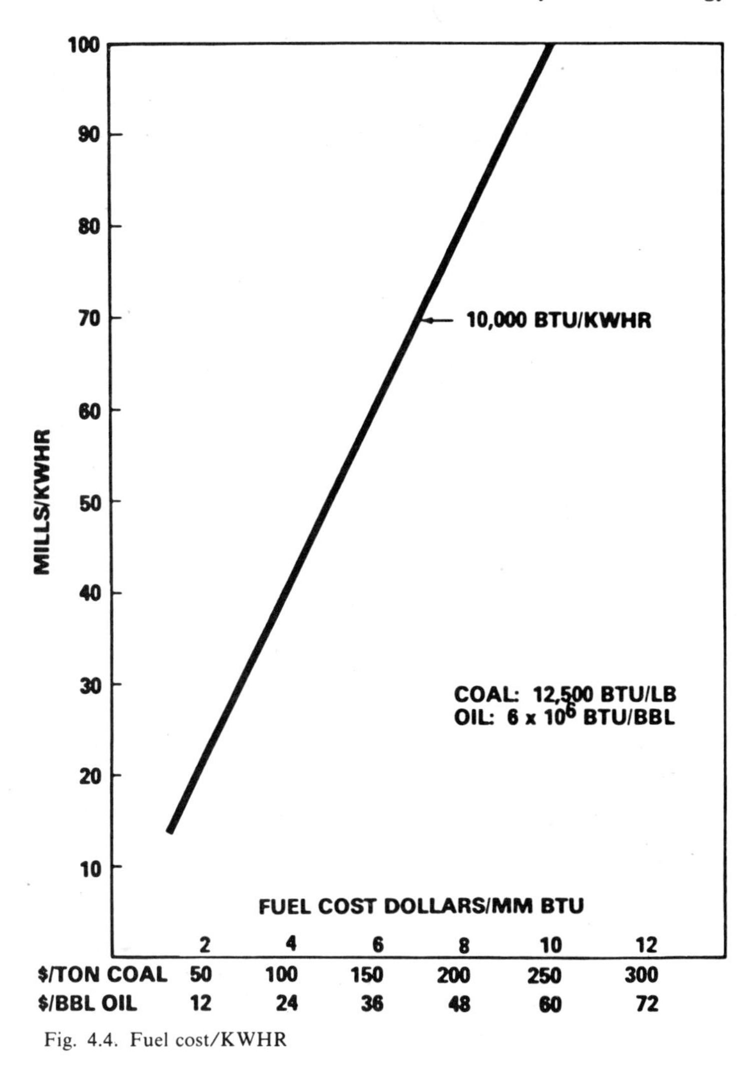

Fig. 4.4. Fuel cost/KWHR

the first year of operation. The capital expenditure for a hydroelectric plant today could be $2680/kW and still produce energy at the same cost as the coal-fired plant.

In addition to the economic benefits, the energy from falling water has a number of other advantages:

1. First and foremost, hydro is a renewable resource and does not require the use of the world's finite supply of coal, oil, and gas.
2. Hydropower is essentially nonpolluting. While we have made great strides in developing means to remove the sulphur and nitrogen oxides from

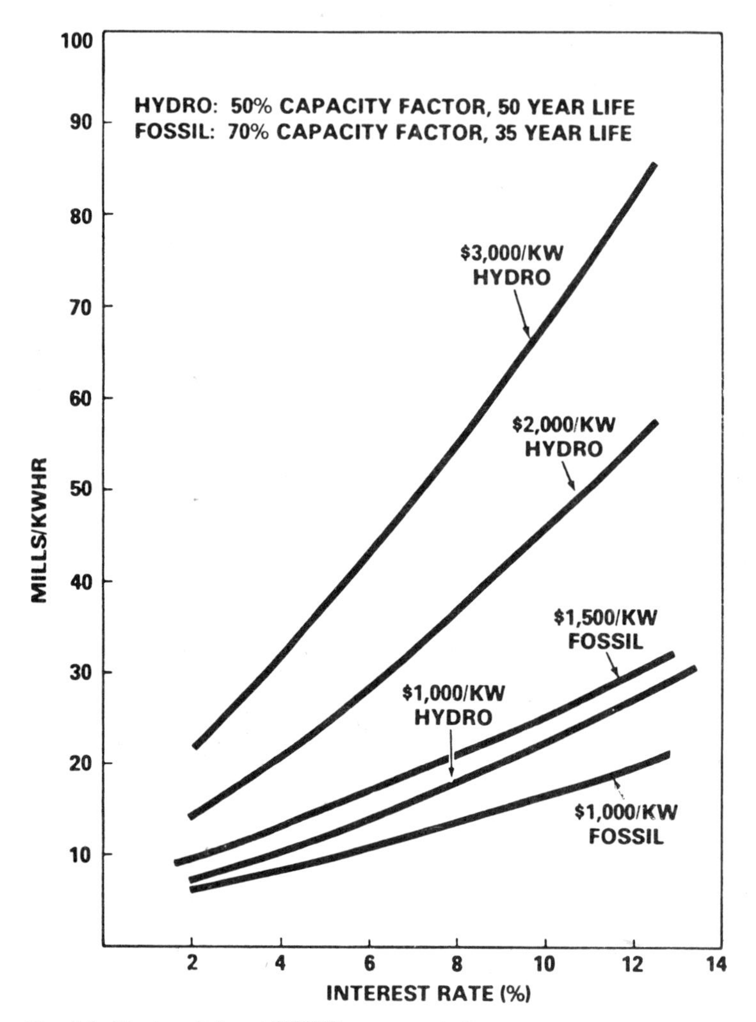

Fig. 4.5. Plant capital cost/KWHR versus cost of money

stack gases, there is still the unresolved question of what happens when we drastically increase the amount of CO_2 in the atmosphere.

3. A multipurpose hydro plant with storage facilities can control floods and can substantially increase the amount of water available for agriculture.
4. The energy reserve from a hydro plant is almost instantly available, which makes it ideally suited for reacting to changing load demands and for providing peaking power.
5. One of the main advantages of hydro power that does not show in a short-range study is that of the long life. We have many plants in the United States now over sixty years old that have become as good as new with moderate refurbishing.

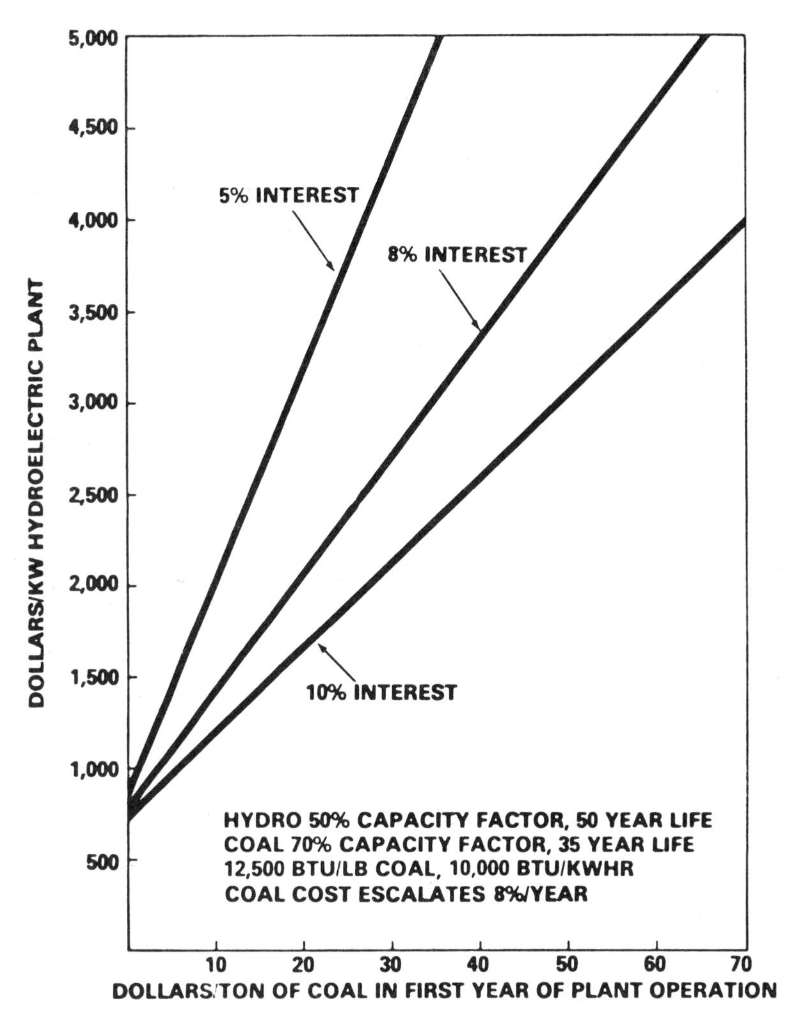

Fig. 4.6. Justified hydroelectric capital cost versus $1,000/KW coal plant

6. A significant advantage for a developing country is that a hydro plant requires fewer and less skilled operating personnel than does a thermal plant.
7. With its large civil component, hydro requires a smaller percentage of foreign exchange and a smaller percentage of imported skilled labor.

With all the advantages that we have cited, hydro power in the developing countries will not progress as it should until the developing countries have a reasonably good knowledge of their potential sites and a rough estimate of the cost of their development. This is an area where the developed countries and agencies such as the World Bank can be of great assistance. Many hydro sites are relatively inaccessible, and many developing countries lack the experience and technology to evaluate an individual site, let alone the country's overall hydro potential. The World Bank, fortunately, has instituted such a program, and one of the first countries studied was Peru.

The results of this study indicate that Peru has an estimated maximum hydro capacity of 195,000 MW, of which 30 percent, or 58,000 MW are feasible considering economic, geological, and topographical restraints. This is 40 times Peru's 1977 installed capacity of 1450 MW, and 8 times Peru's year 2000 projected demand. Most nations are not this fortunate and some nations, of course, have no potential.

In addition to the World Bank and regional development banks, a number of countries such as the United States and Canada have aid programs that provide grants for initial studies and, in some cases, for capital costs, particularly for small hydro installations.

Small Hydro Plants

Many less developed countries have a modest infrastructure and energy requirements. Particularly in isolated areas, small hydro plants may be the only practical means of initially serving such a country.

In Nepal, the village of Namche Bazar is the center of the Sherpa culture. The Sherpas are an industrious and intelligent people whose current situation, nonetheless, is such that they use yak butter for light and yak dung and tree limbs for cooking. There are no roads, only trails and two small airports. The Austrian government, as part of their aid program, is installing a small hydro plant to provide 600 kW of energy for Namche Bazar. This facility will increase the Sherpas' standard of living in many ways. For example chronic eye irritation from the smoky atmosphere will be reduced, as will the denuding of the countryside of trees for cooking fuel.

In many of the poorer countries of the world, the labor of obtaining water and fuel consumes most of the productive energy of the women. Unclean water and lack of fuel are among the chief causes of illness and human misery. Electric power can help alleviate such problems.

The People's Republic of China (PRC) provides an example of a country that has moved aggressively to provide small hydroelectric plants for many isolated areas. Since 1958, the PRC has constructed more than 80,000 small and micro hydro power stations with a total capacity of 5400 MW. Most of these facilities were built by the people's communes at the county level for the use of the commune members. Hydroelectric units with capacities below 500 kW account for more than 60 percent of the total capacity. These units supply the power for local industries and small workshops, grain drying, woodcutting, oil extraction, and lighting. According to a presentation by representatives of the PRC at the International Conference on Water Power held in Washington, D.C., in October 1979, these small hydro facilities have been standardized and are being offered for export. Table 4.1 lists the 16 typical plants.

Large Hydro Plants

At the other extreme of the scale are the very large plants. The largest single plant in the world will be the Itaipu Plant being developed by Brazil and Paraguay on the Paraná River. It will have an installed capacity of 12,000 MW. A project of this size is only possible because Brazil has a rapidly growing load demand and can absorb the greater part of the energy as it becomes available.

Another interesting installation is the 2000-MW Cobora Bassa project in Mozambique. While the plant is in Mozambique, the major portion of the energy goes to South Africa via an 800-mile transmission line. A facility of this size could not have gone ahead without the participation of South Africa, because Mozambique has a very modest demand and could not have financed such a large plant. Now, however, Mozambique has a plant that can ensure its energy supply for a long time and be a major source of foreign exchange as well. In addition, the facility can be very economically enlarged to 4000 MW. The political philosophies of Mozambique and South Africa are diametrically opposed, but this is in no way interfering with the operation of a project that benefits both countries.

India's neighbor, Nepal, is an undeveloped country with few people and a minimum transportation infrastructure. However, it has great hydroelectric potential in rivers such as the Karmali in the west, the Kali Ghandaki in the center, and the Sun Kosi and Arun in the east. There is no way that the people of Nepal can use all of the energy from a large hydroelectric project, nor is it possible to locate much large industry close to these three rivers. However, a substantial part of India's population lives within 400 miles of this source of energy, half the distance from Cobora Bassa to Johannesburg. The benefits to both countries by similarly developing Nepal's hydro potential would be substantial.

On a much smaller scale, there will be opportunities for developed countries to work with developing countries to help locate industrial plants adjacent to hydro plants to absorb a portion of the capacity. This arrangement will give the developing country the economies of size for local consumption and the income to pay for the capital cost of a plant.

Table 4.1. People's Republic of China Water Turbine Generator Sets 12–500KW

Item		Turbine			Generator				
						Speed (rpm)		Voltage (v)	
No	Type	Model	Head (M)	Flow (m^3/sec)	Capacity of generator (kW)	50Hz	60Hz	50Hz	60Hz
1	Axial-flow reaction type	ZD760–LM–60	2~6	0.80~1.80	18. 30. 40. 55. 75	1000	1200	400	460
2		ZD760–LM–80	2~6	1.40~3.50	30. 40. 55	1000	1200	400	460
					75. 100. 125	750	900	400	460
3		GD560–WZ–60	8~14	0.90~2.20	40. 55. 75. 100. 125. 160. 200	750	720 (900)	400	460
4	Mixed-flow reaction type	HL210–WG–20	6~15	0.12–0.19	12. 18	1000	1200	400	460
5		HL310–WG–30	6~18	0.32~0.53	18. 30. 40. 55	1000	1200	400	460
6		HL260–WJ–35	10~25	0.50~0.80	55. 75	750	900	400	460
					100. 125	1000	1200	400	460
7		HL260–WJ–42	8~25	0.65~1.15	75. 100. 125. 160	750	900	400	460
					200	1000	1200	400	460
8		HL110–WJ–42	30~80	0.36~0.59	100. 125	750	900	400	460
					160. 200. 250. 320	1000	1200	400	460
9		HL240–WJ–50	15~40	1.20~2.00	125. 160. 200	600	600	400	460
					250. 320. (400)	750	720	400 (6300)	460 (6300)
					500	100	900	6300	6300

Table 4.1. (Continued)

Item No	Type	Turbine Model	Head (M)	Flow (m^3/sec)	Generator Capacity of generator (kW)	Speed (rpm) 50Hz	Speed (rpm) 60Hz	Voltage (v) 50Hz	Voltage (v) 60Hz
10	Mixed-flow reaction type	HL220–WJ–50	25~59	1.50~1.80	320. (400)	750	720	400 (6300)	460 (6300)
					500	1000	900	6300	6300
11		HL110–WJ–50	35~80	0.56~0.85	200. 250. 320	750	900	400	460
					400. (500)	1000	1200	400 (6300)	460 (6300)
12		HL110–WJ–60	30~70	0.70~1.10	160. 200	600	600	400	460
					250. 320. (400)	750	720	400 (6300)	460 (6300)
13		HL240–WJ–71	23~32	2.90	500	500	514	6300	6300
14	Impulse type	CJ–W–55/1×58	100~260	0.12~0.19	75. 100. 125. 160	750	720	400	460
					200. 250. 320. 400	1000	900	400	460
15		CJ–W–65/1×72	100~260	0.18~0.29	125. 160	600	600	400	460
					200. 250. 320. (400)	750	720	400 (6300)	460 (6300)
					500	1000	900	6300	6300
16		CJ–W–92/1×11	138~145	0.50	500	500	514	6300	6300

For water turbine generator sets having capcities not over 500 kW, customers may make use of the above table for choosing the right specifications. Those not herein specified can also be supplied subject to individual negotiations.

Multipurpose Developments for Energy, Water Supply, and Agriculture

According to the World Energy Conference (WEC) Long-Range Energy Study:

Multipurpose Use of Water: In considering any magnitude of hydroelectric development, planning must consider all water resource needs and the ways to meet the needs. Hydroelectric development cannot be undertaken in isolation from other needs. In many areas of the world, water supply is the controlling factor of present and future human activity. Thus, a comprehensive master plan for a river basin, and perhaps including that of adjacent river basins, is essential for optimum development, especially as water supply becomes a more critical consideration in the region.

The value of hydroelectric energy produced in a comprehensive plan for a river basin can prove to be the economic support that makes the entire development possible, with full consideration of future needs and the facilities to meet them.

The short-term effects, at least on the existing social structures in an area, will depend on the magnitude of the water resource development. Small hydroelectric installations, in providing electricity for house and street lighting, and sometimes an assured water supply as a by-product, can initiate the change from a primitive, poverty-stricken culture to improvement in living standards. A large development that may require the relocation of villages and large numbers of people presents problems that have serious impacts and require full knowledge and sound judgement in determining solutions. There are no easy ways to accomplish these relocations, but often the conditions of those moved can be greatly improved with proper planning. Such problems emphasize the need for multidisciplinary systems analysis of proposed developments, and for decisions made with mature judgments.

Financing Projects

An essential element in the hydroelectric program of a developing country is its ability to finance the projects. Middle- and upper-income developing countries such as Korea, Taiwan, and Brazil are able to attract financing for projects from a wide variety of international financing sources including international financial institutions, export credit agencies, and international commercial banks because their economies have the capacity to generate sufficient foreign exchange revenues to repay loans made on commercial banking terms. Lower-income developing countries frequently do not have the economic capacity to service large amounts of debt on commercial terms and therefore are limited to financing their projects on more concessionary terms.

Three basic sources of international funding are being used to finance the development of hydro power resources in developing countries: the international financial institutions (development banks, development funds, and concessionary financing agencies), export credit agencies, and international commercial banks. The terms for these basic sources of finance are outlined in table 4.2, and a representative listing of the different international financial institutions is contained in Appendix A.

The cost of large hydroelectric projects often exceeds the available resources of any individual source of financing. Consequently, multi-national packages of financing, using credit offered by numerous financial institutions, have become common. This often produces a complex combination of medium- and long-term loans and grants at differing interest rates. Five examples of power projects currently being planned or under construction in developing countries illustrate the multiple sources of financing as well as differences in financing terms available to lower- and middle-income developing countries. These are described in detail in Appendix B. Project A is a 300-MW hydro power plant being constructed in a middle-income developing country. Projects B, C, and D are smaller projects being constructed in progressively lower-income developing countries. The average weighted interest rates for these projects range from 7.4 percent per annum for Example A to 3.1 percent for Example D (including the interest-free grant elements); and the average weighted repayment period ranges from approximately 18 years beginning after a 6-year grace period for Example A to 23 years beginning after an 8-year grace period for Example D. Example A is financed primarily by a combination of development bank loans, export credits, and commercial bank credits, while Example D, located in a lower-income and less credit-worthy country, is financed primarily by highly concessionary loans and grants provided by development funds and bilateral financing institutions. As a basis for comparison, Example E shows the financing arrangements for a 2 × 200-MW oil-fired thermal power project being constructed in a lower-middle-income developing country. The weighted interest rate is 5.1 percent per year (including the interest-free grant element) and the weighted average repayment period is 12 years after a 4-year grace period.

We have included the information on financing to provide a long-term basis for analyzing the potential economic benefits of hydroelectric projects and to illustrate the potential sources of funds for projects in even the poorest developing countries. Hydro power, properly developed, can be a powerful force in eliminating the poverty, ill health, and misery of millions of people in future generations.

Table 4.2. Potential Sources of Finance for Power Projects in Developing Countries

Source	Terms	Special Conditions	Comments
Development Banks (Examples: World Bank, African Development Bank, Asian Development Bank, Interamerican Development Bank)	• 15–25 year term • 6–8% interest	• Application through diplomatic channels • Extensive economic and technical analysis • International tenders	• Lengthy project evaluation process • Participation by the World Bank can often stimulate cofinancing by other institutions
Concessionary Financing Agencies (Examples: International Development Association, European Development Fund, EEC, Special Account, African Development Fund, OPEC Member Funds)	• 15–40 year term • 0–6% interest	• Same as above for multilateral financing • Bilateral financing may be tied to purchases in lender's country	
Export Credit Agencies (Examples: U.S. Eximbank, Japan Eximbank, German HERMES, French COFACE, U.K. ECGD, Canadian Export Development Corporation, etc.)	• Up to 15 year term • 7–8½% interest • One time insurance fees of 0–3½% or guarantee fee depending on agency • Commitment fee	• Financing can cover up to 85% of foreign-sourced goods and services • Each national export credit agency will finance exports sourced from its own country • Export credit agency in one country will often match more favorable terms offered from a competing country for similar goods and services • Some agencies lend directly, others guarantee commercial bank loans	• Financing can be arranged relatively quickly once suppliers are selected • Preliminary expressions of interest can be obtained and evaluation made of competing terms prior to selection of suppliers
International Commercial Banks	• Up to 10 years • Margin over LIBOR* per annum adjustable every 6 months • Management fee • Commitment fee	• Terms negotiable and subject to change with market conditions • Funding available to supplement export credit financing and to cover other procurement, interest during construction, and local costs	• Interim financing can often be arranged to permit project start prior to finalization of other longer term financing

*LIBOR is the interbank funding rate in the London market.

APPENDIX A: SOURCES OF FUNDING INTERNATIONAL FINANCIAL INSTITUTIONS

Development Banks

African Development Bank
Asian Development Bank
European Investment Bank
Inter-American Development Bank
Islamic Development Bank
World Bank
Central American Bank for Economic Integration

Development Funds

African Development Fund (ADF)
EEC Special Account
European Development Fund (EDF)
Inter-American Development Bank—Special Funds
International Development Association (IDA)
United Nations Development Program (UNDP)
Arab Fund for Economic and Social Development
Arab Bank for Economic Development in Africa
OPEC Special Fund

Bi-Lateral Financing Institutions

OECD Members

Australia	Australian Development Assistance Bureau (ADAB)
Canada	Canadian International Development Agency (CIDA)
France	Caisse Centrale de Cooperation Economique (CCCE)
	Fonds d'Aide et de Cooperations (FAC)
Germany	Kreditanstalt für Wiederaufbau (KFW)
Japan	Overseas Economic Cooperation Fund (OECF)
Sweden	Swedish International Development Authority (SIDA)
Switzerland	Direktion für Entwicklungszusammenarbeit und humanitäre Hilfe
United Kingdom	Commonwealth Development Corporation (CDC)
United States	Agency for International Development (AID)

OPEC Members

Abu Dhabi	Abu Dhabi Fund for Arab Economic Development
Iraq	External Iraq Fund for Development

Kuwait	Kuwait Fund for Arab Economic Development
Saudi Arabia	Saudi Development Fund
Venezuela	Venezuelan Investment Fund

APPENDIX B: PROJECT A— MIDDLE-INCOME DEVELOPING COUNTRY 300-MW HYDROELECTRIC PROJECT

Scope of project: Construction of a rock-filled dam, a 25.6-kilometer power tunnel, and a 300-MW surface power house, and provision of associated transmission lines.

Cost of project: $421 million

Amount of foreign financing (approximate): $308.7 million

Sources of financing:

- World Bank—$72 million, 7.5% interest, 15-year repayment after 5-year grace period
- Regional development banks (3 loans)—$68.8 million, 8.0% interest, 1-¼% management commission, 18½-year repayment after 6½-year grace period
- Regional development fund—$45 million, 1% interest for first 10 years, 2% interest through final payment, ½% management commission, 29½-year repayment after 10½-year grace period
- Bilateral assistance—$32.7 million equivalent, 8% interest, 1-¼% management commission, 18½-year repayment after 6½-year grace period
- Syndicated commercial bank credit—$15 million, 1⅜% over LIBOR, 5-year repayment after 2-year grace period. For purposes of analysis, LIBOR assumed to average 9.5% for life of loans.
- Export credits (various export credit agencies)—Approximately $75 million, 7½–8½% interest, 7–10-year repayment after 3–5-year grace period
- Local financing—$112.3 million, 8% interest, 18½-year repayment after 6½-year grace period

Weighted interest rate: 7.4%

Weighted average repayment period: 18.0 years

Weighted average grace period: 6.1 years

PROJECT B— LOW/MIDDLE-INCOME DEVELOPING COUNTRY 40-MW HYDROELECTRIC PROJECT

Scope of project: Construction of a 60-meter high-embankment dam, a water conductor system, and an above-ground power station with an initial capacity of 40 MW which can be expanded to 80 MW.

Cost of project: $54 million

Amount of foreign financing (approximate): $38.4 plus $11.5 million grant

Sources of financing:

- World Bank—$15 million, 7.5% interest, 12-year repayment after 3-year grace period
- Regional development banks (3 loans)—$68.8 million, 8.0% interest, 1-¼% management commission, 18½-year repayment after 6½-years grace period
- European Investment Bank—$14.3 million, 5.4% interest, 15-year repayment after 4½-year grace period
- Bilateral concessionary finance—$9.1 million, 8.75% interest, 20-year repayment after 4-year grace period
- Bilateral aid—$11.5 million grant
- Local development bank financing—$3.6 million, 7½% interest, 20-year repayment after 4-year grace period

Weighted interest rate: 5.5%

Weighted average repayment period: 15.4 years

Weighted average grace period: 3.8 years

PROJECT C—LOW-INCOME DEVELOPING COUNTRY 90-MW HYDROELECTRIC PROJECT

Scope of project: Construction of Stage I of project with installed capacity of 90 MW.

Cost of project: $66.4 million

Amount of foreign financing (approximate): $57 million

Sources of financing:

- World Bank—$9 million, 8.5% interest, 16-year repayment after 5-year grace period
- World Bank (3rd Window)—$8 million, 4.5% interest, 19-year repayment after 6-year grace period
- International Development Association—$8 million, ¾% service fee, 30-year repayment after 10-year grace period
- Regional development bank—$7 million, 7.5% interest, 19-year repayment after 5-year grace period
- Bilateral concessionary finance (various countries)—$25 million, 3% interest, 22-year repayment after 10-years grace period
- Local financing—$9.4 million, 8% interest, 18½-year repayment after 6½-year grace period

Weighted interest rate: 4.8%

Weighted average repayment period: 21.0 years

Weighted average grace period: 7.8 years

PROJECT D—LOW-INCOME DEVELOPING COUNTRY 54-MW HYDROELECTRIC PROJECT

Scope of project: Construction of a concrete diversion dam and provisions of a 54-MW underground power house with ancillary facilities, transmission lines, and technical assistance. Possible later expansion to 108 MW.

Cost of project: $116 million

Amount of foreign financing (approximate): $100 million

Sources of financing:

- International Development Association—$33 million, ¾% service fee, 30-year repayment after 10-year grace period
- Bilateral development aid—$20 million, 3% interest, 30-year repayment after 10-year grace period
- Bilateral development aid—$10 million grant
- Four Arab development funds—$37 million, 4% interest, 15-year repayment after 5-year grace period
- Local financing—$16 million, 8% interest, 18½-year repayment after 6½-year grace period

Weighted interest rate: 3.1%

Weighted average repayment period: 23.0 years

Weighted average grace period: 7.7 years

PROJECT E—MIDDLE-INCOME DEVELOPING COUNTRY THERMAL POWER PROJECT

Scope of project: 2 × 300-MW units oil-fired.

Cost of project: $500 million

Amount of foreign financing (approximate): $389 million

Sources of financing:
- World Bank and IDA—$139 million, 7.9% interest, 15-year repayment after 5-year grace period
- European Investment Bank and EEC—$70 million, 7.9% interest, 15-year repayment after 5-year grace period
- Bilateral assistance—$80 million, 4% interest, 9-year repayment after 3-year grace period
- Bilateral assistance—$100 million grant
- Local financing—$11 million, 5% interest, 9-year repayment after 3-year grace period

Weighted interest rate: 5.1%

Weighted average repayment period: 12.1 years

Weighted average grace period: 4.1 years

Chapter 5
The Role of Nuclear Energy in The Less Developed Countries

Thomas J. Connolly

The projections presented in this report for the deployment of nuclear electric capacity in various countries were derived from a study of world nuclear energy futures.[1] The work was performed at the German Nuclear Research Center at Julich. The analysis was done for the world community as a whole, with no specific emphasis on developing countries.

The starting point for this study was the report of the Conservation Commission of the World Energy Conference (WEC).[2] We accepted the electric demand trajectories to the year 2020 contained in the WEC L4 scenario. This is classified as a low-growth scenario. The world-averaged economic and energy-growth values are given in table 5.1. Regionally disaggregated values for electricity demand growth are given in table 5.2.

The nuclear share of the electricity supply in the WEC L4 scenario was developed independently in the work reported here. First, country-by-country electricity production trajectories were calculated by increasing each country's average annual production in the years 1971–75[3] according to the

Table 5.1. World Energy Conference 1977

L4 Scenario			
		(% Per Year)	
World	1972–85	1985–2000	2000–2020
Economic Growth	3.2	2.9	2.6
Prim. En. Growth	2.0	2.7	2.4
Electricity Growth	3.5	3.9	3.3

Table 5.2. World Energy Conference

L4 Scenario

World Energy Conference Regions	Electricity Growth Rates 1972–2000	2000–2020
1. N. America	3.4	2.9
2. W. Europe	3.7	2.7
3. JANZ	4.1	2.7
4. USSR, E. Europe	5.1	4.0
5. China, C.P. Asia	7.5	4.8
6. OPEC	7.2	5.5
7–11. Non-OPEC Developing Countries	5.5	4.8

growth rates of table 5.2. The growth rate of a region was applied to each country within that region. The nuclear share in these demands was then computed according to the rules of table 5.3. These rules might be said to describe a vigorous nuclear economy in the post-1995 world, preceded by a pause in nuclear expansion prior to that time. This pause is intended to reflect the present hesitant ordering position of utilities. The application of these rules obviously does not constitute an analysis or a prediction. It simply represents an attempt to place world-aggregate or region-aggregate nuclear deployment projections on a basis which has some country-by-country plausibility.

The rule which most specifically affects the developing countries is No. 3, which requires an annual electricity production of 20 terawatt hours (TWh)

Table 5.3. Scenario Development Rules

1. Only Reactors
   currently in operation
   or currently under construction
   or currently announced with site designated
   will be in operation by the end of 1995

2. In countries having at least one reactor qualifying under rule 1, the nuclear share of electricity capacity beginning operation after 1995 will be:
   1/3 if country has high fossil resources
   1/2 if country has intermediate fossil resources
   2/3 if country has low fossil resources

3. The nuclear share of electric generating capacity placed in operation after 1995 in countries not having a single reactor qualifying under rule 1, will follow rule 2, but not until the annual electric generation reaches 20 TWh. At that time, a 500-MWe nuclear plant is placed in operation and the application of rule 2 is started.

Table 5.4. Nuclear Electric Capacity Projections

(GWe)			
	1990	2000	2020
OECD	310	610	1400
Dev. Countries	25	80	380
WOCA	335	690	1770
CPE	50	200	770
WORLD	385	890	2540

prior to operation of a first nuclear plant. The rationale behind this entry threshold value is that a 500 megawatt electrical (MWe) nuclear power plant, operating at a 65 percent capacity factor, would produce 14 percent of the electricity. This value approaches the maximum which seems prudent from a grid-reliability standpoint. This threshold value could be lowered if nuclear plants of less than 500 MWe could be installed or if the grids of some nations might be linked. On the other hand, no allowance was made for the fact that, in many developing countries, a fragmented supply system exists for reasons related to geography and different regional development.

The nuclear electric deployment which results from the application of these rules is shown in Table 5.4, aggregated by conventional political-

Table 5.5. Nuclear Generating Capacity in Developing Countries

World Energy Conference Regions	1990	2000	2010	2020
6. OPEC	2	9	22	47
7. Latin America	9	32	78	154
8. Middle East, North Africa	0	4	14	30
9. Africa South of Sahara	0	0	3	12
10. South East Asia	9	20	43	78
11. South Asia	5	13	31	60
Totals:	25	78	191	381
Comparisons				
Strout (1979)	74	286		
IAEA (1977)	43–80	197–299		
R.J. Barber Assoc. (1975)	132	448		
IAEA (1974)	167	530		

(* All the comparison values quoted in Table 5.5 were extracted from the work of either Strout[4] or Goldemberg[5].)

economic groupings. These values are considerably below the nuclear contribution to the electricity supply projected in the WEC L4 scenario. Our conclusion is that the nuclear supply projections of that scenario will be very difficult to meet.

Tables 5.5 and 5.6 give additional detail with respect to developing countries. In Table 5.5, the values obtained in this work are compared with previous projections.* Goldemberg has already noted the declining trend of the 1990 projections toward stagnation. Only time will tell whether the nuclear revival implicit in our scenario will in fact occur.

The only message I would extract from this exercise is that the deployment of nuclear generating capacity in the developing world is likely to proceed at a much slower pace than that projected in most studies. I suspect that this is a conclusion already reached by many students of the subject. The question which remains is whether other energy resources and forms will substitute for nuclear energy or whether the aggregate demand for electricity in the developing countries will be substantially less than the WEC L4 projections.

Table 5.6. Time Frame of Initial Nuclear Generating Capacity in Developing Countries

WEC Region	1980	1990	2000	2010	2020
6.		Iran	Venezuela	Algeria	Ecuador
				Iraq	
				Kuwait	
				Nigeria	
				Saudi Arabia	
				Indonesia	
7.	Argentina	Brazil	Chile	Cuba	Jamaica
		Mexico	Colombia	Peru	Uruguay
			Puerto Rico		
8.			Egypt		Morocco
			Israel		
			Turkey		
9.				Ghana	
				Rhodesia	
				Zambia	
				Zaire	
10.	S. Korea		Philippines	Hong Kong	
	Taiwan			Malaysia	
				Singapore	
				Thailand	
11.	India				
	Pakistan				

NOTES

1. T. J. Connolly, U. Hansen, W. Jaek, K-H. Beckurts, "World Nuclear Energy Paths," (New York: The Rockefeller Foundation and London: The Royal Institute of International Affairs, 1979).
2. "World Energy Resources, 1985–2020," The World Energy Conference, IPC Science and Technology Press, 1978.
3. "World Energy Supplies, 1971–75," United Nations Statistical Papers, series J, no. 20, table 20, 1977.
4. A. M. Strout, "Prospects for Nuclear Power in the Developing Countries," *Advances in the Economics of Energy and Resources* 1: 291–310, JAI Press, Inc., 1979.
5. J. Goldemberg, "Nuclear Power in the Developing World," May 20, 1979.

Chapter 6
Criteria for Nuclear Electricity Generation in Developing Countries

Eli B. Roth

For a number of months I have been trying to identify criteria by which one might evaluate whether any specific developing country could appropriately and successfully generate its electricity by nuclear power.

By December 1979, the countries listed in table 6.1 had an installed and operable capacity of at least 100,000 kWe.* (A United Nations listing, rather than objective standards, was used here to distinguish developed from developing countries.)

As the study progressed, it became more and more evident that, in terms of the criteria for nuclear power development, there are greater extremes of capability among the various developing countries than there are between some which are classed by the United Nations as developed but are impoverished and others recognized as developing and at the takeoff stage. Of the ten developed countries which were not listed above because they have not yet generated nuclear electric power, three—Austria, Luxembourg, and South Africa—have programs for reactor construction. The remaining seven—Australia, Denmark, Iceland, Ireland, New Zealand, Norway, and Portugal—have so far not chosen to initiate nuclear electric programs. In any case, more than 56 percent of the developed countries already have nuclear electric programs, while fewer than 4 percent of the 154 developing countries do. Or in other terms, 99 percent of all nuclear electricity generating capacity is in developed countries and the Soviet bloc.

In examining those countries, whether developed or developing, that have nuclear programs that have already successfully generated significant amounts of electric power, I have found several crucial common factors:

1. *Alternative sources of energy*: The policy-makers of the country perceived sufficient problems with the conventional sources of electricity generation

to warrant accepting the challenges and potential dangers of nuclear power.

2. *Reactor sizes and grid capacities*: The electric power grid in each country was large enough to accept the generating capacity of the commercially available nuclear electricity generating plants.
3. *Financing capabilities*: The capital for building the nuclear power plants was available.
4. *Technological development*: The technological development of the people and the institutions of the country were sufficiently advanced to deal with nuclear power.

The four common factors that emerged were studied with the analytic purpose of attempting to establish specific criteria that might be applied as guidelines *in advance* by a country's decision-makers when contemplating a nuclear generating program as part of their development planning. I shall take each of those four common factors in turn and summarize the findings.

Table 6.1. Worldwide Nuclear Capacity

	Nuclear Capacity Megawatts(e)
Developed Countries	
Belgium	1,650
Canada	5,476
Finland	1,080
France	6,538
West Germany	8,205
Italy	1,307
Japan	11,524
Netherlands	493
Spain	1,073
Sweden	3,700
Switzerland	1,020
United Kingdom	8,080
United States	52,621
Developing Countries	
Argentina	335
India	602
South Korea	100
Pakistan	125
Taiwan	100
Soviet Bloc Countries	
Bulgaria	880
Czechoslovakia	550
East Germany	1,390
USSR	9,475

* Extracted from information in *Nuclear News* (August 1979), pp. 69–88.

Alternate Sources of Energy

In choosing nuclear energy for electric power generation instead of any other possible primary energy source, a country may have any of several motives. Some of these motives may be political, related to prestige and the issue of possessing nuclear weapons. Others may relate to the strategic consideration of having an assured fuel supply without having to protect shipping lanes for access to bulky fuels such as oil or coal. Still others may be economic, that is, nuclear energy might prove to be the cheapest possible source of electricity.

In any case, as part of the information needed to judge whether or not nuclear electricity generation is appropriate for a country, that nation's decision-makers need to know what other choices are available. The obvious rivals of nuclear energy are coal, oil, natural gas, and hydro power. Solar electric power can be dismissed here because it is not yet being treated as a conventional source of power.

The lack of an alternate source of power with adequate capacity, dependability of supply, or acceptable cost is the most frequent justification for selecting nuclear electricity generation. This is certainly as one would expect, but what is more important is the country-specific nature of the justification. Only after a proper energy survey—an analysis of available resources and capabilities and their associated costs, along with accurate projections of demand for electric power—can this criterion be used properly.

Reactor Sizes and Grid Capacities

Commercial nuclear power plants have had to be large to compete economically with other methods of generating electricity. Figure 6.1 shows the trend in sizes of reactors entering commercial service within supplier countries and in LDCs. Essentially all the reactors entering commercial service are now of about 1000-MWe capacity. Even the Canadian CANDU design, which has not yet reached 1000 MWe, is already at 740 MWe. The British advanced gas-cooled reactor is up to 625 MWe.

This means that any utility desiring to order a nuclear generating plant of modern design which has been proved and tested in a supplier country (and generates power economically) must have a system that can use at least 600 MWe of power. This in turn tends to set the minimum size of the power grid in which a reactor can be installed.

Power plants must be shut down periodically for maintenance, and all commercially available nuclear plants except for the Canadian CANDU design must also be shut down for refueling. No matter how much attention is paid to quality control in power-plant manufacturing and assembly, all plants are subject to unscheduled breakdowns and outages [to say nothing of personnel malfunctions like those so dramatically exhibited at Three Mile

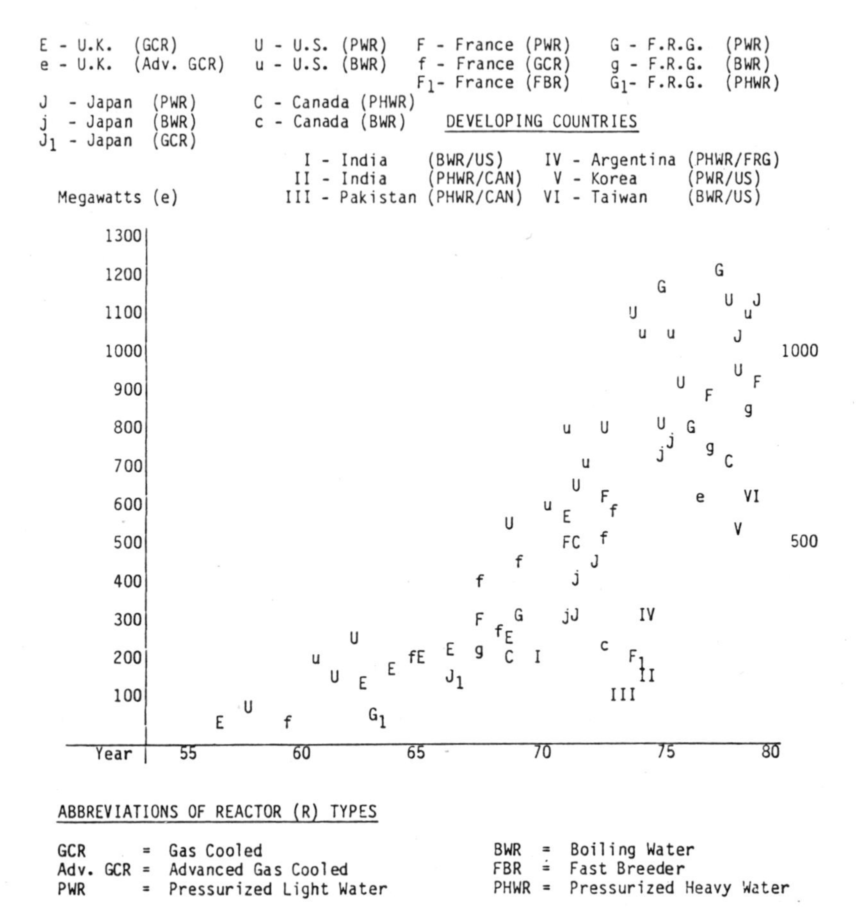

Fig. 6.1. Size (MWe) versus first year of commercial operation for domestic nuclear power plants (by country and reactor type)

Source: Diagram prepared by E. B. Roth from information in *Nuclear News* (August 1979).

Island (TMI)]. If the power plant that must shut down is too large a fraction of the total system's capacity, the reliability of the whole system would be jeopardized.

Basing our calculations on standards that would depend on each country's accepted practices and ability to tolerate power failures, we can fix the upper limit of acceptable power-plant size. One report indicates that it would require at least a 10,000-MWe system to accept a 600-MWe plant, that is, 6 percent of its generating capacity. In International Atomic Energy Agency (IAEA) Surveys, acceptances ranged from 5 percent to 20 percent. International Board for Reconstruction and Development (IBRD) studies have concluded that 15 percent is an average figure. The actual figure selected, however, depends on the specific system. It depends on such local factors as the willingness of customers to accept power interruptions, the nature of the electric load, the age, reliability, and size of other units, and the expected reliability of the nuclear unit.

Very few of the developing countries have national grids with a system capacity as large as 10,000 MWe. Because setting the grid size too high could eliminate candidates prematurely, it is reasonable—acknowledging that their electricity users are generally more tolerant of interruptions of service than users in developed countries—to set 2,500 MWe as the minimum-sized grid to accept a 600 MWe nuclear power plant.

Financing Capabilities

The capital costs of nuclear power plants have risen in recent years at rates almost comparable with the increases in the international price of fuel oil. For estimating purposes, power-plant costs are often calculated on the basis of dollar-per-KWe capacity. But these values usually hold only for a relatively small range of power-plant sizes because of the already-mentioned economies of scale in nuclear power. The dollar values per KWe generally apply to the sizes of plants being delivered at the time of quotation. A 1968 IBRD internal report quoted $120/KWe as a reasonable cost for a conventional oil-fired plant as against $240/KWe for pressurized water reactor (PWR) nuclear units. At that time, we might note from figure 6.1, the PWR units being delivered were of about 500-MWe capacity.

In contrast, *Nuclear News*,[1] reporting on a September 1979 international conference in Copenhagen on the financing of nuclear power plants, stated that, allowing for interest during construction and a seven-year lead time required for nuclear plants, the cost per kilowatt has risen from $226 in 1969 to $1,607 in 1978 (for twin 1,200-MWe sizes). The corresponding figures for coal plants are $183 and $1,266 (but for 800-MWe sizes). As has already been noted, the cost per KWe of 600-MWe size would be higher.

Even more discouraging for nuclear power proponents is the threat, caused by statutory and regulatory delays, of increasing the lead time to 12–14 years. By then the 1978 nuclear figure becomes $2,321 per KWe. About one-third of the increase from $1,607 is attributable to inflation.

In any case, using the shorter optimistic lead time and estimates for 1,200-MWe units, we can calculate the cost of a 600-MWe power plant at $964 million. The first reactor core for such a plant would add about $100 million, so that in round numbers $1 billion in capital would be required. (A coal-fired plant of equivalent total capacity would cost about $760 million, and an oil-burning unit about one-half that, or $380 million.)

Of the total capital required, a certain proportion of the amount, depending on the specific country involved, might be generated within the developing country desiring the plant. Presumably the capital remaining to be borrowed from external markets would be for those parts of the plant and services that would have to be imported and would be paid for with foreign credits. This proportion might be estimated as one-half the total, although that may vary. Thus the minimum sized 600-MWe power plant would require about $0.5 billion in foreign exchange. Of course the nations exporting nuclear systems might, in the present depressed climate for nuclear exports, agree to underwrite a substantial portion of the needed capital, but this would still represent foreign capital for the developing country.

Technological Development

Until the circumstances surrounding the accident at Three Mile Island were sufficiently understood, it appeared to be the consensus, among those who gave the subject much thought, that the problem of getting a country technologically equipped and culturally prepared to deal with the complexities inherent in handling nuclear power was the conventional issue of technology transfer, and that this could be handled simply by augmenting the mechanisms ordinarily used for granting and receiving technical aid. The IAEA, in keeping with its implicit responsibilities associated with Article IV of the Nonproliferation Treaty (NPT), sponsored international meetings, underwrote courses of instruction and training, and issued model guidance documents for countries interested in trying to apply nuclear reactor technology to electricity generation. As a first step, the IAEA approach to this problem of technology transfer has certainly been necessary; whether it is sufficient from the standpoint of the applications of nuclear power to developing countries is a serious question that, following the Three Mile Island accident, has become critical.

The nuclear power industry grew out of the military programs of the 1950s; in fact, the first PWR, Shippingsport, originated as a reactor intended to fuel an aircraft carrier, and that reactor in turn had developed out of the light water technology of the submarine reactor program. This technological lead—established early in the United States through the expenditure of substantial sums on engineering development of the light water reactor, when civilian economics had no relevance—was never relinquished to other types of reactors.

This first-off-the-mark advantage the PWR had was reinforced by U.S. policies initiated under the Atoms-for-Peace programs. The U.S. government agreed to supply the slightly enriched uranium fuel required by such reactors to domestic and foreign manufacturers alike on such favorable commercial terms that several countries changed over from other designs to U.S. PWRs or boiling water reactors (BWRs) even after they had operated the other designs on a commercial scale. This happened in France, Sweden,

Spain, and Japan. In addition, the U.S. manufacturers aggressively marketed their light-water reactor technology through licensing arrangements with large firms in Europe and Japan as well as on their own.

By fortunate coincidence, we were able to contrast a developed with a developing country in their performance of taking advantage of the Atoms-for-Peace program. The starting points were two reports prepared by the National Planning Association, both completed within a ten-month period: *The Outlook for Nuclear Power in Japan* in June 1956 and *Nuclear Power and Economic Development in Brazil* in April 1957.[2]

To summarize these case histories: Japan started in 1954 with an economy and industrial plant still almost prostrate from the aftereffects of World War II. Its GNP was about $200 per person per year (that is, about one-tenth that of the United States then). Lack of foreign exchange was a chronic problem and a hindrance to economic growth. Japan at that time imported about 80 percent of all its raw materials, including 95 percent of its fossil fuels. It had substantial numbers of hydropower sites, potentially economic but as yet undeveloped. Its reserves of coal, although significant, were of poor quality and were becoming increasingly more difficult to mine. Its petroleum reserves were minuscule. Arable land was extremely limited. To manufacture the export products that would pay for its raw materials imports, the country needed additional power at a reasonable price.

In Brazil in 1954, while the economy had not been physically affected by World War II, Brazilian raw material and commodity exports had intermittently generated foreign exchange surpluses. Nevertheless, foreign exchange deficiencies were limiting economic growth. Brazil's GNP was also about $200 per person per year. Its land area, 23 times that of Japan but with substantial portions difficult of access in the tropics, supported a population about 20 percent smaller than Japan's. Brazil also had a substantial number of yet-undeveloped, potentially large hydro power sites. Its coal reserves were minor and domestic petroleum production almost negligible, although there may be oil under some of the huge unexplored interior. By 1954, a significant diversification of Brazilian industry had taken place. There were frequent power shortages, but because the price of power was held down, the major power problem had to do with availability rather than price. The country's major income from abroad came from raw-material and commodity exports.

From a material and financial standpoint, then, it would have seemed that at the takeoff point for Atoms for Peace, 1954, Japan and Brazil were endowed equally, with perhaps a slight advantage for Brazil. But there was a fundamental difference. Japan had spent the previous 100 years becoming an industrialized society. By 1954 Japan had a relative abundance of skilled scientists and—even more important—an abundance of skilled technicians and working people who knew how to exercise their knowledge and diligence in order to earn a tolerable standard of living. The whole fabric of Japanese society—cultural organization, social institutions, government, industry—was geared through long experience to planning and managing material and human resources for the country's overall economic benefit. In spite of its low GNP, Japan was a developed country.

In contrast, Brazil had in 1954 a population untrained and uneducated in the skills an industrialized society requires. Its potentially huge natural resources had seen little systematic investigation. Foreign exchange available for generating capital tended to vary widely and to fluctuate with the changes in the mercurial commodity and raw materials markets. When capital was available, it was frequently invested in relatively unproductive but quick-return enterprises. Government skills and experience in economic planning were, in the word of the report, negligible. Industrial management skills were also in very short supply. There was no tradition of long-term planning by government or industry, and the information needed for such effort was either nonexistent within the country or erroneous. Occasionally political changes would interrupt projects with long lead-times and they simply would not be finished; this would cause multiple economic losses—loss of the investment in that particular enterprise and loss, too, of the needed capital that might have been used profitably elsewhere. In other words, Brazil was then an undeveloped country and has been one of the fastest-developing countries ever since.

Most significant for this study is what happened to nuclear power in the two countries. Both Japan and Brazil started operating research reactors in the same year. The first Japanese research reactor (JRR I) was operated in Tokai Mura from 1957 to 1961, after which it was replaced by the JRR II. The first Brazilian research reactor was installed in the laboratories of the Brazilian atomic energy commission (CNEN) in São Paolo in 1957. From these early concurrent beginnings, the two nations' programs started to diverge.

Japan's first commercial electricity generating reactor plant, which began operating in 1966, was a British designed and built gas cooled reactor (GCR). Subsequent reactors have been light water reactors, either BWR or PWR, because the Japanese decided early that water reactors were the way to go for both economic and technical reasons. Since then, most of the world has concurred. Even the British have been considering changing from their quarter-century-old GCR and the advanced GCR designs to light water reactors. The Japanese also expect to operate their first fast breeder reactor by 1985, even though the United States has been deferring its own fast breeder. The Japanese now report themselves capable of fabricating at home 95 percent of all that goes into a nuclear power plant. This story illustrates the degree of sophistication, confidence, and management drive characteristic of the developed Japanese economy.

In Brazil, the nuclear program has been subject to fits and starts which have been caused sometimes by political considerations and at other times by economic ones. In the early 1970s, because the Brazilian government hoped to become the regional supplier of the large-size nuclear steam supply systems equipment, Brazil contracted with German industry to build a large nuclear steam supply system fabrication plant at the cost of several hundred million dollars. So far the plant has not supplied any nuclear steam supply system and it sits virtually idle while, because of the nuclear-power slowdown, the world's capacity to produce such systems is 2½ times the projected ordering rate for reactors for the foreseeable future.

Meanwhile, the first nuclear power plant in Brazil, a 626-MWe PWR, which, when ordered, was scheduled to begin commercial operation in March 1977, was later supposed to start operation in September 1980. The economic losses that Brazil has suffered are not unique to that country. They are characteristic of the performance of any developing country that attempts to master complex advanced technology (and nuclear technology is among the most complex).

The conclusion that one must draw from these comparisons, as well as others not cited here, is that to consider nuclear electricity generation appropriate now for developing countries is to reverse the order of things. To apply nuclear electricity generation appropriately, a nation must first have certain technological and institutional attributes of a developed country. Contrary to the belief that the road to development is via nuclear power, the road to nuclear power is really via development.

This conclusion need not be the last word. Conceivably, the pressures on world energy resources and the shared dangers of nuclear weapons proliferation might stimulate a new kind of multinational cooperation in the nuclear power field far beyond the scope of current IAEA programs. That organization's resources for conducting hardware-oriented projects would not suffice. What would be needed would be means for cooperation in developing and proving reactor designs with size, cost, and safety attributes suitable for use in developing countries and at the same time economical in the use of world uranium resources.

NOTES

1. *Nuclear News* (November, 1979), p. 9.
2. Michael Sapir and Sam J. Van Hyning, *The Outlook for Nuclear Power in Japan* (June 1956); and Stefan Robock, *Nuclear Power and Economic Development in Brazil* (April 1957).

Chapter 7
Energy Consumption and Decentralized Solutions

José Goldemberg

INTRODUCTION

The seemingly insurmountable problems and frustrations of life in large cities have encouraged the search for alternative solutions that could rescue us from present nightmares. Big cities have large densities of people, who consume large amounts of energy per capita, depend very heavily on the use of fossil fuels, and produce large amounts of waste and pollution. Most of the technologies now in use in cities (including the treatment of refuse and sanitation) are very "hard," in the sense of requiring large and bulky systems for the production and use of energy. This, of course, has much to do with the unavoidable laws of thermodynamics, since there is no way to generate electricity from coal, oil, or uranium without producing a lot of unwanted heat and other polluting products. Although these systems can be optimized (using cogeneration, for example) there are physical limits to what one can achieve.

No wonder, therefore, that many groups of people have looked for ways to supply the energy needed by society from renewable sources, basically from solar energy harnessed directly in solar collectors or indirectly as hydropower, biomass, and wind. In addition to that, some people have associated the use of renewables with small-scale decentralized use, as in rural areas in developing countries, where per capita energy consumption is much less than in large cities (see table 7.1).

As table 7.1 shows, per capita energy consumption in cities does not vary much all over the world. The energy consumption in rural areas in India and São Paulo, however, is more than 10 times less on a per capita basis than it is in most cities. In addition, the energy consumption density (watts per square meter) in urban areas is also pretty much the same in most countries; in rural areas this density is more than 100 times less than it is in cities. The associa-

Table 7.1. Urban and Rural Densities

	Population Density ($people/km^2$)		Energy Consumption (kcal/capita/day)		Energy Consumption Density W/m^2	
Country/City	Urban	Rural	Urban	Rural	Urban	Rural
India	6,000	135	41,600	7,200	12	0.04
New York	560	—	238,000	—	6.4	—
London	1,100	—	108,000	—	5.7	—
Tokyo	980	—	81,000	—	3.8	—
S. Paulo	1,260	13	26,000	9,600	1.6	0.006

1 kW of installed power corresponds to a consumption of 20,800 kcal/day.

tion of renewable, small scale, and decentralized is therefore a fairly natural one because direct solar energy has such low density (~ 100 W/m^2); biomass, hydropower, and wind have even lower densities (see table 7.2).

It is, therefore, an attractive idea to use solar energy as it comes in small dispersed quantities that do not require much processing. The recreation of a long-lost bucolic rural life exerts a strong attraction in the minds of many proponents of this "soft" technology. However, anyone familiar with rural areas in less developed countries (where approximately half of mankind lives) knows that life in these areas is, generally speaking, very difficult, unhealthy, and full of hard work and drudgery. Health and sanitation problems, malnutrition, lack of education, and outright poverty are the rule and not the exception in most of this part of the Third World. No wonder, therefore, that a very strong tendency for urbanization has developed in several parts of the world, and people have been clustering in large towns which grow at rates of 6–7 percent a year, doubling in size every 10 years. As can be seen in figure 7.1, 80 percent of the world's population was rural in 1900, only 60 percent was rural in 1980, and most likely only 50 percent will be rural at the end of the century.

The migration to cities has generated tremendous problems and given rise to slum areas that frequently are larger than the cities themselves. Although life in slums might seem unbearable to the well-to-do urban dweller, it represents progress with respect to life in many rural areas. In slums there are at least some health services, electricity, public transportation, and easy access to some of the niceties of urban life.

In order to counteract the strong tendency to urbanization, it is necessary to improve the conditions of life in small and very small communities of which there are hundreds of thousands in many of the LDCs. For example, in India there are more than 500,000 communities with less than 500 people. Improved living conditions mean, to a large extent, the consumption of more energy. The obvious idea is to match the fact that people are distributed in a decentralized way with the use of decentralized sources of energy. The use of small-scale renewable energy sources seems the obvious choice. (The coun-

Table 7.2. Energy Densities

Source	W/m^2
Wind (north seacoast)	~4.5
Fuelwood plantation	~1.
OTEC (tropical oceans)	~0.8
Wind (continents)	~0.6
Fuelwood (natural forests)	~0.2
Biogas	~0.18
Hydropower	~0.02

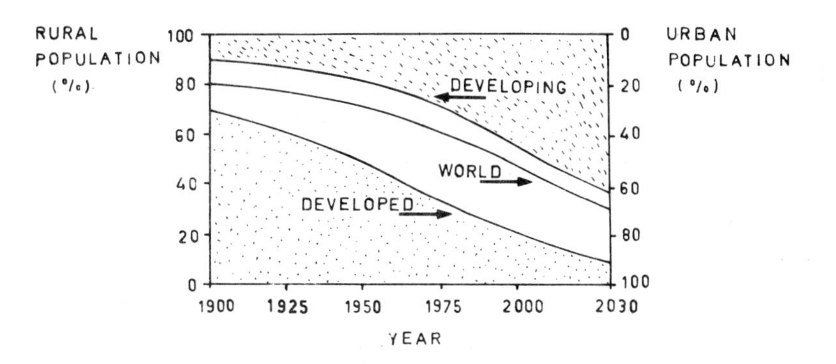

Fig. 7.1. The evolution of urban and rural populations in the twentieth century for developed and developing regions. The global worldwide evolution is also shown.

terpart to this idea is the generation of electricity in large hydroelectric dams or nuclear reactors, that is, in a centralized way, to feed large cities where consumption is centralized.)

For these reasons a large effort is being made to implement small-scale technologies in rural areas around the Third World. In the West it is estimated that approximately $200 million per year is being invested in demonstration projects,[1] an amount that is bound to grow in the near future. In the East one has the example of China that has until recently encouraged, as a matter of national policy, the idea of decentralized solutions that lead to setting up hundreds of thousands of farm-size biogas digestors and almost 100,000 minihydro stations, and also to the production of pig iron by domestic forges.[2] These efforts gave impetus to the idea that "small is beautiful." However, in practice this simple, philosophically attractive, Gandhian idea is facing serious difficulties. We will discuss here some of these problems and try to make the point that *renewable* and *small* do not have to go together, and in fact might not represent any great advantage if taken together.

ENERGY CONSUMPTION IN RURAL AND URBAN AREAS

Starting from first principles, it is not obvious that rural life is less energy-intensive than urban life. Table 7.1 shows that on a per capita basis a city dweller almost anywhere in the world consumes approximately 100,000 kcal/day while a villager in India consumes only 7,200 kcal/day. This does not mean that rural life is less energy-intensive than urban life. It merely illustrates the fact that most villagers lead a miserable life, consuming little more than is needed to stay alive. The correct question to ask is What is the energy consumption for urban and rural life for *comparable* levels of comfort in the same country?

The answer to this question exists for the United States[3] and Norway,[4] where the total energy consumed was calculated as a function of income for rural and urban households (fig. 7.2); both direct and indirect energy expenses were taken into account. In both these developed countries there is hardly any difference between the two categories, with some indication that rural households are actually more energy-intensive than urban. The main reasons for this are the larger use of transportation in rural areas and the higher requirements for heating in the winter in isolated, badly insulated farmhouses.

We did a similar study for Brazil, a large less-developed country, using results of a household expenses survey conducted in 1974 by the Brazilian Statistics Bureau.[5] Results are shown in figure 7.3, for the direct and total energy expenses for rural, urban, nonmetropolitan (small and medium-sized cities) and metropolitan areas, in the State of São Paulo, one of the most populous states of Brazil. As one can see, there is hardly any difference in the total energy expense of households for the same income. (Monetary and nonmonetary incomes were added since the latter are very significant in rural areas.) In our view, this is a very important and rather surprising result. Although presently there are few rural households with large incomes, it seems inescapable that, as their income increases, they will fall into the general pattern of consumption in urban areas.

It could be argued that the quality of life in rural or urban nonmetropolitan areas is better than life in metropolitan areas for the same income. This is

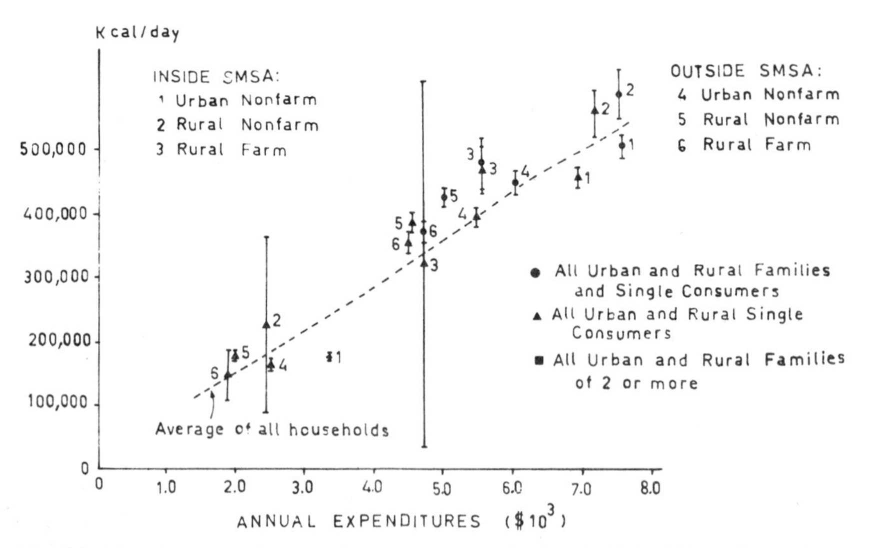

Fig. 7.2. Monetary expenditures and energy consumption for the United States for rural and urban areas. (SMSA means Standard Metropolitan Statistical Area).

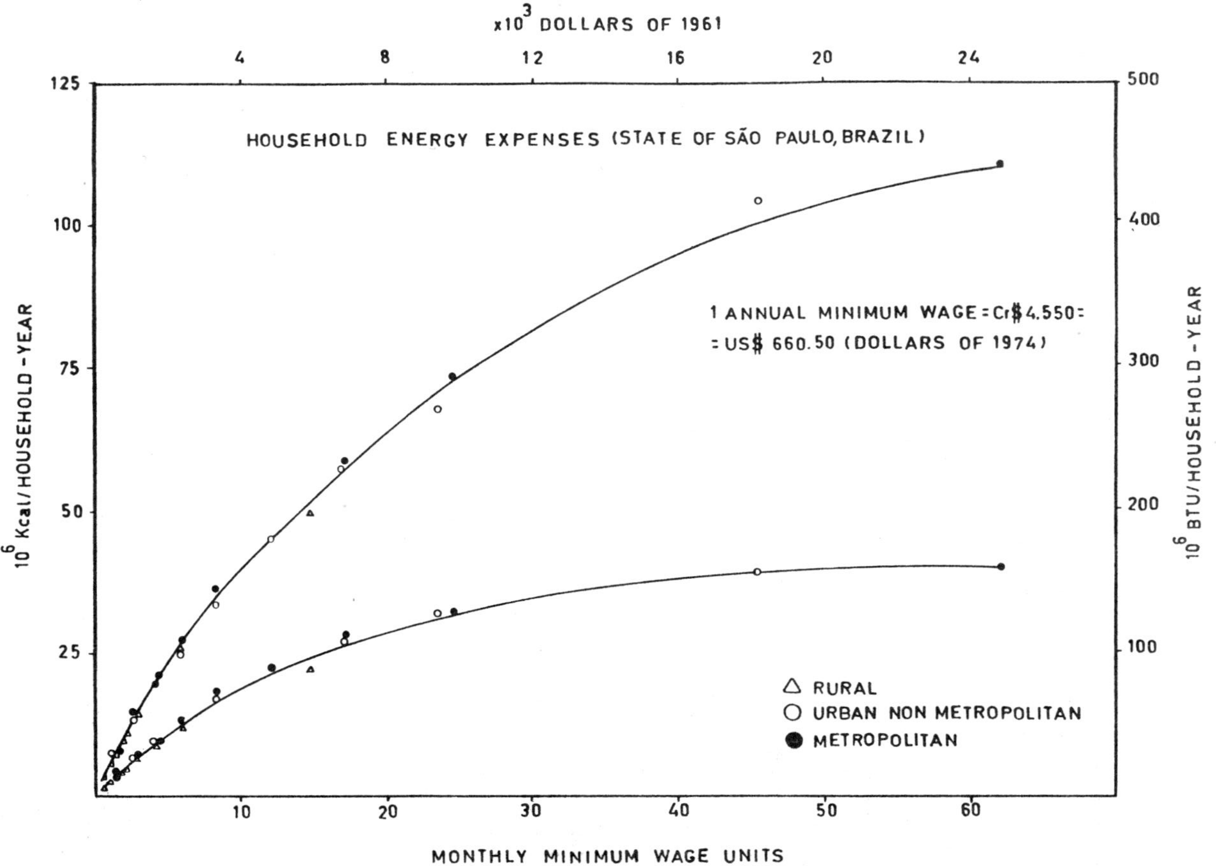

Fig. 7.3. Expenditures (monetary and nonmonetary) and energy consumption in the State of São Paulo, Brazil, for rural, and urban nonmetropolitan, and metropolitan areas.

probably true, but it involves a subjective scale of values. In any case it seems very unlikely that somewhat comparable levels of comfort could be obtained in rural and nonmetropolitan areas with less than half the energy needed in metropolitan areas. As a consequence, if one really wants to improve the living conditions of the poor in most LDCs, one should be prepared to expect very high requirements for energy.

In principle the energy consumption in rural and urban areas in LDCs does not have to be the same, and the development of rural areas—since it lies ahead—could be modulated to be less energy-intensive. Urban areas are already in place with all their infrastructure, and it is more difficult to change them. This is the case for certain commodities such as construction materials: To build high-rise apartments requires much more energy (in steel and cement) than to build flat houses with only one pavement, which are the rule in rural areas. This can be seen in table 7.3, which gives the energy invested in Hong Kong, with its maze of high-rise apartments.[6] Steel and cement account for almost 70 percent of all energy embodied in the buildings. The substitution of wood and other construction materials for these materials could lower this energy appreciably. Other items, however, such as piping, would represent a larger fraction of the total in individual isolated houses than in compact apartment buildings.

It is interesting to note that, in the case of Hong Kong, the per capita energy consumption, (in the form of electricity, town gas, and solid fuels) is 31.5×10^3 MJ per year—2.5 times smaller than the total energy per capita invested in the buildings. It is obvious, therefore, that urbanization is a very energy-intensive process. To compensate for that, however, rural life uses less efficiently certain collective conveniences, such as stores. More transportation might be needed in some cases, as is the case in the United States. In Brazil more energy is spent on food in rural than in urban areas.

Table 7.3. Energy Investment in Hong Kong* (MJ $\times$ 10^3/capita)

Material	In Buildings	In Roads, Paths, and Parks	Total
Iron (piping)	1.7	—	1.7
Steel	32.6	0.1	32.7
Wood	15.8	—	15.8
Bricks	2.3	—	2.3
Cement	15.7	0.8	16.3
Sand	0.5	—	0.5
Crushed rock	0.7	—	0.7
Earthenware (piping)	0.5	—	0.5
Lime	4.0	—	4.0
Bitumen	—	0.3	0.3
Glass	0.6	—	0.6
TOTAL	74.2	1.2	75.4

Although the structure of the energy expenditures changes from rural to urban areas, the total amount of energy per given income ends up being the same. This is true for the United States as a whole and also for the State of São Paulo in Brazil, as can be seen in figures 7.2 and 7.3. It is interesting to notice, however, that the absolute level of energy consumption for the same income is much lower in São Paulo in the United States, as can be seen in figure 7.4.

The reasons for this difference are being investigated. The most obvious is the fact that in a moderate tropical climate such as the one of São Paulo, no heating is needed in the winter. Less transportation is another important reason. However, these differences alone cannot account for the factor of 4 which is evident in figure 7.4 (for high incomes). Life styles and the extravagant expenditures in housing and electric gadgets prevalent in the United States have probably a lot to do with it. This has been noticed already in a comparison between energy consumption patterns of the United States and Sweden made by Schipper and Lichtenberg[7]. For the same GNP, per capita energy consumption in Sweden is approximately half that of the United States.

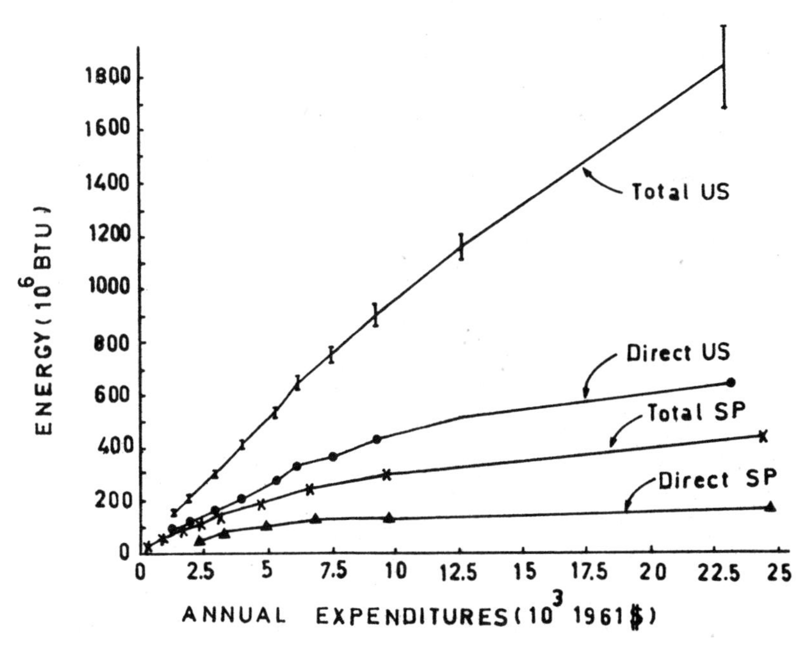

Fig. 7.4. Annual average energy and household expenditures in the United States and the State of São Paulo in Brazil. Direct and indirect energies are indicated.

A positive point that emerges from figure 7.4 is that it will be easier to supply the energy needs of many LDCs than it is to supply those of the United States. We will address ourselves, therefore, to the questions "How to supply the energy needs of LDCs? and Can small decentralized energy sources be enough?

ENERGY AVAILABLE FROM SMALL DECENTRALIZED SOURCES

In a recent rather critical analysis of the successful use of energy sources generated in small units (direct solar collectors, solar cookers, biogas digestors, mini-hydro, etc.), Ashworth points out that the "back lots of communities all over the Third World are littered with the rusted remains of what development experts considered best for the people." He points out, too, that the "humanitarian motives of helping the rural poor are mixed with the desire to create new markets. This is particularly true for expensive technologies such as photovoltaic cells."

Reddy, in India,[8] is particularly sensitive to these points and is trying to mix with the people in villages in India without patronizing them or trying to modernize their consumption habits. He puts a greater emphasis on self-reliance than on technology. Actually a number of people in LDCs view with extreme skepticism any attempt to solve poverty problems with new technologies because at the root of the problem are social inequities that could be solved with present existing technologies if the proper social changes were made. Sri-Lanka and Cuba are important examples—incompletely studied, so far—in which a redistribution of income did a great deal to eradicate illiteracy and proverty.

On the other hand Smil[2] analyzes the case of China, which at one time seemed the ideal testing ground for decentralized energy sources (mainly biogas and minihydro). In Smil's words the Chinese "discovered that small is useful but small is not enough," for the following reasons:

1. Small hydrostations are the first to be affected by the usual cycle of droughts and floods. Since 1975 this has happened to thousands of the 86,000 small hydrostations of China.
2. Family-size biogas digestors work well, but alkalinity and pH must be kept in the proper range; temperature should be higher than 10° C, and improper carbon-nitrogen ratios affect fermentation. This means that technical assistance is needed to have them operating properly, which is not very feasible when there are hundreds of thousands of digestors. Trained personnel can give this assistance, however, in large community-size digestors.

In addition, other products, such as pig iron, cannot be produced efficiently on a small, decentralized scale. The domestic production of pig iron in China was a clear demonstration of this, and accordingly the country shifted rapidly to the use of modern steel production equipment imported from Japan, which is vastly more efficient. What this means is that the use of decentralized renewable sources is desirable and can be quite useful at a low level of development but the use of centralized technologies is unavoidable when a higher level is reached.

The energy needed for the centralized technologies has deservedly received a very bad name because—in developed countries—it originates in coal- or oil-burning thermal stations or in nuclear reactors. This is not an unavoidable path, and it is our contention that one can sustain development at a high level (much above present village levels) using renewable energy sources, on a much larger scale than the use of biogas digestors and small solar collectors suggest. What can be done and is being done in some parts of the world is to use biomass as a significant energy source, although the energy density for the production of biomass is quite low.

ENERGY FROM RENEWABLE SOURCES

The growing of crops is an important example of the use of the low density of solar radiation. Although solar radiation intensity is low and photosynthetic efficiency never exceeds 10 percent (being generally much lower), this problem has been solved successfully by collecting the product of photosynthesis over large areas. The harvesting of crops (or harvesting the solar radiation) allows one to feed millions of people in a city with the product of a few thousand hectares (ha) around the city. Table 7.4 shows typical productivity of agricultural products in Brazil. As a rule of thumb, 1 ha produces the food needed to feed one inhabitant for one year.

The same is approximately true for the production of liquid fuels. One ha of sugar cane can produce 3,500 liters of ethyl alcohol per year;[9] 2 ha

Table 7.4. Agricultural Productivity in Brazil

CROP	Yield (tons/ha)
Rice (nonirrigated)	1–2
Rice (irrigated)	2–4
Black beans	0.8–1.5
Soybeans	1.5–2.0
Corn	1.5–3.0
Wheat (nonirrigated)	0.8–2.0
Wheat (irrigated)	1.0–2.5
Coffee	1.5–2.0

covered by eucalyptus can supply the same amount of alcohol plus an appreciable amount of charcoal. Brazil has embarked on such a program, and at present 50,000 barrels of alcohol per day are being produced with plants planned to produce 200,000 barrels per day in 1985. This will use approximately 4 million ha of land—5 percent of the arable land currently in use in Brazil.[10] Additional expansion of this program will be made using eucalyptus from reforestation projects, thus allowing the use of marginal lands. Since typical plants for the production of ethyl alcohol are not very large (500–1000 barrels per day) they are better suited to some decentralization than are petroleum refineries, which have typical capacities 5 times larger. It is therefore possible for Brazil to have many areas self-sufficient in charcoal and liquid fuel. If one takes Latin America as a whole,[11] the situation can be seen in table 7.5.

Renewable sources already represented 35 percent of the energy consumed in 1975 and will account for 38 percent in 1995. It is interesting to notice, however, that woodfuel (the least sophisticated and least efficient method of using biomass energy) will be replaced by modern methods of using solar energy, such as alcohol fuels, biogas, direct solar collection, and others. In addition to that, the use of hydropower will increase from 14 percent to 21 percent in this period.

CONCLUSIONS

The analysis of the existing data for the United States, Norway, and Brazil shows that larger incomes are closely tied to large energy expenditures in rural and urban areas. A policy of decentralization (and an attempt to supply the energy needs of small comunities from small renewable dispersed sources) might be useful in improving present living conditions from an energy saving point of view, but it does not represent a clear-cut advantage

Table 7.5. Energy Sources in Latin America (%)

	1975	1985	1995
Nonrenewable			
Fossil fuels	64.7	62.5	59.6
Nuclear	0.1 (64.8)	1.8 (64.3)	2.5 (62.1)
Renewable			
Hydropower	14.1	18.9	20.9
Woodfuel	20.3	9.8	6.1
Nonconventional	0.8 (35.2)	7.0 (35.7)	10.9 (37.9)
Total	100.0	100.0	100.0

when compared to a policy of encouraging urbanization. Wealthier people unavoidably will consume more energy regardless of where they live.

It is possible, however, to supply the energy needed in most LDCs from renewable energy sources, avoiding the "cul de sac" in which developed nations currently find themselves. This, not the rural versus urban question, seems to be the central issue in facing energy problems in LDCs. It does not exclude, however, the possibility of saving appreciable amounts of energy in LDCs by including in their development the use of less energy-intensive materials, better architecture and less intensive patterns of consumption. In the cases of Brazil, the United States, and Norway we have not found indications that this is being done.

NOTES

1. J. Ashworth, "Renewable Energy for the World's Poor," *Technology Review* 82, no. 2 (1979): 42.
2. V. Smil, "Renewable Energies: How Much and How Renewable?" *The Bulletin of the Atomic Scientists* 35, no. 10 (1979): 12.
3. R. Herendeen and J. Tanaka, "Energy Cost of Living," *Energy* 1 (1976): 165.
4. R. Herendeen, "Total Energy Cost of Household Consumption in Norway, 1973," *Energy* 3 (1978): 615.
5. G. M. Gil Graça, V. R. Vanin and J. Goldemberg, "Energy Consumption Patterns in Brazil: Metropolitan, Urban and Rural Areas." Unpublished (1980).
6. K. Newcombe, J. D. Kalma, and A. R. Astron, "The Metabolism of a City: The Case of Hong-Kong," *Ambio* 7, no. 1 (1978): 3.
7. L. Schipper and A. J. Lichtenberg, "Efficient Energy Use and Well-Being: The Swedish Example," *Science* 194 (1976): 1001.
8. C. Holden, "Pioneering Rural Technology in India," *Science* 207 (1980): 159.

9. An automobile consumes approximately 3000 liters of alcohol per year, which means that 1 ha of land can supply the energy needs of a family.
10. The production of 1 quad of energy (10^{15} Btu = 252×10^{15} cal) would require approximately 12 million ha, i.e., 120,000 square km of land.

11. Future Requirements of Nonconventional Energy Sources in Latin America," UNDP-OLADE, Quito, Ecuador (1979).

Chapter 8
Energy and Food in The Poor Countries

Roger Revelle

By the year 2025, the earth's population is likely to be 8 billion people—twice today's 4 billion. Over 80 percent of this number will live in the presently developing countries of Asia, Africa, and Latin America. To achieve a leveling-off of population growth, the incomes of most people in these countries must rise substantially as a result of socioeconomic development. If this happens, demands for food and fibers will be much more than twice those of today.

In the tropical or subtropical climates of most developing countries, more food can be grown on existing cultivated land by double- or multiple-cropping and by increasing the yield of each crop, that is, the production per unit area. Both double-cropping and higher yields usually require more water for irrigation and larger doses of chemical fertilizers, as well as the use of farm machinery for seedbed preparation and other farm work. All three of these inputs depend upon increased energy supplies. In most rural areas, modernization of agriculture also depends upon greatly improved transportation to enable farmers to receive higher prices for their crops and to lower their costs. Additional energy is needed to fuel the vehicles used in transportation.

Agricultural modernization requiring increased use of energy is—and will be—essential to provide enough food for growing populations, but it cannot provide sufficient employment for rapidly growing rural labor forces. New, nonfarming jobs must be created to raise incomes and levels of employment. One critically important way to do this is to develop small industries in market towns and small cities throughout the countryside. A very considerable increase in rural energy supplies is a prerequisite for such industrial development.

Energy supplies must grow more rapidly than populations in order to raise the quality and quantity of human diets, increase incomes and employment, and relieve human drudgery. Greater efficiency of energy use, as well as additional sustainable energy supplies, is needed for cooking, space heating, and other domestic activities of rural households.

These generalizations can be made more concrete by considering the 15 largest countries of South and East Asia, which contain half the Earth's population but only 14 percent of its inhabited land area and about one-fourth of the cultivated land. Although these countries were originally heavily forested, the need for agricultural land to feed growing populations has resulted in a reduction of the forest area from perhaps 700 million to 400 million hectares, only 11 percent of the present world total (see table 8.1). Because of their large populations and relatively small cultivated areas, the densities of population on cultivated land are very high—averaging more than five persons per cultivated hectare, compared to about 1.3 persons in the United States. Population densities on cultivated land vary by a factor of 12 from country to country.

Resources of coal, oil, natural gas, and potential hydroelectric power, relative to population, are even more variable (see table 8.2). The People's Republic of China has by far the largest reserves, both in total and per capita. Its estimated coal and petroleum deposits and potential hydro power correspond to about 870 tons of coal equivalent per person and make up 13 percent of the estimated world total. Outside of China, the remaining Asian countries, with 30 percent of the world's population, are endowed with only a little more than 2 percent of the world's fossil fuel and hydropower resources. Even this small fraction is unevenly distributed among the different countries. It is not surprising that these countries contain the earth's largest numbers of people living in extreme poverty—poverty so dire that many of them are able to obtain barely enough food to stay alive.

In recent decades, many less developed market economy Asian countries have experienced significant economic growth, both in agriculture and in industry. Except for Indonesia, this has involved relatively large net imports of petroleum and petroleum products in terms of the size of their economies, even though their total oil imports have been only about 3 percent of world oil production. Prior to the first rapid rise of petroleum prices in 1973–74 (table 8.3) the total cost of oil imports for the poor Asian countries was 6 percent of the value of all imports. But by 1978, the total cost of oil imports had risen to 16 percent of that of all imports, and more than 20 percent for Thailand, India, and the Philippines (table 8.4).

The further sharp rise in prices of petroleum in 1978–79 could be devastating. For example, Thailand's oil import bill in 1980 is likely to be close to $2.7 billion, about 40 percent of the value of all imports, and its trade deficit, plus service on external debt, could be more than $2 billion, on the order of 10 percent of GNP. Thailand may have little choice but to slow drastically or even reverse its recent rapid rate of industrial growth. Agricultural develop-

Table 8.1. Population, Agricultural Land, and Forests in 15 South and East Asian Countries

	POPULATION millions	TOTAL AREA million ha	AGRICULTURAL AREA irrigated[1] million ha	AGRICULTURAL AREA total million ha	FORESTS million ha	POPULATION DENSITY cultiv. ha caput	POPULATION DENSITY forest ha caput
Bangladesh	85	14.4	1.2	9.1	1.5[2]	.11	.02
Burma	32	67.8	0.8	18.9	39.0	.59	1.22
China	900	959.7	76.0	127.0	111.8	.14	.12
India	638	328.0	31.6	165.0	65.8	.26	.10
Indonesia	137	190.4	6.9	18.1	121.8	.13	.89
Japan	116	37.2	2.6	5.3	24.5	.05	.21
Korea	37	9.8	0.8	2.4	6.6	.06	.18
Malaysia	13	33.0	0.3	3.5	23.5	.26	1.81
Nepal	13	14.1	0.2	2.0	4.5	.15	.32
Pakistan	77	80.4	14.0	19.4	2.6	.25	.03
Philippines	46	30.0	1.2	11.1	13.9	.24	.30
Sri Lanka	14	6.6	0.5	2.0	1.3[2]	.18	.09
Taiwan	17	3.6	NA	0.9[3]	NA	.05	—
Thailand	45	54.4	2.5	13.9	16[2]	.31	.36
Viet Nam	50	33.0	0.6	5.3	13.5	.11	.27
TOTAL	2,220	1,862.4	139.2	403.1	446.3	.18	.18
% of estimated world total	50	14		27	11		
% world total, less China	30	7		19	8		

[1] Irrigated area around 1970.
[2] See text.
[3] Author's estimate.
Sources: Population and total land areas: Asian Development Bank, 1979; Land use: *FAO Production Yearbook, 1974*, table 1, except where noted.

Table 8.2. Estimated Fossil Fuel, Hydropower and Forest Energy Resources in South and East Asian Countries in millions of metric tons of coal equivalent

	COAL AND LIGNITE[1]	PETROLEUM & NATURAL GAS LIQUIDS	NATURAL GAS	HYDRO[2]	FORESTS[3]	TOTAL	TOTAL PER CAPUT metric tons/person	
							total	less forests
Bangladesh	1,187		290	10	270	1,753	20	17
Burma	NA	12	NA	NA	6,920	>6,932	>217	?
China	719,000[4]	17,300	26,500	18,000	19,850	800,650	890	868
India	85,800	3,000	9,800	730	11,680	110,010	172	154
Indonesia	2,520	2,300	NA	410	21,620	26,850	196	38
Japan	8,760	11	150	7,950	4,350	21,221	182	145
Korea	1,515	—	NA	615	1,170	3,300	89	58
Malaysia	NA	630	NA	4,915	4,170	9,715	747	426
Nepal	—	—	—	400	800	1,200	92	31
Pakistan	650	42	690	470	460	2,312	30	24
Philippines	920[5]	400[5]	NA	240	2,470	4,030	88	34
Sri Lanka	—	—	—	45	230	275	20	3
Taiwan	450	2.5	—	1,280	NA	>1,733	>102	102
Thailand	235	0.4	190	825	2,840	4,090	91	28
Viet Nam	>12	NA	NA	>335	2,400	>2,747	>55	> 7
TOTAL	821,049	23,687	37,620	36,225	79,230	996,818	449	413
% of estimated world total	16.2	5.3	10.9	30.7	12.4	16.5		
% world total, less China	2.0	1.4	3.2	15.4	9.3	3.2		

[1] Total known resources of coal and lignite; economically recoverable resources may be less.

[2] Hydropower resources taken as 100 times potential annual production of electric energy, converted to coal equivalents, assuming one equivalent ton of coal = 3×10^{10} joules.

[3] Forest resources taken as 100 times potential annual yield (= 22% of estimated net annual primary production of 8.07 metric tons of coal equivalent/hectare).

[4] 50% of total estimated resources, based on assumed recoverability.

[5] Solid fuel and petroleum resources from data in Republic of the Philippines, Ministry of Energy, "Ten-Year Energy Program, 1979–1988." The Philippines also have significant geothermal resources. The planned electric generating capacity for geothermal sources may be nearly 1900 megawatts by 1988, equivalent to annual combustion of 6 million metric tons of coal equivalent.

Sources: Coal and lignites from World Energy Conference, *Coal Resources* (IPC Science and Technology Press, Guildford, England, 1978); Petroleum, Natural Gas, and Hydropower resources from *Survey of Energy Resources 1978* (IPC Science and Technology Press, Guildford, England, 1978), supplemented by *Oil and Gas Journal*, December 5, 1978, pp. 102–103; Richard Nehring, *Giant Oil Fields and World Oil Resources* Report R-2284-CIA (Rand Corporation, Santa Monica, Calif.: 1978); Joseph D. Parent and Henry R. Linden, *A Survey of U.S. and Total World Production, Proved Reserves and Remaining Recoverable Resources of Fossil Fuels and Uranium as of December 30, 1975* (Chicago: Institute of Gas Technology, 1977).

Table 8.3. Trade Balances and Petroleum Imports for Developing Market Economy East and South Asian Countries, 1972–1973 in millions of US dollars

	TOTAL IMPORTS	PETROLEUM IMPORTS		DEBT SERVICE	EXPORTS	EXPORTS less IMPORTS and DEBT SERVICE
		value	% imports			
Bangladesh	875	24	2.7	9	415	−475
India	3,210	264	8.2	652	3,325	−535
Indonesia	2,730	44	1.6	211	3,300	360
Korea	4,240	296	7.0	617	4,115	−742
Malaysia	2,510	156	6.2	85	3,315	720
Nepal	95	13	13.2	0.5	29	−66
Pakistan	980	65	6.6	187	1,130	−37
Philippines	1,800	188	10.4	214	2,465	451
Sri Lanka	430	38	8.8	55	425	−60
Taiwan	3,796	100	2.6	180	5,140	1,168
Thailand	2,050	226	11.0	55	2,110	5
TOTAL	22,712	1,414	6.2	2,266	25,769	790

Source: Asian Development Bank, Annual Report, 1979.

Table 8.4. Trade Balances and Petroleum Imports For Developing Market Economy East and South Asian Countries, 1978 in millions of US dollars

	GNP	TOTAL IMPORTS	PETROLEUM IMPORTS ' value	% of imports	EXTERNAL DEBT SERVICE	EXPORTS	EXPORTS less IMPORTS and DEBT SERVICE
Bangladesh	7,580	1,555	175	11.2	94	550	−1,10
India	112,665	7,400	1,900	25.7	913	6,400	−1,91
Indonesia	48,530	6,650	580	8.7	1,388	11,200	−3,16
Korea	39,000	14,970	2,312	15.4	1,795	17,100	33
Malaysia	14,545	5,920	627	10.6	705	8,000	1,37
Nepal	1,585	225	21	9.3	2.2	81	−14
Pakistan	17,525	3,325	495	14.9	316	1,840	−1,80
Philippines	23,080	5,140	1,030	20.0	646	4,715	−1,07
Sri Lanka	2,720	935	142	15.2	89	970	−5
Taiwan	23,935	11,050	1,590	14.4	633	14,380	2,69
Thailand	21,790	5,355	1,125	21.0	171	5,015	
TOTAL	312,955	62,525	9,997[1]	16.0	6,752	70,251	9
% OF WORLD TOTAL		4.7	3.0[2]			5.5	

[1] volume of petroleum imports = 87.7 million metric tons.
[2] volume of total world crude oil production = 2,903 million metric tons.
Source: Statistical Tables, ***Asian Development Bank 1979 Annual Report*** (Bangkok, 1979). Data Services Limited, *Statistical Outline of India, 1980* (Bombay, 1980).

ment will also be severely hampered by the rising costs of fuel for transportation of farm inputs and harvests and of manufacturing nitrogen fertilizer from oil products or natural gas.

Most available statistics on energy use refer only to so-called commercial energy, that is, energy from fossil fuels and hydroelectric installations. But for the poor countries of Asia, the figures on commercial energy are misleading. On the average, total energy use is about twice the quantity of commercial energy. The major part of the difference is made up of biomass fuels—wood, charcoal, agricultural residues, and cattle dung—and a significant fraction is accounted for by energy expended in human and animal labor. Most fuel energy is devoted to cooking and other household uses. Indeed, fuel for cooking is just as essential as food, and in many cases nearly as expensive. Cereal grains and grain legumes—wheat, rice, maize, sorghum, millets, beans, and lentils—are the principal foods of the poor, and these usually cannot be eaten without being cooked.

Throughout history, the major source of energy for most of mankind has been the photosynthetic conversion by plants of solar into chemical energy. In absolute terms, biomass net primary production (photosynthetic production minus plant respiration) is very large. Most of it takes place in the world's forests.

Wherever fuel wood is available in less-developed countries, wood is "the poor man's oil." It is the principal source of energy for cooking and other domestic uses in the rural areas of most Asian countries. In these countries, the forests are being rapidly destroyed as populations grow—not only because of the need for fuel wood but also because of destructive logging for timber, clearing of land for settled agriculture and shortening of the rotation cycle for "slash-and-burn" agriculture.

The quantity of energy that can be utilized from forests is constrained by several factors: (1) the low efficiency of most forest trees in photosynthetically converting solar energy into the chemical energy of organic matter; (2) the fact that the sustainable yields from forests are relatively small fractions of net primary production; (3) the existence of necessary alternative uses for wood and other forest products; (4) the necessity of using nonbiomass energy to attain an increase in sustainable yields; (5) the requirement for substantial investments, if sustainable yields are to be raised; (6) the great distances between many forested areas and the regions of high energy demand, which require that much energy be used in transportation of harvested wood to the point of use; (7) the necessity to use some energy in harvesting; and, (8) the growing demands for forest land for agriculture and other human uses.

The yield of wood for either timber or fuel from the natural forests of South and Southeast Asia is extremely low. Estimates for Thailand indicate that less than .02 percent of the incoming solar energy is photosynthetically converted into wood that can be used for fuel and timber on a sustainable-yield basis. In Nepal, the efficiency of photosynthetic solar energy conversion into fuel wood and animal forage is probably about .03 percent. The famous Sal forests of India take 80 years or more to mature.

Intensive tree-planting programs are urgently needed to provide fuel wood for cooking and other domestic uses in the rural areas of Asian countries. In the long run, wood could well be a partial substitute for fossil fuels as a source of energy for the generation of electricity and other industrial uses. Because of the pressures on land use for agriculture, the areas planted in trees must be kept as small as possible, and this will require a very considerable increase in the yields per unit area of useful products from trees.

Higher yields can be obtained from several varieties of fast-growing trees that have been discovered in different parts of the world in recent years. One of these, the "Hawaiian Giant", or Salvador variety of the evergreen, tropical legume, *Leucaena leucocephala* is especially well suited to humid and semihumid lowland tropics. In dense plantations in the Philippines, this tall, virtually branchless tree, which can grow up to a height of 20 meters in 6 to 8 years, has produced from 12 to more than 50 tons of wood per hectare per year. At 25 tons per hectare, about .7 percent of the incoming solar energy is converted into the chemical energy of wood.

Leucaena is said not to grow well in acid soils high in alumina or at elevations above 500 meters, and the young trees must be protected against grazing animals, which eagerly browse on the protein-rich foliage. Other varieties of fast-growing trees which may be better suited to particular environments include several species of *Acacia*, *Casuarina*, *Eucalyptus*, *Calliandra*, and *Prosopis*. Many of these, like *Leucaena*, possess symbiotic nitrogen-fixing bacteria, and consequently their leaves provide a high-protein forage for domestic animals.

In Indonesia, Bangladesh, and Sri Lanka, trees for fuel wood and livestock forage are grown together with vegetables, spices, and fruits in home gardens—small plots surrounding the homestead. The plants grown in these areas have a complex, three-dimensional ecology in which the tallest trees partially shade the lower trees, bushes, and crops that grow near the ground. Falling leaves and litter from the upper canopy are beneficial to the smaller plants. Research needs to be done on the most suitable fast-growing trees which can be introduced into these micro-ecosystems.

With the severe pressure on agricultural land in the crowded Asian countries, it will be essential to increase the yields of wood from forested areas that are not suited for agriculture because of steep slopes, irregular topography, rocky, thin soil, or other reasons. Most of the net primary production from forests is in the form of leaves and roots, but it is estimated that about 22 percent could be recovered on a sustainable basis as wood, useful for fuel and other purposes, in forests under intensive silviculture. In the Tropics, this will undoubtedly require replanting of some forested areas with those species of fast-growing trees that are most appropriate for each local environment. Such a transformation of the Asian forests will require a broad, sophisticated research and development program based on tree genetics (including somatic-cell genetics) and physiology, as well as on practical field experimentation. Increases in biomass energy supplies will usually depend upon the use of a relatively small amount of nonbiomass energy. Increasing the sustainable yield of biomass energy will also require considerable technological development and capital investment.

The challenges and the difficulties of modifying existing farming systems to provide a sustainable supply of food and fuel for present and future populations are illustrated by conditions in the hills of Nepal. Here, a tightly knit, virtually closed farming system has developed over many centuries. Until the last few decades of rapid population growth, forests, common pasturelands, cultivated fields, domestic animals, and human beings existed in stable equilibrium. Farm terraces constructed in the valley floors and on the lower hillsides were planted chiefly to rice or maize during the rainy (monsoon) season, and fertilized with the dung of cattle and buffalos. These livestock grazed on the village pasturelands during the rainy season, in effect concentrating in their dung a large fraction of the nitrogen fixed by the soil bacteria of the pasture, together with a portion of the phosphate released by microbial action. During the eight-month dry season, the nutritional values of the pasture grasses rapidly disappeared, and the livestock were fed on leafy branches lopped from the forest trees, plus rice straw and maize stover from the cultivated fields. Forest trees also provided fuel for human cooking and warmth and timber for construction of houses and furniture. There were few external inputs into this integrated system and few exports from it.

With the rapid growth of human populations in the hills, the system has become destabilized. In an attempt to extend the cultivated area, terraces are being cut into the steep, friable upper hill slopes, with the inevitable consequences of rapid erosion and widespread soil destruction by landslides. Further erosion results from overgrazing of the pastureland by the increased cattle and buffalo population, and from destruction of the forests as the needs for livestock forage and for fuel grow beyond the regenerative capacity of the slow-growing forest trees.

Because they must survive from day to day, the hill people are destroying the resources on which their future survival depends. Despite these desperate measures, many are unable to feed themselves, and large numbers migrate during the winter months down to Katmandu or to the northern edge of the Ganges plain in search of work and food.

A radical alteration of the farming system is needed and would be possible if a relatively small amount of nonbiomass energy could be made available. This could be accomplished in many villages by construction of small (15 kW) run-of-the-river hydroelectric plants.

So far as is known, Nepal has no reserves of fossil fuels, and distributing imported fuels would be extremely difficult in the hills, where there are very few roads and nearly all imports and exports must be carried on the backs of men and animals. But Nepal is laced by rivers and streams of all sizes, with a total fall in the hills of about 3000 m and a power potential of 16,000 MW. Construction of 15-kW generating systems in all the 30,000 hill villages of Nepal would correspond to a total installed electric power capacity of 450 MW, about 3 percent of our estimate of the minimum power potential in small perennial streams.

Most of the electric energy produced could be used to fix two to three tons of nitrogen as calcium-nitrate fertilizer in each village, using the almost-forgotten electric arc process. A large part of the remaining energy would be used during the winter months to pump groundwater for irrigation from the valley bottoms up to the stable lower terraces, and the remainder could be used for pumping domestic water supply. The excess heat from the electric arc process might be used for drying hay from the pastures, which cannot now be sun-dried during the monsoon season and, consequently, is left uncut in the pastures, where it rapidly loses its nutritional value for cattle.

The combined availability of sufficient nitrogen fertilizer and irrigation water would make it possible to grow two high-yielding crops per year on the lower terraces, thereby reducing the necessity of cultivating the unstable, highly erodible upper terraces. These could be converted into permanent hay-producing pastures, which would be protected from erosion, partly by their perennial grass cover, and partly by keeping the cattle in pens, thereby preventing erosion damage from livestock paths cutting into the hillsides. Penning the livestock would result in a considerable increase in availability of organic nitrogen fertilizer. Not only could more dung be recovered from penned animals, but the nitrogen in their urine, which makes up about two-thirds of total nitrogen excreted, could be recovered in straw bedding.

With the increase in pasture land and agricultural residues, it would no longer be necessary to use forest foliage as livestock feed, and about 60 percent of the existing forests could be placed in conservation reserves. But fuel wood would still be needed for cooking and for warmth during winter months. This could be obtained by replacing part of the existing forest area with plantations of fast-growing trees. With an estimated annual yield of 5 tons of fuel wood per hectare and a present consumption of 3/4 of a ton of wood per person, an average village of 350 people in the year 2000 would need a community wood plantation of 50 hectares. This assumes a population growth rate for the next two decades of 1.5 percent per year, after allowing for out-migration.

The young trees would have to be protected by fences or other means from destruction by browsing livestock, particularly if the animals were not penned. These and other costs of a 50-hectare village woodlot might be as much as $30,000. To this must be added the estimated cost of the hydroelectric generating system at about $20,000 per village, and another $20,000 for an electric-arc fertilizer unit, irrigation pumps, and related equipment—a total of $70,000 per village. For the 30,000 hill villages of Nepal, around $2 billion would be required. Spread over 20 years, the annual investment would equal 8 percent of Nepal's gross national product. The hill ecosystem in this projection would still be relatively closed, in the sense that few materials for export would be produced, hence amortization of loans from increased agricultural production would be next to impossible.

But the hills of Nepal can be thought of as a priceless resource for mankind, and the investment to save them from destruction might well be thought of as a necessary social cost to be borne by the international community. Moreover, Nepal's giant neighbor, India, should have a direct economic stake in reducing erosion in the hills, in order to ameloriate the ever more damaging floods in the Ganges plain that are caused by siltation and the rivers debouching on the plain from Nepal. At the same time, a determined effort should be made to introduce small-volume, high-value exportable products into the system, such as honey and other products of honeybees, spices, and medicinal plants. This will require considerable applied research.

In order to diminish rural poverty in the Asian countries, new nonfarming jobs must be created. Considerable employment can be provided by the development of small-scale rural industries based on processing and adding value to agricultural products and on meeting local needs for consumer and capital goods.

Rural employment is not an end in itself, but rather a means to attain greater equalization of incomes; to enhance human dignity based on the freedom that comes from self-reliance; and to raise average incomes through fuller use of human resources. Empirical evidence indicates that equalization of incomes acts as a powerful force towards the stabilizing of population size.

Because of its flexibility and ease of use, *electrical energy is probably more satisfactory than any other form to provide shaft power for rural industries.* Many industries suitable for rural areas also require process heat and/or process steam, as well as shaft power. In most cases, these can probably be obtained more economically by direct combustion of fuels rather than by electric heating.

Although electrification may often be a necessary condition for the establishment of small industries in rural areas, experience shows it is far from a sufficient condition. Other factors that are likely to be important are (1) adequate transportation facilities to and from markets and sources of raw materials; (2) an adequate supply of other forms of energy needed for a particular industry or industrial complex; (3) sufficient investment capital, in the form of either credit or savings; (4) trained technicians and managers familiar with problems of marketing, cost accounting, personnel management and business operation; (5) a market town or small city in which a complex of mutually supporting industries can be established; and (6) the availability, skills, discipline, and comparative costs of labor in the rural areas as distinct from the labor force in large cities.

In recent years, numerous small industrial plants have sprung up in the market towns and small cities of the Indian subcontinent. These industries carry on a great variety of activities, from rice milling, cotton ginning, and dairying, to the manufacture of household utensils and diesel engines for irrigation pumps. Kraft paper making, weaving of cotton grey goods on small power looms, gem cutting and polishing, and the construction of ferroconcrete boats and grain storage structures are among other important activities. All depend on electric power, and many are handicapped by the frequent interruptions in power which characterize electric systems in the subcontinent.

The future sources of primary energy for the generation of electricity and for motor vehicles present many of the poor countries with a cruel dilemma. With inadequate hydropower potential and small reserves of fossil fuels, they will be forced to import fuels at very high foreign exchange costs to provide energy for the central-power-station generation of electricity. Yet the alternative of decentralized electricity generation and production of liquid fuels for transportation from locally grown wood and other biomass may seriously limit food production. In very large countries such as Brazil, where only a small fraction of cultivatable land has been required to grow enough food, the production of alcohol from sugar cane as a vehicle fuel may actually raise the real income of rural people by providing increased employment. But in the Asian countries, with their high population densities on cultivatable land, the use of biomass fuels for transportation and electricity will almost certainly result in increased food prices and may reduce the real income of the poor in spite of the added employment provided by rural industry. As Lester Brown has recently pointed out, the food situation is likely to be worsened if the United States and other major food exporting countries develop extensive programs for production of liquid fuels such as gasohol from agricultural products. The 315 million owners of the world's automobiles would be able to compete all too successfully for the agricultural products which might otherwise be used to feed the world's poor.

There can be no single or short-term solution to the energy problems of the poor Asian countries, but only many partial solutions which would gradually bring energy supplies into balance with energy needs. An underlying requirement is a greatly expanded effort in biological research—to increase forest yields and reduce forest losses through research in forestry and agroforestry; to maximize production from household gardens through better understanding of their ecology; to increase the production and widen the uses of natural fibers and rubber; and to obtain more useful microorganisms for production of gaseous and liquid fuels, nitrogen fixation, and mycorrhizal symbiosis with trees. These potential applications of biology to the solution of energy problems in the Asian countries give an added dimension to its already recognized importance in agriculture, food preservation, nutrition, health, and birth control.

Relative to research in physics and chemistry, the Asian countries have a comparative advantage in biological research. It is relatively inexpensive, requiring neither huge research establishments nor massive, highly complex equipment. It is in many ways a "new" science, a science of the future. Most discoveries in biology remain to be made. Hence developed countries have less of a head start than they have in physics and chemistry. At the same time, biological research is a most promising field for international scientific cooperation.

BIBLIOGRAPHY

Brown, Lester. *Food or Fuel: New Competition for the World's Cropland.* Worldwatch Paper No. 35. Washington, D.C.: Worldwatch Institute, 1980.

——. *Losing Ground: Environmental Stress and World Food Prospects.* New York: W. W. Norton & Co., 1976.

Eckholm, Eric. *The Other Energy Crisis: Firewood.* Worldwatch Paper No. 1. Washington, D.C.: World Watch Institute, 1975.

——. *Planning for the Future: Forestry for Human Need.* Worldwatch Paper No. 26. Washington D.C.: Worldwatch Institute, 1979.

Makhijani, A. and Poole, A. *Energy and Agriculture in the Third World*, pp. xv and 168. Cambridge, Mass.: Ballinger Publishing Company, 1975.

Revelle, Roger. "Energy Sources for Rural Development," *Energy*, Special issue: *Renewable Energy Prospects* (October 1979).

——. "Energy Use in Rural India." *Science* 192 (June 1976): 969–975.

U. S. Agency for International Development. *War on Hunger.*

Chapter 9
Hunger, Energy, and Agricultural Development

Vashek Cervinka and Richard DeForest

The problem of hunger is new to many societies in developing countries. In the past, many countries used to export food, and others had no food surpluses, but no food shortages either. What has happened is that our access to fossil fuels, to advanced science, technology, and management, has resulted in many countries experiencing widespread hunger for the first time.

This paper will be based on the authors' years of experience in West Africa and Brazil, where they worked in agricultural development programs. The discussion will be specifically related to the situation in Ghana.

In the past, people in Ghana had no food shortages. Cocoa was a major export commodity, and enough food was produced for internal consumption. After describing some problems and also some negative experience from agricultural policies, several causes for food shortages will be analyzed. Before any measures can be developed to improve food supplies, the causes of the progression from food sufficiency to shortages should be evaluated. Some problems and observations include:

1. The philosophy of agricultural development has been oriented towards intensive mechanization and high fossil energy inputs. Farmers were trained in schools to work with expensive equipment which, for the lack of capital, they could not afford to purchase and operate on their farms. At the same time, farmers were encouraged to adopt food production technologies requiring high fossil energy inputs.

2. Food assistance programs have frequently been offered for political or humanitarian reasons. In many cases, these programs reduced the capacity of local farmers to produce food and develop efficient market infrastructures.

3. The priorities of agricultural development and food policies are established by government officials working in the capital cities or overseas. This is done without any, or with very limited, input from farmers.

4. The needs of farmers are not considered. For example, in a sub-Saharian region, with a single rainy/cropping season, the farming cycle consists of crop planting (April)—harvesting (October)—storage. During the planting period, which should provide a base for future food production, the farmers experience food (i.e., energy) shortages. Well-designed food storages and water management systems would assist in resolving this continuous problem.

5. Farmers' self-reliance was reduced by induced dependence on external production inputs (equipment produced for European and American farms, fertilizers, chemicals, seeds). This dependence has increased the vulnerability and costs of food production in many developing countries.

6. In Europe and North America, the mechanization of farms and industry was encouraged by labor shortages, and still countries in both regions had experienced periods of high unemployment. Any sound planning of economic development should be based upon the essential facts that people are the most important factor and that they should be the main objective. The technology which creates millions of unemployed and frustrated people cannot promote economic development.

The people of Ghana are well-educated and hardworking; the country has good environmental conditions for agricultural production. Why, then, has development taken a path from food sufficiency to food shortages? Two major factors contributing to this situation are (a) philosophy (model of life styles), and (b) energy.

A philosophy is the result of given conditions of existence, as is a biological environment. Ghanaian philosophy could not have flourished in Europe, just as European philosophy is foreign to Africa. Neither is better nor worse. Probably, the most damaging effect of colonialism was to implant a complex of inferiority in the minds of Ghanaians. Instead of using their culture, traditions, and skills as the base for economic development, they were encouraged to follow a model of development which cannot be simply transplanted, and even if transplanted, is not viable in a new environment.

The wrongly established model was both technological and political. Without much regard to traditional Ghanaian economic, social, and cultural institutions, the attempt was made to adopt a system of European democracy. The negative sides of this political system (political power and corruption) sometimes had a stronger influence than the grassroots political movement which is the real base of a democratic system. An agricultural cooperative movement can be given as an example—if ordered by the government, the cooperatives cannot prosper; only those cooperatives which are established and supported by members' initiative have the basic conditions for economic viability.

The twentieth century will be remembered as a period of historically unbalanced energy resources. Past civilizations used those energy resources which were created during their own existence—within the time frame of one or two generations. Our civilization is mostly dependent upon fossil fuels, energy resources created millions of years ago. This causes an extreme economic and social vulnerability in both industrial and developing countries.

A policy of agricultural development designed after the models which require large inputs of nonrenewable (fossil) energy has a low probability of success. Further, while fossil fuels used to be relatively inexpensive in industrial countries, that has not been the case in most developing countries. The transportation of imported farm production inputs from a harbor to inland agricultural regions can illustrate this problem. Considering the typical level of fuel consumption and the price of fuels at $2.00 and $3.00 per gallon, a truck traveling 300 miles uses $240.00 to $360.00 in fuel alone. This significantly adds to the costs of food production and creates the difference between the planned agricultural development policy and reality.

Future agricultural development cannot be based upon fossil fuels, as they represent a nonrenewable energy supply. Their diminishing availability results in the high cost of energy inputs and low supply reliability. The dependence on fossil fuels increases the political and economic vulnerability of both industrial and developing nations. However, this situation should be approached as a challenge rather than as a crisis. All nations will benefit from the development of renewable energy resources which are the only realistic means for future economic development. Further, this creates an opportunity for developing nations to progressively follow the path of renewable energy applied to agricultural development and, thus, to establish sound and lasting foundations for the production of food.

Renewable energy resources will lead to increased energy and food self-reliance and the re-creation of viable political, economic, social, and cultural trends in the countries of Africa, Asia, and Latin America.

Chapter 10
Of Trees and Straws
Vaclav Smil

Although I have not seen any comparisons, I will not hesitate to guess that the quantity of information available on energy in the developed world is at least an order of magnitude greater than it was just seven years ago—and the tide is still rising. (I would also argue that we do not need any more of these studies, surveys, and analyses to initiate, finally, some genuine action.)

In contrast, our knowledge of developmental energetics is dismal—and I do not have in mind input-output matrices, disaggregated final uses by sectors, or temperatures or energy costs of principal products. What we do not know are the bare essentials: How much energy is really consumed throughout the poor world? What are its uses? What is the magnitude of energy resources, renewable and nonrenewable? What share of the renewable potential could be optimally harvested, and in what ways?

I recall my frequent exasperation with what I felt to be a serious lack of data when I started to work on the studies of European and Soviet energetics in the early 1960s—and my complaints continued a decade later with regard to the situation in the United States and Canada. Then I plunged into interdisciplinary work on China's energetics—and quickly discovered what is the *real* lack of information. And yet, in the Chinese case, there are, at least, relatively abundant records from the 1950s; a fragmentary and dispersed, but never completely drying out, flow of bits in the 1960s and 1970s; and now a renewed progression of somewhat systematic information. For the overwhelming majority of the other developing nations, that would constitute a fabulous abundance of hard facts!

If the information considered standard in the Western context is scarce, our understanding of the rural energy resources and uses in the developing world hangs almost solely on approximations that rest largely on arrays of assumptions and outsiders' calculations. To appreciate this key and decisive segment of developmental energetics and to evaluate the appropriateness of alternative modernization paths, we must work within an ecosystem framework, dealing with plants, people, and animals, rather than just with fossil fuels and power generation.[1]

The map is not altogether blank: Currently we have such accounts for China[2], India[3], Bangladesh[4] and Egypt[5]. None is fully satisfactory, and better analyses are under way. But profound limits remain. Any ecological approach attempting to embrace the intertwined triad of energy-environment-food must deal with trees and straws—forest fuels and crop residues—the two principal rural fuel energy resources for the developing nations that provide at least two-fifths of the total energy consumed in the poor world.[6]

What follows is an attempt to impress you with the scarcity of information and the gaps in our knowledge about trees and straws, and with what is a nearly insurmountable difficulty of saying how much is where, and to what extent, and in which form it is to be harvested.

TREES

Mencius said, There was a time when the trees were luxuriant on the Ox Mountain. As it is on the outskirts of a great metropolis, the trees are constantly lopped by axes. Is it any wonder that they are no longer fine? With the respite they get in the day and in the night, and the moistening by the rain and dew, there is certainly no lack of new shoots coming out, but the cattle and sheep come to graze upon the mountain. That is why it is as bald as it is.

Mencius, Book VI, Part I,
Section 8

The lines are nearly two and one-half millennia old, but what they describe is timeless. Forests, by far the largest repository of phytomass, have been disappearing at an increasing pace for several thousand years—first the temperate broadleaf forests of northern and eastern China, the monsoon forests and woodlands of India, and the mixed vegetation of the Mediterranean; then the deciduous and boreal forests of western, central, and northern Europe and their counterparts in North America and Russia, and finally the rich rainforests, extending largely undisturbed over almost all of the lowlands of the humid tropics for 60 million years, started to recede drastically just several decades ago.

The questions are obvious. How much phytomass remains in the world's forests and how fast is it disappearing? What is its spatial distribution? What is its net productivity, and how much of it can be harvested on a sustained basis? How much are the forest fuels contributing to the current energy balance of the developing world? None of these questions, it turns out, can be answered with reasonable accuracy and, consequently, even a general appreciation of the forests' role as potential energy suppliers is exceedingly difficult.

Area and distribution

To begin with, even the simplest measure—the total area of the world's forests and the disaggregated figures for individual continents and countries—is grossly deficient. Global totals in forest inventories and other estimates prepared by the Food and Agriculture Organization (FAO) of the United Nations range from slightly over 37 million km^2 to more than 44 million km^2, a difference of some 19 percent (table 10.1). Appreciably higher totals were put forward by two ecologists working on the International Biological Program: Olson's provisional estimate of woodland and forest is 48 million km^2,[7] while Lieth's figure is 57 million km^2.[8]

On the continental level, the greatest discrepancy exists for Africa, where the previous estimates ranging between 6.3 and 8.0 million km^2 were drastically reduced by Sommer's reappraisal[9] to only 3.3 million km^2; substantial variations are also evident in the case of Latin America (see table 10.1). Similar discrepancies and a lack of solid statistics exist on the national level: Most disturbingly, this is true of all four of the world's most populous developing nations—China, India, Indonesia, and Brazil.

Estimates of China's forest area ranged between 46.5 and 100 million hectares in the late 1950s and early 1960s; however, nearly half of the higher total was at that time classified by the Ministry of Forestry as a secondary growth of low productivity.[10] Latest disclosures put the forested area at 12.7 percent of China's territory, or some 121 million hectares, but this total

Table 10.1. Areas of Forests According to FAO 1953–1977

Area	1953	1958	1963	1971	1976	1977*
North America	6,564	7,334	7,130	7,100	·	6,161
Western Europe	1,090	1,141	1,117	1,220	·	1,244
USSR and E. Europe	7,693	11,579	7,643	7,640	·	9,490
Oceania	857	959	1,021	920	·	1,859
Africa	8,014	7,528	7,000	7,110	3,340	6,396
Latin America	8,900	10,309	9,010	7,940	9,640	10,264
Near East	336	180	263	160	·	139
Far East	4,920	5,019	4,736	5,030	4,170**	5,899
Developed	15,889	20,868	16,650	16,720	·	18,636
Developing	22,485	23,181	21,270	20,400	·	22,816
World	38,374	44,049	37,920	37,120	·	41,452

* Forests plus woodland.

** China excluded.

Sources: Food and Agriculture Organization, *World Forest Resources* (Rome, 1955), pp. 60–69; *World Forest Inventory, 1958* (Rome, 1960), pp. 60–69; *World Forest Inventory, 1963* (Rome, 1966), pp. 48–55; "Supply of Wood Materials for Housing," *Unasylva* 25, nos. 101–103 (1971), p. 29; *Production Yearbook, 1978* (Rome, 1979); and A. Sommer, "Attempt at an Assessment of the World's Tropical Forests," *Unasylva* 28, nos. 112–113 (1976): 5–24.

includes a large, though unknown, area of just recently afforested land.[11] India's official sources classify 73 million hectares of the country (or 22 percent of the total territory) as forests—but as the country has never conducted a nationwide forest inventory it is well known that this figure includes extensive areas that have been completely or largely deforested, such as pastures, scrubland, wasteland, or even cultivated fields, and that a large share of the actually forested area has a very low phytomass density[12]. Not surprisingly, knowledgeable Indian officials admit privately that little more than half of the so-called forested area deserves the title.[13] In Indonesia there are simply no statistics of forest areas available either for the major islands or for the country as a whole; the only reliable figures are for scattered inventoried areas (a requirement for logging concession applicants) that cover so far no more than approximately a quarter of the total estimated forested land.[14] Finally, disparate estimates for Brazil are listed in table 10.2

Obviously, problems of definition are involved: at what point does the density of trees warrant a classification as forest, woodland, scrubland or grassland? And, a more important indicator, what is the actual standing phytomass of trees and other plants per unit area? LANDSAT imagery may be a rather inexpensive and fast tool to determine the simple areal extent of major plant formations, but it is far from satisfactory for a detailed, reliable classification and inventory of forest resources; these tasks must be performed through much costlier and more time-consuming surveys.[15]

The spatial distribution of forests is, obviously, extremely uneven. The admittedly imperfect data in table 10.1 indicate that Latin America has 43 percent of the developing world's forested area while the Near East has a mere 10 percent. Of the nineteen developing nations with populations exceeding 20 million and accounting for four-fifths of the Third World people, 15 nations with over 2.3 billion inhabitants have less than one hectare of forested land per capita; the only exceptions are Brazil, Colombia, Zaire, and Burma (table 10.3).

Table 10.2. Estimates of Brazil's Forested Area 1955–1977

Forested Area (10^6 hectares)	Year of Estimate
480.2	1954[1]
561.67	1958[2]
335.1	1964[3]
461.0	1975[4]
319.12	1977[5]

Sources: [1] Food and Agriculture Organization, *World Forest Resources* (Rome, 1955) pp. 60–69; [2] *World Forest Inventory, 1958* (Rome 1960), pp. 60–69; *World Forest Inventory*, 1963 (Rome, 1966), pp. 48–55; [4] M. K. Muthoo, "Forest Energy and the Brazilian Socio-Economy with Special Reference to Fuelwood" (Paper delivered at the 8th World Forestry Congress, Jakarta, October 1978), and [5] J. Goldemberg, "Brazil: Energy Options and Current Outlook," *Science* 200 (14 April 1978): 158–164.

Table 10.3. Populous Developing Nations (Over 20 Million) and Their Forested Areas per Capita

Country	Population in mid-1978 (million)	Forests (hectares/capita)
China	958	0.1
India	635	0.1
Indonesia	140	0.8
Brazil	115	3.1
Bangladesh	85	0.2
Pakistan	77	0.9
Nigeria	68	0.5
Mexico	67	0.6
Vietnam	49	0.2
Philippines	46	0.3
Thailand	45	0.6
Turkey	42	0.2
Egypt	40	0.0
South Korea	37	0.2
Iran	35	0.3
Burma	32	1.4
Ethiopia	30	0.3
Zaire	27	4.8
Colombia	26	2.7

Sources: *World Population Data Sheet* (Washington, D.C.: Population Reference Bureau (1978); and Food and Agriculture Organization (FAO), *World Forest Inventory* 1963 (Rome, 1966); and FAO, *Forest Resources in Asia and the Far East* (Rome, 1976).

Distribution within countries is similarly uneven. China's northeast, with 10 percent of China's people, has 60 percent of the country's forests, while provinces in the north and in the east, comprising about two-fifths of China's population, contain less than 5 percent of China's forests. In India there are analogical differences between, for example, Arunachal Pradesh on the one hand and Rajasthan on the other. In Indonesia, Kalimantan and West Irian are estimated to have nearly three-fifths of the forested land, while Java has only about 2 percent. Brazil's predicament is shown in table 10.4.

Disappearance

Many processes result in the disappearance of forested areas throughout the developing world. Land hunger is a ubiquitous cause, and in the long run it matters little if those involved are already very well off beef ranchers in Brazil who, encouraged and often handsomely subsidized by the government, convert vast tracts of intact forest with the help of bulldozers—or landless, unemployed, illiterate Filipino peasants who so frequently move illegally into a patch of forest, burn it down, clear the stumps, and establish

Table 10.4. Spatial Distribution of Brazilian Forests and Population in 1975

	Amazonia	Northeast	Midwest	Southeast	South	Brazil
Forests						
10^6 hectares	321	40	72	21	7	461
(percent)	(70)	(9)	(16)	(4)	(1)	(100)
Population						
10^6 people	4	32	7	45	19	106
(percent)	(4)	(31)	(6)	(42)	(18)	(100)

Source: M. K. Muthoo, "Forest Energy and the Brazilian Socio-Economy with Special Reference to Fuelwood" (Paper delivered at the Eighth World Forestry Congress, Jakarta, October 1978).

themselves as squatters for just a few years of meager crop harvests. Often it is not a matter of survival but of social status; for example, much forest is converted to pasture in Latin America because many crop farmers desire nothing else but to become cattle ranchers.

New roads, settlements, industrial enterprises, and water reservoirs are the other consumers of forests associated with the still relentless population pressure. Available figures for officially authorized deforestation in Brazil—the country which, in absolute terms, is stripping its vegetation faster than any other developing nation—show the relative shares of the main categories (table 10.5). "Mining" of tropical and subtropical forests for timber is, obviously, another major cause of vegetation disappearance: It spans a wide spectrum of methods, from relatively controlled selective logging to ecologically disastrous clear-cutting; in some major lumber-producing nations, there are signs of stricter control being established through limitations on logging licenses and the establishment of protected forest areas.

Nobody keeps any solid figures for any seriously affected developing nation. Clearly, the process of disappearance over the past few decades has been most intensive in the tropical rain forests of Latin America, Africa, and Southeast Asia. By the middle of this century, these forests, the world's most

Table 10.5. Officially Sanctioned Deforestation in Brazil 1966–1975

Activity	Area (hectares)	Percent of the total
Cattle raising	4,375,271	38.0
Colonization	3,519,480	30.7
Highways	3,075,000	26.8
Forestry	500,000	4.4
Total:	11,469,751	100.0

Source: Brazilian Institute of Forestry Development (IBDF) cited in S. H. Davis, *Victims of the Miracle* (Cambridge: Cambridge University Press, 1977).

important storehouse of biomass, had already receded about 42 percent from their climax area, and the officially reported roundwood logging is diminishing their still intact natural stands by at least 0.6–1.0 percent annually.[16] Although it is clearly impossible to come up with a good global estimate, it is evident that the addition of legal and illegal fuelwood harvesting and forest conversion to cropland, settlements, roads, and reservoirs would easily double this rate. Simple extrapolation is not particularly meaningful except to point out that the continuation of the current global trend could destroy virtually all the remaining natural rain forests within two to four generations. For more than a dozen developing nations, there are published estimates (guesstimates is, in most cases, a better term) of annual rates of forest disappearance which are summarized in table 10.6. Some of these figures may be nearly an order of magnitude off and should be viewed merely as indicators of a very alarming process.

Table 10.6. Disappearance of Forests in Some Developing Nations During The Late 1960s and 1970s

Country	Annual Rate of Disappearance (10^3 hectares)	Approximate Share of the Forested Area (percent)
Bangladesh	10	0.4
Bolivia	46	0.1
Brazil (officially authorized)	1150	0.3
Colombia	250	0.4
Costa Rica	60	2.0
Ghana	50	4.8
Haiti	22	5.0
India	113	0.2
Indonesia	150	0.1
Ivory Coast	400	3.3
Laos	300	2.0
Madagascar	300	1.8
Malaysia (peninsular)	150	1.7
Papua New Guinea	20	0.1
Philippines	33	0.3
Thailand	300	1.1
Venezuela	50	0.1

Sources: Food and Agriculture Organization, *Forest Resources in Asia and the Far East* (Rome, 1976); A. Sommer, "Attempt at an Assessment of the World's Tropical Forests," *Unasylva* 28, nos. 112–113 (1976): 5–24; A. Rowley, "Forests: Save or Squander?" *Far Eastern Economic Review* 98, no. 48 (2 December 1977): 53; P. Bowring and R. Tasker, "Philippines in Decline," *Far Eastern Economic Review* 98, no. 48 (2 December 1977): 63; J. Josephson, "Building the Third World," *Environmental Science and Technology* 11, no. 9 (September 1977): 852; D. Lamb, "Conservation and Management in Tropical Rain-Forest: A Dilemma of Development in Papua New Guinea," *Environmental Conservation* 4, no. 2 (Summer 1977): 121–129; and R. Stolz, *Plan de Desarrollo del Sector Forestal en Bolivia* (MACA/CDF, La Paz).

Productivity

Even if we had very good figures for areal extent and disappearance, we would still be confronted with the problem of the different concepts that can be applied in evaluating the forest biomass and its productivity. The traditional forestry approach has been to use the merchantable bole concept within which the utilization of phytomass is limited to the bole from the stump to some commercial height and to trees of certain size, species, and density.[17] This neglects roots and green parts, which, depending on the forest type, account for 23–29 percent of total phytomass (table 10.7), as well as a part of the perennial above-ground growth (bark and branch wood). Share of the utilized phytomass would then range from above two-thirds of the standing mass to as low as 36 percent. In the case of climax stands of tropical rain forests, where among scores or hundreds of species, only a handful are commercially interesting, the difference is still greater.

Annual production (increment) within the bole concept is again limited to merchantable wood and bark, and its global averages are correspondingly low—just 1.8 cubic meters (m^3) of wood with bark per hectare for conifers and 2.5 m^3 per hectare for nonconifers.[18] Using standard conversion values of 625 and 750 kg/m^3 respectively,[19] this translates into 113 and 188 g/m^2. In the ecosystem approach, all vegetation—above and below ground—is accounted for, and the average values for forest biomass range between 20 and 45 kg/m^2 and net primary productivities (increment) between 200 and 2,200 g/m^2.[20]

Very large differences between the two concepts are illustrated on a global level in table 10.8. There is more than a fourfold difference in biomass estimates and nearly a tenfold difference in annual increment values. While it is true that the ecosystem concept is not relevant to evaluate commercial exploitability of forests for industrial lumber, it is highly appropriate in assessing energy potential of forests. After all, branches, leaves, bark, stumps, dead trees, shrubs, and even grasses already constitute a large portion of

Table 10.7. Component Breakdown of Biomass in Selected Forests (All figures are in percent)

Phytomass Category	Tropical Rain Forest	Subtropical Forest	Oak Forest	Fir Forest
Green parts	8	3	1	6
Perennial above ground parts	74	77	75	71
Roots	18	20	24	23

Source: I. E. Rodin and N. I Basilevic, "World Distribution of Plant Biomass," in F. E. Eckardt, ed., *Ecosystems at the Primary Production Level* (Paris: UNESCO, 1968), pp. 46–47.

Table 10.8. Comparison of Two Forest Phytomass Concepts

A Merchantable Bole Concept		B Ecosystem Concept		Ratio B/A
Area (10^6km^2)	37.1	Area (10^6km^2)	37.1	1.0
Growing stock (kg/m^2)	6.8	Phytomass (kg/m^2)	30.1	4.4
Growing stock (10^9t)	252.3	Phytomass (10^9t)	1,116.7	4.4
Increment (kg/m^2/year)	0.13	Net primary productivity (kg/m^2/year)	1.2	9.2
Increment (10^9t/year)	4.8	Production (10^9t/year)	44.5	9.3

Sources: Food and Agriculture Organization (FAO) *World Forest Inventory 1963* (Rome, 1966), pp. 48–55; FAO, "Supply of Wood Materials for Housing," *Unasylva* 25, no. 101–103 (1971): 29; and R. H. Whittaker and G. E. Likens, "The Biosphere and Man," in H. Lieth and R. H. Whittaker, eds., *Primary Productivity of the Biosphere* (New York: Springer-Verlag, 1975), pp. 305–328.

forest phytomass collected for fuel in developing countries and, as will be seen shortly, neglect of that fact leads to considerable underestimates of actual harvest.

On the other hand, the net primary productivities do not represent the actually available mass increments. As Evans properly remarks, net primary productivity is "an academic concept which ignores herbivore, insect, and decay consumption and . . . can be more than twice as great as actual material storage."[21] Yet if our knowledge of net primary productivity, especially in tropical rain forests, is far from satisfactory, our understanding of that more realistic measure, the net ecosystem production, is even less so, and any generalization on a global, regional or national scale is just a gross estimate.

Calculation of forest biomass energy equivalents is further complicated by differences in mean combustion value of plants. Average dry matter energy content of forest ecosystems ranges from 4.1 kcal/g for tropical rain forests to 4.9 kcal/g for chaparral.[22] That of wood averages 4.7 kcal/g for deciduous species and 4.8 kcal/g for conifers, with the highest individual values exceeding 5 kcal/g. By necessity, mean values must be used to calculate the storage and production totals.

Given all of these difficulties, it is clearly impossible to offer any accurate estimate of forest phytomass and production, be it on a global, continental, or national basis. Plainly, we do not know what are the forest energy resources in any major developing nation. Nor do we know much about how they are harvested and consumed as fuel.

Harvesting and consumption

Depending on needs, accessibility, and the state of the forests, harvesting of phytomass for fuel (I am avoiding, purposefully, the term fuelwood or firewood because, as already mentioned, forest fuels in many parts of the developing world contain, besides stem wood and branches, twigs, bark, and stumps, and nonwoody tissues as well may be done every day and fairly easily by children and adults in the immediate neighborhood of a settlement—or it may require periodical, tedious, and lengthening trips with pack animals or headload teams of adults. Availability of other biomass fuels, such as crop residues after the harvest time, and the duration and intensity of the rainy period will determine the time of the sole or dominant reliance on forest fuels and the need for seasonal storage.

Naturally, the bulk of this harvest is completely beyond the reach of regular statistical accountability and hence the worldwide conservation of forest fuels can be estimated with only very poor reliability. Consequently, the best available estimates of the total fuelwood use in the developing world in the mid-1970s range from around 1.3 billion,[23] to about 2.5 billion m^3.[24]

Special surveys on fuelwood and/or charcoal use were carried out in some locations and for limited periods of time in about a dozen developing nations, including India, Thailand, Lebanon, the Sudan, Gambia, Tanzania, Kenya, Uganda, Upper Volta, and Indonesia.[25] Results suggest that the estimates regularly published by the Forestry Division of the FAO underrate the actual use of forest phytomass in most of the poor countries. On the other hand, there are instances where the FAO estimates are much higher than the alternative figures (table 10.9). It would appear that, with the exception of a handful of developing nations with reasonably reliable and consistent fuelwood statistics, nationwide forest fuel consumption estimates throughout the poor world are nearly useless for any serious assessment of phytomass harvesting potentials and limits.

Only two essential points are clear: First, forest fuels are the most important source of energy for most of the people throughout the developing

Table 10.9. Fuelwood Consumption Estimates for Selected Developing Nations

Country (Year)	FAO Estimates	Alternative Estimates (million cubic meters)
India (1975)	115.285	184.00
Bangladesh (1975)	14.012	3.25
Thailand (1972)	15.050	47.49
Ghana (1975)	10.106	5.60
Zambia (1972)	0.180	0.79

Sources: Food and Agriculture Organization, *1975 Yearbook of Forest Production* (Rome, 1977); J. W. Powell, "Wood Waste as an Energy Source in Ghana," in N.C. Brown, ed., *Renewable Energy Resources and Rural Applications in the Developing World* (Boulder: Westview Press, 1978), pp. 115–128; P. Argal, "Role of Wood Energy in the Rural Economy of India," (Paper delivered at the Eighth World Forestry Congress, Jakarta, October 1978); K. Openshaw, "Woodfuel: A Time for Reassessment," *Natural Resources Forum* 3 (1978): 35–51; and R. Tyers, "Optimal Resource Allocation in Transitional Agriculture: Case Studies in Bangladesh," Dissertation, Harvard University, 1978.

world; and second, with increasing populations and the growing needs of local industries, consumption of fuelwood and charcoal has been climbing and contributing—with timber logging, colonization, and industrial projects—to relatively rapid deforestation, especially in the tropical zone, a process that is likely only to accelerate during the rest of this century.

Our knowledge is grossly inadequate to properly evaluate and to rationally affect the process. While we have a fair understanding of the standing biomass, primary production levels, and dynamics of the temperate forests, our understanding of tropical forest ecology is incomparably weaker. Little information is available on the structure and function of natural forests, almost nothing on modified forests or combined agrosilviculture systems, and appreciation of long-term changes caused by variations in management and environmental conditions is pure conjecture.[26]

Consequently, we do not know the total energy stored in tropical and subtropical forests, the share that could be safely harvested, and the best practices to do so. However, we have a growing body of observations and indications pointing toward an alarming conclusion with unpredictable environmental consequences: Throughout most of its extent, the rain forest seems to be incapable of regeneration under present land-use practices.[27]

Ecosystem functioning, especially in the case of the biome containing most of the planet's tree phytomass, is vital to human welfare, though certainly not easy to quantify.[28] Those who are not disturbed by the disappearance of natural forest stands and talk glibly about putting huge plantations of fast-growing species in their place are ignoring this key reality of man's existence in the biosphere.

Silviculture could be practiced carefully, cautiously, and responsibly, but a massive blanketing of areas previously covered by a profusion of multi-layered species (rooted in meager soils yet functioning well owing to the large and continuous flux of energy and matter, with intimate bonds between scores of auto- and heterotrops) with a handful of 'desirable' plantation trees—be it *Eucalyptus*, *Gmelina* or *Leucaena*—that should yield cheap and plentiful wood for fuel, pulp, and construction, would be a dubious strategy.

The local results are already known—increased storm flow and soil erosion and reduced low water flow, two important contributions to devastating flood-drought cycles.[29] The long-term regional and global effects of dust loading of the atmosphere, changed albedos, energy fluxes, and CO_2 and water cycles are still largely a matter of speculation—but they are sure to follow.[30]

To sum up, I view the developing world's dependence on forest fuels as one of the most dangerous environmental problems of our time. The process is a major contributor to the destruction of precisely those terrestrial ecosystems—tropical rain forests, raingreen forests, and summergreen forests—that are potentially least suited to man-influenced restoration, be it attempts at total regeneration or ecologically impoverished and potentially injurious silvicultural practices.

STRAWS

Although the global phytomass is overwhelmingly concentrated in forests and only a fraction of a percent is in seasonal and permanent crops, the easily accessible residues—straws, stalks, leaves, vines, even roots—have always been a very important source of energy, not only in deforested regions but, seasonally, also in the areas with adequate forest cover.

In contrast to the evaluation of forest primary production, satisfactorily reliable calculations of residue yields could be done rather easily, usually by applying appropriate multipliers to the known crop harvest. This is a simple exercise for smaller areas where the residue yield per unit surface or per unit yield can be easily sampled, but the application of approximate multipliers to regional, national, or global totals will almost inevitably result in major errors. This is because the residue yields of the same crop differ and fluctuate considerably according to the plant varieties, maturation periods, soil characteristics, fertilization, use of pesticides, irrigation, and weather conditions of different regions.

A summary of the ranges of residue coefficients for major crops compiled from nearly a dozen reports from various parts of the world makes clear the very wide margin of error caused by indiscriminate application of a general multiplier (table 10.10). Just to arrive at continental figures of the right order of magnitude, I multiplied the mid-1970s production totals for major crop groups by modal estimates—1.5 for cereals, 0.2 for tubers, 1.5 for legumes, and 0.2 for sugar cane (table 10.11). Although the actual values may easily be off by as much as ± 25 percent, the worldwide yield looks, in any case, quite impressive: somewhere around 2.5 billion dry tons, the equivalent of about 950 million tons of crude oil.

This is a huge total, nearly equal to three-fifths of the current OPEC crude oil production, but only about half of it is available in the developing nations, and a very large portion of this potential residue harvest is not—and, in fact, should not become—available for direct combustion or for further conversion to gaseous or liquid fuels. It has more important competitive uses as feed, as fertilizer, and for erosion protection.

In densely populated regions of the developing world, where grazing land or fodder crops are almost nonexistent, most of the cattle—the principal motive power in the countryside—as well as other domestic animals, are stall-fed with residues ranging from rice straw to sweet-potato vines. In China and India, approximately three-quarters of all harvested crop residues are consumed as animal feed and bedding.[31] The crop wastes play a similarly

Table 10.10. Residue Coefficients for Major Crops

Crop	Residue/crop ratio
Rice (paddy)	0.75–2.51
Wheat	1.10–2.57
Corn	0.55–1.30
Barley	0.82–1.50
Oats	0.95–1.75
Rye	1.20–1.75
Sorghum	0.85–1.99
Potatoes	0.20–0.30
Soybeans	1.10–2.60
Other legumes	1.20–1.50
Sugar cane	0.20–0.25
Cotton	1.40–3.00
Rape	1.85–2.00

Sources: J. L. Buck, *Land Utilization in China* (Nanking: University of Nanking Press, 1937); O. L. Dawson, *Communist China's Agriculture* (New York: Praeger, 1970); N. Parathasarathy and S. S. Rajan, "Plant Breeding Strategy for the Seventies with Special Reference to Developing Countries in the Tropics," *International Rice Commission Newsletter* 21, no. 4 (December 1972): 1–17; A. Tanaka, "Methods of Handling the Rice Straw in Various Countries," *International Rice Commission Newsletter* 22, no. 2 (June 1973): 1–20; R. C. McGinnis, "Potential Biomass Energy Crops," in *International Biomass Energy Conference* (Winnipeg: Biomass Energy Institute, 1973), pp. xviii–6); A. Poole, "The Potential for Energy Recovery from Organic Wastes," in R. H. Williams, ed., *The Energy Conservation Papers* (Cambridge: Ballinger Publishing, 1975), p. 288; E. Owen, "Straw and Other Fibrous Materials in Human Food Chains and Nutrient Cycles," in A. N. Duckham et al., eds., *Food Production and Consumption* (Amsterdam: North Holland, 1977); Z. O. Müller, "Recycling of Organic Waste for Production Purposes," UNDP/FAO working paper, 1978; and Food and Agriculture Organization, *Agricultural Residues: Quantitative Survey* (Rome, 1979).

irreplaceable role throughout Southeast Asia and in parts of Latin America and Africa.

Cereal straws, by far the most important crop residues, also have an amazing variety of other uses as construction materials (in thatching and fencing, and for irrigation gates) and as a raw material for manufacturing paper, boards, kitchen utensils, and household and personal items, such as mats, baskets, hats, sandals, ropes, and bags. Very little quantitative information is available on the partitioning of straw use. Judging from the recently available facts on China's rural energetics, the results of Buck's classic study (table 10.12) seem to be still basically valid in many of the country's villages.[32]

Currently I know of only one nation, Taiwan, that has detailed and reliable figures on all the uses to which its main crop residue, rice straw, is put (table 10.13). Partial accounts are available for a few other countries. For example, in South Korea nearly 25 percent of the rice straw goes for roofing; in Sri Lanka it is split between feed and fuel; in the Philippines the principal uses are feed, mulch, and fuel in pottery making.[33] In other cases even the

Table 10.11. World Production of Major Crop Residues in 1975

	Cereal Straws	Tuber Vines	Legume Vines	Sugar Cane, Stalks, and Leaves	Total residues		
		(million tons)			(10^6t)	(10^6TJ)	(percent)
North America	430	3	2	5	440	7.38	19
Europe	330	22	4	—	356	5.96	15
USSR	203	18	8	—	229	3.83	10
Latin America	118	9	7	58	192	3.21	8
Africa	99	15	8	10	132	2.21	6
Asia	836	44	38	54	972	16.29	41
Oceania	29	—	—	5	34	0.57	1
Developed	1,035	46	14	13	1,108	18.56	47
Developing	1,009	66	53	119	1,247	20.89	53
World	2,044	112	67	132	2,355	39.45	100

Sources: author's calculations using the crop production totals from Food and Agriculture Organization, *Production Yearbook 1978* (Rome, 1979). residue multipliers of 1.5 for cereals and legumes and 0.2 for tubers and sugar cane and 16.75GJ/t of dry residue.

Table 10.12. Traditional Uses of Crop Residues in China

Crop Residues	Feed	Fuel (percent of total)	Manufacture
Barley straw	17	66	11
Buckwheat straw	10	74	13
Corn stalks	20	67	9
Cotton stalks	1	90	6
Kaoliang stalks	1	70	18
Millet straw	58	26	1
Oat straw	38	52	7
Potato vines	61	23	6
Rice straw	46	32	11
Wheat straw	18	59	13

Source: J. L. Buck, *Land Utilization in China* (Nanking: University of Nanking Press, 1937).

gross availability estimates are widely apart: While a recent study of Bangladesh energy supply estimates total crop residue use at 112.9 peta joules (PJ), the other puts it at 235.1 PJ.[34]

Even if the crop residues used for feed and manufacture could be replaced by fodder crops and substitute materials, complete removal of residues from fields is highly undesirable. Part of the residues should be incorporated into the soil, either directly or as a compost, not only to recycle valuable nutrients that would have to be otherwise replenished by energy-intensive synthetic fertilizers, but also to activate microbial soil processes, to improve the tilth, to maintain proper carbon ratios in the soil, and to reduce the soil and water losses.[35] Reduction of soil erosion in this fashion is especially desirable in row crops. A recent survey has shown that the soil losses on sloping land planted with row crops are incompatible with permanent agriculture.[36]

Table 10.13. Rice Straw Uses in Taiwan

Final Use (percent of total)	First Crop 3,446 kg of straw per hectare	Second Crop 2,832 kg of straw per hectare
Mushroom growing	16	9
Paper making	15	17
Fuel	12	11
Feed	7	4
Fertilizer composting	13	15
Fruit mulching	6	15
Ploughed-in	14	14
Other uses and waste	17	15

Source: unpublished data, "Joint Commission for Rural Reconstruction" (Taipei), personal communication (1978); the figures refer to the year 1977.

The amount of residue to be left in the field (for decomposition, ploughing-in, or burning) or returned to it after feeding and fermentation depends, naturally, on the availability of chemical fertilizers and on the soil type and field slope. Appropriate balances between residue removal and retention can be established only by detailed local studies. Considering the cost of synthetic nutrients as well as the pace of worldwide soil erosion and desertification, crop residues should be recycled as much as practicable. A recent interdisciplinary appraisal of biomass energy potential in the United States concluded that owing to the agricultural, environmental, and energetic aspects, "little or none of these remains should be used for biomass energy conversion."[37] Independently, I had to conclude the same in the case of the developing nations—except that here the case is even stronger.

There does seem to be a way to obtain both fuel and fertilizer from the residues—through widespread anaerobic fermentation. However, the numerous environmental constraints on the process, and in this instance, the overriding necessity to maintain an appropriate C/N ratio of material fed into digesters, limit the mass of residues put through in most circumstances.[38] Moreover, the problems associated with crop residues reduce their desirability as a fuel for direct combustion: They are bulky, often high in moisture content at harvesting time, difficult to store, seasonally available, relatively low in energy density, and difficult to collect and transport.[39]

To conclude, although the resource quantification of crop residues is easier than that of the forest fuels, it is impossible to offer any reliable values for optimal amounts to be harvested for energy in the developing world. Nonetheless, I would argue that, as with the forest fuels, the smaller such use the better.

This, indeed, seems to me the most important, the most dangerous, and, unfortunately a not easily eradicable fact of the developmental energetics: So many poor nations who are barely able to feed themselves rely for their energy supply so heavily on those resources whose destruction and use are diminishing and, in the long range, destroying their very potential to grow food.

NOTES

1. V. Smil, "Energy Flows in the Developing World," *American Scientist* 67, no. 5 (September–October 1979): 522–531.
2. Smil, *China's Energy* (New York: Praeger, 1976); and "China's Energetics: A System Analysis," in *Chinese Economy Post-Mao* (Washington, D.C.: United States Government Printing Office, 1978), pp. 323–369.
3. R. Revelle, "Energy Use in Rural India," *Science* 192 (4 June 1976): 969–975.
4. R. Tyers, "Optimal Resource Allocation in Transitional Agriculture: Case Studies in Bangladesh" (Dissertation, Harvard University, 1978).
5. G. Stanhill, "A Comparative Study of the Egyptian Agro-Ecosystem," *Agro-Ecosystems* 5 (1979): 213–230.

6. Smil, "Energy Flows."
7. J. S. Olson, "Carbon Cycles and Temperate Woodlands," in D. E. Reichle, ed., *Analysis of Temperate Forest Ecosystems* (New York: Springer-Verlag, 1973), p. 234.
8. H. Lieth, "Primary Productivity of the Major Vegetation Units of the World," in H. Lieth and R. H. Whittaker, eds., *Primary Productivity of the Biosphere* (New York: Springer-Verlag, 1975), p. 205.
9. A. Sommer, "Attempt at an Assessment of the World's Tropical Forests," *Unasylva* 28, nos. 112–113 (1976): 5–24.
10. S. D. Richardson, *Forestry in Communist China* (Baltimore: Johns Hopkins Press, 1966), pp. 259–261.
11. Smil, "Environmental Degradation in China," *Asian Survey* 20, no. 8 (August 1980): 777–788.
12. L. C. Sharma, *Forest Resources of India* (New Delhi: Eastern Economist, 1972), and Food and Agriculture Organization, *Forest Resources in Asia and the Far East* (Rome, 1976). 1976).
13. E. Eckholm, *Losing Ground* (New York: W. E. Norton, 1976).
14. Food and Agriculture Organization, *Forest Resources.*
15. P. J. Newbold, *Methods for Estimating the Primary Production of Forests* (Oxford: Blackwell Scientific, 1967).
16. Sommer, "Attempt at an Assessment of the World's Tropical Forests."
17. J. E. Young, "The Enormous Potential of the Forests," *Journal of Forestry* 73, no. 2 (February 1975): 99–102.
18. Food and Agriculture Organization, *World Forest Resources* (Rome, 1955), pp. 60–69).
19. Food and Agriculture Organization, *Production Yearbook, 1977* (Rome, 1979).
20. R. H. Whittaker and G. E. Likens, "The Biosphere and Man," in H. Lieth and R. H. Whittaker, eds., *Primary Productivity of the Biosphere* (New York: Springer-Verlag, 1975), pp. 305–328.
21. R. S. Evans, *Energy Plantations: Should We Grow Trees for Power-Plant Fuel?* (Vancouver: Canadian Forestry Service, 1974), p. 7.
22. Lieth, "Primary Productivity."
23. J. E. M. Arnold, "Wood Energy and Rural Consumption," (Paper delivered at the Eighth World Forestry Congress, Jakarta, October 1978).
24. K. Openshaw, "Woodfuel: A Time for Reassessment," *Natural Resources Forum* 3 (1978): 35–51.
25. T. H. Digernes, "Wood for Fuel: Energy Crisis Implying Desertification: The Case of Bara, the Sudan," (Dissertation, University of Bergen, 1977); Openshaw; E. Ernst, "Fuel Consumption Among Rural Families in Upper Volta, West Africa," (Paper delivered at the Eighth World Forestry Congress, Jakarta, October 1978); Arnold; and H. Soesastro (personal communication, 1979).
26. E. F. Brünig, "The Tropical Rain Forest: A Wasted Asset or an Essential Biospheric Resource?" *Ambio* 6 (1977): 187–191; J. S. Olson, H. A. Pfuderer, and Y. H. Chan, *Changes in the Global Carbon Cycle and the Biosphere* (Oak Ridge: Oak Ridge National Laboratory, 1978).
27. A. Gomez-Pompa, C. Vazquez-Yanez, and S. Guevara, "The Tropical Rain Forest: A Nonrenewable Resource," *Science* 177 (1 September 1972): 762–765.
28. W. E. Westman, "How Much Are Nature's Services Worth?" *Science* 197 (2 September 1977): 960–964.
29. Brunig, "The Tropical Rain Forest."
30. G. M. Woodwell et al., "The Biota and the World Carbon Budget," *Science* 199 (13 January 1978): 141–146; Y. H. Chan, J. S. Olson, and W. R. Emanuel, *Simulation of Land-Use Patterns Affecting Global Carbon Cycle* (Oak Ridge: Oak Ridge National Laboratory, 1976).

31. Revelle, Energy Use in Rural India; Smil, *China's Energy*.
32. Smil, "China's Energy Balances," (Manuscript in preparation, 1981).
33. A. Tanaka, "Methods of Handling the Rice Straw in Various Countries," *International Rice Commission Newsletter* 22 no. 2 (June 1973): 1–20; and Food and Agriculture Organization, *Agricultural Residues: Quantitative Survey* (Rome 1979).
34. Montreal Engineering et al., *Bangladesh Energy Study for the Government of Bangladesh*, 1976 and Tyers, Optimal Resource Allocation.
35. J. V. Mannering and L. D. Meyer, "The Effects of Various Rates of Surface Mulch on Infiltration and Erosion," *Soil Science Society of America Proceedings* 27 (1963): 84–86; and W. D. Shrader, "Effect of Removal of Crop Residues on Soil Productivity," in *Biomass: A Cash Crop for the Future*? (Columbus: Midwest Research Institute and Battelle, Columbus Laboratories, 1977): pp. 49–74.
36. R. A. Brink, J. W. Densmore, and G. A. Hill, "Soil Deterioration and the Growing World Demand for Food," *Science* 197 (12 August 1977): 625–630.
37. D. Pimentel et al., "Biological Solar Energy Conversion and U.S. Energy Policy," *BioScience* 28, no. 6 (June 1978): 377.
38. *Methane Generation from Human, Animal and Agricultural Wastes* (Washington, D. C.: National Academy of Sciences, 1977); and Smil, "Renewable Energies: How Much and How Renewable?" *Bulletin of the Atomic Scientists* 35, no. 10 (December 1979): 12–19.

Chapter 11
Aid to Nonconventional Energy Development in the Third World

James W. Howe

INTRODUCTION

As used in this paper, *nonconventional energy* is synonymous with *decentralized renewable energy*, except that it includes fuelwood, which, when grown in large forests, may be considered centralized. Thus, even though ocean thermal, tidal, and solar satellite energy generation are usually considered nonconventional, they are not treated in this paper because of their limited applicability to the Third World. The definition in this paper does include small-scale energy technology that converts direct sunshine, water flow in small streams, wind, and biomass into useable energy, as well as equipment to make efficient use of that energy. It also includes production of fuelwood and other fibers used for their energy content. For example, it would include solar water distillers, photovoltaic cells, wind pumpers, mini-hydroelectric systems, biodigesters, improved charcoal kilns, improved stoves, and forestry and woodlots, among other technologies.

More than three years ago, the Overseas Development Council (ODC) was asked to review and evaluate the potential of decentralized renewable energy for African villages. We were struck then with the fact that, although there are many passionate advocates—and many equally passionate critics—of decentralized renewable energy (DRE) there was very little hard information on the performance of DRE even in the laboratory, and virtually none on the performance of such technology in African (or other Third World) villages. With the possible exception of China (on which few quantitative data were available to us) there had been very few occasions where DRE equipment had been tested in Third World villages and where the results had been measured, recorded, evaluated, and reported. Consequently, the ODC concluded that a high priority for Third World policymakers and aid agencies was to site-test such equipment in a variety of villages in order to introduce some objective evidence into the escalating debate. We outlined a procedure for such tests and, on behalf of the Agency for International Development (AID), worked with a number of African countries to identify and describe some specific projects to conduct such tests. In the process, we gained several important insights.

1. We learned that the physical performance in village sites of the equipment being tested, although an important factor, is by no means the only one. We identified at least five other factors, discussed below.
2. If an aid agency is going to undertake a testing program jointly with Third World countries, it is normally essential and useful for the project to be directed by an agency of the host country (with advice and support from the aid agency).
3. The project needs to include support to strengthen the host country institution that has project responsibility.
4. Active participation in the site tests by villagers themselves will normally improve the relevance of the tests. We developed a hypothesis (still not adequately tested) that such participation in the planning stages of the test would increase the likelihood that the equipment being tested would succeed.
5. Many aid institutions are doing research on DRE in many Third World countries, and a number of Third World laboratories are conducting research on their own. The amount of such work appears to be proliferating. We were reminded of the proliferating research on food of two decades ago and of the successful efforts by private and public aid programs to build networks that linked such researchers together so that information on research being undertaken or on research breakthroughs could be quickly communicated to all of the researchers in the networks. The idea of similar efforts to create networks of researchers on DRE seemed to us to be worth exploring.
6. In subsequent work that we have done on fuelwood in Africa, the importance of village participation has become even more manifest than in testing DRE equipment.
7. There is room even in the poorest Third World villages for improving the efficiency of energy use.

Many of the ideas we struggled with in those early days, three years ago, have been made much more tangible, lucid, and sophisticated in two admirable pieces by John Ashworth and his associates at the Solar Energy Research Institute[1] (SERI). Indeed, it is gratifying to see how much progress has been made in so short a time. The Africa Bureau of AID has done important pioneering work. They are preparing tests of some 21 energy technologies, including fuelwood, in 10 countries, in from 1 to 115 villages (or other) sites each. According to our unofficial count, there are currently some 15 bilateral aid programs and 14 multilateral programs with renewable energy and fuelwood projects in 41 African countries, with a total of 347 assistance projects. We do not have such an inventory on project activities in other continents.

DECENTRALIZED RENEWABLE ENERGY: ACTION AGENDA

Learn What Tasks Villagers Want to Energize

Anyone who has lived or travelled outside the cities in the Third World has seen the ruins of biodigesters or the skeletons of windmills long since abandoned. From questioning those involved in, or familiar with, these unsuccessful efforts, one is led to the hypothesis that in addition to the normal quota of bad management and faulty machinery, one major factor is the failure of the villagers to become personally committed to the project.[2] We are told of cases where outsiders—either foreigners or nationals of the host country—came to a village with a preselected technology to do a preselected task (for example, an imported steel windmill to pump water for household and livestock use), erected the equipment, attempted for several weeks or months to get the local people interested in it, and then left. Soon after the departure of the outsiders, a component part may have broken or someone may have needed a piece of metal to repair a leaky roof (it being the rainy season, the windmill was not urgently needed). Once the equipment was inoperative, it was soon abandoned. Memories of the outsiders' intervention faded, and the old ways of obtaining water were resumed.

We know that villagers do not always resist new ways. Sometimes they adopt them with impressive speed (for example, the high yield varieties of grain in Asia). The way to ensure that villagers accept a new technology as part of the economy and sociology of the village—something they will make great efforts to keep operating—is to let them participate in the early planning for the project. Instead of presenting them with a solution to a problem identified by an outsider, let them pick the problem they wish to resolve (e.g., a shortage of cooking fuels, darkness, grain that needs grinding, or the lack of clean water nearby). Describe for them, perhaps with models or pictures, a range of technologies that might help resolve the problem. Discuss with them the advantages and disadvantages of each and, if possible, enlist their interest in helping the outsiders conduct a series of tests to see which ones are best for a given task in that village. Make sure there is something in it for the villagers and that they understand that the goal is to find out which technologies work best so there will not be too much frustration at learning that some of them are failures. In the words of one exceptionally experienced pioneer in this work, "Let the technology talk its way into the village even if it takes time."[3]

Availability of Local Energy Resources

The range of local energy resources might include fuelwood, crop residues, animal wastes, a local year-round stream, sunshine, and wind. Precise and detailed measures taken over a long time period are available, but where time and funds are short, it may be necessary to compromise. Local villagers can supply a great deal of data about such things; for example, how many animals there are, whether the stream ever runs completely dry, how long the millet stalks last before they are gone, and how far one must go to gather wood for cooking. Other measures such as wind speeds and insolation may be inferred roughly from national central data but, before wind equipment is installed, one should have speeds in the particular village at various times of the day throughout the year. Whether one should have village preference before measuring primary energy availability or vice versa, is open to debate. In some countries, there may be merit in launching a program to gather primary energy data in a large sampling of villages, using science teachers, headmasters, domestic service corps, U.S. Peace Corps people, or any other village residents qualified to make and keep records of the measurements.

Testing the Equipment

This matching process is described thoroughly in Ashworth's forthcoming paper.[4] The hardware tests should be designed to answer a series of questions:[5]

1. How well does a given device perform technically within the physical conditions—including the all-important variable of settlement pattern—of the village?
2. How do its costs compare with those of alternative energy technologies?
3. How does it fit the local culture?
4. How does it match existing or prospective village institutions that could own, operate, and maintain it?
5. What are the best techniques for introducing the technology into the villages?
6. What is the effect of the increase in available energy upon such indicators of community well-being as literacy, infant mortality, income, migration, and birth rates?

Each of these tests deserves further comment.

Physical Conditions. The selection of technology must match the physical circumstances of the village. A windmill requires a certain wind velocity for a reliable period of time in order to function; sandstorms may damage the glass on a photovoltaic cell; hydroelectric generators will not work if the stream is dry for several months during the year; humidity or salt spray may corrode metal parts; biodigesters are less effective at low temperatures and require the right mix of organic wastes and water; and a pyrolitic converter requires a ready supply of woody material to make charcoal, chargas, and charoil. A test of equipment should match various kinds of hardware with a range of physical conditions and measure and record technical performance, maintenance problems, safety record, breakdowns, and the like.

Costs. Initial capital costs are easiest to measure. Other costs include those of maintenance, repair, and operating personnel. Solar equipment tends to have relatively high costs for purchase and installation, but in many instances, very low costs for maintenance and operation over the remainder of its life. It is, therefore, important to know as soon as possible the approximate life of the project in order to compare it with diesel equipment, which costs less to install but a great deal more to operate.

Local Culture Factors. Some devices may be unacceptable to local residents because they offend local beliefs. For example, the design of a device may be reminiscent of an evil omen. Some devices may stir up family problems. For example, men may resent it if the first device introduced is one that helps only women. If the men are dominant in the locality, they may kill the prospect of a successful experience with the device. The device may require the handling of cow dung, which in some societies may be unacceptable.

How Does It Fit Village Institutions? Some technologies can be operated by an individual family. This might be true, for example, of a biodigester in a culture where families own several cattle each, pen them up at night, and are willing to handle dung. A solar cooker that cooks by concentrating the sun's rays on a cooking pot could also be operated by an individual family in those cases where cooking is done in a pot and where there is no objection to cooking out of doors when the sun is high overhead. Other technologies are not amenable to management by single families. Thus a technology that produces electricity calls for some kind of village public utility. In some villages, there is a tradition of carrying on activities on a village-wide basis. In others, there may be no such tradition so that it may be necessary either to avoid such a technology, have it operated by a unit established from outside the village, or experiment with untried forms of village organization. Some technological applications may lend themselves readily to private enterprise. Thus, for example, a photovoltaic cell-driven grinder might be operated by a village entrepreneur who would sell the service of grinding corn or millet to people in the village. Similarly, a pyrolitic converter to make charcoal out of fibrous waste, such as wood, sawdust, coconut husks, or peanut shells might in some cultures be best managed by a small industrialist.

The range of applications is great, and the cultural-institutional variations within and among villages are almost limitless. In order to give the technology the best chance of performing well, at low cost, and being well maintained, a great deal of thought needs to be given to the best institutional arrangement within the village for owning and operating the device. A great deal of field research has been done in many cultures and individual villages by sociologists and anthropologists. Techniques need to be found to draw upon this knowledge while the project is being designed. The village test should make provision for observing and recording the performance of various kinds of village institutions in utilizing various kinds of hardware.

Introducing the Technology. We believe the technique for introducing a project into a village will significantly affect the success of the project. Involving innovative villagers in the very early stages of a project will increase the enthusiasm and involvement they have over the life of the project. It will increase the likelihood that the project will be incorporated into the life of the village, and that the villagers will maintain and operate the new device well and get the most out of it. Moreover, involving villagers will improve the design of the project because villagers often have the most reliable opinions on what task to select for energizing (e.g., pumping, grinding, or cooking), what primary energy source to use (e.g., sunshine, organic waste, animal power, wind generator, solar pumps, or solar cell), and what village institutional arrangements to make for the operation of the device.

On the other hand, involving villagers in project planning has some drawbacks. It may be time-consuming and hence unattractive to impatient Americans (or other foreign) aid officials or to host country operators. It calls for skills not always possessed by energy technicians. In many countries the normal method of dealing with villagers is to impose change from above. This normal technique, of course, requires the government office that imposes the change to supply a corps of people to operate and maintain the new technology. It is a much greater budgetary burden than if villagers could be motivated to take over maintenance, operation, and simple repairs.

It is for each government to consider these advantages and disadvantages and decide whether to impose technology from above or attempt to involve villagers in designing and implementing the changes. The fifth purpose of a well-designed test would be to furnish such governments with objective data on the advantages and disadvantages of each technology transfer technique. Hence the test should attempt to determine whether involving villagers in the early stages of the project enhances the prospects for success.

Impact on Community Well-Being. The tests should be able to tell policy-makers of developing nations something about the impact of new energy sources on the life of the villagers (or urban slum dwellers) and therefore, ultimately, upon the nation.

What is the impact of energy on agricultural production? Is a substantial portion of the labor released from other tasks devoted to raising additional crops or livestock? Does employment occur as a result? Are small industry or handicrafts stimulated by the coming of energy to the village? Does the provision of reading lights improve the pace or quality of education? What effect does energy have on health services or the incidence of disease (for example, from clean drinking water)? What effect is there on the role of women? Is there an increase, a decrease, or no change in the number of babies born each year? The AID-financed Evaluation Study of the MISAMIS Oriental Rural Electric Service Cooperative in the Philippines concluded that "The sharp decline in the crude birth rates (in the wake of electrification) is one of the most interesting phenomena uncovered by the study." Is there a change in infant mortality? What changes, if any, take place in the pattern of migration? Does energy set up a demand for imported items that drain limited foreign exchange? These and other evidences of the impact of energy on village life should be measured and analyzed in order to better understand the importance of village energy, to anticipate problems and opportunities it may create, and to differentiate among the effects of different uses of energy (lighting, clean water, cooking fuel, for example, contrasted with irrigation) and different forms (electricity versus gas, charcoal, or mechanical energy).

Evaluating the Results

As we noted above, hundreds of projects involving several hundred villages are getting under way in various parts of the Third World. A large fraction of these projects are experimental, designed to gain knowledge about what works. Many individual projects have a provision built into them for evaluating the results of that particular experiment. No effort has been made to coordinate this work to ensure that the kinds of questions being researched on these many projects will produce information that can be accumulated to provide answers about which technologies work best under given conditions. Nor is there any provision to take stock across the board from time to time about what has been learned.

Clearly, some international leadership is called for. It could be a job for the Development Assistance Committee (DAC), the donor nations's watch-dog organization. Perhaps even better equipped—and more acceptable to the developing countries because they are not members of DAC—would be the World Bank. With vigorous leadership, the tremendous effort in renewable energy experimentation in Third World laboratories and villages could be made to yield answers as to the advantages and disadvantages of the several technologies far earlier and more reliably than without such an integrative effort.

Extending Hardware to Third World Rural Energy Users

Once answers become available on the performance, cost, acceptability, and other characteristics of various technologies, the policymakers in each nation need to consider what sort of system is needed to deliver, install, operate, maintain, and repair such technologies.[6]

A local support system is required. From the analysis given above of the sequence of testing, one may identify the components of such a system. First, the system must help villagers develop an energy strategy for their particular village, taking into account what their local raw energy sources are, what tasks they would like to use them for, and what devices they would like to employ. Second, the system must provide a qualified person to work with the villagers to set in place the chosen technology. This calls for helping villagers to order the necessary equipment and supplies and find the financing that may be needed, training local persons for operation, maintenance, and repair, and supervising the actual building of the energy facility. Third, there is a need to monitor the use of the local energy facilities, trouble-shooting when there is a physical breakdown, counseling when cultural, personal, or political problems arise, arranging for training of additional village people, and bringing information from the world of energy researchers that might improve on the local technology. Fourth, there is need for the experience of the village with the energy device to be recorded and made known to the research world so that its work can be kept relevant to the energy needs at the village level. In particular, careful records on energy costs could provide indispensible information to researchers.

In the early stages of a program of installing small-scale renewable energy in selected villages, and while nations are gathering experience on which to judge whether village source energy is feasible for nation-wide application, the energy extension system to perform these four steps will of necessity be elementary and somewhat informal. A full-scale national extension service would not be warranted. However, once a country becomes convinced of the feasibility of village-scale technology and determines to apply it on a wide scale, it must face the need for some kind of energy extension system.

That system may take one of three forms. In some cases, private manufacturers and distributors might produce and market one or more devices, install them, train local attendants to operate and maintain them, and provide such central servicing as may be necessary. It might be necessary at first for a government agency to help the villages design an energy strategy. In other cases, where poor prospects of profits, the ideology of the government, or the fragility of the private sector make it unlikely that the private sector can do the job, a governmental extension service, such as the existing agricultural extension system, may have to take over. And in still other cases, it may be necessary for a wholly new national energy extension service to be established.

At this point, we can safely say it will be necessary for all countries to have a system for extending village source energy technology and for supporting that technology if village source energy is to succeed. But it is up to each country to decide whether to rely on the private sector, to turn to an existing agricultural extension system, or to develop a new system for energy.

Building Third World Energy Institutions

Each of the above actions should be conducted insofar as possible under the direction of a Third World institution. Each assistance project should be designed with two purposes in mind: First to learn something that Third World energy policymakers will need to know; and second, to help develop Third World institutions capable of carrying on the investigative work and of applying the knowledge to bring energy to Third World villages. Thus, if the task at hand is to test some technologies in a series of villages or to measure village preferences or available primary energy, a host country institution should be the project manager, supported as needed by outside experts who report to and work for the host country institutions. In addition to efforts to strengthen specified host country institutions, outside aid agencies should also engage in general programs to train cadres of energy professionals.

R&D on Decentralized Renewable Energy Technologies in the Industrialized Countries

The vast bulk of research and development funds in the world originate in the northern industrialized countries. By a large margin, most of that is spent on nonrenewable centralized systems. A small but growing part is spent on renewable energy, but that tends to emphasize large-scale, sophisticated systems of doubtful applicability to the Third World. Industrialized countries could do a great deal to help with the rural and urban slum energy problems of the Third World by increasing their outlays for R&D on decentralized renewable energy. To the greatest extent possible, increasing portions of that research should be done cooperatively with Third World research institutions and located physically in Third World countries.

General Surveys of Noncommercial Energy Use, Needs, and Availabilities in the Third World

A number of comprehensive energy assessments have been done and many more are in prospect for Third World countries. The U.S. Departments of State and Energy, the World Bank, and others are involved. These assessments tend to neglect the noncommercial energy scene because there are no good data on noncommercial energy to use in making an assessment. Clearly an assessment that neglects this important part of the energy picture is inadequate. Several initiatives have been undertaken recently to remedy this problem, including a conference in January sponsored by AID on the subject of surveys, and a joint AID–Peace Corps–ODC project in which Peace Corps volunteers in selected countries will assemble data over a period of time about the energy regime of the community in which they live.

FORESTRY AND FUELWOOD: AGENDA FOR ACTION

Forests and Energy

The forests of the world are threatened by competition between the need for food and the need for fuelwood. While the demand for agricultural land may be the greater threat in many areas, in other critical areas, deforestation is in large measure the result of a heavy demand for fuelwood. Whatever the cause, the problem is enormous and growing more serious. In Africa, for example, it has been estimated that annual new fuelwood plantations of about 1 million hectares a year will be required for the next twenty years in order to meet rural and urban fuelwood needs.[7] The World Bank believes that in the nine African countries it surveyed, the rates of afforestation need to be increased by from 8 to 50 times to meet domestic fuelwood needs between now and the end of the century.[8]

Barriers to Resolving the Problem

The first barrier to resolving the problem is that Third World authorities and donor agencies are only just beginning to recognize the extent of the problem. In every country, some people recognize the problem fully, but this recognition needs to be disseminated throughout the policy community. Secondly, costs are a major barrier to dealing adequately with the problem. Estimates of the costs per hectare of planting trees vary widely. In the Sahel, experience suggests a cost of about $725 per hectare. Other costs have ranged from a low of $66 where labor was volunteered to as much as $2460 per hectare of surviving trees.[9] If we apply the Sahelian cost of $725 to the estimated need of 1 million hectares of new plantings a year for Africa, the resulting estimate of $725 million per year illustrates the point that costs may become a major barrier.

In some cases, conflict over land use will present a barrier. Where villagers believe land is needed for grazing or cropping, they may pull up seedlings planted by forestry authorities.[10] This suggests the importance of fully involving villagers in any fuelwood or forestry projects. Land tenure questions can also be important. If a villager's right to ultimately use the wood being planted is in doubt, it may not be possible to get his cooperation in planting or protecting trees.

Lastly, administrative barriers may stand in the way of resolving the problem. In order to carry out a successful forestry or woodlot program, there must be an effective forestry department with good people, vehicles, gasoline for the vehicles, funds for labor, and many other ingredients. These are frequently lacking.

Actions Needed

Action is needed on a number of fronts on a massive scale. There is ample room for relevant donor agencies to cooperate. There is no need to argue whether to emphasize large forest reserves, small village commercial woodlots, or private profit-making plots. All should be emphasized. There is a need to conserve wood by using better stoves and making of charcoal in improved kilns. There is also a need to move ahead with renewable substitutes for wood (e.g., solar cookers or precookers or solar crop or fish dryers).

A final word on method: Whether in testing decentralized renewable energy, in planting and protecting trees, or in gathering data, we believe the best results come when local persons are involved as full participants. This includes involvement during the early planning stages of the project. It is especially important to involve those in the village who stand to gain if the project succeeds or to be hurt by some of its side effects. In cultures where cash crops are concerned, this may be the men. Where domestic energy or subsistence crops are concerned, it is likely to be the women.

NOTES

1. John H. Ashworth, *Renewable Energy Sources: Current International Development Assistance Programs*, (Golden, Colorado: Solar Energy Research Institute (SERI), October 1979).
2. James W. Howe and Francis Gulick, *Fuelwood and Other Renewable Energies in Africa*, (Washington, D.C.: January 16, 1980).

3. Richard Stanley, *Arusha Appropriate Technology Project, Annual Report, 1977–78*, private conversation, Spring, 1978.
4. John H. Ashworth and Jean Neuendorffer, *A Process for Matching Village-Level Energy Needs and Renewable Energy Systems*, (Golden, Colorado: Solar Energy Research Institute [SERI], January 1980).
5. James W. Howe, "New Village Uses of Renewable Energy Sources," in Priscilla Reining and Barbara Lenkerd, eds., *Village Viability in Contemporary Society*, (Boulder, Colorado: Westview Press, 1979), pp. 16–20.
6. James W. Howe and staff of the Overseas Development Council, *Energy for the Villages of Africa*, (February 25, 1977), pp. 35–36.
7. Reidar Persson, *Forest Resources of Africa: An Approach to International Forest Resources Appraisals, Part II: Regional Analysis*, (Stockholm: Royal College of Forestry, 1977), pp. 128–32, 144–45. Part I, published in 1975, provides a country-by-country summary analysis of the status of national forests and man-made forests, including details on inventories.
8. John S. Spears, "Wood as an Energy Source: The Situation in the Developing World," (Paper presented at the 103rd Annual Meeting of the American Forestry Association, October 8, 1978).
9. James W. Howe and Francis Gulick, "Fuelwood and Other Renewable Energies in Africa, a progress report on the problem and the response, January 16, 1980, p. 18.
10. Howe and Gulick, p. 19.

Chapter 12
Coordination of Renewable Energy Projects: An Agenda for Action

John H. Ashworth

The number of foreign assistance projects using renewable energy technologies in developing countries has increased rapidly in the last five years. However, the amount of useful information derived from these projects has not kept pace with the information needs of program managers, project leaders, and applied researchers in the developing and developed world. These information needs range from actual data on system performance, and the economic costs of installing and operating systems to the social acceptance of new production systems based on renewable energy systems.

One solution to this lack of data is the sharing of project information. As a first step toward the development of standard data collection and exchange, the U.S. Agency for International Development (USAID) and the U.S. Solar Energy Research Institute (SERI) sponsored a Workshop on Evaluation Systems for Renewable Energy Technologies held on February 20–22, 1980. Twenty-five program managers and energy planners from bilateral and multilateral donor agencies, private foundations, and developing countries met to discuss what information on renewable energy systems they would like to receive from other organizations. The second half of this paper reports the preliminary results of that meeting, along with the recommendations of workshop participants for further steps to continue the process of establishing common information elements and evaluation criteria.

NATURE OF THE ENERGY PROBLEM

In the past five years, the international assistance community has mounted a major effort to remove the roadblock to the economic and social development of the Third World caused by the sudden disappearance of assured supplies of low-cost energy. Concessionary financing mechanisms have been developed to manage the balance-of-payments deficits caused by soaring oil prices. Several major lending organizations, led by the World Bank, have focused on the location and rapid exploitation of indigenous conventional fuels in non-OPEC nations. Potential sources range from peat in Burundi to coal and natural gas in India.[1] The nuclear option received a great deal of attention during the 1970s, despite concerns over the proliferation of nuclear weapons, and is still the cornerstone of the development programs of several of the most advanced, middle-tier Third World nations.[2] For the poorest nations of the Third World, almost totally dependent on imported petroleum for their conventional fuels,[3] neither the nuclear option nor domestic hydrocarbon development are major short-term alternatives. Their domestic markets are too small for large-scale conventional plants, and they lack the capital resources, infrastructure, and distribution network to substitute nuclear power for the small, inefficient, decentralized diesel- or oil-fired generating plants that they currently use. The stiffening U.S. position on reprocessing facilities has also altered calculations even for those nations large enough to use a conventional nuclear facility, and has raised concerns about future fuel availability. Many of the most oil-dependent developing nations do not have promising hydrocarbon-bearing geological formations, although a number of large-scale resource assessments are currently underway to seek out previously overlooked oil and gas fields, coal deposits, and hydroelectric generating sites.

For most developing nations, renewable energy sources are both part of the immediate problem and the centerpieces for the long-term solution. Expanding populations, combined with spiralling world oil prices, have brought increasing pressure on supplies of traditional fuels, particularly firewood, charcoal, animal dung, and agricultural residues. Associated with the increasing use of these fuels has been deforestation, increased siltation of riverbeds down-stream from deforested mountain slopes, and soil depletion from the combustion as fuel of traditional manure fertilizers.[4] For many applications, renewable energy systems seem to offer alternatives both to dependence on expensive and unreliable imported petroleum and to depletion of natural resources for fuel. Simple solar thermal systems can provide crop drying and water heating, while wind and small-scale hydroelectrical and hydromechanical systems are well matched to the production of electricity and shaft power. Other decentralized systems, such as biogas generators, photovoltaic arrays, and flat-plate solar systems coupled with rankine-cycle engines seem to offer possible sources for water pumping and the provision of potable water.

THE DONOR RESPONSE

To determine if these renewable energy systems offer real alternatives to fossil or traditional fuels, a great deal of field testing as well as technology modification will be required. Thus far, much of the initiative for this testing and installation in developing country sites has come from the foreign assistance community. A small group of bilateral donors, led by France, the United States, Canada, the Netherlands, and Sweden, began the process, along with regional organizations such as the Organization of American States and the various organs of the UN systems. More recently, these initial donors have been joined by other major bilateral, multilateral, and private assistance agencies, such as the government of West Germany, the Rockefeller Foundation, the aL Dir'iyyah Institute, the Inter-American Bank, and the European Economic Community. The World Bank currently is contemplating a major renewable energy lending program, as are other organizations traditionally associated with infrastructure-creation projects.

The result of all this interest has been a rapid increase in the number of renewable energy systems currently being installed or in the advanced stages of planning. A survey which SERI published in late 1979 located projects totalling over $225 million dollars from a variety of foreign assistance donors.[5] More recent publications of the Overseas Development Council[6] and the International Institute for Environment and Development[7] have located a great many more projects, focusing on the African continent in the first case and major European and North American donors in the second case. This proliferation of projects is accelerating as projects under consideration for the past several years receive funding and as agencies prepare programs to present to the 1981 UN Conference on New and Renewable Sources of Energy.

INFORMATION NEEDS FOR PLANNING AND TECHNOLOGY SELECTION

This tremendous spurt of interest in renewable energy projects has not been spurred by the concrete results of earlier experiments but has taken place *in spite* of the lack of useful data on project outcomes and system performance. Most renewable energy projects are seen by their sponsors and recipients as initial experiments, despite the fact that other organizations may have already installed similar or identical systems for the same end-use energy need. Unfortunately, there is a good possibility that little will be learned after the pilot projects are completed, since information on project success and system performance is not being systematically collected. This problem is particularly serious for systems that fail to produce the expected output, that break down, or that are rejected by local users. This vital information is not reaching researchers and manufacturers so that they can seek solutions for design and materials defects. It is also failing to reach the program managers who must make decisions on what systems to install to meet particular energy needs.

Part of this lack of information on field performance and on problems encountered in existing installations is due to the commercial rivalries among several of the industrial nations seeking Third World export markets for their renewable energy systems. This rivalry currently is most intense in photovoltaic systems, with large demonstration projects being installed in Mexico, in sub-Sahara Africa, and in the Middle East by the United States, West Germany, and France. In some cases, such as France, West Germany, and Japan, long-term export promotion and short-term foreign assistance are tightly intertwined, making it difficult for the foreign assistance agency to report objectively on problems encountered with a system that its government is actively marketing as the solution to rural Third World energy needs. In other industrial countries, export promotion and foreign assistance are proceeding parallel to one another, with little information flowing back and forth between the agencies charged with running each effort. The United States is the obvious example of this second case.

The lack of information, both on system successes and on field problems and failures, is most serious for the development manager considering a proposed renewable energy project. Without sufficient field data, decision-makers are unable to use their traditional tools for allocating scarce resources among competing project proposals. For example, how can one use social benefit-cost analysis when one does not have even rough estimates of the major components of the stream of cost and benefits from one system? This uncertainty is compounded by two other features: the high degree of site-specificity of the performance of most renewable energy systems, and the intermittent nature of the output of the technologies. Site-dependence requires a careful assessment of the resource base for the location and sizing of the technology. Until this is done, only the roughest figures on the cost of a delivered unit of energy can be developed. For example, the output of a wind turbine is proportional to the cube of the wind speed. The potential output of a system installed in a location with a steady 15-mile-per-hour wind regime is over three times that of the same system installed at a site 100 meters away with a constant 10-mile-per-hour wind velocity. The intermittent nature of most renewable energy systems also dictates that some form of storage will be needed if the local demand for energy does not occur when the resource is available. In a recent methodological paper that a colleague and I developed on matching renewable energy systems to rural energy needs,[8] we found that the requirement to store the energy produced or to transform it from one energy form to another were central determinants of the cost-competitiveness of renewable energy systems. We located a total of 12 criteria that are necessary to determine which of an array of renewable energy technologies would be most appropriate and closely matched to the characteristics for a particular village-level energy need. These are listed in table 12.1.

It is the closeness of the match of the temporal, thermal, climatic, social, and cultural characteristics of the output of the technology with the characteristics of the need that determines whether or not the system will be used and adopted by the potential end-user. If the transfer and adaptation of the technology is to proceed unassisted after the first trial installations, the energy source must be compatible with local energy usage patterns and must be capable of being locally operated and maintained. The development planner must have information on operational procedures and maintenance problems in order to make an intelligent choice among alternative technologies. This will be particularly important when technologies begin to make the transition from one-time experimental installation to widespread dissemination. The current practice of having a contract engineer fly in from Paris or Cleveland to repair a minor component failure in an imported photovoltaic or solar thermal pumping system may be acceptable on a single demonstration installation (although wasteful of donor funds), but not for a commercially viable system. Not only does this practice make the cost of replacing a defective $50 controller close to $5,000, but it makes the system inoperative for weeks or even months at a time.

Lack of information on field results can also be a major problem for the manufacturer or researcher producing a renewable energy system, who needs to know precisely what went wrong (if anything) in the field installation. Were there problems with materials degradation or durability once the system was installed? If so, were these caused by environmental factors (blowing sand, high ambient temperatures, humidity, boring insects, etc.), design deficiencies, or improper maintenance? Researchers and manufacturers also need detailed information on system output compared to both the resource base and the pattern of demand. This would assist in sizing the next system and in determining the amount of storage required. Lastly, the researcher and manufacturer need reliable data on market acceptance. How did the final user feel about the system? Would changes in system design make it more compatible with existing social and cultural practices? Who within the local setting should be charged with operation and maintenance, and what differences does this choice make in the pattern of usage?

BRIDGING THE INFORMATION GAP

To meet the information needs of both the planner and the energy system producer, two steps must be taken. First, the information required must be collected in a form that is useful for the decision-maker. Program managers seeking to assess the potential contribution of a small-scale hydroelectrical generator in a specific location need different information than production engineers redesigning that generator for production economies, ease of manufacture, and minimization of maintenance. Second, all available information on alternative or similar systems should be collected and compared. This last step is particularly important for new or experimental systems, since the decision-maker and organization may not have an accumulated body of knowledge on system sizing and performance, practical problems, and low-cost solutions that might have been gathered through experience with more time-tested systems (such as large-scale hydroelectric generation facilities).

Recognizing that both of these steps have not yet been taken for many renewable energy technologies and that a great many installations were under consideration within their own organizations, foreign assistance agencies have begun to discuss the necessity of sharing information on systems installed in developing-country sites. Led by USAID and other major bilateral donors, some tentative steps have already been taken. As part of the follow-up to the summit meeting of western leaders held in Bonn in July 1978, representatives of most of the OECD nations met periodically in 1978 and 1979 to discuss increased energy assistance for developing countries. The findings of those representatives were published in the McPhail report, as it was called after its Canadian chairman, and were approved by the OECD Council in May 1979. The report endorsed an increase in renewable energy foreign assistance projects, and agreed in principle to the concept of common data collection techniques. To this impetus was added, in 1979, the approval and funding of the major United Nations Conference on New and Renewable Sources of Energy, to be held in 1981. The attention given to and preparation for this meeting have ensured that more renewable energy projects will be funded in the future, with the concomitant need for more information.

Based on the favorable response from other major foreign assistance donors and developing country planners to the OECD agreements, USAID decided to hold a workshop to focus on the question of the common information elements required for planning and evaluating renewable energy systems. In order to ensure that the discussion would focus on information useful to decison-makers, it was decided to invite a small group of individuals charged with program management, the selection of new initiatives, and energy planning for the major bilateral, multilateral, and private donor agencies. A number of senior managers from developing country govern-

Table 12.1. Characterization Criteria

Criteria	For Basic Needs	For Renewable Energy Technologies	Unit of Measure
		I. Discrimination Criteria	
Type of output	Form of energy that can satisfy demand	Form of energy produced	Not applicable
Temperature of output	Level of heat to perform required work	Range of temperature of energy system output	°C
Spatial dispersion	Number of locations per village needed for the performance of the basic need task	Capability to distribute the energy output produced by the technology	Number of sites per village required
		II. Site-Specific Temporal & Climatic Criteria	
Seasonality	Time of year when the energy demand occurs	Time of year when the resource produces useful energy output	Growing season, nongrowing season, or all year long
Time of day	Time of day when energy is required to perform the basic need task	Time of day when the useful energy is produced	Morning, daytime, night, or 24-hr day
Duration	Length of time per day required to perform the basic need task	Length of time the technology provides useful energy during the day	Number of hours per day
Sensitivity to interruption	Length of time the performance of the task can be halted	Variability of output of the energy source	Can be interrupted or cannot be interrupted / Variable or not variable

Table 12.1. (Continued)

Criteria	For Basic Needs	For Renewable Energy Technologies	Unit of Measure
	III. Site-Specific Social/Cultural/Environmental Critiera		
Usage by type of person	Persons participating in the basic need task affected by changing the energy source	Persons likely to be involved in operating the renewable energy technologies	By sex, age, and class
Historical, social, and religious influences	Historical, social, and religious requirements/customs that affect how basic needs are met	Traditional patterns that could create resistance to the use of the energy technology and energy use	Description of the historical, social, and religious customs that affect basic needs
Traditional energy sources used	Sources of energy used to satisfy village requirements	Traditional use of renewable energy sources. Traditional technologies used to protect renewable energy soruces	Units consumed per capita or per task (kilograms of firewood, charcoal, dung, etc.)
Environmental and ecological factors	Climatic and resource conditions that limit local ability to satisfy needs or that alter relative importance of basic human needs	Factors that influence energy system performance, durability, maintenance requirements, etc.; also factors that are affected by the installation of an energy system or the reallocation of resources	Qualitative descriptions
	IV. Cost Considerations		
Cost	Cost limits for new energy technologies (given the monetary, labor, and social costs of traditional and conventional energy used for basic need requirements and the financial resources available to the village)	Costs of the technology's local application	Costs given in dollars (and/or person days) per unit of work or per unit of output; social cost qualitatively described

ments were also invited, as were several specialists in information systems for development planning. The Workshop on Evaluation Systems for Renewable Energy Technologies was cosponsored by USAID and SERI, using funds provided by USAID. It was held at SERI February 20–22, 1980. The USAID and SERI planners of the meeting deliberately structured the meeting as an informal series of working sessions, with the participants themselves actually developing and ranking the information components in small working groups then reporting back to the larger sessions. By limiting participation to 25 individuals, each representing one organization, it was hoped that specific practical recommendations for data items could be developed rapidly, thus providing a framework for subsequent efforts at information coordination and exchange.

The workshop completed, the SERI staff have begun the task of pulling together the information developed in the working groups and plenary sessions. The final proceedings will be coordinated with the participants and will be available in a few months. What I would like to report today are some of the problems encountered in the process of developing the common information elements, some of the information elements for project planning on which there was general agreement, and the results of several case studies developed by the participants.

PROBLEMS ENCOUNTERED DURING THE WORKSHOP

During the two-and-one-half days of working sessions, several conceptual and practical problems surfaced repeatedly, both in the small groups, where the participants sought to identify common information elements and in the plenary sessions, where the findings of the groups were discussed and combined. They will be central to future efforts to develop common criteria and must be addressed from the very beginning of the effort.

What Is the Purpose of the Project?

The participants found that the information required and expected from an installation depends a great deal on why the system is being installed. Experimental projects, designed to assist an ongoing research and development program, have a much larger information component than straightforward development assistance projects or projects designed to commercialize the production of a technology. This distinction is even more confused in many of the foreign assistance projects currently underway, because the renewable energy system may be an experimental component in an otherwise routine assistance project aimed not at energy development but at a different development objective (integrated rural development, small-farmer assistance, expansion of health facilities, etc.).

Who Is the Audience for the Information?

While information is required for every stage of the normal project cycle, the emphases are different among information elements depending on the final user of the data. The program planner needs a different level of information than does the project administrator in the field, the evaluation specialist, or the researcher engaged in technology adaptation and modification.

What Level of Detail?

This is closely linked to the two preceding questions. The collection of information costs money, and the collection of detailed and comprehensive information costs more money and requires more time than a simple site visit. But, conversely, in the early stages of learning about a technology, more information is needed to inform later decision-making.

How to Generalize from Site-Specific Data

Renewable energy systems must be tailored both to local physical conditions and to local usage patterns. How much information about this tailoring process is essential? In particular, what data on local institutional and social conditions are necessary to understand the success or failure of a project?

APPROACHES ADOPTED DURING THE WORKSHOP

To meet these methodological questions on the purpose, audience, and scope of information to be gathered and shared, the USAID/SERI Workshop leaders decided to focus in the small working groups first on the general categories of information that the participants considered necessary for the planning of any renewable energy project. This information was then presented to the group as a whole and discussed. There was surprising agreement on the elements, although each of the different working groups place different emphasis on individual components. A general outline of those findings is given in table 12.2.

Each working group was then assigned a particular application of a technology and asked to develop a list of the information elements they would like to receive from other organizations, ranking them in order of importance. In order to address the questions of purpose and audience, one application considered to be experimental was selected (photovoltaic-powered water pumping), along with an application where the purpose was the diffusion and adaptation of a known techology (fuelwood stoves). Lastly, the third small working group was asked to focus on a single-technology (biogas generation), and to develop what information might be required regardless of application.

Table 12.2. General Planning Data Elements: Consensus of Workshop on Evaluation Systems for Renewable Energy Systems

I. ECONOMIC

Market Price of Currently Available Fuels to Consumer (both fossil and renewable sources)

*Capital Costs** (include foreign exchange percentage for each component)
- equipment and materials
- installation and labor
- engineering
- financing costs

Operating Costs
- operation
- training
- maintenance

Existing Government Incentives and Taxes

Benefits from Project
- value of energy produced, measured by most economical alternative

Indirect Benefits
- employment and local income generated
- import substitution
- market provided for local industries
- costs of not doing the project (environmental degradation, etc.)

II. RESOURCE BASE

Inventory of Physical Resources Available

Inventory of Human and Animal Resources Available

III. TECHNICAL SYSTEM PERFORMANCE

Local Need for Energy
- quality and quantity of energy required
- time, duration, and seasonality of energy need

System Output (vs. resource availability in each case)
- type and quality of energy produced
- temporal pattern of output
- seasonal pattern of output
- availability of output (system reliability)

Maintenance and Repair
- skills required
- local materials
- spare parts required

Lifetime of System
- degradation of components/materials

Optimization of System
- changes made to system
- local adaptations required
- additional research and development required

Table 12.2 (Continued)

IV. SOCIAL/CULTURAL FACTORS

Patterns Prior to System Introduction
- energy use
- energy materials collection

Effects of Project
- beneficial/losers
- family structure
- traditional roles
- migration-settlement patterns
- employment
- income distribution

Local Receptivity to Innovation and Change
- compatibility of project with cultural patterns
- replacing existing system or introducing new supply

Community Participation in Planning, Operation, and Maintenance

V. INSTITUTIONAL FACTORS

Existing Decision-Making Organizations and Management Capability
- decison-making in village/household
- outreach/enforcement from government to village
- communication/representation from village to government

Existing Support Infrastructure
- extension and education organizations
- repair and service facilities
- transport and communications facilities
- financing mechanisms

Patterns of Ownership of Local Facilities

Institutional Linkage between Energy and Other Sectors
- agriculture
- transport
- health and education

VI. ENVIRONMENTAL

Impacts of Previous Energy Consumption Patterns
- water quality and availability
- air quality
- soil quality

Direct Impacts of Renewable Energy Technology Operation
- by-products produced
- land taken out of production

Indirect Impacts of System Operation
- displacement of other fuel sources
- increase in total economic production

* More important for demonstration and commercialization projects than for experimental R&D installations.

Table 12.3. Findings of Case Studies Performed by Working Groups of Workshop on Evaluation Systems for Renewable Energy Technologies

<table>
<tr><th></th><th>Group #1</th><th>Group #2</th><th>Group #3</th><th>Consensus</th></tr>
<tr><td>Technology
Applications
Type of Project</td><td>Photovoltaics
Small-Scale
Water Pumping
Experimental</td><td>Fuelwood Stoves
Cookings
Demonstration/Diffusion</td><td>Biogas Generator
Old Applications</td><td></td></tr>
<tr><td colspan="5">Major Criteria</td></tr>
<tr><td>Technical</td><td>Site Data
• insolation
• pumping lift
• extraordinary costs
• System configuration
Operation and maintenance of pump and array
System output/durability
• actual vs. rated</td><td>Design specifications
Material components
Flexibility of fuel sources
Energy efficiency
Simplicity of installation and maintenance
Maintenance requirements
Reliability
Safety</td><td>Design of digester/gasholder
Construction materials for digester/gasholder
Biomethanation process
Characteristics of change
• water content
• toxic components
• carbon/nitrogen
• volatile solids
Temperature
Retention time
Gas production rates
Maintenance and cleaning
System down-time</td><td>Design specifications
Construction materials
Resource base/inputs
Output of system
Installation and maintenance
• actual vs. expected
System reliability
• down-time—actual vs. expected</td></tr>
<tr><td>Economic</td><td>Operation/maintenance
Training costs
Capital costs
• equipment
• engineering
• financing
Backup system</td><td>Operation and maintenance
Cost of woodfuel saved
Social cost/benefits
Beneficiaries</td><td>Equipment/materials
Labor
Operation and maintenance
Value of byproducts
Value of fuel displaced
Social benefits/costs</td><td>Operation/maintenace
Capital costs
• equipment
Labor/training
Social costs/benefits
Value of fuels displaced</td></tr>
</table>

Table 12.3. (Continued)

	Group #1	Group #2	Group #3	Consensus
Technology Applications Type of Project	Photovoltaics Small-Scale Water Pumping Experimental	Fuelwood Stoves Cookings Demonstration/Diffusion	Biogas Generator Old Applications	
		Major Criteria		
Socio/Cultural	Impacts • migration/settlement • employment generated • inside vs. outside income generated Beneficiaries/losers • equity of distribution	(included under economic criteria)	Beneficiaries/losers • sex/age • class status Increased agricultural production Changes in health/sanitation	Beneficiaries/losers Identification of specific impacts
Institutional	Support provided by national, regional, and local organizations Project Management	Needs for education or extension service Local skills and fabrication potential	Institution in charge Location of expertise in R&D and extension Supply of materials/equipment • village organization • source of financing	Support available for local organizations Education and extension services Local sources of skills, materials, and fabrication
Environmental	Land use/availability	Impact on deforestation	Reduction of deforestation • reduction of pollution • pollution from pit run-off	

Each working group spent the better part of a day developing the information required for its case and placing it in order of importance. All this information was presented to the group as a whole. Table 12.3 is a preliminary cross-tabulation of the findings of the three groups, with some minor editing to make the language compatible across the groups. As can be seen, there is not as clear a consensus across the individual technologies and applications as there was in the data elements for planning in table 12.2. This is in keeping with one of the findings of the last plenary session, to be discussed below, that future efforts at coordination and establishment of common information elements should focus on specific applications of specific technologies. However, the proceedings of the meeting will provide some analysis of these case study findings and the group reasoning that went into the formulation of the lists. It is interesting to note that there was consensus on certain items that might not have been expected. For example, most of the participants were vitally interested in complete data on operation and maintenance of the technologies, but were split on the question of the importance of capital costs (this split depended on the purpose of the installation). All the panels were concerned with system reliability, as measured by the amount of down-time, meaning the number of hours the system was expected to be producing useful energy and did not. The working groups were uniformly interested in a careful accounting of the village beneficiaries and losers as the result of the project, as well as indicators of the social and cultural impacts of the renewable energy project. Most of the participants concurred on the importance of institutional consensus on what these factors were across the specific applications.

WORKSHOP FINDINGS AND RECOMMENDATIONS

The following are some of the major points presented during the final morning of discussion, after the working groups had completed their case studies. More detailed presentations, along with references to earlier portions of the workshop, will be included in the final proceedings.

1. Common information elements for renewable energy projects are both valuable and feasible. Further steps should be taken to develop common data elements, based on the initial categories and approaches developed at this working session.
2. The development of common information categories would speed the exchange of information among development organizations, researchers, and project field managers.
3. Information elements are most useful when they are specific to a particular technology (although some elements may be common to all systems). It is possible that the data elements should be further divided according

to each application of each technology, but this can only be determined after several attempts at developing the individual data elements and after these data elements have been reviewed by potential users.

4. The refinement of the information elements is the first priority. After the detailed sets of data elements have been developed, they then should be reviewed by experts familiar with data collection, by practitioners familiar with the particular technology, and by the program managers who would ultimately use the data.
5. The need to specify data elements for each technology suggests the utility of developing one or more data handbooks. This could be completed by late 1980 by one donor organization or by individuals drawn from cooperating organizations. This effort should be coordinated with the efforts of the technical panels of the 1981 UN Conference on New and Renewable Sources of Energy.
6. A mechanism for information exchange should be provided. In order to establish such a mechanism, several issues need to be resolved, including what type of organization is needed and what type of collection and dissemination system would be most accessible for the large number of donor agencies and developing countries. Existing data centers should be consulted on the best method for implementing such an international information exchange. There is also the possibility of coordinating or dividing the tasks required among the existing centers, focusing either on a technology-specific division or on an application-by-application division.
7. There is an immediate need to catalogue all the developing and developed country centers currently collecting data on renewable energy systems. This would allow these centers to be involved from the outset in the development and review of the common data elements.
8. When the organizations developing the data elements and exchange mechanisms have examined the alternatives, they could report their suggestions to an international forum, which would then make the final decision on what institutional mechanisms to adopt. It was suggested that the 1981 UN Conference on New and Renewable Sources of Energy might serve as that forum, and that these efforts on developing common elements and exchange mechanisms be coordinated with the preparation for the conference.

NOTES

1. R. Vedavilli, "Petroleum and Gas in Non-OPEC Developing Countries, 1976–1985," World Bank Staff Working Paper #289 (April 1978).
2. For an overview of one part of the problem, see John Gray et al., *International Cooperation on Breeder Reactors*, (New York: The Rockefeller Foundation, May 1978).
3. See Palmedo et al., *Energy Needs, Uses and Resources in Developing Countries*. (Brook-

haven: Brookhaven National Laboratory Center for the Analysis of Energy Systems, 1978).

4. Arjun Maklijani and Alan Poole, *Energy and Agriculture in the Third World* (Boston: Ballinger Press, 1975).
5. John Ashworth, "Renewable Energy Sources for the World's Poor: A Review of Current International Development Assistance Programs," Solar Energy Research Institute, TR-195 (October 1979).
6. James Howe and Frances Gulick, "Fuelwood and Other Renewable Energies in Africa," Overseas Development Council, January 1980.
7. Thomas Hoffman et al., unpublished working paper for the World Bank.
8. John Ashworth and Jean Neuendorffer, "Matching Renewable Energy Systems to Village Level Energy Needs," Solar Energy Research Institute, TR-744-514, June 1980.

Chapter 13
Energy for Rural Development

John Hurley

As the world faces the implications of less abundant and more expensive energy resources, it is easy to focus on geopolitical relationships and economic strategies, and to overlook the human problems involved. These problems are especially troublesome for the rural areas of most developing countries.

Despite the rapid industrialization and urbanization that is taking place in many developing countries, rural populations are still vast and closely tied to agricultural pursuits. It is estimated that 74 percent of the total population of developing countries in 1970 was in rural areas, and that in 2000, despite increasing migration to urban areas, 57 percent of their total population will still be rural.[1] For low-income developing countries, about 73 percent of the total labor force was engaged in agriculture as of 1977. The comparable figure in the middle-income countries was 46 percent.[2]

A World Bank report suggests that about 85 percent of all absolute poverty (defined by income levels below which minimum standards of nutrition, shelter, and personal amenities cannot be maintained) in developing countries is in the rural areas. In the mid-1970s, the report estimates, this percentage amounted to about 550 million people.[3]

For people in rural areas, the effects of poverty frequently include high levels of morbidity and mortality, inability to sustain hard work on a regular basis, and lack of access to education and a variety of services such as safe water supply.

The Overseas Development Council has devised a "Physical Quality of Life Index" (PQLI) that compares countries on the basis of rates of infant mortality, literacy, and life expectancy. A 1979 report of the Council indicates that, in comparison to an average PQLI of 93 for the high income countries, the low-income countries, with generally large rural populations, had a PQLI of only 40.[4]

Rural development must involve strategies to increase production and raise productivity as a means of improving the conditions of life for large numbers of people. At the same time, these improvements also will enable the rural sector in developing countries to contribute to the wider process of national economic and social development.

The task of rural development, however, is complex and difficult. Rapid population growth puts increasing pressure on available resources, and especially on the need for increased food supplies. Since good farming land is limited, the production of food per hectare will need to be increased. There is also a great need to increase the level of nonfarm employment in rural areas.

Energy is a critical aspect of the rural development process. It is expended in agricultural operations, in food processing and transportation, in the production of fertilizers, pesticides, and farm equipment; it is necessary for industrial operations that provide jobs, and it is required for cooking, for household light and heat, and for the construction and operation of the infrastructure needed for schools, health centers, and water supply.

The growth rate of energy consumption in developing countries is greater than in industrialized countries. From 1960 to 1974 the average annual energy consumption growth rate for low-income countries was 5.7 percent and for middle-income countries 7.6 percent. This compares to a consumption growth rate of 4.9 percent for the industrialized countries. From 1974 to 1976, the consumption growth rate for low-income countries was 4.6 percent, for middle-income countries 5.2 percent, and for the industrialized countries only 1.3 percent.[5]

Rural development is affected by the severe problems faced by many developing countries that rely heavily on imported petroleum. As prices rise, governments face the necessity of using increasingly large sums of foreign exchange for imported oil, thus diverting resources that might be used for other development activities.

At a 1980 conference on energy surveys in developing countries, several papers illustrated the problems of rising energy costs. In Kenya, for example, 85 percent of commercially traded energy comes from imported petroleum at a cost of 25 percent of the country's foreign exchange earnings.[6] In Jamaica, 90 percent of the total energy supply comes from petroleum imports.[7] Although it is estimated that traditional fuels (firewood, animal dung, crop residues, etc.) supply 73 percent of the total energy consumption of Bangladesh, the cost for petroleum imports in 1977–78 amounted to 33 percent of the country's total export earnings.[8] About 36 percent of the 1979 export revenues of the Philippines were allocated to oil imports, which account for close to 92 percent of total energy requirements.[9]

Although the energy situation is very troublesome in many developing countries, especially as it affects rural development, there are alternatives to the scenario that shows rising dependence on imported oil at rising prices. There is, for example, the possibility of discovery and developing of new fossil fuels and other indigenous resources in the developing countries. Although it is uncertain whether this will add major new resources to the world supply, such development certainly could help relieve the situation in specific countries. Several developing countries have relatively unexploited coal reserves. Peat represents a source of fuel for certain countries, such as Rwanda, Burundi, and Jamaica. Many developing countries have had relatively little exploration for new petroleum sources. And it is estimated that developing countries have 70 percent of the world's major hydroelectric capacity, but only 20 percent of the installed capacity.[10]

Conservation represents another important way in which developing countries can hold down the rate of energy consumption. Cooking, for example, is a major use of energy in households in developing countries. If more efficient stoves were widely adopted for cooking, water heating, and other domestic uses, considerable savings might be made both in energy use and in expenditure of income. It should be noted, however, that this sort of innovation must be based on cultural and behavioral understanding to be effective; technological change often generates new and unanticipated behavior.

The traditional sectors of most developing countries make extensive use of noncommercial or nonconventional energy sources. Firewood, charcoal, plant and animal residues, human and animal power, and direct and indirect solar energy are used in the household and for a variety of agricultural and rural industry tasks. It is estimated that while these forms of energy supply only about 5 percent of total world energy consumption, they represent about half the total energy production of developing countries that import oil, and more than 85 percent of energy production in many of these countries.[11]

Many developing countries have important characteristics that make the increasing use of renewable and non-conventional energy sources attractive. These have been summarized in one report as follows:

> For the most part, they do not have the massive investments in petroleum-oriented energy infrastructure that mark the industrialized countries; many of them have more sunlight than the industrialized countries, so solar energy is more reliable for them; and their climatic conditions permit faster growth of vegetation for firewood and biogasification. Finally, the decentralized nature of renewable energy makes it an ideal energy source for programs of small-scale rural development.[12]

Nonconventional sources will not in themselves provide enough energy to power rural development in most countries. They can, however, substitute for imported fuels and can play an increasingly important role in national energy balances.

The importance of wood, charcoal, and crop and animal residues for cooking and other domestic energy uses in rural areas indicate a need to find ways of using these fuels more efficiently. Charcoal yields, for example, can be greatly increased by use of better kilns. At the same time, careful planning for afforestation and reforestation is required to ensure that wood supplies are available: in many countries, there is a grave shortage of fuelwood, and a consequent severe problem of soil erosion and depletion as the demand for wood increases. Similarly, the use of crop and animal residues has implications for soil fertility.

Rural applications of nonconventional energy resources and technologies now receive considerable attention, but will benefit from continuing research and development. Direct solar energy can be used heating water, heating (or cooling) buildings, drying agricultural and animal products, and for salt production. Wind energy can be harnessed by small-scale wind machines for use in water pumping and irrigation, in tasks such as grinding, threshing, cutting, and winnowing, and in generating electricity for various purposes. Small-scale, low-head hydroelectric units can generate electricity or power grain mills, pumps, wood- and metal-working machinery, and other types of machines. Fermentation of plant products or animal wastes can be used to produce alcohol fuels or methane gas.

None of these alternative energy sources and technologies will offer a complete solution to the energy requirements of rural development. It is clear, however, that all countries must use the broadest possible range of available energy resources and must devote considerable effort to identifying, developing, and disseminating these resources.

Careful planning is necessary if nonconventional energy sources are to be used effectively; economic and behavioral factors can be as important as technological feasibility. In some cases, for example, more efficient stoves have not been accepted because they were not at the height to which local women were accustomed. In some cases, initial capital costs of a technology may be out of reach of most potential users, even though the technology would save money over its lifetime. Some energy sources or technologies which are not economically attractive at present prices but will become competitive as other energy sources rise in price or become more scarce warrant continuing research and development.

The collection of basic energy-related information is essential for sound planning related to national energy needs, especially for the complicated task of rural development. A study of energy in the developing countries has characterized the information situation by saying, "Information on the *supply* of noncommercial energy is not good." And further, "For *consumption* the situation is even worse. Very few actual field studies have been done so estimates are largely based on guesses and suppositions."[13]

The need for better energy information in developing countries led the USAID to ask the National Academy of Sciences to organize a conference in 1980 on the topic. A Workshop on Energy Survey Methodologies in Developing Countries was held at Jekyll Island, Georgia, January 21–25, 1980, with about seventy-five participants, including many from developing countries. Information was exchanged on surveys that have been carried out or are underway in a number of countries. The special information needs pertaining to the urban, rural, transportation, and industrial sectors were discussed in working groups, and recommendations were made. The proceedings of the workshop has been published and includes background papers, reports of the working groups, general conclusions and recommendations, and a directory of energy surveys in developing countries.

The workshop participants concluded that in most developing countries there is a lack of basic information that spans all uses and sectors of energy consumption, and that surveys are necessary to provide a firm informational base for analyzing energy decisions. Surveys must be as diverse in depth and extent as the energy issues and problems of a country, and should closely involve the institution which needs the information. Although for the most part there are no general survey methods that have universal applicability, there is a great need to provide mechanisms for exchanging survey information and experience among developing countries. Because of the critical importance of improving the basis for decisions on energy problems, a high priority should be given to assistance to developing countries for information gathering and analysis.

The energy outlook in most developing countries is uncertain and complex, particularly for rural areas which are the predominant sector of many nations. The difficulties posed, however, make it all the more imperative to increase and intensify the effort being devoted to energy planning and the development of alternative energy sources and technologies. If progress cannot be made in the task of rural development, then a dangerously divisive gap will continue to grow between the developing and the industrialized countries of the world.

NOTES

1. Morton M. McLaughlin et al. (Overseas Development Council), *The United States and World Development: Agenda 1979* (New York: Praeger Publishers, 1979) p. 178.
2. The World Bank, *World Development Report, 1979*, (New York: Oxford University Press, 1979), p. 162.
3. The World Bank, *The Assault on World Poverty* (Baltimore: The Johns Hopkins University Press, 1975), p. 19.
4. McLaughlin et al., p. 151.
5. The World Bank, *World Development Report, 1979* (New York: Oxford University Press, 1979), pp. 138–139.
6. Oyuko O. Mbeche, "Energy Use in Kenya," *Proceedings, International Workshop on Energy Survey Methodologies for Developing Countries* (Washington, D.C.: National Academy Press, pp. 160–164.
7. W. R. Ashby, "Energy Assessment in Jamaica: A Progress Report," *Proceedings, International Workshop on Energy Survey Methodologies for Developing Countries* (Washington, D..C: National Academy Press, 1980), pp. 155–159.

8. M. Nural Islam, "Energy Use in Bangladesh," *Proceedings, International Workshop on Energy Survey Methodologies for Developing Countries* (Washington, D.C.: National Academy Press, 1980), pp. 150–154.
9. Fernando R. Manibog, "Data Need Identification, Collection and Utilization Problems in the Commercial and Traditional Energy Sectors: The Case of the Philippines," *Proceedings, International Workshop on Energy Survey Methodologies for Developing Countries* (Washington, D.C.: National Academy Press, 1980), pp. 165–170.
10. McLaughlin et al., p. 68.
11. The World Bank, *World Development Report, 1979* (New York: Oxford University Press, 1979, p. 41.
12. McLaughlin et al., p. 68.
13. Philip F. Palmedo, Robert Nathans, Edward Beardsworth, and Samuel Hale, Jr., "Energy Needs, Uses and Resources in Developing Countries" (Brookhaven National Laboratory: Policy Analysis Division, National Center for Analysis of Energy Systems, March 1978), pp. 77–78.

Chapter 14
Developing Countries' Energy Prospects: A Long-Term, Global Perspective

Paul S. Basile

This paper presents research conducted at the International Institute for Applied Systems Analysis (IIASA) in Laxenburg, Austria, and is adapted from Part IV ("Balancing Supply and Demand: "The Quantitative Analysis") of the IIASA Energy Systems Program final report, "Energy in a Finite World: A Global Energy Systems Analysis," to be published in 1980. While the author takes responsibility for much of the selection and direction of the research, and for the presentation of it here, others at IIASA have contributed major pieces to the whole. Dr. Arshad Khan performed most of the detailed energy consumption analyses reported here; the work of Dr. Jyoti Parikh contributed as well, as did the analyses of Alois Hoelzl, Malcolm Agnew, and Jack Eddington. The whole effort was part of the larger research program directed by the Energy Systems Program Leader, Professor Wolf Haefele.

A full survey of the assumptions, approach, and results of the IIASA Energy Systems Program can be found in its final report.

The regions of the world covered in the IIASA Energy Systems Program are defined in figure 14.1. This paper will discuss primarily, the long-term (to the year 2030) energy prospects for Regions IV—Latin America (LA), V—Africa and South and Southeast Asia (Af/SEA), VI—the Middle East and Northern Africa (ME/NAf), and to a limited extent, VII—China and centrally planned Asian economies (C/CPA). The High and Low scenarios which were constructed and analyzed by the IIASA Energy Systems Program are summarized in table 14.1. Many of the results for developing regions presented here are part of these scenarios.

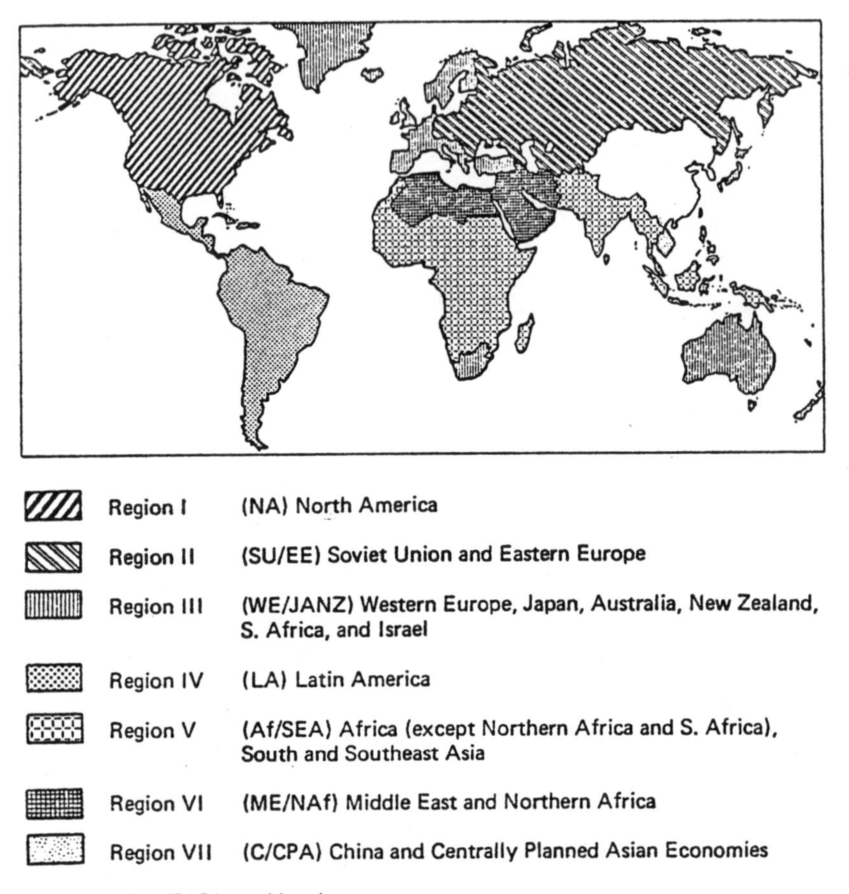

Fig. 14.1. The IIASA world regions.

A DOMINANT THEME

The integrative analysis of the IIASA Energy Systems Program, reported in the eight-chapter Part IV ("Balancing Supply and Demand: The Quantitative Analysis") of the final report, repeats a dominant theme: over the next five decades, even with rigorous conservation measures in industrialized regions, increasing need for liquid fuels throughout the world may exceed the capabilities of global energy supply systems. The energy problem, viewed from a sufficiently long-term and global perspective, is not an energy problem, strictly speaking; it is an oil problem or, more precisely, a liquid fuels problem.

A great deal of information and a good many insights lie behind this analysis of the globe's long-term energy future. Different regions of the

Table 14.1. Two Global Supply Scenarios: Definitions of Main Parameters

	High Scenario	Low Scenario
Assuming, by 2030:		
Population		
factor increase	2x	2x
total (10^6)	7,976	7,976
Gross world product		
factor increase	6.4x	3.6x
growth	declining rate	
average growth (%/yr) 1975–2030	3.4	2.4
Final energy demand		
factor increase	4.0x	2.5x
2030 annual rate (TWyr/yr)	22.8	14.6
Results in, by 2030		
Primary energy		
factor increase	4.3x	2.7x
2030 annual rate (TWyr/yr)	35.65	22.39
average per capita (kWyr/yr)	4.5	2.8

Further details about these scenarios can be found in "Energy in a Finite World, A Global Systems Analysis."

world face widely different economic and energy prospects, with widely different potentials and constraints. Some energy end-use markets can be more easily supplied than others, and different energy supply options carry different economic, environmental, political, and institutional implications.

The developing country analyses to be presented here should be viewed in the global context, a few points of which are summarized here.

- The supply analyses show world crude oil production peaking in about 2010, at nearly 95 million barrels per day in the High scenario (or in 2015 at 75 mbd in the Low scenario), and declining thereafter. Only large-scale exploitation of oil shales and tar sands seems capable of reversing this decline, because these unconventional sources are the only ones of sufficient global magnitude identified so far.
- Indeed, by 2030, only 61 percent of liquid fuel demands (under the assumptions of the High scenario) could be met by produced crude oil. The remainder (39 percent) must come from synthetic liquid fuels made from coal. In combination with this, the ceiling of 33.6 mbd assumed on oil production in Region VI (ME/NAf) would have a great impact on the extent and timing of oil alternatives: As imports become restricted, alternatives appear sooner and to a greater extent.
- Coal would be required in tremendous quantities—7–13 billion tons per year in 2030 under the conditions of the scenarios. Some 5 percent of these

amounts would be required for the production of synthetic fuels. Whether this amount of coal can be mined in the world is open to serious question.

- Efficiency improvements and substitutions of less energy-intensive economic activities for more energy-intensive ones would reap substantial energy savings in the developed regions. There, technical and economic sophisticated and saturation effects, could reduce growth in final energy from the 3.8 percent-per-year values of the past two or three decades to 1.1–1.7 percent per year over the next five decades.
- Development needs, expanding populations, and increasing urbanization would lead to continued energy growth in the developing regions. The prospects for saving energy—reducing energy growth relative to economic growth—seem dim for these regions; there, energy-to-GDP (gross domestic product) elasticities are expected to be greater than 1, while they are likely to be less than 1 in the developed regions.

Graphic presentations of the last observation are given in figures 14.2 and 14.3. Energy use per capita and GDP per capita show no fixed or even clearly definable relationship. Yet the development needs for countries at $250 to $500 per capita and less are great; simple multipliers will dictate high energy needs for these regions in the future. Note that these projections assume neither that LDCs will repeat the development paths of today's industrialized countries nor that they will chart drastically new courses.

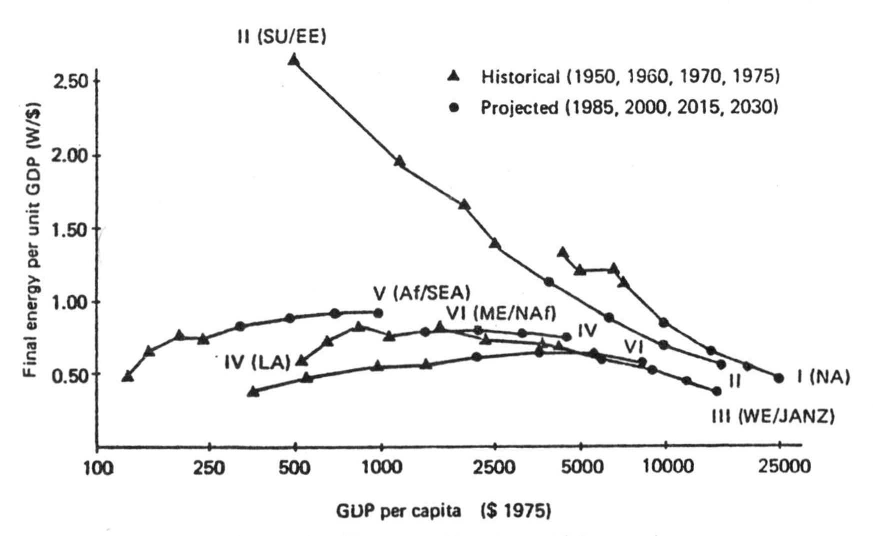

Fig. 14.2. Energy intensiveness in different world regions—high scenario.

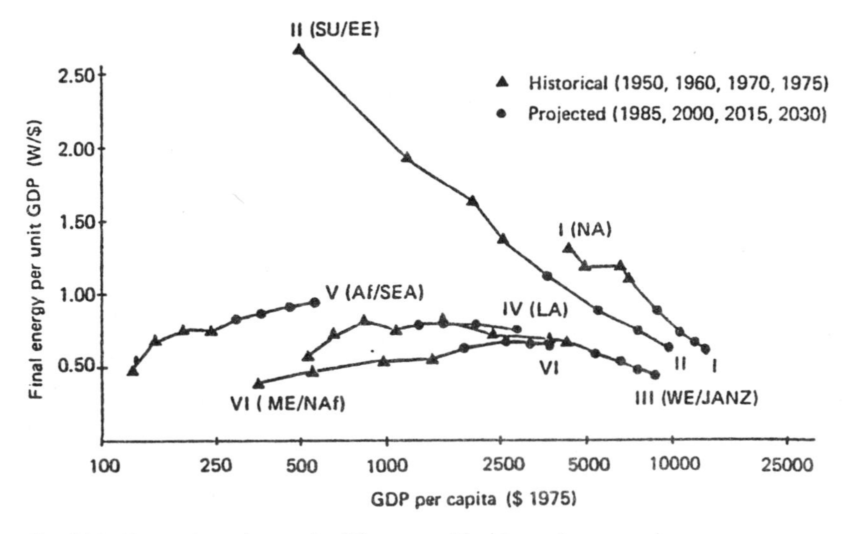

Fig. 14.3. Energy intensiveness in different world regions—low scenario.

ENERGY CONSUMPTION IN DEVELOPING REGIONS

In economic terms, the developing regions typically have low per capita income levels, high consumption and low investment rates, and relatively large agricultural sectors. They usually have large and life-giving imports and foreign aid, and supply-constrained (rather than demand-driven) growth patterns. Most of the developing countries seem to be heading toward steady industrialization and rapid urbanization, although there may be some slackening of the latter trend in some places. In energy terms, national total and per capita energy consumption levels are very low, the share of oil among commercial energy uses is often high, and noncommercial energy plays an important baseline role in rural areas and, to a certain extent, in households, in urban areas.

These are generalizations, to be sure. But they convey some impressions; they imply, almost certainly, the considerable likelihood of major increases in energy demand in the future.

Low Energy Consumption

The exceedingly low level of commercial energy consumption in many developing countries is a difficult fact for many to grasp. In 1975 North America consumed 11kW/cap, the USSR and Eastern Europe about 5kW/cap, Western Europe about 4kW/cap. In developing regions IV (LA), V (Af/SEA), VI (ME/NAf), and VII (C/CPA) in 1975, consumption was 1.1, 0.2, 1.0, and 0.5

kW/cap, respectively. Consider some typical examples: in 1975 Brazil had 0.7 kW/cap, India 0.2 kW/cap, and about 30 African and Asian countries—including Bangladesh, Burma, Ethiopia, Nigeria, and Uganda—used primary energy at the rate of less than 0.1 kW/cap.

High Oil Share

Current oil prices are high, as the whole world knows. But they have been low, and oil is easy to transport. Even without expensive high voltage lines or underground pipelines, oil in tin cans reaches even the remotest of villages. As a result, the oil shares of total primary commercial energy in 1975 were 67 percent in Latin American and 49 percent in Africa and South Asia. Many countries—Peru, Thailand, and Kenya among them—have 80 and 90 percent or more (of commercial primary energy) oil shares. In developed Regions I (NA) and III (WE/JANZ) the shares were 44 percent and 56 percent, respectively; in the oil-rich Soviet Union it was 39 percent. But oil must be imported by many developing countries, and this strains hard currency reserves which are also needed in no small measure for imports of food, equipment, and spare parts. The oil payments outflow in many developing countries is serious; in some countries 30 percent and even more of export earnings go to oil purchases.

These figures motivate a closer look at the possible future uses of oil in developing countries—and in the context of the oil-tight world foreseen in the global scenarios already described. (See "Transportation Sector and Oil Use," below.)

Rural Electrification Backlog

Vast rural areas in many developing countries are without electricity. Only 15 percent of the rural population in Latin America and Asia has electricity; less than 4 percent of Africa's 80 percent-rural population does. At the same time, the degree of electrification, and in particular its consumption density in urban areas in these countries is comparable to that in the urban areas in developed countries.

Agriculture

Agriculture in the developing regions based largely on traditional farming practices, is currently far less energy-intensive than that in the developed regions. According to the economic projections of the scenarios here, the agricultural GDP in Regions IV(LA), V (Af/SEA), and VI (ME/NAf) is expected to increase by a factor of 3.7–4.5 over the next 50 years; the expected increase would be 2.2–2.5 times in Regions 1 (NA), II (SU/EE), and III

(WE/JANZ). The implications of these projections in energy terms can be seen in the parameters of table 14.2.

Consider arable land in the developing regions. There is not much potential for expanding arable land in Regions IV, V, and VI, where the present per capita availability of arable land is about 0.34 ha compared to 0.62 ha in developed Regions I, II, and III. If no significant new area is brought under cultivation, the per capita arable land availability would decrease over the next 50 years to .14 ha in the developing regions and to 0.46 ha in the developed regions. These limits on arable land expansion imply that essential agricultural productivity improvements must come from increases in the use of fertilizers, irrigation, and farm mechanization. But surface water is in short supply, and precipitation is not adequate in most areas; increasing use would therefore have to be made of underground water.

Taking these factors into account, the energy intensiveness of agriculture (including mechanization and irrigation but not including energy used to produce fertilizers) in Regions IV, V, and VI is assumed to increase by a factor of 10 over the next 50 years. Thus, by 2030, the average energy intensiveness in these regions would be about the same (2.8 kWh per dollar value added) as the present average value for the developed regions. The final energy used in agriculture would increase for the High and the Low scenarios by about 45 and 37 times the 1975 level in the developing regions and by just 2.4 and 2.0 times in the developed regions.

NONCOMMERCIAL ENERGY AND RENEWABLE RESOURCES

The role of noncommercial energy in rural areas in the developing countries is difficult to measure. In 1975, these sources provided probably about 24 percent of total energy in Region IV (LA) and as much as 51 percent in Region V (Af/SEA). But these shares have been declining—from 35 percent and 61 percent, respectively, in 1965. Obviously, this trend is important for estimates of future consumption of commercial fuels.

Wood continues to play a major noncommercial energy role in the developing regions, and its use has been growing in Africa and Asia at over 4 percent per year for the past two decades. About 70 percent of noncommercial energy use today is met by fuelwood—village women trudging seveal kilometers each day, gathering and carrying. The resulting long-term deforestation effects have led to recognition of a possibly imminent fuelwood crisis.

Still, the so-called traditional sources of energy in developing regions may have a useful future. Wise policy may well dictate an aggressive posture in developing regions, of saving on fossil fuels by making extensive use of

Table 14.2. Agricultural Patterns

Region	Arable Land Per Capita (ha/cap)	Irrigation (% of Arable Land)	Mechanical Appliances (per 1,000 ha)	Fertilizer Use (kg/ha)
I (NA)	1.07	7	22	80
II (SU/EE)	0.77	7	15	96
III (WE/JANZ)	0.34	9	45	117
IV (LA)	0.45	9	7	32
V (Af/SEA)	0.32	14	1	14
VI (ME/NAf)	0.33	25	4	27
VII (C/CPA)	0.15	61	2	50

All data refer to arable land including land under permanent crops. Mechanical applicances included here are tractors and harvester-threshers. Fertilizer use refers to consumption in terms of N_2, P_2O_5, and K_2O.

renewable resources. A portion of substitutable fossil fuel demand in the future in each region may be replaced by the commercial supply of charcoal/wood, agricultural wastes, and biogas, without disrupting the normal evolution of various activities in the industrial and household/service sectors.

Table 14.3. Assumed Penetrations of Renewable Sources of Energy, Regions IV (LA) and V (Af/SEA)

Demand Sector	Nature of Demand	Projected Penetration of Renewables[a] (% of Useful Heat Demands) 2000		2030
		Region IV	Region V	Regions IV and V
Households				
Cities	Cooking, space and water heating	0.40	0.45	0.60
Towns	Cooking, space and water heating	0.60	0.70	0.80
Villages	Cooking, space and water heating	0.75	0.85	0.90
Service Sector				
Cities	Space and water heating		0.20	0.40
Towns	Space and water heating		0.30	0.60
Manufacturing				
	Low temperature steam/hot water		0.40	0.80
	High temperature steam		0.30	0.60
	Furnace heat		0.06	0.12

The analysis of the maximum renewable contribution (see table 14.3) takes into account the evolution of settlement patterns and the likely distribution of various energy-consuming activities among rural areas, towns, and large cities.

On the basis of the assumptions in table 14.3 and the level of use of noncommercial fuels assumed, it is expected that biogas generation in Region V may substitute for as much as 15 GW of fossil fuel by 2000 and 35 to 40 GW by 2030. The substitutable fossil fuel demand to be met by commercial supplies of charcoal/wood and biogas would be 100–121 GW by 2000 and 273–424 GW by 2030. These are discouragingly small figures, on a regional scale, but imply vigorous widespread and numerous local efforts.

Table 14.4. Resource Utilization[a] of Renewable Energy Sources in Regions IV (LA), V (Af/SEA), and VI (ME/NAf) in the High and Low Scenarios

	Maximum Production Capacity (GWyr/yr)	Capacity in 2030	
		High Scenario (GWyr/yr)	Low Scenario (GWyr/yr)
Hydroelectricity[b]:			
IV (LA)	583	355	355
V (Af/SEA)	761	426	426
VI (ME/NAf)	68	12	12
Wood from forests[c]			
IV (LA)	2,090	704	458
V (Af/SEA)	1,880	673	604
VI (ME/NAf)	55	0	0
Wood from plantations[c]			
IV (LA)	[e]	0	0
V (Af/SEA)	[e]	340	0
VI (ME/NAf)	[e]	0	0
Agricultural and animal wastes[d]			
IV (LA)	291–374	0	0
V (Af/SEA)	1,054–1,355	67	58
VI (ME/NAf)	98–126	0	0

[a] Commercial energy use only.

[b] The figures refer to primary energy equivalents at an efficiency of about 37 percent.

[c] These data refer to dry wood above ground; it is assumed in the scenarios that the harvested wood will be converted to charcoal with an efficiency of about 0.45 before it is used as a replacement of fossil fuels.

[d] The agricultural and animal wastes are assumed to increase by 2030 to 3.5 to 4.5 times the 1975 value. The commercial use of these wastes is assumed through biogas generation, for which the conversion efficiency is taken as 0.60

[e] Not available.

TRANSPORTATION SECTOR AND OIL USE

As noted earlier, developing regions today typically rely heavily on liquid fuels. Yet their transportation sectors, (liquid-fuel intensive) remain woefully underdeveloped. If development mandates increased transportability of people and things—and many contend that economic development (specialization and efficiency) requires just exactly that—then we can expect even greater liquid fuel needs in LDCs in the future.

Comparing data on automobile ownership and use among regions (table 14.5), one can see the great disparity between the most and least developed regions now and according to the assumptions made at IIASA. One could hardly say that these figures imply excessive automobile use in today's developing regions by 50 years from now. Still, oil use in the LDCs would grow rapidly in these scenarios.

And the oil would not be used indiscriminately. As table 14.6 indicates, motor fuel and feedstock uses would represent increasingly large shares of all oil use in the future. Oil should be saved for its premium uses, in both industrialized and developing regions (although the potentials for such savings are likely to be greater in the developed world).

Table 14.5. Assumptions for Automobile Ownership and Usage in Six World Regions

	Base Year	High Scenario		Low Scenario	
	1975	2000	2030	2000	2030
Auto. Ownership (auto./1000 pop.) Region:					
I (NA)	500	526	526	526	526
II (SU/EE)	25	63	100	50	67
III (WE/JANZ)	192	305	450	240	313
IV (LA)	39	100	230	72	144
V (Af/SEA)	4	11	38	9	22
VI (ME/NAf)	17	59	177	42	79
Intercity and Urban Distance Travelled (10^3 km/auto./yr) Region:					
I (NA)	15.9	16.1	16.5	15.9	15.9
II (SU/EE)	18.5	17.1	17.6	16.2	16.5
III (WE/JANZ)	9.2	10.2	10.3	10.1	9.9
IV (LA)	13.2	13.8	13.1	15.4	15.0
V (Af/SEA)	25.2	21.3	19.3	24.4	26.0
VI (ME/NAf)	16.4	16.6	18.1	19.3	24.4

Table 14.6. Final Energy Demands for Liquid Fuels, 1975–2030

Region		Base Year 1975	High Scenario 2000	High Scenario 2030	Low Scenario 2000	Low Scenario 2030
I	(NA)					
	Liquids[a] (TWyr/yr)	0.951	1.14	1.67	0.99	1.13
	% motor fuel+feedstock[b]	(74)	(86)	(94)	(84)	(91)
II	(SU/EE)					
	Liquids (TWyr/yr)	0.438	0.70	1.31	0.64	0.89
	% motor fuel+feedstock	(65)	(90)	(100)	(88)	(100)
III	(WE/JANZ)					
	Liquids (TWyr/yr)	0.979	1.65	2.06	1.32	1.44
	% motor fuel+feedstock	(52)	(69)	(86)	(64)	(76)
IV	(LA)					
	Liquids (TWyr/yr)	0.190	0.68	1.68	0.49	1.05
	% motor fuel+feedstock	(69)	(79)	(90)	(79)	(89)
V	(Af/SEA)					
	Liquids (TWyr/yr)	0.138	0.53	1.57	0.42	1.02
	% motor fuel+feedstock	(58)	(74)	(91)	(74)	(88)
VI	(ME/NAf)					
	Liquids (TWyr/yr	0.070	0.31	0.88	0.23	0.47
	% motor fuel+feedstock	(74)	(85)	(94)	(85)	(91)
World[c]	Liquids (TWyr/yr)	2.85	5.46	10.36	4.37	6.69
	% motor fuel+feedstock	(64)	(79)	(92)	(77)	(88)

[a] Liquids mean total final energy demand for liquid fuels.
[b] % motor fuel and feedstock means the % of total liquids demand that goes to "nonsubstitutable" liquids: for motor fuel (in transportation, agriculture, mining, and manufacturing) and petrochemical feedstock uses requiring liquids.
[c] World includes Region VII (C/CPA), assumed here simply to be the average of other developing regions.

HYDROELECTRIC

Developing regions generally have large and undeveloped hydroelectric capacities. As figure 14.4 illustrates, the analyses at IIASA have made use of these estimated potential resources in the scenario analyses. Further gains beyond those estimated here would become increasingly expensive, and long-distance transportation problems would multiply. By 2030 in the scenarios, an average of about 35 percent–40 percent of LDC electricity needs could be met by hydroelectric capacity.

REGION IV, LATIN AMERICA

In Latin America, the dominant energy source of the past, present, and future is oil. Adding the recent increased oil resource estimates for Mexico to

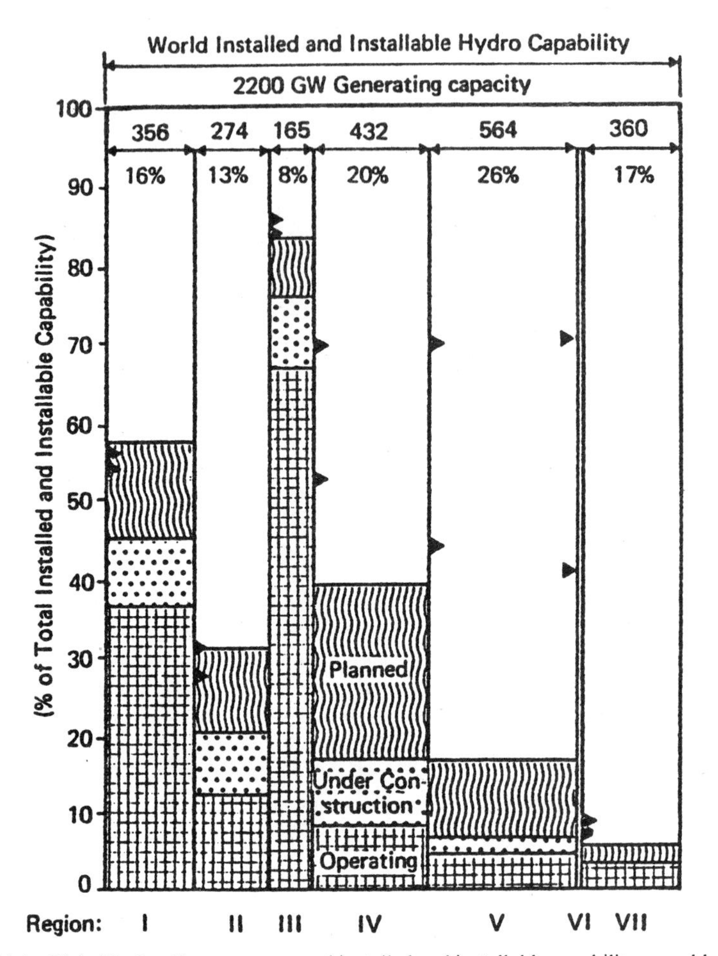

Fig. 14.4. Global hydraulic resources, total installed and installable capability—world regions. Top arrow in each regional column is the assumed maximum realizable. Bottom arrow is the actual realized by the integrated analysis in 2030, both high and low scenarios; area of column proportional to total potential installable capacity.

the vast heavy crude oil potential in Venezuela and Colombia produces total oil resource figures which seem capable of supporting even the high growth prospects of this region.

LA, like other developing regions, is projected to have a more rapid GDP growth than that of the developed regions. The projected range is 3.5–4.4 percent per year (average) to 2030; these are high by global standards, but still are reduced rates from the recent past. In the High scenario, per capita GDP in 2030 would exceed that in WE/JANZ today. But the challenges to such development would be formidable; LA would hold some 800 million people in 2030—two and one-half times its 1975 population. Providing sufficient services and jobs for so many people would require all the energy and other wealth that the region could generate.

Assuming that some fraction of heavy crude oil can be produced at relatively low cost, then all of the oil production necessary for domestic needs in the scenarios to 2030 could be provided from resources at $16 per barrel or less. Total LA oil production would grow from 4.6 million barrels per day in 1975 to 16–25.5 mbd by 2030 in the scenarios. These are 2.3–3.2 percent per year annual increases in output; from 1975 to 1978 the growth was 2.6 percent per year. But can 3 percent per year growth be sustained for 55 years?

The assessment here is that LA can increase its oil output to meet domestic needs, but would choose not to export oil (in net terms). Too rapid rises in income can be destabilizing; there are signs already that the oil-rich countries of the region will build up their petroleum industries only at a carefully measured pace. Also, oil production solely for internal purposes would be enormous, aside from any production for export. The oil production rates for LA in 2030 in the scenarios equal 30–45 percent of the total global production in 1975.

Major nonoil energy sources in the supply mix of the future in LA are expected to be hydroelectricity and some other renewable source—notably commercially distributed wood (e.g., charcoal for industry and households) and ethyl alcohol from plantation crops (extensive planning is underway in Brazil).

The primary energy source mix in the High scenario for LA (figure 14.5) shows that oil, while growing in magnitude, would be about constant as a share of total primary energy—while renewables and nuclear would grow to account for some 18–23 percent of total primary energy by 2030 (from essentially 0 percent in 1975). Natural gas would be somewhat market-limited in this region, but the resource base could conceivably allow gas to play a larger role than that in the scenario projections.

REGION V, AFRICA (EXCEPT NORTH AND SOUTH AFRICA) AND SOUTH AND SOUTHEAST ASIA

Af/SEA may face the bleakest energy future of any region. Endowed with neither energy resource riches nor capital wealth (skills, technological know-how), while having large and rapidly growing populations, the favorable energy options for the long term seem few. Still, development objectives for the region are legitimate. While the situation is challenging and somewhat discouraging, it is not hopeless.

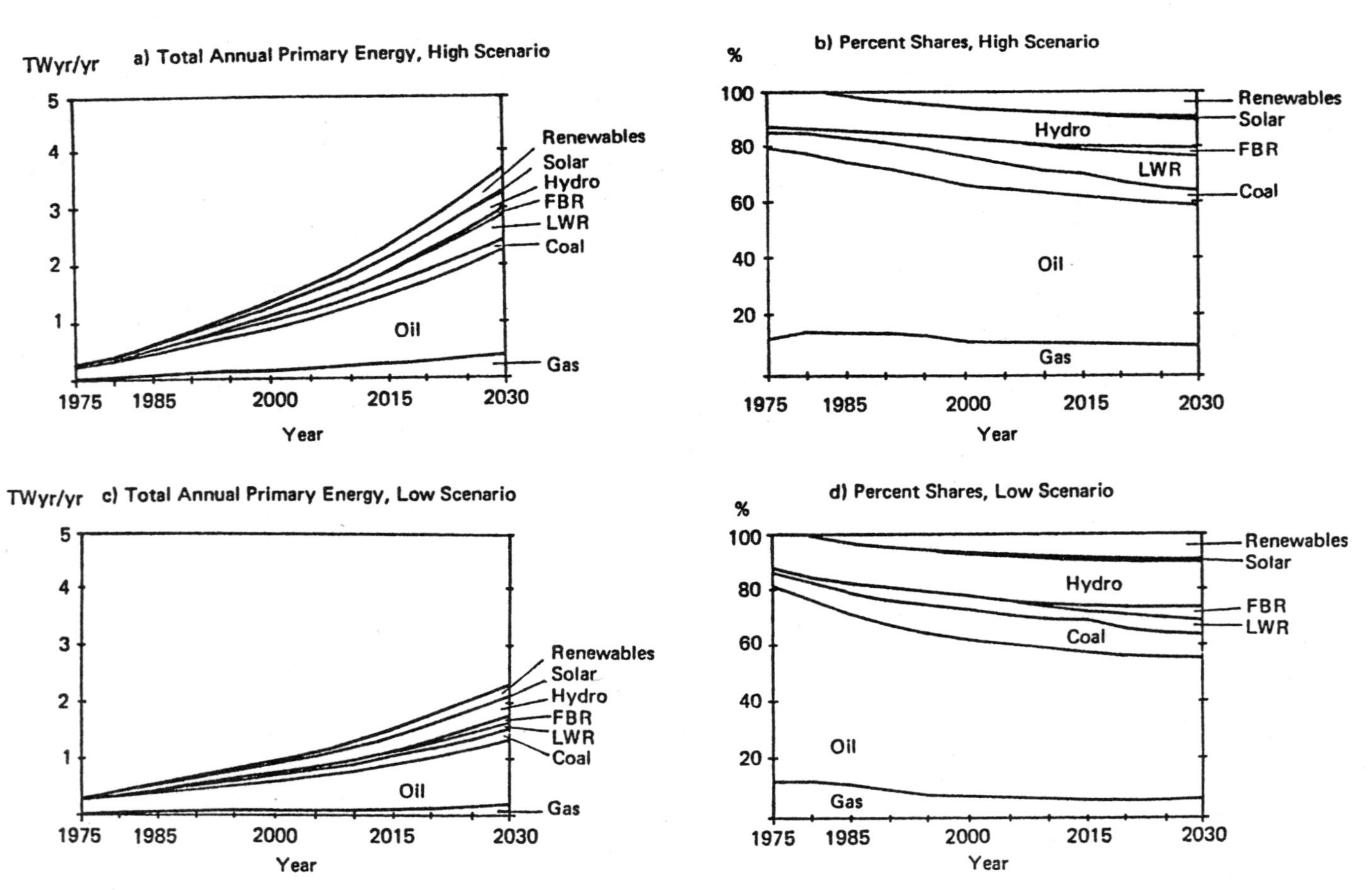

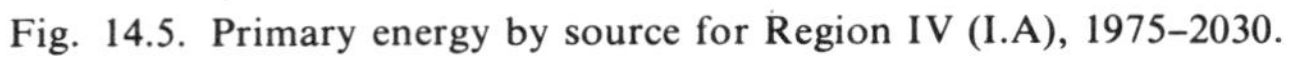
Fig. 14.5. Primary energy by source for Region IV (I.A), 1975–2030.

Af/SEA is the poorest in per capita GDP of all the seven world regions. It would still be the poorest by 2030 in the scenarios, in spite of vigorous assumptions made about the region's economic and energy prospects. Real GDP growth rates would average 3.3–4.3 percent per year from 1975 to 2030—higher than the rates in the developed regions. The development path for this growth would continue in the presently observed shifts towards the industry, service, and energy sectors, and a decline of the agricultural share (from 36 percent of GDP in 1975 to around 20 percent by 2030 in the High scenario).

Energy savings, as noted earlier, generally require sophistication, and so should not be expected to any appreciable extent in Af/SEA.

Today, the region is a net oil exporter, because Nigeria, Gabon, and Indonesia are exporters and because aggregate liquid fuel demands are quite low. The scenarios project Af/SEA to become a net oil importer within the next decade, and by 2010 imports would become greater than domestic crude oil production. Coal liquefaction then is expected to provide a small, but growing domestic source of liquid fuels.

Although liquid fuels offer a convenience and flexibility of use that makes them highly desirable for broad development objectives, they will inevitably become more expensive. Af/SEA would therefore be wise to tap other sources where possible. Over the long term, renewables would play a growing role in the scenarios here, with biogas and charcoal assumed (in the High scenario) on an aggressive policy to meet up to 16 percent of household and industrial final energy needs. All renewables combined (including hydroelectric power) could account for some 20 percent of total primary energy by 2030 in the High scenario, or even 30 percent in the Low scenario (figure 14.6). Electricity needs (growing at 4.7–5.9 percent per year in the scenarios) would be met by coal, nuclear, and hydroelectric power.

REGION VI, MIDDLE EAST AND NORTHERN AFRICA

This region, as a whole, would want to constrain its long-term oil output—and would be wise to do so since oil resources would be seriously depleted by 2030 with a sustained production rate of just 33.6 mbd, as assumed here. (An extensive discussion of the perceived character and expected behavior of this region will be given in Part IV of "Energy in a Finite World: A Global Systems Analysis.")

GDP growth rates can be expected, of course, to be high for quite some time. Development would come with the growth, reaching levels comparable to those in NA and WE/JANZ today. Development also signals growing shares of manufacturing and services in GDP, and a relative decline in the energy production sector share.

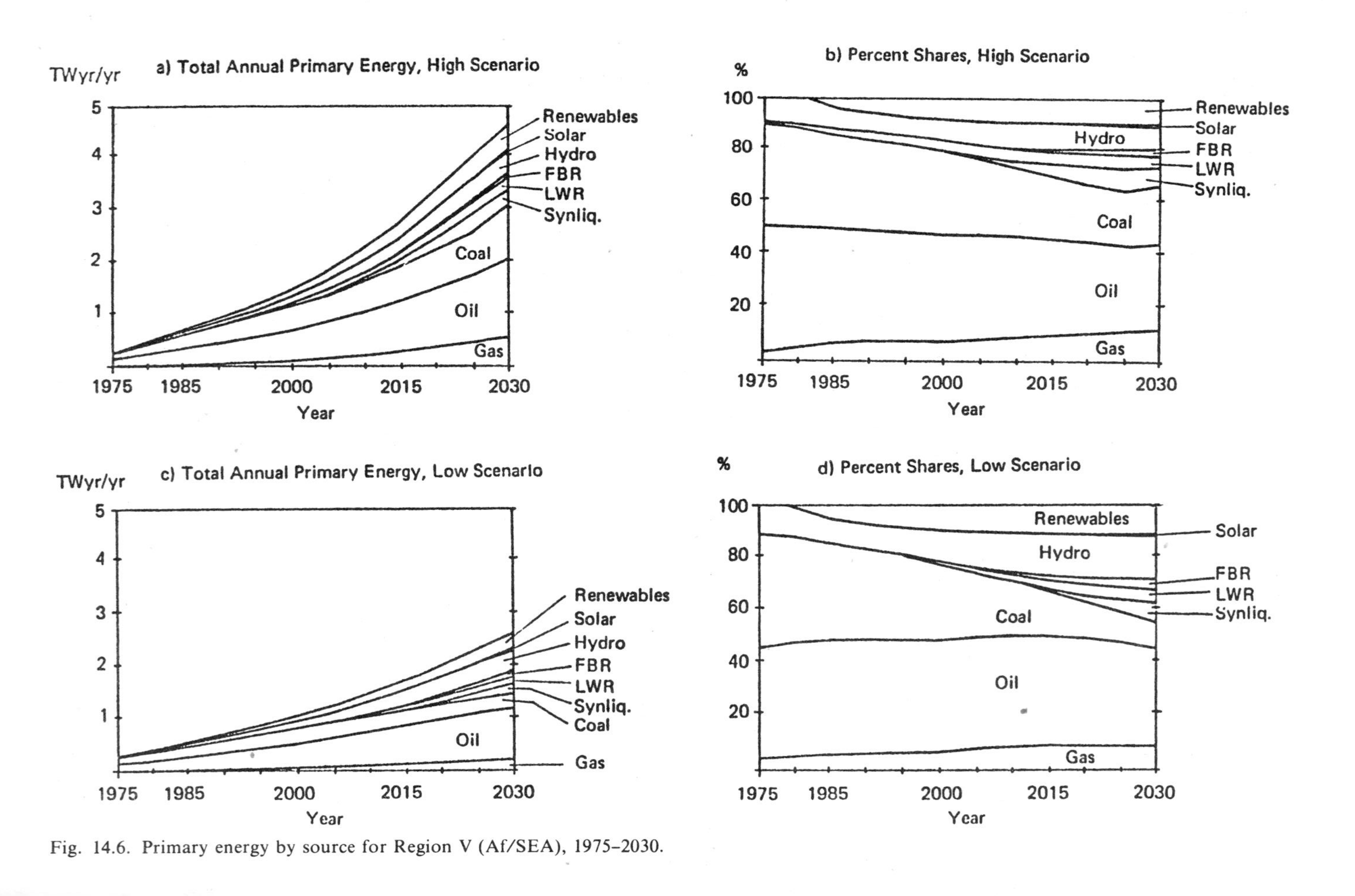

Fig. 14.6. Primary energy by source for Region V (Af/SEA), 1975–2030.

Oil and gas together provide about 90 percent of primary energy in ME/NAf over the 55-year scenario period in the High scenario. An assumed quantity of cheap natural gas, used for domestic purposes, leads to gas taking at least 50 percent of the primary energy market by 2030 in the scenarios. Gas could even be used rather extensively for electricity production.

REGION VII, CHINA AND CENTRALLY PLANNED ASIAN ECONOMIES

The primary observation about the energy future of C/CPA is that very little is known. Thus, the summary here is confined to a few key assumptions and resulting generalizations about the energy prospects of Region VII:

- GDP growth seems likely to be high, but so does population growth. With the prediction of average GDP growth rates of 2.6–3.8 percent per year and of an average population growth rate of 1.2 percent per year, per capita GDP levels would reach, in 2030, levels comparable to those of LA today.
- It is assumed that C/CPA would be neither a net exporter nor an importer of energy. But domestic oil resources are not likely to be able to keep pace with rising demands; coal liquefaction would be required post-2000.
- Coal production, required for liquid fuels synthesis and electricity generation, would reach nearly 3.5 billion tons per year by 2030 in the High scenario, from 0.48 billion tons per year in 1975; over 2 billion tons per year of this would be needed simply for synthetic liquid fuel production. This coal production would exceed that of NA in 2030.
- The main primary energy source for the future in these projections would be coal. Natural gas and, to a small extent, nuclear would grow noticeably, but still represent less than 20 percent of primary energy by 2030 in the scenarios.

A FINAL WORD

The search for effective and speedy means of meeting human needs and hopes is accentuated in the developing countries by their present plight—the great distance between practice and potential, realities and aspirations. The challenges to progress in the developing regions are substantial; they reflect individual and societal distresses that call for fresh thinking and innovative solutions. Distinctive characteristics of economic structure, cultural patterns, and political goals in the developing regions cannot be ignored.

In 1975, the developing regions–Regions IV (LA), V (Af/SEA), VI (ME/NAf) and VII (C-CPA)—accounted for 15.7 percent of total global primary (commercial) energy. By 2030, under the assumptions of the High and the Low scenarios (see table 14.1), they would represent 38–43 percent. By 2030, per capita levels of consumption in the developing regions would have grown substantially, but in most instances they would not have reached 1975 West European (4 kW/cap) levels. Only Region IV, (LA) and Region VI (ME/NAf) would be nearing, by 2030 (especially in the High scenario), a status that could be called "developed".

Yet of course no one knows the future. This cannot be said too often or too strongly—particularly after summarizing, above, a great deal of detailed thinking about the world and its energy systems in the year 2030. Yet there are, perhaps, a finite number of truly important uncertainties—a few large imponderables.

Will there be—can there be—a breaking of the link between economic growth and transportation activity? Can the world grow, can the developing regions develop, and can the developed regions spend new disposable income without large increases in transportation (and hence in liquid fuel demands)?

Will the Middle East cap its oil production at a level allowing a smooth transition to alternatives, or will the tensions caused by their oil ceilings create global instabilities?

Will all of the energy problems restrain economic growth in the world to the extent of causing a long-term global recession?

Will the developed regions succeed in reducing their dependence on liquid fuels—by savings and by developing alternatives—so that the world's finite oil supplies can be left for the developing poor?

REFERENCES

Part IV, "Balancing Supply and Demand: The Quantitative Analysis," in *Energy in a Finite World: A Global Energy Systems Analysis*, the final report of the IIASA Energy Systems Program, to be commercially published in 1980.

Chapter 15
Energy Demand and Structural Change in Developing Countries

Lutz Hoffman and Matthias Mors

INTRODUCTION

Since the quadrupling of oil prices in 1973/74, the world has learned that energy is a scarce resource like many others. The reaction to this lesson has been two-fold. Engineers became preoccupied with the questions of how the supply of energy might be increased and whether energy could be used more efficiently in production and households. Economists who had rarely thought about energy discovered that energy use differs widely among countries and that changes over time also show wide variations. In particular it was found that energy demand increases much faster in countries at low levels of economic development than in the advanced industrial countries. This led to the question of what actually determines energy demand and whether an answer to this question would improve the ability to foresee with reasonable accuracy possible future demand. The present paper deals with this issue.

DETERMINANTS OF ENERGY DEMAND AND THEIR QUANTIFICATION

Four basic factors determine the development of energy demand over time. These are changes in

- the general level of economic activity, usually measured by gross domestic product (GDP)
- the sectoral or product composition of an economy

- the technology of energy use
- the efficiency of energy use

The crucial questions which have to be answered for an assessment of the future growth in energy demand are, first, What can be said about the changes of these factors over time? and second, What is the quantitative impact of these factors on energy demand?

With regard to the first question, the economist's skills are rather limited and probably will always remain so. Within certain limits, economists can make conditional forecasts for these factors, but, being honest, they can never claim to prophecy certain developments with a quantifiable degree of reliability. The economist ought to be able to provide answers to the second question, but even here the profession's knowledge is still scanty. The basic problem seems to be that the factors mentioned work simultaneously to determine energy demand. It is therefore rather difficult to assess correctly the impact of one factor as long as it has not been possible to quantify the impact of the remaining factors.

The first studies on energy demand attempted to estimate simple demand-GDP relationships. The results differed widely not only among countries but also for different time periods for a particular country. The reason is obvious. If factors other than GDP are not explicitly accounted for, the estimated parameters of an energy demand function with GDP as the only explanatory variable reflect the impact of all factors. For a conditional forecast of the change of energy demand due to a change of GDP, such a relationship is of rather little use because of its instability. It can only be expected to be relatively stable if all the other factors change exactly as in the past. This is however very unlikely.

In the light of this, it was certainly an improvement when real energy prices were introduced into the estimated relationship. The idea apparently was that the efficiency of energy use depends on the energy price. This holds for consumption of energy in households as well as in industry. It might also have been thought that the energy price captures part of the impact of technology. If all other factor prices were constant, and the economy were functioning like a perfectly competitive equilibrium system, changes in the energy price would indeed induce technological changes and could therefore be considered representative of these changes. However, in reality we neither have perfectly competitive economies nor constant factor prices. Even if we could not account for insufficient competitiveness, what at least ought to be considered if the impact of technology is to be captured, are changes in nonenergy input prices.

This has not been done up to now. The introduction of the real energy price into the demand function, therefore, takes the impact of efficiency changes out of the income parameter. As real energy prices were falling during the 1960s, for which most estimates were made, the income parameter was generally estimated lower with the price variable than without it. It may be noted that the cost share equations, derived from translogarithmic production functions, are a rather different approach from the one discussed here. Nevertheless they gave some indication of the order of magnitude one could expect from the impact of technological change.

The sectoral or product composition was added to the demand function by the present authors in demand estimates for developing countries. As structural change in developing countries typically favor energy-intensive industries, the result was again—as with the introduction of the price variable—that the parameter of the income variable was lowered. In most estimates the structural variables contributed more to a reduction of the income elasticity than did the price variable. Since structural change is less pronounced in the industrial countries than in the developing countries, the most striking result was that the wide disparities between the income parameters of these two country groups largely disappeared.

A SIMPLE ECONOMETRIC ENERGY DEMAND MODEL

The Theoretical Model

Energy demand of an industry (EI) may be described by the following function:

$$EI = f_1 (X, P_E, P_L, P_R) \tag{1}$$

where X is real output and P_E, P_L, and P_R are the real prices of energy, labor, and raw materials in terms of capital (see Appendix I). Let us partition the independent variables into one subset with output and the energy price and a second subset with the two other factor prices. If we assume weak separability with respect to this partition, we can write

$$EI = f_2 (X, P_E). \tag{2}$$

By definition we have

$$X = \frac{X}{Y} \cdot Y = u \cdot Y. \tag{3}$$

(2) can then be written as

$$EI = f_3 (Y, u, P_E). \tag{4}$$

Similarly for households, the following energy demand function may be defined:

$$EH = f_4 (Y, P_{EH}), \quad (5)$$

where Y is real income and P_{EH} the real energy price in terms of consumer goods. If we assume that the real energy prices in (4) and (5) are expressed as indices and that the indices move parallel over time, we may use just one price index (P). A function for total energy demand (E), being the sum of demand by n industries and by households, can then be written as

$$E = f_5 (Y, u_1, u_2, \ldots u_n, P). \quad (6)$$

Econometric Specification

The econometric specification is primarily a matter of convenience. For the estimation of production and demand functions, it has become common practice to use logarithmic specifications because they can account for non-linearities and therefore tend to produce better statistical results than un-logged forms. If the specification is log-linear, the functions directly estimate elasticities which are convenient to interpret.

The advantages of logarithmic specifications must, however, be weighted against their shortcomings. One problem is that for the larger values, the deviation of the observed from the estimated values tends to be understated and the coefficient of determination therefore tends to be overstated (see Appendix II). In time series analysis, where one usually deals with increasing functions, this has the consequence that deviations in the most recent years are generally understated. This is undesirable if it is believed, for example, that the more recent developments are at least as important, if not even more important than the earlier values for the purpose of making projections. Theoretically, the problem could be solved by weighting the observations. In practice, however, it is difficult to determine the correct magnitude of the weights.

An aggregation problem arises if it is postulated that the demand function for total energy can be disaggregated into the functions of the user categories in the sense that the parameters of the latter functions can be identified in the first (see Appendix III). Such a postulate seriously constrains the range of permissible specifications. In particular, it excludes all logarithmic specifications, because such functions can generally not be disaggregated.

In spite of these shortcomings, a log-linear specification was chosen in the present study, primarily because the structural variables certainly change in a nonlinear fashion. From the work of Kuznets (1971), Fels, Schatz, and Wolter (1971), and Chenery and Syrquin (1975), it is known that the sectoral shares in GDP approach certain ceilings with rising per capita income and eventually reverse from there later on.

The function actually estimated is

$$\ln \frac{E}{B} = a_0 + a_1 \ln \frac{GDP}{B} + a_2 \ln (R4S) + a_3 \ln (R5S+R6S) \tag{7}$$

$$+ a_4 \ln (R7S+R8S) + a_5 \ln (R9S) + a_6 \ln PR + u.$$

R4S = share of agriculture in GDP
R5S = share of mining in GDP
R6S = share of construction in GDP
R7S = share of manufacturing in GDP
R8S = share of utilities in GDP
R9S = share of transport in GDP
PR = index of real energy price

In formula (7), the shares of mining and construction as well as those of manufacturing and utilities are lumped together, because for several countries they cannot be obtained separately. Energy consumption and GDP were divided by population (B) in order to avoid having very large or very small values of these variables dominating the results.

Data and Estimation Method

The estimation of a function like (7) requires a substantial amount of data, most of which are not readily available for developing countries. In our case, the data for all exogenous variables except prices were obtained from the data banks of the World Bank. The data for energy consumption are from a United Nations tape.[1]

The Data. The price data were partly obtained from B. J. Choe of the Economic Analysis and Projections Department of the World Bank and partly collected by ourselves. Energy price indices were constructed from the price series for individual fuels by using the 1970 fuel consumption shares of the respective country as weights. 1970 was also chosen as base year. Because the income variables are expressed in real terms, the nominal price indices also had to be deflated. For this deflation, either the consumer price index, the retail price index, the wholesale price index, or the GDP deflator were used, depending on data availability.

The price series was converted into indices because it appeared impossible to make an average energy price of one country comparable to that of another. As a result, the estimated price elasticities reflect only the impact of price changes over time and not the impact of inter-country differences.

A Two-Stage Estimation Method. In order to estimate the income, structural, and price elasticities represented by the parameters of (7), the equation is first applied to a pooled time-series cross section sample, in which time-series, ranging from 1960 to 1975, of individual countries are lumped together. Theoretically, the parameters estimated from such a sample reflect differences between countries as well as changes over time. However, in the case under investigation, the cross-country effects appear to dominate the estimated parameters.

The application of equation (7) to a country pool instead of individual countries was necessary for several reasons. First, for a multiple regression with six independent variables, the available time series with a maximum of 16 observations for the individual country is simply too short for estimating statistically significant parameters. Second, the structural change within countries over a period of 16 years is generally too small and the spread of the respective variables therefore too narrow, to expect statistically significant parameters. Third, structural change is a long-run phenomenon which best is explained by cross section estimates which indicate long-run adjustments much better than time series.

The situation is somewhat different with respect to the income variable. Income changes over time are certainly much more pronounced than structural changes. Furthermore, reactions of energy demand to income changes can be expected to be more immediate than reactions to changes in the economy's structure and the energy price. Hence, time-series estimates of the energy income relationship for individual countries are likely to produce statistically satisfactory results. In addition, by estimating the energy income relationship for each country individually, it is possible to bring out the trends prevailing over time in the various countries.

These considerations led to the following second step in our estimation procedure. After having estimated the parameters of equation (7), a reduced energy variable was calculated by deducting the structural and the price effect from the original variable:

$$\text{(8)} \quad \ln \frac{E^*}{B} = \ln \frac{E}{B} - \hat{a}_2 \ln (R4S) - \hat{a}_3 \ln (R5S + R6S) - \hat{a}_4 \ln (R7S + R8S) - \hat{a}_5 \ln (R9S) - \hat{a}_6 \ln PR .$$

The reduced energy variable was then regressed for every country against the income variable:

$$\text{(9)} \quad \ln \frac{E^*}{B} = b_0 + b_1 \ln \frac{GDP}{B} + \gamma$$

The explained variance of the two step estimation can be calculated as follows:

$$(10) \qquad R^2_{TOTAL} = 1 - \frac{s_\gamma^2}{s_{EB}^2}$$

where s_γ^2 is the variance of the error term in equation (9) and s_{EB}^2 is the variance of the original energy variable as it appears on the left side of (7).

In order to understand the relationship between the income elasticities in (7) and (9), equations (7), (8), and (9) may be rewritten as follows:[2]

$$(11) \qquad \ln \frac{E}{B} = a_0 + a_1 \ln \frac{GDP}{B} + \sum_{i=2}^{6} a_i \ln w_i + \epsilon$$

$$(12) \qquad \ln \frac{E^*}{B} = \ln \frac{E}{B} - \sum_{i=2}^{6} \hat{a}_i \ln w_i$$

$$(13) \qquad \ln \frac{E^*}{B} = b_0 + b_1 \ln \frac{GDP}{B} + \gamma$$

w denotes all independent variables other than per capita income.

The ordinary least squares method then gives the following estimate of b_1:

$$(14) \quad \hat{b}_1 = \left[\Sigma\left(\ln \frac{GDP}{B} - \ln \overline{\frac{GDP}{B}}\right) \times \left(\hat{a}_0 + \hat{a}_1 \ln \frac{GDP}{B} + \hat{\epsilon} - \left(\overline{\hat{a}_0 + \hat{a}_1 \ln \frac{GDP}{B} + \hat{\epsilon}}\right)\right)\right] \Big/ \left[\Sigma\left(\ln \frac{GDP}{B} - \ln \frac{GDP}{B}\right)^2\right]$$

$$= \hat{a}_1 + \frac{\Sigma\left(\ln \frac{GDP}{B} - \ln \overline{\frac{GDP}{B}}\right)(\epsilon - \bar{\hat{\epsilon}})}{\Sigma\left(\ln \frac{GDP}{B} - \ln \overline{\frac{GDP}{B}}\right)^2}$$

$$= \hat{a}_1 + k$$

It can be seen that b_1 is the sum of $\hat{a}_1$ and k where k is an error-term related item from the point of view of equation (11). Figure 15.1 illustrates the result:

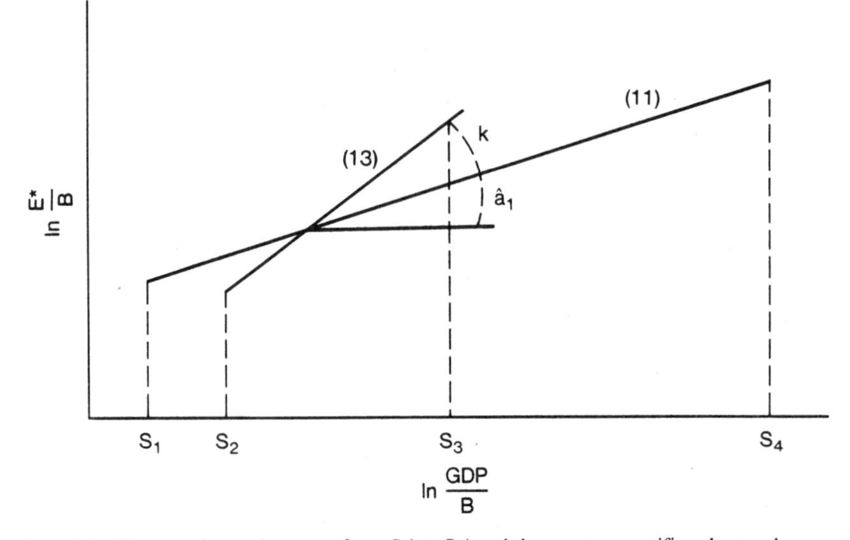

Fig. 15.1. If the total sample ranges from S 1 to S 4 and the country specific sub-sample ranges from S 2 to S 3, the estimated regression lines (11) and (13) can be drawn. The gradient of (13) apparently is the sum of the gradient of (11) and k. If the gradient of (13) is smaller than that of (11), then k is negative.

Application Problems. Estimating (7) and (9) for every individual country is a time-consuming task. If one is primarily interested in estimates and projections for country groups, one may therefore wonder whether the above functions could not be applied to country aggregates.

The easiest procedure is to apply (7) directly to the respective time series of a country group, assuming that the income elasticity estimated for the pooled time series cross section sample adequately reflects the group's energy-income relationship over time. Unfortunately, that procedure is not possible. It can produce very peculiar results. In the group 10 figure at the end of the chapter the observed energy consumption of a country group is plotted against the estimated consumption. It is seen that the first always lies above the latter, though the time trends are very similar.

The reason for this result is an aggregation problem (see Appendix IV). If the variables of (7) are aggregated for a country group, the implicit weights of the group averages differ from one variable to another. For per capita consumption, for instance, the population shares are the weights, whereas the structural variables are weighted by income shares.[3] Let us assume for a moment that per capita consumption is regressed against the structural variables only. If, then, the population shares of countries with high per capita energy consumption are relatively larger than their income shares, one obtains a result like that in the figure.

Statistically better results are obtained if, instead of applying (7) directly, (a) is estimated for the aggregated data. Also, in the cases tested, the projections did not differ much from those obtained by adding up the country projections. However, this approach is also methodologically defective, because in calculating (8), one makes the same aggregation error as in the first approach. The better fit is due to the fact that in estimating (9), the errors made in (8) are largely ironed out.

The Projection Method

The basic idea in designing the projection method was to construct a model that could easily be linked to the World Bank macro model. By taking the exogenous variables projected by the World Bank macro model as inputs to our energy model, it was possible to avoid the enormous task of developing our own consistent world macro model. Unfortunately, the linking process was substantially aggravated by the fact that the output of the World Bank model is on a higher level of aggregation than the input needed for our energy model. This made it necessary to disaggregate the exogenous variables as supplied by the World Bank.

Given the exogenous variables, projections can be made by applying the structural and price elasticities, a_2, to a_6, as estimated by equation (7), as well as the constant and the income elasticity estimated by equation (9).

Two versions are produced for every country and country group—one projection with a constant real energy price over the entire projection period, and another where it is assumed that the real price increase between 1975 and 1990 amounts to 50 percent or an average of 2.7 percent annually.

The projections for the country groups are obtained by simply adding the projections for the individual countries. In the case of per capita energy consumption, the energy and population projections were first added over the various countries and then divided by each other. The aggregates of the structural variables were obtained by calculating weighted means of the country-specific structural variables with the income shares serving as weights. In the case of the group's per capita income, one simply had to divide total income as supplied by the World Bank by total population.

ESTIMATION AND PROJECTION RESULTS FOR TWO COUNTRY GROUPS

Estimates and projections with the model described above have been undertaken for almost 100 countries, which were divided into 14 country groups. The results obtained are described below for two country groups. For every country and country group we have a set of five tables. However, only those for the country group are reproduced at the end of this chapter.

Table 1 lists the parameters estimated by equation (7). The coefficient of determination, the number of observations, and the F-value for that function are also shown. As the coefficient of determination is biased in logarithmic regressions (see above), we have added the sum of the squared deviations in absolute terms in relation to the sum of the squared observations in absolute terms, as a more neutral measure of the goodness of fit. This ratio is denoted by SUM 1/SUM 2. The numerator and denominator are also shown. The table also lists all countries included in the group.

Table 2 gives the total estimated energy consumption in the first column of its upper half. These data are obtained by multiplying the estimated per capita values (column 3) by the population data (column 5). The second column gives actual energy consumption, and the fourth gives the respective per capita values. The lower part of the table shows the time series for the remaining exogenous variables. The prices series is calculated as a weighted average of the country-specific price indices, with the current energy consumption values being the weights.

Table 3 plots actual, estimated, and projected energy consumption per capita.

Table 4 and *Table 5* plot the exogenous variables, except for population, because these plots may facilitate assessing the projections of these variables.

Group 1: Southern Europe

The first step regression produced an income elasticity of 0.4, which is considerably lower than one would expect from other estimates of income elasticities of energy demand. On the other hand, the price elasticity is −0.5, quite in line with results obtained in other studies.

Among the structural variables, the share of agriculture has a negative sign, indicating that the generally observed decline in the share of agriculture affects energy demand positively. This is because the energy intensity of agriculture is rather low in less developed countries.

High positive elasticities are estimated for mining and construction, manufacturing and electricity, and the transport sector. Whereas this result has little significance for the transport sector, because the share of transport in GDP usually does not change very much over time, the highly positive elasticity of manufacturing and electricity certainly is a very important determinant of the changes in energy demand, because this sector typically expands its share in the process of economic development.

The negative impact of agriculture and the positive impact of the other sectors implies that one partly offsets the other. However, the positive impact appears to be dominant. The net result is a reduction of the income elasticity.

For the individual countries, the income elasticities net of structural change and price effects, as estimated in the second step, vary widely. The lowest values are zero and 0.12 for Portugal and Israel. The highest values are estimated at 1.02 and 0.81 for Greece and Cyprus.

Country	Income Elasticity	t-ratio	SUM1/SUM2 × 10
Cyprus	0.81	6.3	1.7
Greece	1.02	8.0	3.2
Malta	0.57	8.2	4.6
Portugal	0	0	0.1
Spain	0.70	7.8	0.5
Turkey	0.60	3.5	1.3
Yugoslavia	0.75	7.3	0.9
Israel	0.12	0.8	1.4

Group 5B: Low-Income East Africa

This group is one with a relatively poor data situation and a very low level of economic development. As can be seen from Table 2, missing data made estimates for several years impossible. Most likely the quality of the data is also not good. In spite of the data problem, the results are quite reasonable. The income elasticity estimated in the first step is again quite low. For countries at this level of development, one usually would expect elasticities above 1.4.

On the other hand, the price elasticity of almost −0.9 is quite high. The reason could be that in very poor countries, price changes for commercial energy induce the substitution of noncommercial energy. During periods of falling energy prices, the switch from noncommercial to commercial energy sources is accelerated, whereas with rising prices, a resubstitution of noncommercial energy sources takes place. In more developed countries, these possibilities do not exist anymore so that one could expect to find lower price elasticities.

Agriculture again shows a negative sign. The elasticity value is considerably higher than in group 1. This is of course not surprising because the countries in this group are predominantly agricultural economies with an extremely low energy intensity of agricultural production.

Manufacturing and electricity as well as transport again show positive signs. Mining and construction were omitted because the estimated parameter was not significantly different from zero.

Although the elasticities of manufacturing and electricity and of the transport sector are lower in absolute terms than that of agriculture, their combined impact on energy demand is probably much larger. The reason is that in relative terms, the changes of the respective shares are higher for these two sectors, which start from a rather low level. For the future this will continue to be so for manufacturing and electricity, whereas the share of transport will change much less.

The income elasticities estimated in the second step again differ widely for individual countries. In a number of cases, the elasticities net of structural change and price movements even became negative. In the projection we put those values at zero.

Country	Income Elasticity	t-ratio	SUM1/SUM2 × 100
Burundi	0	0 4.1	
Ethiopia	2.35	5.9	1.7
Kenya	0	0	1.5
Malavi	0	0	7.7
Rwanda	0	0	1.5
Tanzania	1.00	3.5	2.6
Uganda	2.11	5.0	0.5
Zaïre	0	0	4.1

In summary it can be said that in this group, structural change and price movements account for most of the changes in energy demand. This is understandable if it is realized that real per capita income has hardly changed in these countries and has even declined in some of them.

Increasing Prices

Next we show Table 2 for each of the two country groups for the case of a 50 percent real price increase. It is seen that the overall impact on projected demand in 1990 differs considerably between the two groups. The reason is, of course, the difference in price elasticity. Whereas for Southern Europe total demand in 1990 is about 20 percent lower than with constant prices, the reduction in low income East Africa is about 30 percent. The annual growth rate between 1975 and 1990 reduces in the case of Southern Europe from 5.4 to 3.8 and for low income East Africa from 5.3 to 3.0. As a large portion of the assumed price increase has already taken place, the latter growth rates appear to be more likely for these country groups than those with constant prices.

THE IMPACT OF STRUCTURAL CHANGE

Although it has already become clear from the above discussion that structural change is an important determinant of energy demand, in this section we will demonstrate its importance by means of simulations. We have arbitrarily assumed that the share of agriculture declines by 3 percentage points more than in the original projection and that the share of manufacturing and electricity is 4 percentage points higher.

As the next two tables show, the effect is in group 1 the projected value in 1990 becomes 25.8 percent higher whereas the increase in group 5B is only 18.5 percent. The annual growth rates of energy demand between 1975 and 1990 increase to 6.9 percent for group 1 and 6.6 percent for group 5B.

In the final two tables we have combined the increased structural change with the real energy price rise. For group 1 the result is that the estimated demand for 1990 is only marginally higher than in the original version. In group 5B on the other hand, demand is reduced by 16.4 percent. The latter result obviously is due to the high price elasticity found for this country group. The effect on the rates of growth of energy demand for country group 1 is zero. In the case of country group 5B, the growth rate comes down to 4.2 percent.

CONCLUSIONS

Income elasticities of energy demand vary much less among countries and over time if the impact of prices and structural change is taken into consideration. For projection purposes it is therefore advisable to consider these three elements separately. Countries which encourage industrialization by economic policy are bound to generate larger increases in energy demand than countries where agricultural development is emphasized. An assessment of economic policies should therefore be an important element of energy demand forecasting.

A word of warning with respect to our results appears, however, to be appropriate. Our estimaion methods give preference to the impact of structural change and prices. It is therefore quite possible that the impact of these factors is overestimated. We have experimented with an alternative approach which estimates the country-specific income elasticities, the average structural parameters, and the price elasticity simultaneously. For reasons which cannot be explained here, this approach tends to underestimate the impact of structural change. Nevertheless, one also obtains with this approach significant parameters for the structural variables, although their value is usually lower. The true values may lie somewhere in between.

APPENDIX I: DERIVATION OF AN ENERGY DEMAND FUNCTION OF PRODUCERS

Assume the following Cobb-Douglas type production function:

$$X = E^{\alpha} \cdot L^{\beta} \cdot C^{\gamma} \tag{I-1}$$

where E, L, and C are the energy, labor, and capital inputs. Let p_e, p_l and p_c be the respective prices, and X and p be the output and output price. Maximizing profit function

$$\pi = X \cdot p - E \cdot p_e - L \cdot p_l - C \cdot p_c \tag{I-2}$$

yields the first order conditions

$$p \cdot \alpha \frac{X}{E} = p_e \tag{I-3}$$

$$p \cdot \beta \frac{X}{L} = p_l \tag{I-4}$$

$$p \cdot \gamma \frac{X}{C} = p_c \; . \tag{I-5}$$

From (I–3) to (I–5) one can derive

$$C = \frac{p_e}{p_c} \cdot E \left(\frac{\alpha}{\gamma} \right)^{-1} \tag{I-6}$$

$$L = \frac{p_e}{p_c} \cdot \frac{\beta}{\gamma} \cdot E \left(\frac{\alpha}{\gamma} \right)^{-1} \cdot \left(\frac{p_l}{p_c} \right)^{-1} \; . \tag{I-7}$$

Inserting (I-6) and (I-7) into (I-1) and solving for E yields

$$E = z \cdot X^{\frac{1}{\lambda}} \cdot \left(\frac{p_e}{p_c} \right)^{-\frac{\beta+\gamma}{\lambda}} \cdot \left(\frac{p_l}{p_c} \right)^{\frac{\beta}{\lambda}} \tag{I-8}$$

where

$$\lambda = \alpha + \beta + \gamma \text{ (scale elasticity)}$$

and

$$z = \left(\frac{\beta}{\gamma}\right)^{-\beta} \cdot \left(\frac{\alpha}{\gamma}\right)^{(\beta+\gamma)}$$

The energy demand function (I–8) is equivalent to function (1) in the main paper. It depicts the points of tangency of isoquants, representing X, with isocost planes. In a two-factor space it is the following familiar optimum situation:

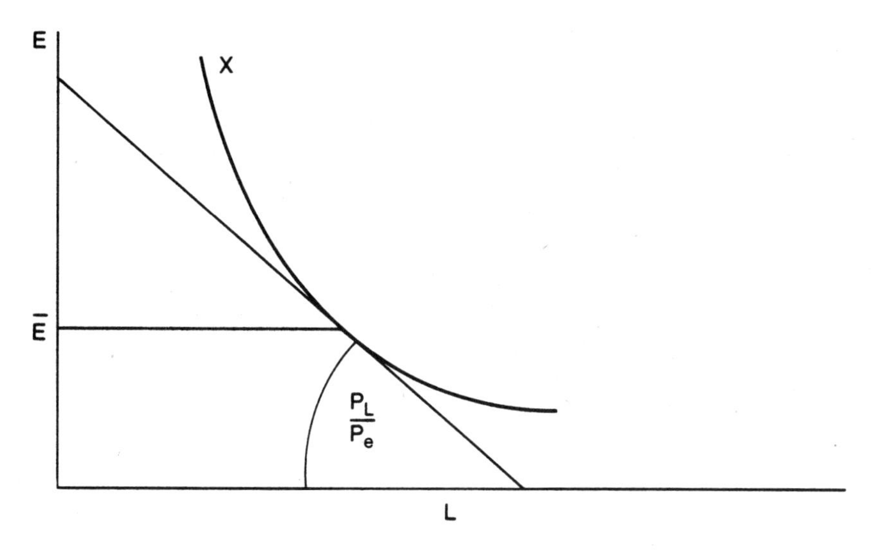

Fig. 15.2

APPENDIX II: MEASURE OF DETERMINATION FOR NONLINEAR REGRESSIONS*

Assume the following nonlinear function:

(II–1) $$Y_t \simeq a x_t^{b}$$

and its logarithmic transformation,

(II–2) $$\ln y_t \simeq \underbrace{\ln a}_{\alpha} + b \ln x_t.$$

*We owe this section and Appendix IV to a note by W. Oberhofer.

One can then estimate

$$\text{(II–3)} \qquad y_t^* = \alpha + bx_t^* + u_t^*$$
$$\text{with } y_t^* = \ln y_t \text{ and } x_t^* = \ln x_t \,.$$

This implies for (II–1) a multiplicative approach:

$$\text{(II–4)} \qquad y_t = ax_t^b \cdot v_t$$
$$\text{with } v_t = \exp(u_t^*) \,.$$

Equation (II–3) can be estimated by ordinary least squares method. The estimated parameters are $\hat{\alpha}$, $\hat{b}$. It is then postulated that $\hat{\alpha} = \exp(\hat{\alpha})$.

For a linear equation like (II–3), the coefficient of determination is usually employed:

$$\text{(II–5)} \qquad B = \frac{s_{\hat{y}^*\hat{y}^*}}{s_{y^*y^*}} = 1 - \frac{s_{\hat{u}^*\hat{u}^*}}{s_{y^*y^*}}$$

This is a measure of the explanation of $y^* = \ln y$ by $x^* = \ln x$. It ranges between 0 and 1. What one really wants, however, is a measure of the explanation of y by x. One could think of the following approach: We calculate

$$\text{(II–6)} \qquad \hat{y}_t = \hat{a}x_t^{\hat{b}}$$

and define

$$\text{(II–7)} \qquad y_t - \hat{y}_t = \hat{u}_t \,.$$

and then obtain

$$\text{(II–8)} \qquad B_1 = \frac{s_{\hat{y}\hat{y}}}{s_{yy}} \quad \text{or} \quad B_2 = 1 - \frac{s_{\hat{u}\hat{u}}}{s_{yy}} \,.$$

Apparently, the properties of (II–8) are not the same as those of (II–5). In general, B_1 will be different from B_2. B_1 can be larger than 1 and B_2 smaller than 0. B_1 as well as B_2 are therefore not easy to interpret, and comparisons between different estimates also become rather difficult.

We therefore propose to use as a measure of determination

$$\text{(II–9)} \qquad B_3 = 1 - \Sigma\hat{u}_t^2/\Sigma y_t^2 \,.$$

As $\bar{\hat{u}}$ usually is different from 0, one should not use $s_{\hat{u}\hat{u}}$, but $1/T\Sigma\hat{u}_t^2$ instead. $\Sigma\hat{u}_t^2$ is minimal with the least squares method. However, as this magnitude is dependent on the dimension, it is standardized as in (II–9).

B_3 approaches 1 in case of a good explanation. If the explanation is very poor, it approaches zero and can even become negative. A negative value implies that the residuals have a larger dispersion than the original data. This indicates that the model is incorrect.

APPENDIX III: AGGREGATED DEMAND FUNCTIONS

Specifications which meet the postulate that total demand can be disaggregated into the demands of different user categories have to be linear in absolute terms. Equation (4) in the text could, for instance, be specified for each of the four sectors as follows:

$$EI_i = c_{0i} + c_{1i}u_i + c_{2i}Y + c_{3i}P_{Ei} \quad (i = 1, 2, \ldots . 4) \tag{III–1}$$

A linear specification of (5) would be

$$EH = b_0 + b_1Y + b_2P_E \, . \tag{III–2}$$

Assuming a uniform energy price (P_{TE}), one can aggregate the two functions as follows:

$$\begin{aligned} E &= \sum_i EI_i + EH \\ &= \sum_i c_{0i} + b_0 + \sum_i c_{1i}u_i + (\sum_i c_{2i} + b_1)Y + (\sum_i c_{3i} + b_2)P_{TE} \, . \end{aligned} \tag{III–3}$$

If not all of the sectoral shares appear in (III–3) as independent variables, as is the case in the present study, the parameters of these two functions are somewhat different. By definition we have:

$$\sum_i u_i = 1 \, . \tag{III–4}$$

Assume that industry 1 is omitted. One may then derive the output of that industry as the difference between one and the output share of the remaining industries. Inserting into (III–3) yields

$$\begin{aligned} E = {} & \sum_i c_{0i} + c_{11} + \sum_{i \neq 1} (c_{1i} - c_{11})u_i \\ & + (\sum_i c_{2i} + b_1)Y + (\sum_i c_{3i} + b_2)P_{TE} \, . \end{aligned} \tag{III–5}$$

From (III–5) it can be seen that the parameter of the income variable can be expected to have a positive sign and that of the price variable a negative one if the respective parameters in (III–1) to (III–2) have the theoretically correct signs. The parameters of the share variables in (III–3) will have either a positive or a negative sign, depending on whether the respective industry has a high or a low energy intensity. A parameter in (III–5) can have a negative (positive) sign, even if the respective parameter in (III–3) is positive (negative).

APPENDIX IV: APPLICATION TO AGGREGATED DATA

a. Unweighted Aggregation

We assume the following simple linear relationship for n sectors:

(IV–10) $$y_{ti} = a_i + bx_{ti} + u_{ti} \qquad 1 \leq i \leq n\,.$$

For the aggregated values one obtains the same regression equation:

(IV–11) $$\sum_{i=1}^{n} y_{ti} = \sum_{i=1}^{n} a_i + b \sum_{i=1}^{n} x_{ti} + \sum_{i=1}^{n} u_{ti}\,.$$

It may now be asked whether one can aggregate nonlinear relationships in a similar way. Starting from equation (II–1), one can write for two sectors

(IV–12) $$y_{ti} \simeq ax_{ti}^{b} \qquad 1 \leq i \leq 2$$

For $b > 1$, (IV–12) may be represented as follows:

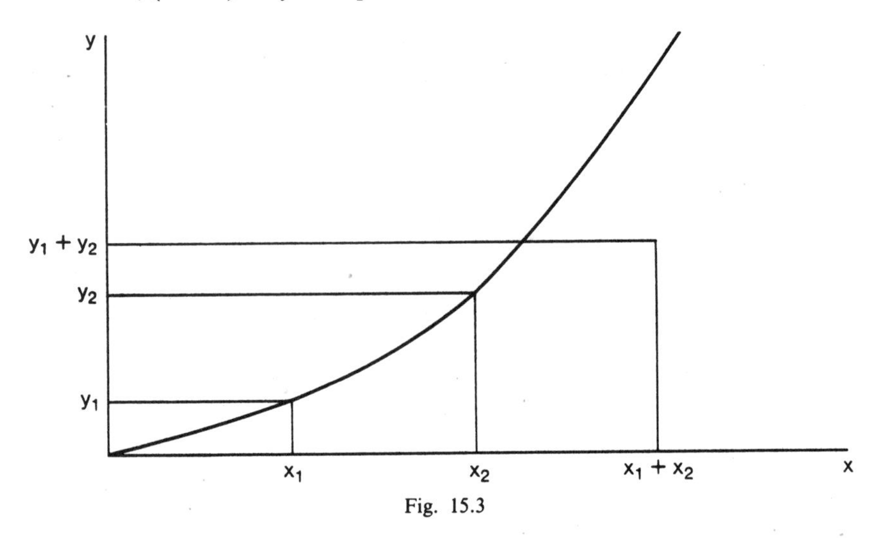

Fig. 15.3

Apparently, the following relationship does not hold:

(IV–13) $$y_{t1} + y_{t2} \simeq a(x_{t1} + x_{t2})^b.$$

Similar considerations apply in the cases $0 < b < 1$ and $b < 0$).

b. Weighted Aggregation

Let us start again from equation (IV–10), where y_{ti} and x_{ti} are now assumed to be ratios:

(IV–14) $$y_{ti} = \frac{a_{ti}}{b_{ti}} \text{ and } x_{ti} = \frac{c_{ti}}{d_{ti}} .$$

If we assume constant shares

(IV–15) $$\frac{b_{ti}}{\sum_i b_{ti}} = \beta_i \text{ and } \frac{d_{ti}}{\sum_i d_{ti}} = \delta_i ,$$

we obtain by aggregation

(IV–16) $$\sum_i y_{ti} = \sum_i \frac{a_{ti}}{b_{ti}} = \frac{\sum_i a_{ti} \frac{\sum_j b_{tj}}{b_{ti}}}{\sum_i b_{ti}} = \frac{\Sigma a_{ti} \frac{1}{\beta_i}}{\Sigma b_{ti}}$$

and

(IV–17) $$\Sigma x_{ti} = \sum_i \frac{c_{ti}}{d_{ti}} = \frac{\sum_i c_{ti} \frac{\sum_j d_{tj}}{d_{ti}}}{\sum_i d_{ti}} = \frac{\Sigma c_{ti} \frac{1}{\delta_i}}{\Sigma d_{ti}} .$$

By summation in (IV–10) we have

(IV–18) $$\Sigma y_{ti} = \frac{\Sigma a_{ti} \frac{1}{\beta_i}}{\Sigma b_{ti}} = \Sigma a_i + b \underbrace{\frac{\Sigma c_{ti} \frac{1}{\delta_i}}{\Sigma d_{ti}}}_{\Sigma x_{ti}} + \Sigma u_{ti} .$$

If the ratios y_{ti} and x_{ti} are aggregated, we obtain a linear relationship. One then has a weighted aggregation of the nominator divided by the aggregate of the denominator.

For the nonlinear case, we start again from equation (IV–12) under assumption (IV–14). At first, we aggregate numerator and denominator:

$$\text{(IV–19)} \qquad y_t = \frac{a_{t1} + a_{t2}}{b_{t1} + b_{t2}} = \frac{a_{t1}}{b_{t1}} \frac{b_{t1}}{b_{t1} + b_{t2}} + \frac{a_{t2}}{b_{t2}} \frac{b_{t2}}{b_{t1} + b_{t2}}$$

$$= y_{t1}\,\beta_1 + y_{t2}\beta_2$$

$$\text{(IV–20)} \qquad x_t = \frac{c_{t1} + c_{t2}}{d_{t1} + d_{t2}} = \frac{c_{t1}}{d_{t1}} \frac{d_{t1}}{d_{t1} + d_{t2}} + \frac{c_{t2}}{d_{t2}} \frac{d_{t2}}{d_{t1} + d_{t2}}$$

$$= x_{t1}\,\delta_1 + x_{t2}\delta_2$$

Assuming, for example, $\beta_1 = 1/4$, $\beta_2 = 3/4$ and $\delta_1 = 1/2$, $\delta_2 = 1/2$, we can draw for $b > 1$ the following figure:

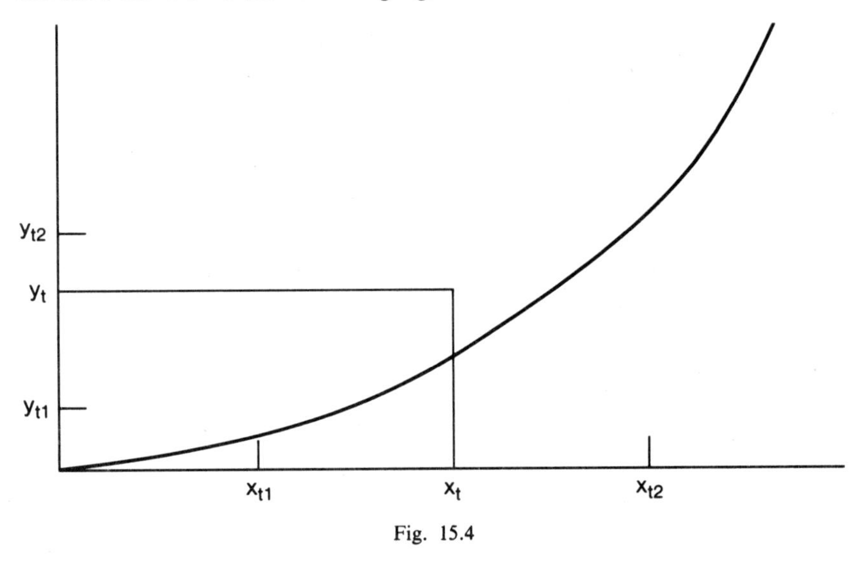

Fig. 15.4

In general, the aggregate will not behave like the individual sectors. The aggregated values can therefore not be expected to lie on the regression line. This will only accidentally be the case.

Now we aggregate, as in the linear case, the ratios

$$\text{(IV–21)} \qquad y_t = y_{t1} + y_{t2}$$

and

$$\text{(IV–22)} \qquad x_t = x_{t1} + x_{t2}.$$

Also in this case, the aggregate generally does not behave as the individual sector does:

(IV–23) $$y_{t1} + y_{t2} \simeq \alpha(x_{t1} + x_{t2})^b$$

For $b > 1$ we obtain the following figure:

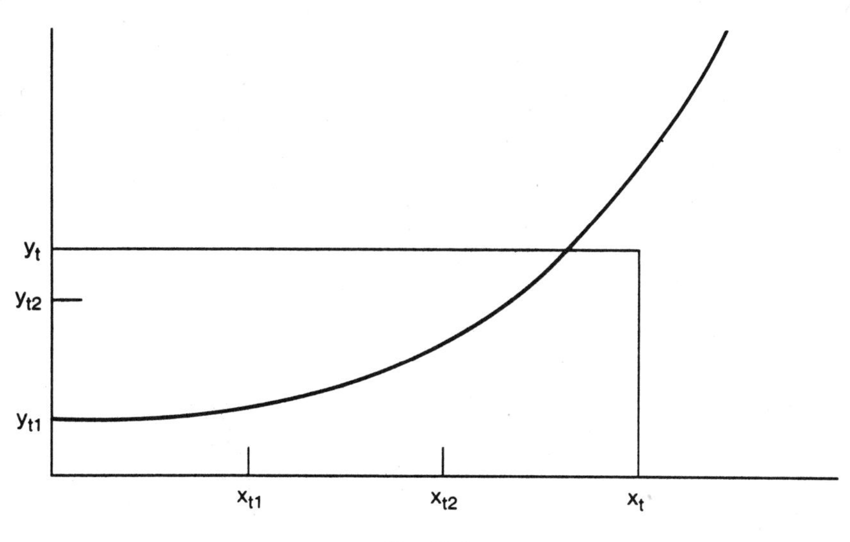

Fig. 15.5

NOTES

1. United Nations, Statistical Papers, Series J World Energy Supplies, 1977.
2. We owe this derivation to a note by Y. K. Wen of the World Bank.
3. For the aggregate of two observations we have

$$\frac{E_1 + E_2}{B} = \frac{E_1}{B_1} \cdot \frac{B_1}{B} + \frac{E_2}{B_2} \cdot \frac{B_2}{B}$$

and

$$\frac{X_1 + X_2}{Y} = \frac{X_1}{Y_1} \cdot \frac{Y_1}{Y} + \frac{X_2}{Y_2} \cdot \frac{Y_2}{Y}$$

Table 1

****************************** GROUP 1 ***********************

ESTIMATED PARAMETERS OF THE POOLED CROSS SECTION / TIME SERIES REGRESSION

REGRESSORS		COEFFICIENTS	T-VALUES
CONSTANT	:	2.7975	8.086
LN(GDP/H)	:	0.4013	7.835
LN(R4S)	:	-0.4081	-8.208
LN(R5S+R6S)	:	0.7424	7.970
LN(R7S+R8S)	:	0.6255	8.848
LN(R9S)	:	0.3136	2.410
LN(PR)	:	-0.5445	-3.979

R SQUARE : 0.9047

OBSERVATIONS : 117

F-VALUE : 174.0372

SUM (E-EEST)**2: 1113405849.340964
SUM E**2 : 268295040846.281231
SUM1 / SUM2 : 0.004150

COUNTRIES : CYPRUS
GREECE
MALTA
PORTUGAL
SPAIN
TURKEY
YUGOSLAVIA
ISRAEL

Table 2

****************************** GROUP 1 ************************

AGGREGATED DATA OF THE COUNTRY GROUP AND SUMMATION OF THE COUNTRY BY COUNTRY ESTIMATES

YEAR	E EST	E	E/B EST	E/B	B
1960	58192.893127	62814.226400	0.600986	0.648713	96829.000000
1961	65463.616247	68433.148800	0.668282	0.698597	97958.000000
1962	69564.553353	71092.153800	0.700126	0.715501	99360.000000
1963	78858.138405	78319.596400	0.782688	0.777343	100753.000000
1964	92591.235707	85291.297700	0.906424	0.834961	102150.000000
1965	95200.689717	91992.506700	0.916879	0.885981	103831.200000
1966	99267.926845	99983.488300	0.944837	0.951648	105063.500000
1967	107736.257629	107533.757400	1.011042	1.009142	106559.600000
1968	120617.357963	116299.663100	1.116597	1.076626	108022.300000
1969	137791.929351	126028.829500	1.258159	1.150752	109518.700000
1970	153154.942319	136867.940900	1.374023	1.227905	111464.600000
1971	158482.908074	155672.832400	1.402887	1.378013	112969.100000
1972	169324.275017	167044.726800	1.477785	1.457890	114579.800000
1973	176587.176694	183155.150700	1.519043	1.575542	116249.000000
1974	180136.252538	195436.974600	1.524653	1.654157	118149.000000
1975	180456.872227	197888.171100	1.502254	1.647364	120124.100000
1980	227867.163332	0.0	1.774666	0.0	128399.999998
1985	314578.795272	0.0	2.301235	0.0	136699.999999
1990	428159.159435	0.0	2.950787	0.0	145100.000002

YEAR	GDP/B	R4S	R5S+R6S	R7S+R8S	R9S	PR
1960	0.840340	0.246675	0.080303	0.266650	0.066947	1.206977
1961	0.901514	0.238939	0.081585	0.269784	0.065736	1.179772
1962	0.951047	0.238843	0.083697	0.265352	0.065762	1.132471
1963	1.024444	0.236494	0.088334	0.263704	0.065971	1.109161
1964	1.078463	0.211244	0.095841	0.271478	0.066201	1.123201
1965	1.125937	0.204660	0.094388	0.267537	0.066954	1.127512
1966	1.196142	0.205181	0.092408	0.261896	0.067362	1.094642
1967	1.232479	0.192988	0.092105	0.259884	0.067279	1.049962
1968	1.300297	0.181132	0.095780	0.261431	0.068856	1.044100
1969	1.387718	0.171436	0.099636	0.264085	0.070618	1.006785
1970	1.449649	0.162200	0.099784	0.267249	0.072875	1.000000
1971	1.533651	0.162298	0.096917	0.266030	0.073558	1.009692
1972	1.636671	0.156503	0.097336	0.271909	0.071819	1.023051
1973	1.729188	0.159547	0.095993	0.275895	0.069981	0.992947
1974	1.795303	0.151467	0.091511	0.284433	0.069165	1.052360
1975	1.806736	0.148407	0.088220	0.276715	0.069908	1.071196
1980	2.085592	0.138773	0.097311	0.263939	0.072997	1.062941
1985	2.556408	0.123001	0.103936	0.274065	0.072854	1.068345
1990	3.117974	0.110000	0.110818	0.284184	0.072270	1.073282

* = FOR THIS OBSERVATION DATA ARE MISSING

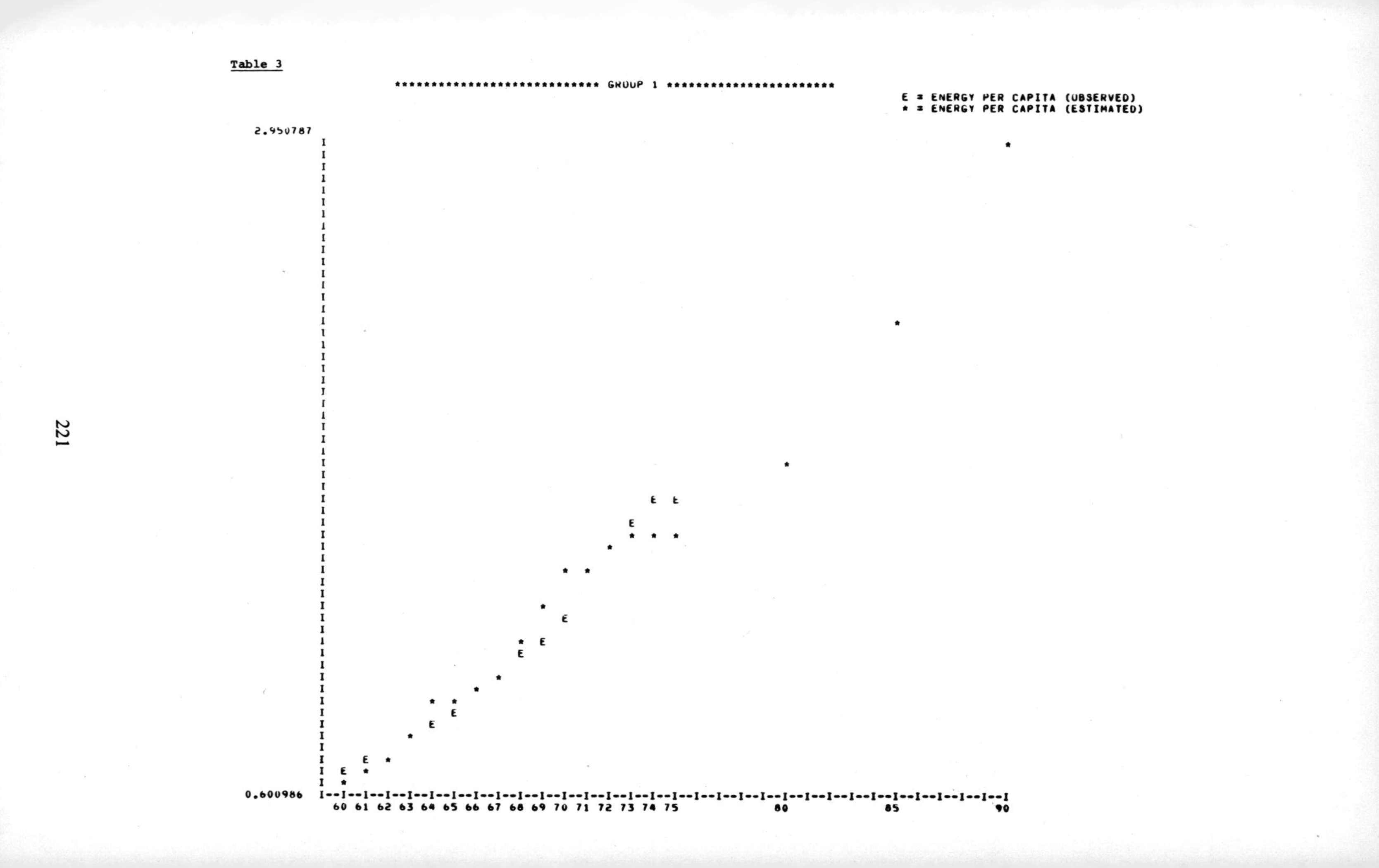

Table 3
GROUP 1
E = ENERGY PER CAPITA (OBSERVED)
* = ENERGY PER CAPITA (ESTIMATED)
2.950787
0.600986
60 61 62 63 64 65 66 67 68 69 70 71 72 73 74 75
80
85
90

Table 4

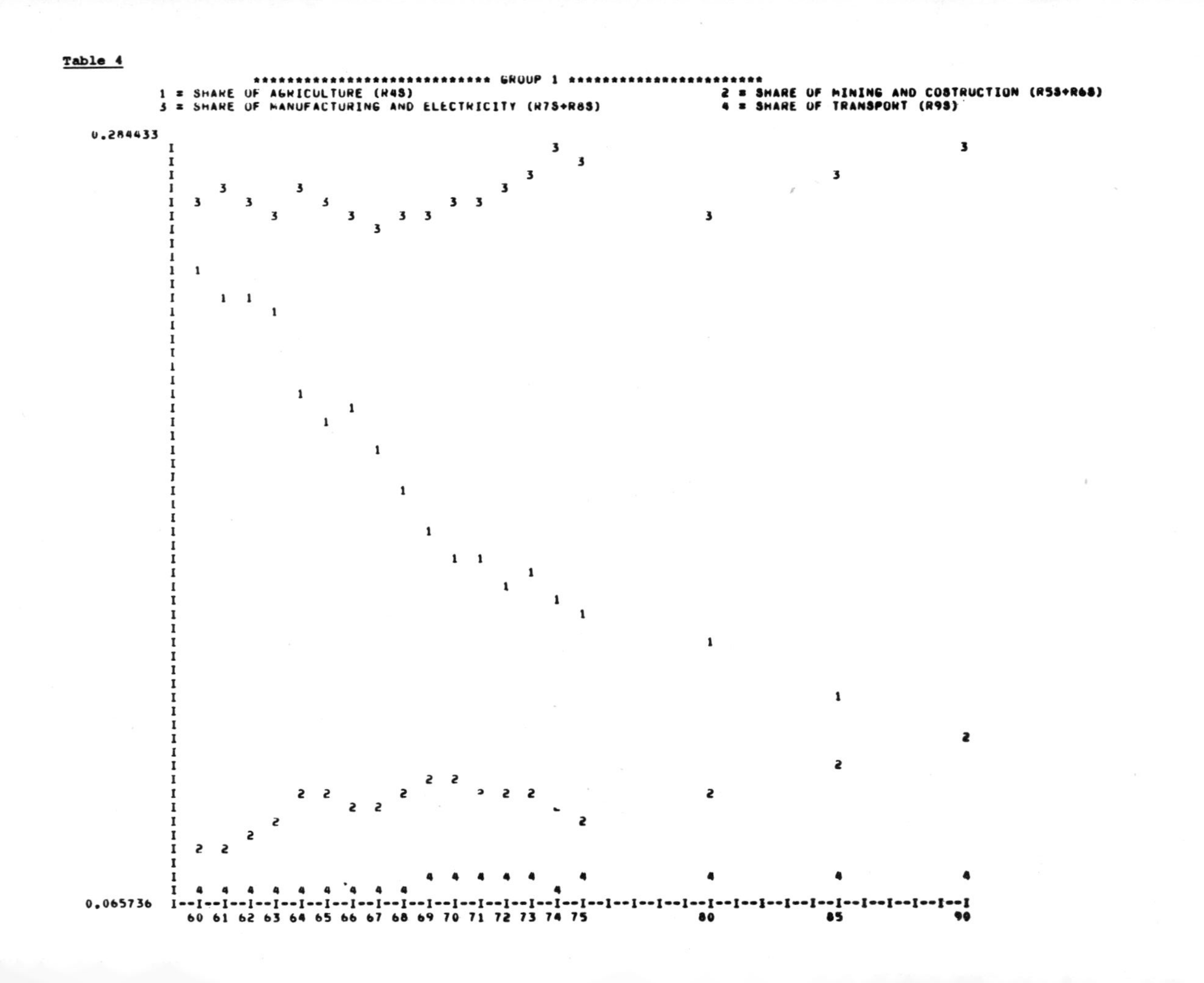

****************************** GROUP 1 ************************
1 = SHARE OF AGRICULTURE (R4S)
2 = SHARE OF MINING AND COSTRUCTION (R5S+R6S)
3 = SHARE OF MANUFACTURING AND ELECTRICITY (R7S+R8S)
4 = SHARE OF TRANSPORT (R9S)
0.284433
0.065736
60 61 62 63 64 65 66 67 68 69 70 71 72 73 74 75 80 85 90

Table 5

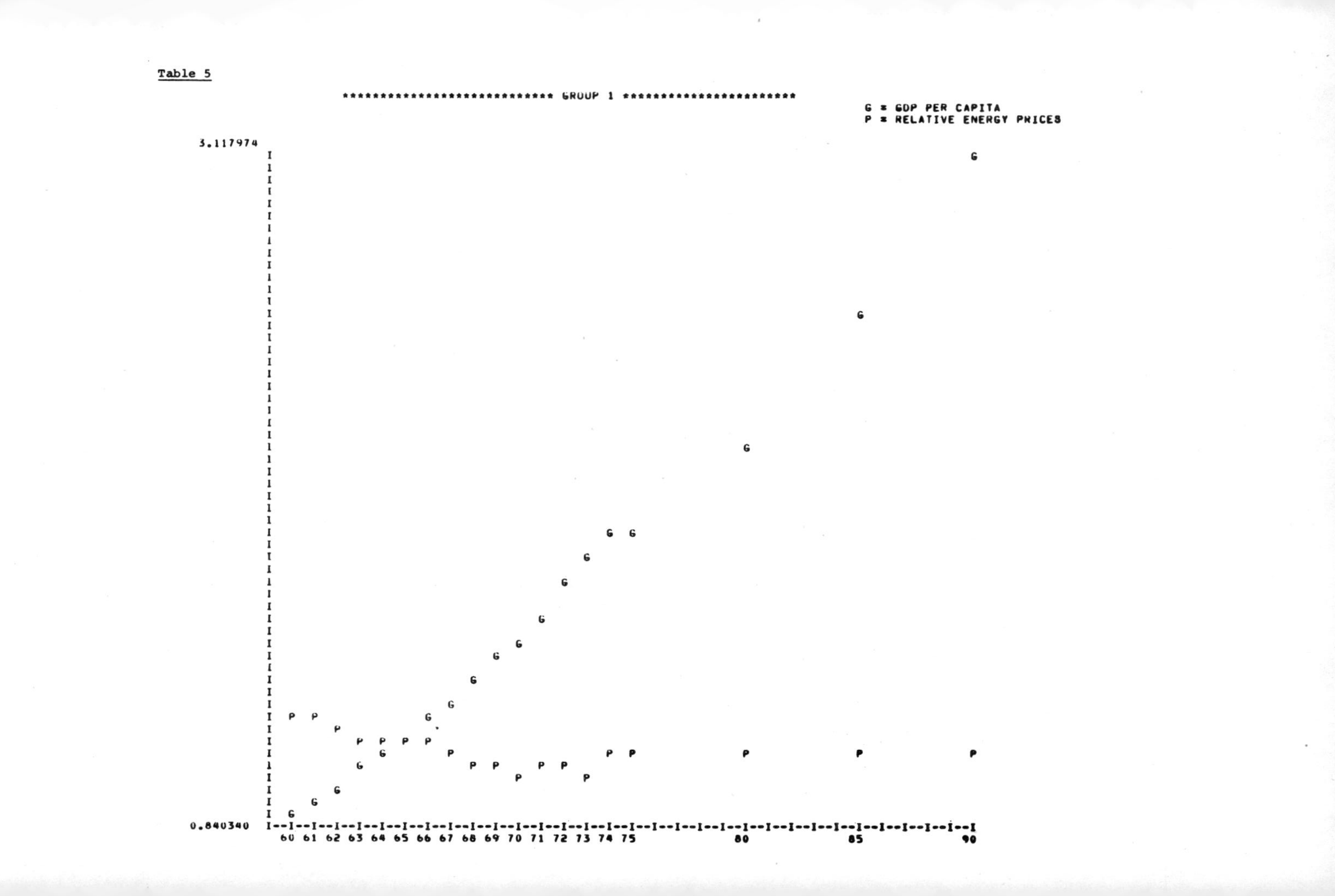

GROUP 1
G = GDP PER CAPITA
P = RELATIVE ENERGY PRICES
3.117974
0.840340
60 61 62 63 64 65 66 67 68 69 70 71 72 73 74 75 80 85 90

Table 1

****************************** GROUP 5B *********************

ESTIMATED PARAMETERS OF THE POOLED CROSS SECTION / TIME SERIES REGRESSION

REGRESSORS		COEFFICIENTS	T-VALUES
CONSTANT	:	-0.0561	-0.158
LN(GDP/B)	:	0.8941	10.137
LN(R4S)	:	-0.8225	-6.345
LN(R7S+R8S)	:	0.3739	3.209
LN(R9S)	:	0.2875	4.361
LN(PR)	:	-0.8858	-4.348

R SQUARE : 0.8895

OBSERVATIONS : 96

F-VALUE : 144.9445

SUM (E-EEST)**2: 1360553.256686
SUM E**2 : 354190658.631405
SUM1 / SUM2 : 0.003841

COUNTRIES : BURUNDI
ETHIOPIA
KENYA
MALAWI
RWANDA
TANZANIA,UNITED REP.
UGANDA
ZAIRE

Table 2

****************************** GROUP 5B **********************

AGGREGATED DATA OF THE COUNTRY GROUP AND SUMMATION OF THE COUNTRY BY COUNTRY ESTIMATES

YEAR	E EST	E	E/B EST	E/B	B
1960	0.0	3831.490600*	0.0	0.061799*	70858.000000
1961	0.0	3840.784300*	0.0	0.060362*	72718.700000
1962	0.0	4127.612000*	0.0	0.058225*	74534.700000
1963	0.0	4325.612000*	0.0	0.059534*	76397.500000
1964	0.0	4434.386300	0.0	0.056624	78313.300000
1965	0.0	4845.347500	0.0	0.060319	80328.200000
1966	0.0	5212.415500	0.0	0.063235	82428.800000
1967	5475.644587	5477.922800	0.064658	0.064685	84686.400000
1968	5565.940222	5802.750500	0.063997	0.066720	86971.900000
1969	5991.333238	6845.685000	0.067107	0.076677	89280.000000
1970	6991.125916	7154.833100	0.076217	0.078002	91726.100000
1971	8024.261385	7662.061300	0.085045	0.081207	94352.600000
1972	8518.183046	7878.316200	0.087827	0.081230	96988.300000
1973	8550.924354	8466.464300	0.085798	0.084950	99663.900000
1974	0.0	8557.784700	0.0	0.083548	102429.000000
1975	0.0	8650.158500	0.0	0.082132	105320.600000
1980	11244.358087	0.0	0.093469	0.0	120300.000000
1985	14578.783868	0.0	0.105643	0.0	137999.999998
1990	19124.507543	0.0	0.120583	0.0	158600.000001

YEAR	GDP/B	R4S	R5S+R6S	R7S+R8S	R9S	PR
1960	0.116852	0.488578*	0.066591*	0.096311*	0.055933*	0.914882*
1961	0.110001	0.484813*	0.060375*	0.096448*	0.053826*	1.110619*
1962	0.119000	0.469186*	0.057811*	0.096748*	0.053528*	0.894433*
1963	0.123149	0.465626*	0.053378*	0.098855*	0.052781*	0.852246*
1964	0.125650	0.454927*	0.054905*	0.103816*	0.052852*	1.007298*
1965	0.127070	0.443477	0.058825	0.111583	0.053961	0.957637*
1966	0.133884	0.453659	0.056815	0.111461	0.055508	1.006235*
1967	0.135437	0.444061	0.060982	0.114552	0.056697	1.063230
1968	0.137414	0.428580	0.093230	0.098448	0.058377	0.987082
1969	0.143417	0.416517	0.105253	0.100455	0.056924	1.008807
1970	0.145557	0.426599	0.111847	0.112221	0.064455	1.000000
1971	0.149510	0.413450	0.094621	0.113842	0.065562	0.948832
1972	0.150176	0.400363	0.091532	0.113152	0.065962	0.931487
1973	0.152537	0.399504	0.113026	0.110395	0.066918	0.925568
1974	0.153427	0.383578	0.114382	0.111097	0.069879	0.935171*
1975	0.149655	0.380938*	0.088895*	0.116399*	0.075070*	0.963226*
1980	0.157002	0.381307	0.112681	0.112681	0.070799	0.895028
1985	0.166678	0.353999	0.106399	0.117599	0.075961	0.896657
1990	0.175815	0.331004	0.110656	0.135246	0.076158	0.897974

* = FOR THIS OBSERVATION DATA ARE MISSING

Table 3

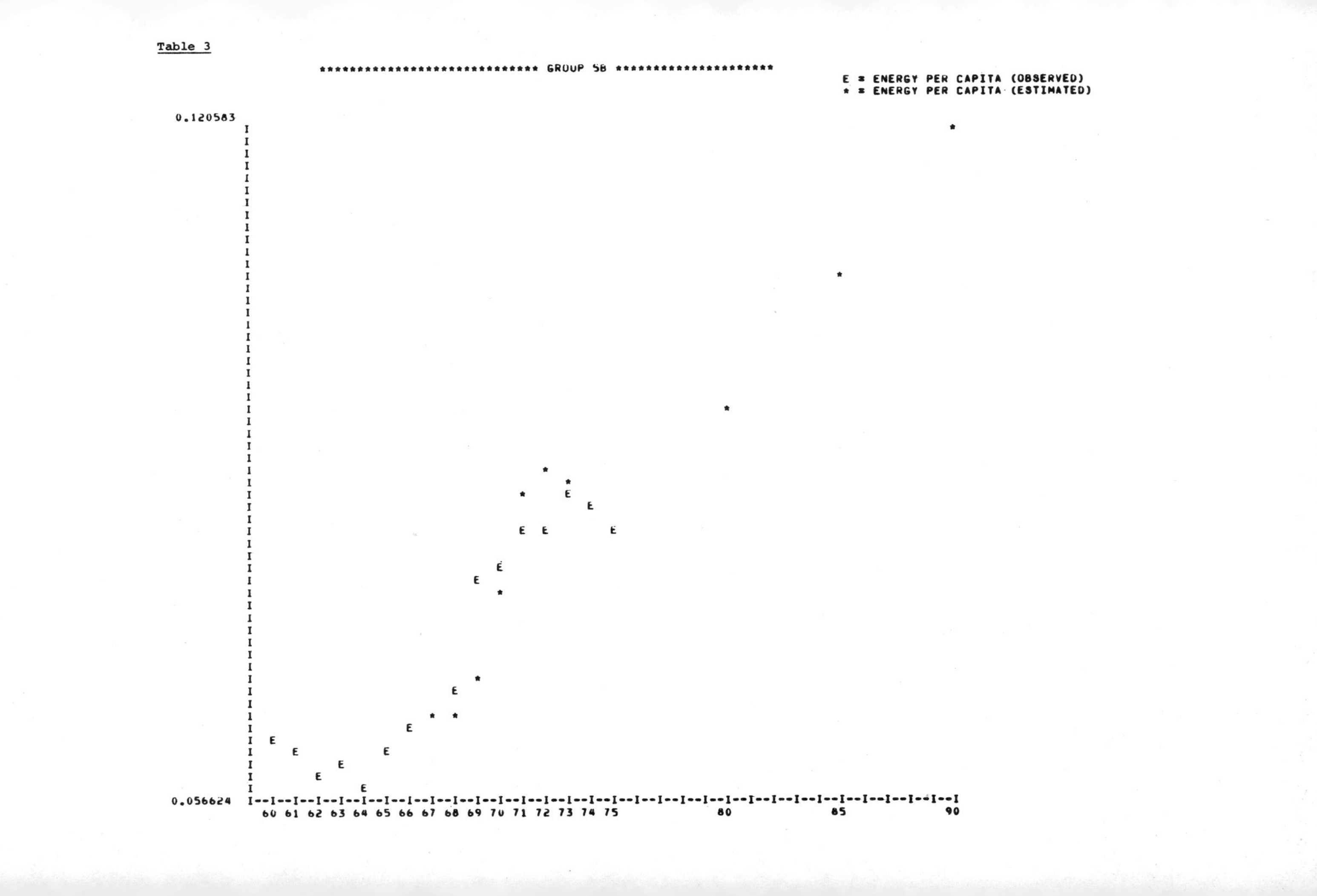
******************************* GROUP 5B **********************
E = ENERGY PER CAPITA (OBSERVED)
* = ENERGY PER CAPITA (ESTIMATED)
0.120583
0.056624
60 61 62 63 64 65 66 67 68 69 70 71 72 73 74 75 80 85 90

Table 4

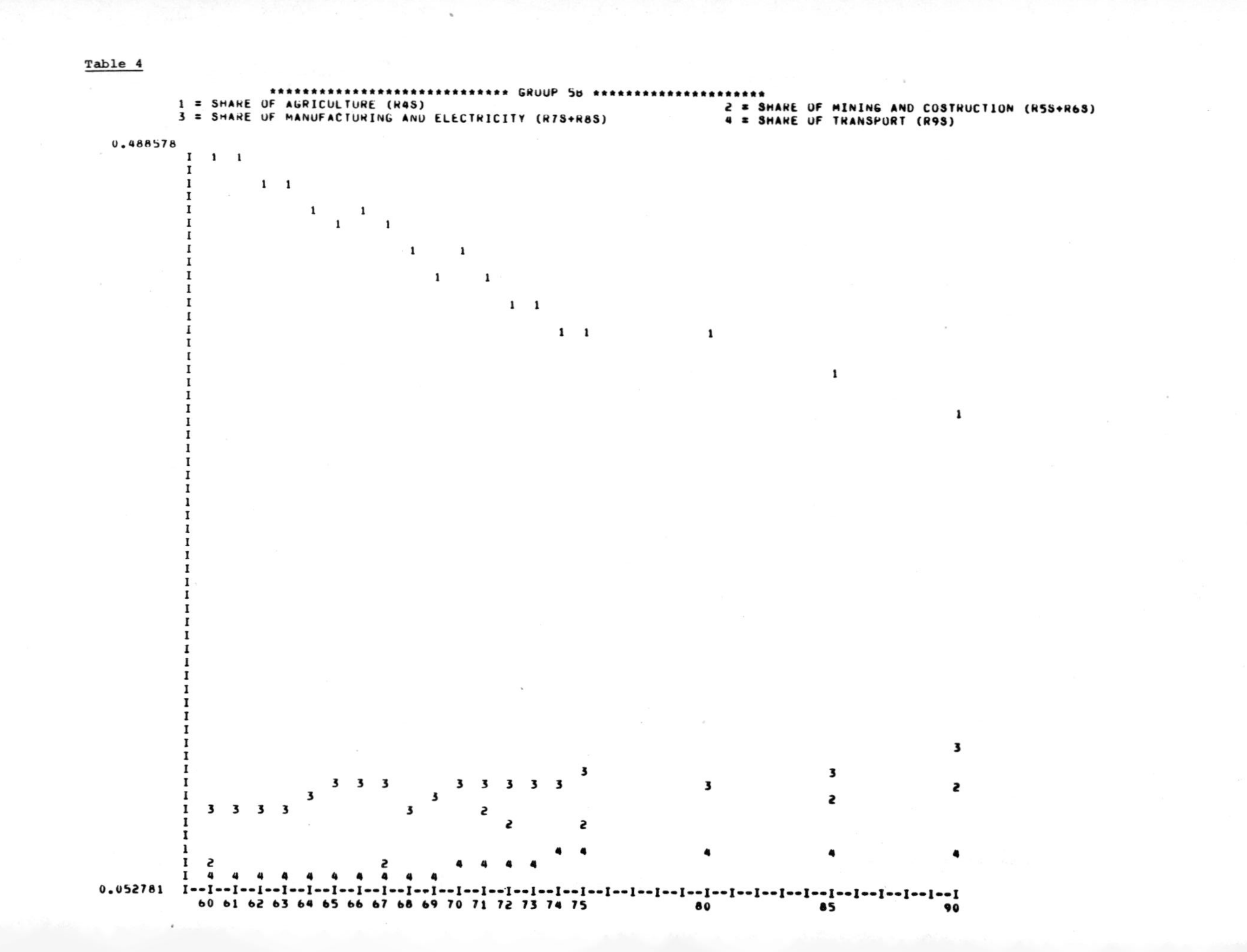

****************************** GROUP 5B **********************
1 = SHARE OF AGRICULTURE (R4S)
2 = SHARE OF MINING AND COSTRUCTION (R5S+R6S)
3 = SHARE OF MANUFACTURING AND ELECTRICITY (R7S+R8S)
4 = SHARE OF TRANSPORT (R9S)
0.488578
0.052781
60 61 62 63 64 65 66 67 68 69 70 71 72 73 74 75
80
85
90

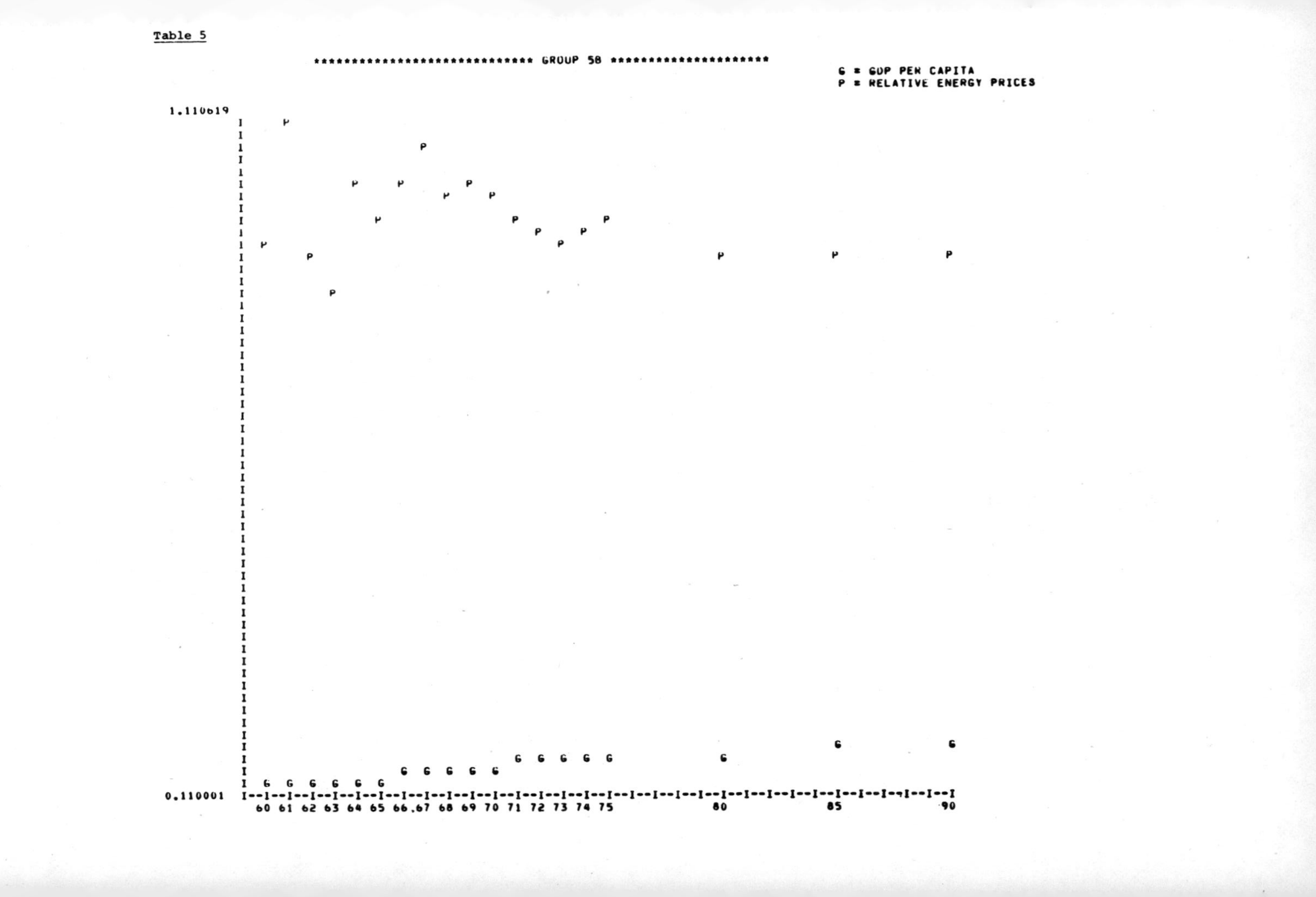
Table 5
GROUP 58
G = GDP PER CAPITA
P = RELATIVE ENERGY PRICES
1.110619
0.110001
60 61 62 63 64 65 66 67 68 69 70 71 72 73 74 75
80
85
90

Table 2

****************************** GROUP 1 ************************

AGGREGATED DATA OF THE COUNTRY GROUP AND SUMMATION OF THE COUNTRY BY COUNTRY ESTIMATES

YEAR	E EST	E	E/B EST	E/B	B
1960	58192.893127	62814.226400	0.600986	0.648713	96829.000000
1961	65463.616247	68433.148800	0.668282	0.698597	97958.000000
1962	69564.553353	71092.153800	0.700126	0.715501	99360.000000
1963	78858.138405	78319.596400	0.782688	0.777343	100753.000000
1964	92591.235707	85291.297700	0.906424	0.834961	102150.000000
1965	95200.689717	91992.506700	0.916879	0.885981	103831.200000
1966	99267.926845	99983.488300	0.944837	0.951648	105063.500000
1967	107736.257629	107533.757400	1.011042	1.009142	106559.600000
1968	120617.357963	116299.663100	1.116597	1.076626	108022.300000
1969	137791.929351	126028.829500	1.258159	1.150752	109518.700000
1970	153154.942319	136867.940900	1.374023	1.227905	111464.600000
1971	158482.908074	155672.832400	1.402887	1.378013	112969.100000
1972	169324.275017	167044.726800	1.477785	1.457890	114579.800000
1973	176587.176694	183155.150700	1.519043	1.575542	116249.000000
1974	180136.252538	195436.974600	1.524653	1.654157	118149.000000
1975	180456.872227	197888.171100	1.502254	1.647364	120124.100000
1980	216343.302089	0.0	1.684917	0.0	128399.999998
1985	272930.085885	0.0	1.996562	0.0	136699.999999
1990	344253.312868	0.0	2.372525	0.0	145100.000002

YEAR	GDP/B	R4S	R5S+R6S	R7S+R8S	R9S	PR
1960	0.840340	0.246675	0.080303	0.266650	0.066947	1.206977
1961	0.901514	0.238939	0.081585	0.269784	0.065736	1.179772
1962	0.951047	0.238843	0.083697	0.265352	0.065762	1.132471
1963	1.024444	0.236494	0.088334	0.263704	0.065971	1.109161
1964	1.078463	0.211244	0.095841	0.271478	0.066201	1.123201
1965	1.125937	0.204660	0.094388	0.267537	0.066954	1.127512
1966	1.196142	0.205181	0.092408	0.261896	0.067362	1.094642
1967	1.232479	0.192988	0.092105	0.259884	0.067279	1.049962
1968	1.300297	0.181132	0.095780	0.261431	0.068856	1.044100
1969	1.387718	0.171436	0.099636	0.264085	0.070618	1.006785
1970	1.449649	0.162200	0.099784	0.267249	0.072875	1.000000
1971	1.533651	0.162298	0.096917	0.266030	0.073558	1.009692
1972	1.636671	0.156503	0.097336	0.271909	0.071819	1.023051
1973	1.729188	0.159547	0.095993	0.275895	0.069981	0.992947
1974	1.795303	0.151467	0.091511	0.284433	0.069165	1.052360
1975	1.806736	0.148407	0.088220	0.276715	0.069908	1.071196
1980	2.085592	0.138773	0.097311	0.263939	0.072997	1.169235
1985	2.556408	0.123001	0.103936	0.274065	0.072854	1.386712
1990	3.117974	0.110000	0.110818	0.284184	0.072270	1.602083

* = FOR THIS OBSERVATION DATA ARE MISSING

Table 2

******************************* GROUP 5B **********************

AGGREGATED DATA OF THE COUNTRY GROUP AND SUMMATION OF THE COUNTRY BY COUNTRY ESTIMATES

YEAR	E EST	E	E/B EST	E/B	B
1960	0.0	3831.490600*	0.0	0.061799*	70858.000000
1961	0.0	3840.784300*	0.0	0.060362*	72718.700000
1962	0.0	4127.612000*	0.0	0.058225*	74534.700000
1963	0.0	4325.612000*	0.0	0.059534*	76397.500000
1964	0.0	4434.386300	0.0	0.056624	78313.300000
1965	0.0	4845.347500	0.0	0.060319	80328.200000
1966	0.0	5212.415500	0.0	0.063235	82428.800000
1967	5475.644587	5477.922800	0.064658	0.064685	84686.400000
1968	5565.940222	5802.750500	0.063997	0.066720	86971.900000
1969	5991.333238	6845.685000	0.067107	0.076677	89280.000000
1970	6991.125916	7154.833100	0.076217	0.078002	91726.100000
1971	8024.261385	7662.061300	0.085045	0.081207	94352.600000
1972	8518.183046	7878.316200	0.087827	0.081230	96988.300000
1973	8550.924354	8466.464300	0.085798	0.084950	99663.900000
1974	0.0	8557.784700	0.0	0.083548	102429.000000
1975	0.0	8650.158500	0.0	0.082132	105320.600000
1980	10334.012294	0.0	0.085902	0.0	120300.000000
1985	11633.775804	0.0	0.084303	0.0	137999.999998
1990	13491.665129	0.0	0.085067	0.0	158600.000001

YEAR	GDP/B	R4S	R5S+R6S	R7S+R8S	R9S	PR
1960	0.116852	0.488578*	0.066591*	0.096311*	0.055933*	0.914882*
1961	0.110001	0.484813*	0.060375*	0.096448*	0.053826*	1.110619*
1962	0.119000	0.469186*	0.057811*	0.096748*	0.053528*	0.894433*
1963	0.123149	0.465626*	0.053378*	0.098855*	0.052781*	0.852246*
1964	0.125650	0.454927*	0.054905*	0.103816*	0.052852*	1.007298*
1965	0.127070	0.443477	0.058825	0.111583	0.053961	0.957637*
1966	0.133884	0.453659	0.056815	0.111461	0.055508	1.006235*
1967	0.135437	0.444061	0.060982	0.114552	0.056697	1.063230
1968	0.137414	0.428580	0.093230	0.098448	0.058377	0.987082
1969	0.143417	0.416517	0.105253	0.100455	0.056924	1.008807
1970	0.145557	0.426599	0.111847	0.112221	0.064455	1.000000
1971	0.149510	0.413450	0.094621	0.113842	0.065562	0.948832
1972	0.150176	0.400363	0.091532	0.113152	0.065962	0.931487
1973	0.152537	0.399504	0.113026	0.110395	0.066918	0.925568
1974	0.153427	0.383578	0.114382	0.111097	0.069879	0.935171*
1975	0.149655	0.380938*	0.088895*	0.116399*	0.075070*	0.963226*
1980	0.157002	0.381307	0.112681	0.112681	0.070799	0.984531
1985	0.166678	0.353999	0.106399	0.117599	0.075961	1.156834
1990	0.175815	0.331004	0.110656	0.135246	0.076158	1.331480

* = FOR THIS OBSERVATION DATA ARE MISSING

Table 2

****************************** GROUP 1 ************************

AGGREGATED DATA OF THE COUNTRY GROUP AND SUMMATION OF THE COUNTRY BY COUNTRY ESTIMATES

YEAR	E EST	E	E/B EST	E/B	B
1960	58192.893127	62814.226400	0.600986	0.648713	96829.000000
1961	65463.616247	68433.148800	0.668282	0.698597	97958.000000
1962	69564.553353	71092.153800	0.700126	0.715501	99360.000000
1963	78858.138405	78319.596400	0.782688	0.777343	100753.000000
1964	92591.235707	85291.297700	0.906424	0.834961	102150.000000
1965	95200.689717	91992.506700	0.916879	0.885981	103831.200000
1966	99267.926845	99983.488300	0.944837	0.951648	105063.500000
1967	107736.257629	107533.757400	1.011042	1.009142	106559.600000
1968	120617.357963	116299.663100	1.116597	1.076626	108022.300000
1969	137791.929351	126028.829500	1.258159	1.150752	109518.700000
1970	153154.942319	136867.940900	1.374023	1.227905	111464.600000
1971	158482.908074	155672.832400	1.402887	1.378013	112969.100000
1972	169324.275017	167044.726800	1.477785	1.457890	114579.800000
1973	176587.176694	183155.150700	1.519043	1.575542	116249.000000
1974	180136.252538	195436.974600	1.524653	1.654157	118149.000000
1975	180456.872227	197888.171100	1.502254	1.647364	120124.100000
1980	255692.696100	0.0	1.991376	0.0	128399.999998
1985	370145.847255	0.0	2.707724	0.0	136699.999999
1990	538510.231279	0.0	3.711304	0.0	145100.000002

YEAR	GDP/B	R4S	R5S+R6S	R7S+R8S	R9S	PR
1960	0.840340	0.246675	0.080303	0.266650	0.066947	1.206977
1961	0.901514	0.238939	0.081585	0.269784	0.065736	1.179772
1962	0.951047	0.238843	0.083697	0.265352	0.065762	1.132471
1963	1.024444	0.236494	0.088334	0.263704	0.065971	1.109161
1964	1.078463	0.211244	0.095841	0.271478	0.066201	1.123201
1965	1.125937	0.204660	0.094388	0.267537	0.066954	1.127512
1966	1.196142	0.205181	0.092408	0.261896	0.067362	1.094642
1967	1.232479	0.192988	0.092105	0.259884	0.067279	1.049962
1968	1.300297	0.181132	0.095780	0.261431	0.068856	1.044100
1969	1.387718	0.171436	0.099636	0.264085	0.070618	1.006785
1970	1.449649	0.162200	0.099784	0.267249	0.072875	1.000000
1971	1.533651	0.162298	0.096917	0.266030	0.073558	1.009692
1972	1.636671	0.156503	0.097336	0.271909	0.071819	1.023051
1973	1.729188	0.159547	0.095993	0.275895	0.069981	0.992947
1974	1.795303	0.151467	0.091511	0.284433	0.069165	1.052360
1975	1.806736	0.148407	0.088220	0.276715	0.069908	1.071196
1980	2.085592	0.120001	0.096827	0.290480	0.072919	1.062941
1985	2.556408	0.100000	0.104518	0.309416	0.072424	1.068345
1990	3.117974	0.079999	0.112409	0.328409	0.071877	1.073282

* = FOR THIS OBSERVATION DATA ARE MISSING

Table 2

******************************** GROUP 5B **********************

AGGREGATED DATA OF THE COUNTRY GROUP AND SUMMATION OF THE COUNTRY BY COUNTRY ESTIMATES

YEAR	E EST	E	E/B EST	E/B	B
1960	0.0	3831.490600*	0.0	0.061799*	70858.000000
1961	0.0	3840.784300*	0.0	0.060362*	72718.700000
1962	0.0	4127.612000*	0.0	0.058225*	74534.700000
1963	0.0	4325.612000*	0.0	0.059534*	76397.500000
1964	0.0	4434.386300	0.0	0.056624	78313.300000
1965	0.0	4845.347500	0.0	0.060319	80328.200000
1966	0.0	5212.415500	0.0	0.063235	82428.800000
1967	5475.644587	5477.922800	0.064658	0.064685	84686.400000
1968	5565.940222	5802.750500	0.063997	0.066720	86971.900000
1969	5991.333238	6845.685000	0.067107	0.076677	89280.000000
1970	6991.125916	7154.833100	0.076217	0.078002	91726.100000
1971	8024.261385	7662.061300	0.085045	0.081207	94352.600000
1972	8518.183046	7878.316200	0.087827	0.081230	96988.300000
1973	8550.924354	8466.464300	0.085798	0.084950	99663.900000
1974	0.0	8557.784700	0.0	0.083548	102429.000000
1975	0.0	8650.158500	0.0	0.082132	105320.600000
1980	13372.550251	0.0	0.111160	0.0	120300.000000
1985	17352.454063	0.0	0.125742	0.0	137999.999998
1990	22654.296648	0.0	0.142839	0.0	158600.000001

YEAR	GDP/B	R4S	R5S+R6S	R7S+R8S	R9S	PR
1960	0.116852	0.488578*	0.066591*	0.096311*	0.055933*	0.914882*
1961	0.110001	0.484813*	0.060375*	0.096448*	0.053826*	1.110619*
1962	0.119000	0.469186*	0.057811*	0.096748*	0.053528*	0.894433*
1963	0.123149	0.465626*	0.053378*	0.098855*	0.052781*	0.852246*
1964	0.125650	0.454927*	0.054905*	0.103816*	0.052852*	1.007298*
1965	0.127070	0.443477	0.058825	0.111583	0.053961	0.957637*
1966	0.133884	0.453659	0.056815	0.111461	0.055508	1.006235*
1967	0.135437	0.444061	0.060982	0.114552	0.056697	1.063230
1968	0.137414	0.428580	0.093230	0.098448	0.058377	0.987082
1969	0.143417	0.416517	0.105253	0.100455	0.056924	1.008807
1970	0.145557	0.426599	0.111847	0.112221	0.064455	1.000000
1971	0.149510	0.413450	0.094621	0.113842	0.065562	0.948832
1972	0.150176	0.400363	0.091532	0.113152	0.065962	0.931487
1973	0.152537	0.399504	0.113026	0.110395	0.066918	0.925568
1974	0.153427	0.383578	0.114382	0.111097	0.069879	0.935171*
1975	0.149655	0.380938*	0.088895*	0.116399*	0.075070*	0.963226*
1980	0.157002	0.351308	0.112779	0.152583	0.068999	0.895029
1985	0.166678	0.324001	0.106918	0.157078	0.074161	0.896657
1990	0.175815	0.301005	0.110073	0.175830	0.074357	0.897974

* = FOR THIS OBSERVATION DATA ARE MISSING

****************************** GROUP 1 ************************

AGGREGATED DATA OF THE COUNTRY GROUP AND SUMMATION OF THE COUNTRY BY COUNTRY ESTIMATES

YEAR	E EST	E	E/B EST	E/B	B
1960	58192.893127	62814.226400	0.600986	0.648713	96829.000000
1961	65463.616247	68433.148800	0.668282	0.698597	97958.000000
1962	69564.553353	71092.153800	0.700126	0.715501	99360.000000
1963	78858.138405	78319.596400	0.782688	0.777343	100753.000000
1964	92591.235707	85291.297700	0.906424	0.834961	102150.000000
1965	95200.689717	91992.506700	0.916879	0.885981	103831.200000
1966	99267.926845	99983.488300	0.944837	0.951648	105063.500000
1967	107736.257629	107533.757400	1.011042	1.009142	106559.600000
1968	120617.357963	116299.663100	1.116597	1.076626	108022.300000
1969	137791.929351	126028.829500	1.258159	1.150752	109518.700000
1970	153154.942319	136867.940900	1.374023	1.227905	111464.600000
1971	158482.908074	155672.832400	1.402887	1.378013	112969.100000
1972	169324.275017	167044.726800	1.477785	1.457890	114579.800000
1973	176587.176694	183155.150700	1.519043	1.575542	116249.000000
1974	180136.252538	195436.974600	1.524653	1.654157	118149.000000
1975	180456.872227	197888.171100	1.502254	1.647364	120124.100000
1980	242761.622102	0.0	1.890667	0.0	128399.999998
1985	321140.329225	0.0	2.349234	0.0	136699.999999
1990	432979.015034	0.0	2.984004	0.0	145100.000002

YEAR	GDP/B	R4S	R5S+R6S	R7S+R8S	R9S	PR
1960	0.840340	0.246675	0.080303	0.266650	0.066947	1.206977
1961	0.901514	0.238939	0.081585	0.269784	0.065736	1.179772
1962	0.951047	0.238843	0.083697	0.265352	0.065762	1.132471
1963	1.024444	0.236494	0.088334	0.263704	0.065971	1.109161
1964	1.078463	0.211244	0.095841	0.271478	0.066201	1.123201
1965	1.125937	0.204660	0.094388	0.267537	0.066954	1.127512
1966	1.196142	0.205181	0.092408	0.261896	0.067362	1.094642
1967	1.232479	0.192988	0.092105	0.259884	0.067279	1.049962
1968	1.300297	0.181132	0.095780	0.261431	0.068856	1.044100
1969	1.387718	0.171436	0.099636	0.264085	0.070618	1.006785
1970	1.449649	0.162200	0.099784	0.267249	0.072875	1.000000
1971	1.533651	0.162298	0.096917	0.266030	0.073558	1.009692
1972	1.636671	0.156503	0.097336	0.271909	0.071819	1.023051
1973	1.729188	0.159547	0.095993	0.275895	0.069981	0.992947
1974	1.795303	0.151467	0.091511	0.284433	0.069165	1.052360
1975	1.806736	0.148407	0.088220	0.276715	0.069908	1.071196
1980	2.085592	0.120001	0.096827	0.290480	0.072919	1.169235
1985	2.556408	0.100000	0.104518	0.309416	0.072424	1.386712
1990	3.117974	0.079999	0.112409	0.328409	0.071877	1.602082

* = FOR THIS OBSERVATION DATA ARE MISSING

Table 2

****************************** GROUP 5B **********************

AGGREGATED DATA OF THE COUNTRY GROUP AND SUMMATION OF THE COUNTRY BY COUNTRY ESTIMATES

YEAR	E EST	E	E/B EST	E/B	B
1960	0.0	3831.490600*	0.0	0.061799*	70858.000000
1961	0.0	3840.784300*	0.0	0.060362*	72718.700000
1962	0.0	4127.612000*	0.0	0.058225*	74534.700000
1963	0.0	4325.612000*	0.0	0.059534*	76397.500000
1964	0.0	4434.386300	0.0	0.056624	78313.300000
1965	0.0	4845.347500	0.0	0.060319	80328.200000
1966	0.0	5212.415500	0.0	0.063235	82428.800000
1967	5475.644587	5477.922800	0.064658	0.064685	84686.400000
1968	5565.940222	5802.750500	0.063997	0.066720	86971.900000
1969	5991.333238	6845.685000	0.067107	0.076677	89280.000000
1970	6991.125916	7154.833100	0.076217	0.078002	91726.100000
1971	8024.261385	7662.061300	0.085045	0.081207	94352.600000
1972	8518.183046	7878.316200	0.087827	0.081230	96988.300000
1973	8550.924354	8466.464300	0.085798	0.084950	99663.900000
1974	0.0	8557.784700	0.0	0.083548	102429.000000
1975	0.0	8650.158500	0.0	0.082132	105320.600000
1980	12289.905535	0.0	0.102160	0.0	120300.000000
1985	13847.146839	0.0	0.100342	0.0	137999.999998
1990	15981.806478	0.0	0.100768	0.0	158600.000001

YEAR	GDP/B	R4S	R5S+R6S	R7S+R8S	R9S	PR
1960	0.116852	0.488578*	0.066591*	0.096311*	0.055933*	0.914882*
1961	0.110001	0.484813*	0.060375*	0.096448*	0.053826*	1.110619*
1962	0.119000	0.469186*	0.057811*	0.096748*	0.053528*	0.894433*
1963	0.123149	0.465626*	0.053378*	0.098855*	0.052781*	0.852246*
1964	0.125650	0.454927*	0.054905*	0.103816*	0.052852*	1.007298*
1965	0.127070	0.443477	0.058825	0.111583	0.053961	0.957637*
1966	0.133884	0.453659	0.056815	0.111461	0.055508	1.006235*
1967	0.135437	0.444061	0.060982	0.114552	0.056697	1.063230
1968	0.137414	0.428580	0.093230	0.098448	0.058377	0.987082
1969	0.143417	0.416517	0.105253	0.100455	0.056924	1.008807
1970	0.145557	0.426599	0.111847	0.112221	0.064455	1.000000
1971	0.149510	0.413450	0.094621	0.113842	0.065562	0.948832
1972	0.150176	0.400363	0.091532	0.113152	0.065962	0.931487
1973	0.152537	0.399504	0.113026	0.110395	0.066918	0.925568
1974	0.153427	0.383578	0.114382	0.111097	0.069879	0.935171*
1975	0.149655	0.380938*	0.088895*	0.116399*	0.075070*	0.963226*
1980	0.157002	0.351308	0.112779	0.152583	0.068999	0.984532
1985	0.166678	0.324001	0.106918	0.157078	0.074161	1.156834
1990	0.175815	0.301005	0.110073	0.175830	0.074357	1.331479

* = FOR THIS OBSERVATION DATA ARE MISSING

Chapter 16
The Demand for Oil and Energy in Developing Countries

Charles Wolf, Jr., Daniel A. Relles, and Jaime Navarro

How large will the oil and energy demand of the non-OPEC less developed countries (NOLDCs) be in the next decade? Will it be small and relatively insignificant in relation to world demand, or large and relatively important? How will demand be affected by the economic growth of the NOLDCs?

This report tries to develop a reasonable range of NOLDC demand forecasts, concentrating on the demand for oil, but also devoting some attention to the total demand for commercial energy. It tries to be explicit about the uncertainties associated with the forecasts, and with the income and price elasticities on which they are based. Finally, we wish to consider the implications of these estimates for U.S. policies concerning NOLDCs, North-South relations, the world oil market and for future research on energy issues relating to NOLDCs.

In 1976, total commercial energy consumption (including oil, gas, coal, and primary electricity) by all NOLDCs amounted to 9.3 million barrels per day (mbd) in oil equivalent, or about 11.5 percent of global consumption, excluding the centrally planned economies. Oil consumption (excluding petrochemical feedstocks) by all NOLDCs in 1976 was 5.6 mbd, or about 14 percent of the global figure. Oil imports by NOLDCs in 1976 were about 15 percent of world imports.

Our forecasts of NOLDC demand for the next decade cover an extremely wide range:

		Previous Forecasts
• Oil consumption:	5.7–17.4 mbd	10.0–13.5 mbd
• Share of world consumption:	7.8–32%	14.4–24.5%
• Share of oil imports:	8.3–34.6%	15.5–26.5%

There is more than a three-fold difference between our minimum and maximum estimates of NOLDC oil demand in 1990—between 5.73 and 17.45 mbd. As a share of world oil demand, these figures correspond, respectively, to a minimum between 7.8 percent and 10.6 percent, and a maximum between 23.5 percent and 32.2 percent.* The circumstances under which the extremes of the interval would occur are, admittedly, quite unlikely, especially for the lower end of the interval. As a share of forecasted world imports of oil in 1990, the NOLDC proportion may be as small as 8.3 percent, or as large as 34.6 percent, assuming that the 1976 relationship between the oil imports and oil consumption still prevail in 1990.

The wide range of these forecasts for 1990 depends in part on the several different scenarios that we have assumed for NOLDC growth in real income (3 percent, 5 percent, or 7 percent per year), and oil prices (increasing at 3 percent, or 5 percent per year in real terms).

	3%, 5%	5%, 5%	7%, 3%
• Oil consumption:	5.7–15.3 mbd	6.4–16.0 mbd	7.5–17.4 mbd
• Share of world demand:	8.0–21%	9–22%	10–32%
• Share of oil imports:	8.2–30.5%	9.3–31.8%	10.9–34.6%

The range of the estimates narrows somewhat if we confine our attention to a single scenario, one that perhaps may be considered the most reasonable and likely case—namely, NOLDC income growth of 5 percent per year for the decade of the 1980s, and oil price increases at the same rate.

Table 16.1. NOLDC Energy and Oil Consumption and Imports, 1976 (mbd Oil Equivalent)

	Commercial Energy Consumption	Oil Consumption	Oil Imports
World[a]	80.5	39.2	30.6
NOLDCs	9.3 [11.5%]	5.6 [14.3%]	4.7 [15.3%]
Upper-income NOLDCs	4.8 (6.0%)	3.7 (9.4%)	2.9 (9.6%)
Middle-income NOLDCs	1.8 (2.2%)	1.2 (3.1%)	1.1 (3.6%)
Lower-income NOLDCs	2.7 (3.3%)	0.7 (1.8%)	0.7 (2.1%)

[a] Excluding communist countries.

*The range of the minimum and maximum percentage shares depends on whether we use the high (74 mbd) or low (54 mbd) estimates of world oil consumption in 1990 that have been made in other studies.

Table 16.2. Previous Forecasts, 1980–1990[a] (mbd Oil Equivalent)

Oil Consumption	1980	1985	1990
OECD (1977)	5.3 (10.8%)	6.2 (10.9%)	...
WAES (1977)	7.5–8.3 (15.2%)	9.5–10.7 (16.9%)	11.7–13.5 (18.2%)
Energy Research Group (1977)	9.0 (19.0%)	11.4 (21.4%)	13.3 (24.5%)
World Bank I (1976)	6.2–6.8	7.8	10.0

[a] 1976 NOLDC consumption = 5.6 mbd (14.3%).

- Oil consumption: 6.1–16.0 mbd
- Share of world consumption: 29.6%
- Share of oil imports: 8.8–31.8%

In this case, the range of our 1990 NOLDC demand forecasts lies between 6.1 and 16/0 mbd, or between 8.2 and 21.6 percent of forecasted world oil demand, respectively.

In any event, the range of our estimates still covers an extremely wide interval, reflecting a greater degree of uncertainty than that displayed in previous studies by the World Bank, OECD, and other institutions. Our report includes a comparison and an explanation of these differences.

The models on which our forecasts are based express current demand as a function of current income and price, measured in constant units, and of demand in the immediately preceding period. Data used in fitting the models cover 77 NOLDCs, which accounted in 1976 for 79 percent of total oil consumption by all 124 NOLDCs.

Our demand forecasts are associated with a correspondingly wide range of income and price elasticities shown below.

Income Elasticities of Demand for Oil[a]		
Adjustment to	Short-run	Realized by 1990
Delayed (model 1)	.012 to .144	.024 to .663
Rapid (model 2)	.053 to .249	.187 to .653
	Price Elasticities[b]	
Delayed (model 1)	−.041 to −.081	−.089 to −.756
Rapid (model 2)	−.037 to −.077	−.092 to −.761

[a] Coefficients significant at $p \leq .025$.
[b] Coefficients significant at $p \leq .005$.

The apparently simple question, "What is the income (or price) elasticity of demand for oil or energy in the NOLDCs?" admits of neither a simple nor a singular answer. Four different types of elasticity are calculated from the regression equations. They span an extremely wide range, varying by a factor of five or more, across the four types of elasticities. In general, the income elasticities we have calculated are appreciably lower than those calculated in previous studies, while our price elasticity estimates are similar to those arrived at in previous studies.

There are three principal explanations for the wide range of our forecast, and for the uncertainties it reflects—(1) variations in the scenarios assumed for NOLDC economic growth and for world oil prices, (2) variations in model specifications, and (3) variations in definitions and measurement of the price and income variables employed in the models. Among the three sources of uncertainty, differences in scenario assumptions and differences in the definition and measurement of variables have about equally large effects, while differences in model specification have the smallest effect. The scale and explanation of these uncertanties, as well as those associated with the more familiar standard errors of the estimating equations, are discussed in the Appendix A.

What policy implications follow from our forecasts?

The first and most obvious implication is that policy plans and pronouncements should recognize the inevitably large degree of uncertainty that must accompany efforts to forecast NOLDC oil and energy demand this far into the future. From the standpoint of U.S. energy policy, prudence may warrant relatively greater attention to those cases with high NOLDC income growth rates, high income elasticities of demand, and low price elasticities, because of the possibly serious consequences this outcome would entail for the U.S. If supplies are tight because of high OECD demand or restrictive OPEC policies, high NOLDC demand growth would further increase the tightness, pushing prices upward, assuming that other influences on world oil markets are unchanged. The durability of the OPEC cartel would tend to be strengthened to the extent that its cohesion is helped by higher world demand. (These results would tend to materialize if rapid economic growth occurs in the NOLDCs, if prices in world oil markets are high, and if income elasticities of demand in the NOLDCs exceed their price elasticities.) On the other hand, if economic growth in the NOLDCs is slow, with low income elasticities and high price elasticities, pressure on world oil supplies and prices would be eased, and the cartel's cohesion would probably be strained, especially if OECD demand growth is also moderate or low. However, in this case, the political effects of slower growth in the NOLDCs would be uncongenial to them, as well as to the United States.

U.S. policy therefore faces a dilemma. Accelerated growth and economic development in the NOLDCs—a general aim of U.S. *foreign* policy—is like'" to mean a significantly higher demand for oil, hence further pressure on world oil prices and supplies—circumstances which U.S. *energy* policy would prefer to avoid. Attempts to resolve the dilemma (for example, by encouraging the development of soft energy technology, or nuclear technology in the LDCs) are likely to be viewed by the NOLDCs as disingenuous and self-serving. Thus there is a real conflict, often unrecognized, between the aims of U.S. foreign policy in the arena of North-South relations, and U.S. international energy policy.

However, the relationship between the NOLDCs and U.S. international energy policies may also be viewed from a different, and more congenial, standpoint. Instead of looking at the effect on world markets of NOLDC economic growth and oil demand, one may consider the effect of oil markets on the NOLDCs. From this standpoint, the interests of the United States and those of the NOLDCs are highly compatible. The NOLDCs and the United States, as well as other developed countries, share strong interests in increased world oil supplies and lower, or constant, world oil prices. The developed countries and the oil-importing LDCs are on the *same* side of this North-South issue, not opposite sides—a point which is often missed or obscured in the conventional view of North-South issues.

By 1990, the oil import bill of the NOLDCs can be roughly estimated at between $47 billion and $88 billion (in 1979 dollars), over the central range of our forecasts for NOLDC oil demand in 1990. Thus, the incremental costs to the NOLDCs of their annual oil imports will almost surely be considerably larger than the benefits they might plausibly receive from any of the measures of international economic reform these countries have sought: for example, the proposed stabilization fund for LDC commodity exports, or the debt-service reduction that might result from a rescheduling of LDC international indebtedness, or the increases in foreign economic assistance sought by the NOLDCs and the UN Conferences on Trade and Development.

To alleviate the large and growing burden of oil import costs, an impediment to the economic development of the NOLDCs, we suggest that consideration be given to the possibility of obtaining a concessional oil price from OPEC sellers on petroleum sales to NOLDC buyers based on her historical reports. Notwithstanding the numerous and serious obstacles and drawbacks to such a proposal, we suggest that further consideration should be given to it. The potential attractiveness of a two-tier pricing proposal lies in the possibility it may entail of securing supplementary assistance for the NOLDCs from the OPEC countries, as well as providing a concrete way in which the United States individually, and the other countries of the North as a group, might collaborate with the NOLDC South in furthering NOLDC interests.

Several research suggestions growing out of our analysis of NOLDC energy demand are also discussed.

Forecasting energy demand frequently encounters a familiar methodological issue: What relative importance and attention should be accorded to refinements and sophistication in model specifications, as against more mundane concerns, such as exogenous assumptions or scenarios that are adopted, the definition and measurement of variables, the quality and comparability of the data used to measure them, and the inclusion or exclusion of specific dummy variables? Our results bear on the answer to this question in only a limited way, but one that is nonetheless interesting and significant.

In terms of their relative quantitative effects on the minimum and maximum demand forecasts for 1990, the "mundane concerns" have a much greater impact than do additional model refinements and sophistication.

While the pattern of price and income elasticities for several separate country NOLDC categories—new industrial countries, upper income, middle income, and lower income countries—seems to be similar to that of the all-country grouping, a few interesting differences are noted in the text. Consequently, we suggest that more attention should be focused in future research on the income and price elasticities, and their associated demand forecasts, for the separate country subgroupings.

Such further work would be particularly useful in the case of the oil-importing new industrial countries (NICs), Korea, Taiwan, Brazil, Argentina, Hong Kong, and Singapore. Among the NOLDCs, these are the countries that have successfully surmounted the problem of limited oil supplies and rising oil prices while dramatically advancing their own economic development. To what extent have these countries relied on market forces and price changes, or on direct controls and rationing, in allocating scarce energy supplies? To what extent, and through what means, have they shifted production toward less energy-intensive output and technology? And to what extent have they been able to pass on to others their increased oil import costs simply by raising prices of their exports? Analyzing the policies and patterns of energy use and adjustment in the NICs should be particularly instructive and useful for the other NOLDCs.

Our study of energy and oil demand, like other similar studies, has looked at only one side of the relation between economic development and energy use, namely, the effect of economic growth and of increases in oil energy prices on demand. Econometric research should also focus on the reverse relationship: the effect of changes in energy use and energy prices on economic growth. Such research should consider the relationship between increases in real oil prices as an independent variable, and economic growth in the NOLDCs (as a group and for various country subgroups, especially the NICs), as the dependent variable. The aim should be to provide an answer to this question: Do increases in international oil prices of X percent cause (or contribute to) a decrease in economic growth of Y percent in the importing countries, after proper allowance is made for the effects of other variables?

Finally, a number of research issues are discussed connected with the two-tier oil price proposal mentioned above. One aspect of this research should be an investigation of the extent to which the oil prices actually paid by some NOLDCs in recent years may have been *below* quoted world prices, e.g., due to rebates or concessional devices of various kinds. This investigation should cover information sources (including intelligence sources) beyond those used in the present study.

APPENDIX A

Estimation Method

1. Lagged adjustment model:
   - *Desired* (target) level of oil (energy) demand depends on income and price
   - *Actual* demand adjusts to target gradually, with lag
2. Reduced form estimating equation:

$$Q_{i,t} = a_o(Y_{i,t}^{\alpha} P_{i,t}^{\beta})^{1-\lambda} Q_{i,t-1}^{\lambda}$$

3. Variation in model specification:
   - Delayed *versus* rapid adjustment to income changes

Dimensions of Variation in Regression Analysis

1. Adjustment to income changes (delayed, rapid)
2. Country groupings (all NOLDCs, NICs, high-middle-low income LDCs)
3. Country dummy variables
4. Definition (measurement) of variables
   - Price
   - GDP
   - Demand
5. Per capita and aggregate

DATA SET AND DATA USE

1. 77 countries, 1967–76
2. Data quality and problems
3. Pooling of cross-section and time series data
4. Estimation depends on strong data assumption

Chapter 17
Energy Demand and Conservation in Kenya: Initial Appraisal

Lee Schipper

INTRODUCTION

The Kenya Bureau of Statistics has supplied the International Energy Agency with vital data on the supplies of energy during the past 14 years. Missing from these data, but discussed by other researchers[1] is information on noncommercial fuels—wood and charcoal, dung, waste paper, agricultural wastes, and even solar energy. We have been able to estimate part of the impact of the use of solar energy in Kenya, and do so below in discussing end-use of energy. The rough energy supply balance for Kenya is shown in table 17.1

The demand for energy has been heretofore studied only in the aggregate. It is clear, for example, that a country like Kenya uses very little space heating, and under certain circumstances, need not indulge very heavily in air conditioning except at the coast. Other end-use demands can be roughly estimated from the known output of the East African Oil Refinery (EAOR) as well as net imports of oil products. Electric power production is controlled principally by the East African Power and Light Company (EAPL), and aggregated data on sales have always been generally available.

In this research, however, we take a more detailed approach to accounting for energy end-uses. Following earlier work[2], we disaggregate important

Based upon an earlier work, "Energy Demand and Conservation in Kenya," by L. Schipper and O. Mbeche, presented at the conference "Energy and Environment in East Africa," Nairobi, May 5–11, 1979. This work was sponsored by the Beijer Institute, Stockholm, and was initiated when the author was a guest researcher at the Royal Swedish Academy of Sciences. Support from the Energy Information Administration of DOE through the Lawrence Berkeley Laboratory is also acknowledged.

Table 17.1. 1977 Kenyan Energy Balance (all units GWh = 10^6 kWh)

Primary Supplies	Transformations		Uses							
	Losses*	Net Available	Res.	Bldgs., Sm. Indus.	Lg. Indus.	Auto	Truck, Bus	Rail	Air	Marine, other
Electricity										
Hydro. incl. import 1060	200 El	1210	305	310	640^{++}					
Thermal elec. 350	□ 1450.oil									
Coal 360										
Other 700										
(Bagasse)										
Oil										
Crude & prod. 34500										
Exports, stock 15800										
Net 18700	1200	17425								
Lighting, cooking		750	405	345						
Kerosene										
LPG (bulk)			(--90--------)							
(cylinders)		185		(--95------------)						
Motor spirit—reg./prem.		3145				3145				
Diesel fuel gas oil		3550				(--1440-------)				
heavy		380			1920		(---570-----)			
Jet fuels, av gas		3440							3440	
Heavy oils										□ 1450
		6020		(---3520---------)				1050		El. GE
Totals (do not add vertically)										

Table 17.1. (Continued)

* Losses: 1,2000 GWh oil (used in EAOR); 200 GWh transmission losses, + unaccounted from Elec. Sectors; 1450 □ GWh oil consumed to produce 350 GWh Thermal Electricity; ++50 GWh Self-generation not included.

Notes: We have given here the approximate allocation of various energy forms to various end-use sectors. Figures in parentheses should not be added to row or column totals as they are composite figures. Information from EAPL, the Bureau of Statistics, the oil companies, and the International Energy Agency was not always easy to reconcile, particularly in view of the many different units, both energy and quantity, employed by various organizations. We counted the contribution of hydro power at its direct thermal equivalent, rather than counting it as if made in thermal powerplants, the IEA convention. We used an average value for the energy content of a ton of oil product, typically there should be a 15% variation from this average at the most. For LPG (Liquid Petroleum Gas), however, we took its true energy value per kg, 13.4 kWh or about 4×10^6 Btu/barrel.

Note that we have accounted for the use of electricity by refining under "industry" while crude oil lost in refining appears under "losses." Losses of oil in the conversion to electricity appear under "other."

Unfortunately the various statistics provided by the government do not break down oil use satisfactorily for us to be able to allocate each type of product to each use. Thus "industrial and commercial uses" of oil products as given by the 1978 survey of energy use by the Bureau of Statistics does not tell which kinds of oils were used for which activities. Dotted lines indicate where the allocation of energy over sectors is uncertain.

energy end-uses into various economic or physical activity levels (miles driven, tons produced, households) and into energy intensities (joules/mile, joules/ton, kWh/household). To do this is to pay particular attention to the economic and demographic structure of the country: How many autos are there, and how far are they driven? How much steel is produced by individual plants, at what energy intensity? How many people visit a given hotel in a given year? In this way we can relate energy demands to specific economic activities that are often directly related to the degree of economic development in Kenya, particularly in the cities.

Why should developing countries worry about energy efficiency and conservation. It is often argued that their per capita use among those people and institutions actually coupled to the market economy is so little, that there is literally nothing to conserve. We found the opposite to be the case. Many factory managers and buildings experts were concerned about the cost of energy. Government officials and oil company planners as well were worried about the cost to Kenya of importing increasing amounts of increasingly expensive oil. Ironically the EAOR was a profitable earner of export dollars before the embargo, since a large portion of the crude refined there was reexported, the profits paying for the net outflow of hard currency to buy all the crude. The oil embargo and subsequent price rise changed that situation.

The other important concern voiced in Kenya is over the commercial/noncommercial interface among energy supplies. Most world statistics only count commercially sold energy, particularly that used in activities that are accounted for in the nominal GNP. Of course, there is intense competition among these two kinds of energy sources—deforestation and high cost charcoal may make commercial gas cylinders or solar cooking the only viable option for rural families who cook; low cost commercially sold wood replaced oil (until recently) in one of the manufacturing firms we visited; bark could serve as a firing fuel for the paper mill except that it proved to be cheaper in the past to debark trees where they were cut, by hand, and leave the residue behind. Oil is used instead to raise steam at the mill.

Ultimately, then, there is much interaction and potential substitution among commercial and noncommercial or renewable energy sources. The problem is just that the average Kenyan, whether rural or urban, has little income with which to buy equipment that would make electric or gas cooking possible; has little choice in how efficient higher cost wood is turned into charcoal, and must interact with a market economy that is more or less dependent upon the inflow of commercial imported oil for its health. Better understanding of the efficiency of end-use energy in Kenya, and recognition of the many ways in which commercial and noncommercial energy sources, efficiently deployed, could complement each other, may be the key to Kenya's energy future. For as oil prices rise internationally, Kenya, like the countries of the OECD, may find that as all other energy sources are also rising in cost, the most effective weapon against these costs remains energy conservation, the effective use of all forms of energy.

In addition to officially published statistics, we relied on a certain number of key institutions for data.[3] Oil companies, (EAPL), architectural engineering firms, industrial plant engineers who were interviewed, producers of solar heating equipment, and transportation experts all provided data, which we reference whenever possible (and not proprietary). Our interviews led us to discuss energy use with fifteen of the largest industrial firms in Kenya, representing production of steel, paper, cement, foodstuffs, beer, trucks, tires, and energy itself—the oil refinery and the Kenya pipeline.

END-USE ANALYSIS

Industrial Energy Use

There exist to date no detailed surveys by firm or product of industrial energy use in Kenya. However, we found very quickly that the requisite data exist, given the relatively small number of firms listed in the Directory of Industries (1974 and 1977 editions). We did not have time to survey electricity and fuel use of each type of producer, but were able to sample data from individual firms, oil companies, and EAPL. Finally, we conducted on-site interviews with engineers responsible for heat and power in over a dozen important firms. Eventually we plan to completely classify energy use in Kenyan industries, measure energy intensities, and thereby measure the potential for increased energy efficiency in industry.

The outlook for conservation was mixed. We found one firm where the engineers complained that no one would spend a small amount to fix obvious leaks, improve boiler efficiency, or "optimize" a process, even if the proceeds for such investments were large. An engineer at a metal processing firm told us he was happy if the equipment would simply work at start-up.

On the other hand, a major tire manufacturer pointed out that their factory had reduced energy intensities, taking part in a world wide competition among other firms owned by the same parent company. A manufacturer of food and household items (like detergents) has just hired an engineer who plans to make important process modifications to reduce energy use; in addition this firm is eliminating the use of firewood. Firewood to them was cheapter than oil, while scarce to rural people who can't use oil anyway. Government policy now aims to conserve wood for uses other than process heat.

Included in the section on industry is the Kenya oil pipeline, which supplies the Nairobi area from Mombassa with around 1.2 million tons of oil each year. We give data on pumping energy consumed per cubic meter of oil

Table 17.2. Some Industrial Intensities

TYPE	PLACE	ELECTRICITY	FUEL	OUTPUT UNIT	YEAR
Lorries	Kenya	173.4 kWh	1.85×10^6 Btu	Lorry	1978
Beer	Kenya	.12 kWh	2200 Btu	Liter	1978
Tires	Kenya	1.09 kWh	15000 Btu	Pound of tire	1975
	Kenya	.74 kWh	9650 Btu		1978
	World average, same company,	13,400 Btu (total)			
	U.K., same company	19,000 Btu (total)			
	S. America,	11,110 Btu (total)			
Oil refining	Kenya	16 kWh	1.48×10^6 Btu	Ton	1977
		11.5 kWh	1.37×10^6 Btu	Ton	1973
	Typical refinery, 20% gasoline 3.75×10^6 T/YR	1.75×10^6 Btu/ton			
Cement	(1)	90 kWh + 6.0×10^6 Btu (same plant)		Ton	1975
	(1)	90 kWh + 5.3×10^6 Btu			1977
3 processes in in 2 plants	(2)	105 kWh + 3.1×10^6 Btu			1977
	(3)	95 kWh + 3.6×10^6 (Btu same plant, different kiln)			1977
Other countries	South America	4.5×10^6 Btu total			1978
	Sweden	4.8×10^6 Btu total			1975
Paper (cogeneration)	Kenya	5.1×10^7 Btu total			1977
	Sweden	3×10^7 Btu total			1973
Kenya pipeline	Kenya	10 kWh or 870 Btu		Cubic meter of oil 10^6 Btu of oil	Data
	Lorry Mombassa–Nairobi		15,400 Btu	10^6 Btu	Estimate

These figures were obtained directly from representative firms. Where possible, we give the electricity-fuel breakdown. The comparative figures have electricity figured in, in the case of tires, electricity is counted at 10,000 Btu/kWh.

pumped in table 17.2. Additionally, we estimate the amount of fuel required to haul a ton of oil uphill from Mombassa to Nairobi, and return the empty lorry. Using data from a report from the Oak Ridge National Laboratory, we estimate the energy intensity of tanker transport from U.S. data to be around 2000 Btu per ton mile (certainly a lower limit for Kenya);[3] using a round trip of 500 miles, we found that the consumption of oil for this mode was around 20 times greater than the consumption of electricity for pumping a given quantity of oil. Moreover, the presence of tankers on the road causes apparently serious problems for the road surface. While the pipeline certainly has impacts that we have overlooked, these environmental and energy advantages cannot be dismissed.

What of the tanker drivers who might now be unemployed? This is a problem that surfaces in every country as energy development proceeds. However, one of us (LS) observed a yard in Mombassa where the tanks were being removed from tankers that were then converted to flatbed trucks. This suggests that the impacts of energy development can be mitigated, given time and careful planning. The reemergence of Uganda as a potentially friendly nation and user of EAOR products increases greatly the prospects for the northwest extension of the pipeline; road interests support this as a way of clearing up traffic—and potholes—on the vital route from Nairobi to Kampala.

Table 17.2 gives an overview of energy intensities in key firms. For some, several years' data or comparative figures from other parts of the world are given. The EAOR increased its size after 1973 but did not increase output. Consequently efficiency fell. A conservation program gained back some of these losses. As output increases, the energy intensity will fall again. Given the rapid pace of growth of industry in Kenya, we expect to see new technologies in other industries that will allow output to increase considerably faster than energy use, particularly as energy prices rise.

Commercial Buildings

Commercial buildings include many kinds of enterprises; the most important for Kenya are public services (schools, hospitals), hotels and restaurants, office buildings (including government), and stores. Most of these are classified under the old EAPL tariff system into Tariffs 3 and 4. Similarly some of the classes of users are broken out from the overall oil company data; where possible we indicate in table 17.3 how much total energy goes to various sectors.

We have gone further by using estimates of building energy use for key kinds of buildings—major hotels, office buildings, schools, and hospitals. Some of these buildings are described in table 17.4. In the case of one major hotel, we found that overall electric power use per guest per year had been

Table 17.3. Provisional Breakdown—Other Electricity Uses

	Customers	Price/kWh KE Cents	Total Sector kWh	Remarks
Commercial				
Small[a] (T3)	20,000	55	132×10^6	1976
Large[b] (T4)	400	55	140×10^6	1976
Large industry (T5)	490	27.3	356×10^6	1976
Agriculture[c] (large estates)				
Nairobi	50	40	13×10^6	
elsewhere	50			
Nairobi only				
Total T4	251	—	95×10^6	1977
	270		103×10^6	1978
among which				
Hotels	24	35	19.6×10^6	1977
	25	45	14.4×10^6	1978
(coastal region)	40	—	23.9×10^6	1978
Hospitals			15.7×10^6	1977
Offices, banks (excluding gov't)	22	50	11.4×10^6	1977
	22	60	12.6×10^6	1978
(new customers only)	3	60	1.9×10^6	1978
Kenyatta Center			2.7×10^6	1977
			2.2×10^6	1978

[a] Mostly shops.
[b] Mostly large buildings, schools, some light manufacturing.
[c] Agriculture includes farms and estates; total included in T4 and T5.

decreased substantially from 1976 to 1978, in part due to the recent initiation of a conservation campaign.

On the other hand, energy use in large structures in Kenya depends critically on the building shell design. Most, but not all, of the new office buildings in Nairobi, for example, employ elaborate systems of shading to keep direct sunlight off the windows. Energy conserving buildings have their axes

Table 17.4. Major Hotel, Nairobi, Air Conditioned

	Electrical, kWh/guest	Oil Btu Guest	Gas Btu/ Guest
1976	37.5	3×10^5	—
1977	35.5	2.9×10^5	2×10^2
1978	29.4	2.9×10^5	2×10^2

oriented east-west, with few windows on the east-west ends and shading on the north-south sides. This design maximizes free (day) light, minimizes cooling needs (if any) and requires the use of small electric heaters only sometimes, in some rooms, during cold months. We have not yet surveyed lighting levels in offices, but we found no evidence of the overlighting common in the United States.

Homes

Commercial energy use in the residential sector is characterized by extreme concentration into a small fraction of all households. As the figures in table 17.5 suggest, only 6 percent of the population in Nairobi, and less elsewhere, was connected to the EAPL grid. Complementing this picture is the relatively minor but growing use of kerosene for lighting and cooking and the somewhat greater use of gas cylinders. For domestic cooking, the majority of Kenyans, of course, use charcoal or firewood. Openshaw[5] estimates this as great as 5 times the sum of all commercial energy use!

We have been able to break down electricity use further, using data from EAPL. The largest residential consumers register their hot-water electricity consumption on a special tariff (with an electronic signal interrupter). In 1976 there were 27,000 hot water customers, 51,000 regular residential customers (including the hot water) and 56,000 customers using very little electricity, most living in rural areas or in low-income estates. Typical figures for consumption in these groups are given in table 17.6. It should be noted that the designated income group for each housing tract does not always reflect the income of the people who actually live there, due to subleasing. Similarly figures of wealthy households include use of energy in servants' quarters.

What is missing from this electricity use picture is the complementary use of fuels. While electric cooking probably dominates in those homes on Tariff 1 or 6, it is clearly absent from those on Tariff 2, charcoal or in some cases

Table 17.5. Residential Energy, 1978

	Customers	Price, KeC	use/yr kWh
Regular electricity	51,000	33	3,000 per customer
Small users (elect.)	56,000	125	250
Hot water (elect.)	22,000	18	4,815
Gas cylinders (total country)	—	—	90 GWh (1977)
			103 GWh (1978)
Cooking and lighting oil (total country)	—	—	405 GWh (1977)
			645 GWh (1978)

* Shell estimates 1 liter/week/family for lighting, cooking. Fuel estimates subject to revision.

Table 17.6. Residential Electricity by Income, 1977

Description	Sample Size	Average for Six Months	Tariff
"Lower class estate"	20	10 kWh/mo.	T2
"Middle class estate"	12	258 kWh/mo.	T1
Executive housing	12	565 kWh/mo. normal	T1
	11	444 kWh/mo. hot water	T6
Large estates	18	752 kWh/mo. normal	T1
	17	462 kWh/mo. hot water	T6

Sample of electricity billing data, the descriptions of the housing are only approximate and reflect the outward appearance or purpose of the housing project. Note that the last two samples reflect hot water and cooking for servants. In Nairobi, roughly half of all customers receive Tariff 1 and/or 6, the normal and hot water tariffs, while half receive Tariff 2. The average for "middle class estate" probably should be used for forecasting purposes. Data source: EAPL. See also McGranahan, et al., in *Proceedings of the Workshop*. For prices, see tables 17.5 and 17.8.

gas being more important. One house we visited had switched from gas to charcoal since the gas stove exploded. We obtained estimates of country-wide kerosene and LPG consumption from Kenya Shell. This estimate includes sales of small lots of packaged kerosene and bulk kerosene sold for resale, as well as lots of small gas cyclinders. We may have overestimated consumption in the residential sector of these two fuels since small stores or restaurants may use small quantities of these fuels as well.

The prospects for solar water heating in Kenya are bright. We examined the records of one of the major assemblers and suppliers of solar water heaters. Using an estimate of 2000 M^2 of collectors he has installed thus far, each M^2 providing about 9000 Btu per day of hot water, we find that installed residential and commercial hot water systems save Kenya about 1.5×10^6 kWh/yr that would have been required in the form of electricity (more if gas or oil) for heating this water. Moreover, a great deal of this electricity would be under normal commercial tariff, being used in schools or hospitals. The total investment cost for these collectors has been approximately 3×10^6 Kenya shillings (KSh). If normal tariff electricity cost 50 K-cents/kWh in 1979, then the yearly savings to Kenya from this investment is approximately 750,000 KS.[6] The manufacturer we interviewed pointed out that business was booming, and provided us with examples of new projects (a school, a hospital, a condominium) that he expected to complete soon (table 17.7).

Competing with commercially sold energy, as we pointed out above, is the use of wood, charcoal, and other renewable fuels for cooking and possibly water heating. Unlike electricity, these fuels (and gas cyclinders) are available in the rural regions; thus the choice between them and electricity tends to be biased because of the cost of electrifying villages. That is, both noncommercial fuels and electricity (or gas plus the necessary stove) are scarce resources, if for different reasons. This means that the choice between them, if there is to be a policy favoring one or the other, is difficult. However, sales of kerosene by the largest oil company have increased much over the past few years, suggesting that the scarcity of charcoal is indeed putting pressure on the use of commercial fuels.[7]

Transportation

Statistics on transportation are often well known in the aggregate, because motor vehicles are registered, traffic is often surveyed, and most motor fuels are taxed in one way or another. On the other hand, there are many ways in which sales of fuels do not correspond uniquely to one class of vehicles, or where types of vehicles may provide two kinds of service. Light Diesel fuel, called gasoil, can fuel automobiles or light trucks; heavy diesel fuel can fuel trucks, railways, or some buses. Matatus usually run on motor gasoline, but ordinary trucks can be used as matatus. Thus it is difficult to assign fuel use to specific tasks, and therefore difficult to analyze the fuel efficiency of each vehicle or service.

Nevertheless we give the breakdown of fuels used for transportation, and their end-uses in table 17.1. In table 17.8 we present another view of transportation, the share of vehicles in each class and the share of vehicle miles in each class as estimated from actual road surveys.[8] In general the following rules apply, relating vehicle type to fuel:

Private cars—Up to nine-passenger vehicles, except Landrovers and minibuses: Premium fuel

Medium commercial vehicles—Two-axled goods vehicles, weighing more than 1524 kg, with more than one tire in each axle: Regular fuel

Light commercial—As above, but less than 1525 kg: Regular fuel

Table 17.7. Solar Water Heating*

Type of building	Investment	M^2	Electricity Replaced Yearly	Cost**
Group of 42 flats	360,000 ks	112	4000 kWh/flat	32¢
Luxury condominium, 104 units	510,000 ks	250	3000 kWh/flat	25¢
School, 300 students	100,000 ks	38		(50 ks/students)
Medical center (80% solar)	330,000 ks	128		

* Beasley collectors produce about (32 litres/day per m^2 hot water with 80° F temperature rise). Other supplies in Kenya have not been surveyed. Data from K. Mousley, Instrumentation Ltd., Nairobi and Beasley, Ltd., Australia, fact sheets.

** Cost estimated as ratio of 7.5%/annum fixed charge to electricity produced, i.e., ¢/kWh.

Table 17.8. Amount of Travel by Class of Road and Vehicle Type, 1978 Average Yearly Vehicle per Kilometer of Road Class (VEH/km.)

Class of Roads	% of Total Network*	Cars	Light Commercial Vehicles	Medium Commercial Vehicles	Heavy Commercial Vehicles	Buses	Total (VEH/km)	(%)
Trunk	13	1,185	501	501	178	220	2,970	(62)
Primary	18	339	533	300	12	71	1,275	(27)
Secondary	23	71	258	95	1	53	278	(10)
Minor	46	1	8	2	—	—	11	(10)
All roads	100	1,596	1,705	878	191	344	4,734	(100)
Percent VEH/km	—	(34)	(36)	(19)	(4)	(7)	(100)	

* Excludes special purpose roads.
Source: Development plan 1979–1983.

Heavy commercial—More than two axles: Diesel fuel

Bus—Other passenger vehicles, including minibuses, and dual purpose vehicles: Diesel fuel (except matatus)

Note that the share of vehicles in each class *roughly* corresponds to the share of vehicle miles in each class. That commercially motivated vehicle miles exceed that classification's share of vehicles, when compared with private vehicles is not surprising; owners of capital, such as vehicles, try to maximize the utilization of their often substantial investments. On the other hand, most developed countries reveal clear patterns of growth: The use of private automobiles rises from somewhat faster to much faster than private incomes, while the use of commercial vehicles, particularly for freight, tends to scale only with total output. Armed with more detailed fuel sales statistics, we could closely couple fuel and efficiency to transportation services.

Now we couple to the greatest extent possible the use of energy to the activities of transportation. In table 17.9 some of these relationships are shown. First we show the number of passengers actually embarking from Nairobi International Airport for international destinations, and the amount of fuel loaded there. While this measure of efficiency, fuel per passenger, is crude, its decrease over a period when additional transcontinental flights from Europe and Africa were being added to schedules, thus lengthening the average trip away from Nairobi, suggests conservation. In fact, the 1973–77 era saw the replacement of most narrow-bodied aircraft by jumbo jets and an increase in charter flights from Europe. These changes increase energy efficiency. On the other hand, emergency conditions often dictate that planes cannot take on a full load of fuel in Nairobi, but must bring in fuel. Therefore our figures must be seen as provisional until surveys are arranged to show the exact amount of fuel used in Kenya to transport a passenger—whether a Kenyan or a tourist returning home—to an overseas destination.

In the case of private autos, we give amounts of gasoline sold to autos as well as the number of autos registered. This gives an approximate measure of intensity of use, though not efficiency, since we have not obtained data on actual miles traveled, nor on auto weight or load factor. Moreover, we suspect that a few surveys among auto dealers, registration statistics, and fuel sales records would reveal many of these measures. We hardly need to point out the phenomenal growth in the ownership of autos.

Finally we give a measure of the use of trucks and buses in the table, showing also fuel used. Here we take the sales of gasoil as reported by Kenya Shell, since the official statistical abstract lists light diesel fuel that is also used for some stationary applications. We give this figure separately but warn against any strict interpretation of these figures.

There are other important uses of fuel that we have not covered here but show in table 17.1. Among those are inland and overseas shipping, passenger

Table 17.9. Some Transportation Energy Uses

Jet Aviation (Nairobi only)	Activity Passengers	Gross Energy	Intensity kWh/passenger
1973	730,000	3570 GWh	48.2
1974	790,000	3540	44.2
1975	920,000	4020	43.1
1976	960,000	4140	42.5
1977	—	3860	< 42.5 assuming
1978	—	3800	< 42.5 increase in traffic

Source: Statistical Abstract, Kenya Shell.

Private Autos	Registrations[a]	Gross Energy[a]	Gross Energy[b]
1973	70,000	2710 GWh	3106 GWh
1974	78,000	2625 GWh	3015 GWh
1975	83,680	2730 GWh	3145 GWh
1976	88,700	2800 GWh	3230 GWh
1977		3150 GWh	
1978		3250 GWh	

Source: Bureau of Statistics, "Statistics of Energy and Power" for [a]Statistical Abstract for Registrations, [b]Gross Energy which may include gasoil.

Other Vehicles Trucks, Buses	Registrations[a]	Gross Energy[a]	Gross Energy[c]	Gross Energy[b]
1973	67,750	2950 GWh		3250 GWh
1974	76,460	2910 GWh	1220 GWh	3120 GWh
1975	83,825	2965 GWh	1250 GWh	3260 GWh
1976	91,790	3350 GWh	1360 GWh	3660 GWh
1977		3550 GWh	1490 GWh	

Source: As above—[c]Kenya Shell Gasoil figures. Differences due to definition of product; do not always reconcile with table 1.

and freight rail, buses, and matatus. The most significant development is the switch, by Kenyan Railway Corporation, from heavy fuel oil to lighter diesel oil, which reduces markedly the consumption of fuel per unit of output. Taken together, all this transportation data, combined with economic and demographic projections of incomes, mobility, and location of people, would provide an excellent base for a careful forecast of energy demands for future transportation in Kenya.

ENERGY PRICES

The increase in world energy prices was felt in Kenya. First, the EAOR, which used to turn a foreign exchange profit by reselling oil products amounting to about half of the throughput of the refinery, saw foreign demand drop off somewhat, the domestic share increasing. Worse, the gross profit margin on a unit of product became very small in relation to the price of crude. Thus the situation after 1974 was very different from the situation in 1973 and earlier.

In table 17.10 we present some representative energy prices in Kenya, from data gathered by Bikro Consult, Kenya, and from the Bureau of Statistics. We have converted all amounts to units of 10^6 Btu or kilowatt hours. We remind readers that transportation charges for fuel, and demand charges for heavy electric power users must be added to these figures.

The data presented are in current KS. While the price increases seem dramatic, the GNP deflator for the period 1974–77 inclusive is approximately 1.6, according to the difference between the monetary GDP in constant and current dollars as given in the 1978 Economic Survey. We give the approximate value of these fuels in 1974 prices, and the change is dramatic. While basic fuel prices, with the exception of LPG, have increased considerably; electricity prices have not, due of course to the dominance of hydropower in the supply picture. On the other hand, preliminary data from 1978 indicate that substantial price increases over 1977 occurred, and a new billing system and tariff were introduced in 1979.

Two factors determine the use of energy from an economic point of view. Energy prices determine in part the marginal cost of using certain equipment

Table 17.10. Energy Prices. (Units—Kenya Shillings/10^6 Btu or kWh)

Fuels (KS/10^6Btu)	1973	1974	1977	1977 at 1974 Prices	Remarks
Fuel oil	14.6		39.0	24.4	FOB Nairobi
Diesel oil (heavy)	17.0		46.3	28.9	"
Gas oil (light motor diesel)	24.3		61.0	38.7	"
Lpg, 15 kg cylinders	55.8		97.6	61.0	"
Lpg, bulk	65.5		87.8	54.9	"
Motor gasoline	40.8	64.9	72.3		average of super/regular

Electricity ¢/kWh	1974	1976	1977	1977 at 1974 Prices	
Regular domestic	25.8	35.4			
Special domestic	112.5	128.9	36.4	22.8	1977 is average for all 3 tariffs
Interruptible domestic	22.6	18.6			
Small commercial	41.5	54.8	60.0	37.5	
Large commercial	24.6	34.0	40.0	25.0	
Industrial	17.8	27.3	31.0	19.31	

or enjoying certain amenities, provided the user processes the capital equipment in the first place. Prices also play an important role in the choice of equipment, particularly where solar water heating or most industrial uses are concerned. On the other hand, incomes and income growth play a great role both in the affordability of equipment and in the ability to use that equipment. In Europe, for example, gasoline use is climbing steadily because family incomes are increasing and Europeans are buying their first or second autos. Accordingly gasoline use will be rising there for some time in spite of high prices, though not as fast as the increase in autos, because new autos in Europe may become more energy-efficient now. In Kenya, however, the use of autos by expatriates and diplomats as well as the Kenya upper class, is fully saturated. Growth in the future depends on the rate of increase of the middle class. This kind of analysis must be performed on all sectors of energy use in Kenya in order for us to be able to derive meaningful price–income–intensity-of-use relationships. Moreover, we must be able to measure the use of commercially sold fuels compared to noncommercial fuels, capital intensive renewables like solar hot water, and of course the nonmarket income of many Kenyans.

CONSERVATION IN KENYA?

We noted in several places that sites we visited indicated that energy conservation programs were in progress. In every case the person responsible cited higher prices for fuels and electricity as the primary movitation. As to our pessimism over the lack of interest on the part of some firms, it is well known from economic observations that the response to a price increase, be it steady or one-time, takes between a few and tens of years, to take effect. The reason is simply that the greatest changes in energy use take place with the least cost when new equipment is built. Thus the evidence we have seen so far indicates that conservation is beginning to take place in Kenya. But we found many opportunities worth investigating.

We noticed several buildings that could be retrofitted to reduce solar gain and hence air conditioning, including the building housing the American Embassy, the Hilton Hotel, and even the Kenyatta Center. We mention the names of these buildings, not to single out their owner or managers, but to show that a wide variety of enterprises could take part in energy conservation as energy prices climb. We also note that homeowners can add insulation to hot water heaters (in the United States some utilities now provide this as a service), shade windows, keep refrigerator coils clean, and make a conscious effort to reduce the number of miles they drive.

FOREIGN TRADE AND EMBODIED ENERGY

An extremely important source of energy often omitted from national data is the energy bound up in imports and exports. That is, a unit of goods or services required energy for its fabrication, including the process energy used to make the raw materials and so on. Elsewhere we examined the balance of trade for this embodied energy and found that while the United States imported a small amount (about 1 percent of its 1973 gross energy use), Sweden and other countries in Europe were significant exporters. That is, significant quantities of the oil imported by many nations leave their borders bound up in steel, paper, and other energy intensive goods. Agricultural products tend not to be as energy-intensive on an energy per ton or energy per monetary-unit basis when compared with raw materials. One important energy-intensive export from Kenya is refined oil products, energy for which is consumed at the EA Refinery. This embodied energy amounts to nearly 5 percent of the actual heat content of the fuels exported. Another may soon be paper.

We have not evaluated specific energy intensities for the many materials and products that Kenya deals with. However, we note that three significant categories besides trade in actual fuels show a great import surplus: These are industrial supplies besides food, machinery, and other capital equipment, and transport equipment. These are listed here in the approximate order of greatest to least energy intensity. In 1974 these goods amounted to KSh 240×10^6 imports, KSh 85×10^6 exports, and in 1977 (deflated by 1.6 to 1974 currency) KSh 212×10^6 imports and 56×10^6 exports. Estimating average energy intensity for these kinds of products at about 100,000 Btu per 1974 US$ ($1 = KSh 7 approximately), this amounts to about 40×10^{12} Btu in 1974 and a similar amount in 1977. These figures appear to be greater than half of the recorded energy use in Kenya. While our estimate is rough, this hidden energy is known to form a significant fraction of energy use in other countries, as much as 20 percent in Denmark. We expect that our estimate is correct to within a factor of 2, and point out that the major export from Kenya, food products, tends to be far less energy intensive than the goods we have considered here. However, it would be useful for Kenyan energy planners to look carefully into this hiden energy flow since by any account it is significant in the overall energy balance, particularly as rising world energy prices push up the costs of energy-intensive materials and products.

Units Used: (See also Ref. a). 1 Kilowatt-hour (kWh) of heat or electricity contains 3412 British Thermal Units. 1 kilocalorie (kcal) contains 4.18 kJ.

Oil energy content is often measured in metric tons of oil equivalent, which is given an arbitrary value of 10^7 Kcal, or 11.63 thousand kWh. We have used an average value for all oil products, since true values vary by only about 10% except for LPG. One barrel contains by definition 42 U.S. gallons or about 160 liters, and as a standard of oil contains 5.8×10^6 Btu.

We count electricity at its direct thermal content, 3412 Btu, rather than include fictitious losses in thermal powerplants for hydropower, as is often done.

NOTES

1. See the proceedings of the workshop, "Energy and Environment in East Africa," Nairobi, May 5–11, 1979, to be published by the UN Environmental Program and the Beijer Institute of Energy and Human Ecology.
2. L. Schipper and A. Lichtenberg, "Efficient Energy Use and Well-Being: The Swedish Example," *Science* 194 (December 3, 1976): 1001.
3. Statistics of Energy and Power 1969–77 (Nairobi, Kenya: Central Bureau of Statistics, 1974, 1977, 1978); *Directory of Industries*, Nairobi, Kenya; *Economic Survey*, 1978. Estimates of industrial energy intensities from other countries from Schipper and Lichtenberg; L. Schipper, "1978 Energy Use and Conservation in Industrialized Countries," in J. Sawhill, ed., *Energy Conservation and Public Policy* (New York: Prentice-Hall, 1977) and private communications with other researchers and firms interviewed. In each case we obtained energy intensities of firms listed directly. Sample energy intensities of goods in foreign trade are found in L. Schipper, *Annual Review of Energy* 1 (Palo Alto: Annual Reviews, Inc., 1976). Additional information on energy use and the economy can be found in House and Killick, on urban and residential energy use, in McGranahan et al., in *Proceedings of the Workshop*. See also "Basic Energy Statistics and Energy Balances for LDCs" (Paris: International Energy Agency, January 1978—with updates for 1977 for Kenya supplied by the IEA).
4. Other estimates compiled by O. Mbeche from information of transportation companies in Kenya put the cost much higher.
5. In *Proceedings of the Workshop*.
6. This calculation assumes no standby losses for either system. We count only the hot water actually made available by solar systems. If all this were produced from the low-cost interruptible tariff, the savings would be considerably less, on the order of 300,000 per year. Either way the rate of return is greater than 10 percent.
7. D. French, USAID, points out that the increase in commercial fuel use may also be a result of higher incomes.
8. These figures do not truly represent vehicle miles, but we use them as a proxy for the rough division of traffic into various classes or modes.

Chapter 18
Energy and International Trade Issues

Jean Louis Waelbroeck

The world has now gone through two sharp increases in oil prices. The second of these increases has not yet worked out its effects, which it is desirable to anticipate. We must also live with the idea that the sharp price jump last year may not be the last one. A dangerously large share of the world's energy supply is concentrated in a politically unstable area, and there may be other upheavals around the Persian Gulf. These will then cause other price shocks, compared to which last year's may seem mild, and we must be prepared to adjust to these.

Alan Manne has asked me to discuss the economic logic of this adjustment process. This afternoon Helen Hughes will give you her evaluation of prospects from the unique vantage point provided by her experience in the World Bank. My approach will be more analytic, and I will try to describe and appraise the mechanisms through which energy affects trade and the world economy.

I will distinguish between two kinds of mechanisms and problems, dwelling mainly on those which are relevant in the fairly short run—the three to five years which follow an oil price increase. I will then discuss the long-run adjustment problems of developing countries.

ADJUSTMENT IN THE FAIRLY SHORT RUN

What got talked about at the beginning of 1974, and again in 1979, was a true physical shortage of oil. It was feared that whole industries or activities would have to close down that millions of workers would be unemployed; in table talk people even discussed giving up the automobiles. The solutions advocated were quantitative—conservation measures, and freely accepted efforts to use energy more intelligently. Little of this happened. A few gas queues which could have been avoided by good policies, and a few perfunctory conservation guidelines. The real adjustment has been accomplished through prices. The price mechanism made it impossible for OPEC to embargo Holland and Denmark effectively in 1974. In 1979 any country which was running short went to Rotterdam, and Rotterdam prices attracted enough supplies to satisfy these demands. The adjustment was brutal of course: The loss of something like 2 percent of the world's energy supplies caused the world price to double. With this background it is frightening to imagine the consequences of serious political difficulties in Saudi Arabia.

Sharp changes in prices mean sharp changes in the national and international distribution of income. The value of OPEC's exports went up by over $100 billion last year. The cumulative amount of the taxes levied in last year's energy bill exceeds $200 billion, and the bill gave roughly as much to the oil companies. These are large sums, and it is understandable that a fierce tug-of-war, national as well as international, has developed to pull this way and that the amounts gained and lost by different groups of people.

Internationally, what is at stake is the ability of the governments of oil-importing countries to cut consumption as they promised to do at the Tokyo summit and elsewhere. You will have gathered that I have little faith in soft solutions like conservation. Cutting consumption means either raising higher taxes or (as in the United States), eliminating the implicit subsidies from which consumers have benefited since 1974.

As strikingly illustrated by Alan Manne's model, to increase taxes on oil is paradoxically the best way to drive its price down. I am oversimplifying only a little bit in saying that, if the price which balances the supply and demand of oil in the world is $30, then this price can be achieved either by levying a $15 tax in all consuming countries and paying $15 to OPEC, or by levying no tax and paying $30 to OPEC. Taxing energy is therefore the rational response to shortages; this should of course be done in a way which does not discourage the expansion of non-OPEC supplies. Of course, taxing energy is not what governments have done. We Europeans feel rightly critical of the energy price policy followed until recently by the United States, but we have been only too happy to allow inflation to erode the excise duties on gasoline and heating oil on our side of the Atlantic.

What has worked against this policy is in part that it is based on a reasoning which is too abstract to be understood by voters. It is hard to put across the idea that taxing oil makes it cheaper from the country's point of view; it is even harder to persuade well-informed people that the two oil price crises were more than tricks cleverly engineered by the Seven Sisters. Also—more importantly perhaps—though it is in the collective interest of importers to tax oil, it is very tempting for each individual government to try to get a free ride on the efforts of others, to spare itself the political ill will generated by an oil tax, while benefiting from the easier world energy market which results from taxes imposed by other countries. As a result, each country has tended to wait for the others to do something, and no effective policy has evolved.

At home, the tug-of-war is between governments, consumers, and energy producers. As explained above, higher energy prices lead to large income transfers. For consumers the result is a perceptible cut in the standard of living. Producers gain a lot if the price mechanism is allowed to work and if their gains are not taxed away. This gain seems very unfair given that other people are suffering. As consumers try to squirm away from the threatened cut of their purchasing power, a strong political pressure is generated to keep the price of energy at an artificially low level. As in the United States, this feeds the balance-of-payments deficit and leads to foreign borrowing. In effect, what happens then is that consumers squirm away from an immediate sacrifice by transferring the burden to their children—or to themselves a few years later.

This brings us back to the international tug-of-war, for there is one way to avoid paying back what has been borrowed, and that is to reduce debts by inflation and devaluation. This is a game which countries have played with much gusto in recent years. The slide of the dollar, for example, has wiped out debts which greatly exceed the foreign aid given by the United States since the war. Such tactics cannot be continued forever, however, and it now takes an 18 percent prime rate to persuade foreigners to hold dollars.

It would be possible to go on in this way, but I think that the basic point has been made: Increases in world oil prices mean a large income transfer, which is almost as inevitable as the phases of the moon. But there exist rich possibilities to pass the buck from one country to another, from one social group to another, even from one generation to another within a particular country.

This leads me to my next topic, which is the contribution of energy prices to inflation. Here I will distinguish between two effects, a small one, and large one. It is trivial to note that if gas prices go up, the cost of living index also increases. This direct impact—what in French we call the mechanical impact of inflation—is, however, rather small—on the order of 1 percent last year for developed countries.

What appears to be far more important is the wage-price escalation triggered by this mechanical impact. As I explained, groups of consumers are keen to squirm away from a cut of living standards, the inevitability of which they do not want to understand. Thus as prices rise, wage earners—and other groups, such as doctors and farmers—try to adjust wages, fees, and farm prices to offset the price increase. The result is a vicious circle of inflation, which appears to multiply several-fold the initial impact of energy prices on the price level.

Perception of this inflation multiplier provides a strong temptation to governments to insulate consumers from the increase in energy prices—and so we come back to the difficulty of designing an effective international strategy to deal with the energy situation.

I now come to another multiplier effect, of a more classical sort, the Keynesian multiplier mechanism, to which anyone who has followed an introductory course of economics has been exposed. Keynes has shown that governments can expand or contract demand by increasing or reducing the budget deficit. The induced change in demand is a multiplier of the change of public saving which caused it. If the oil price increase is looked at from a world point of view, it will be seen that it is akin to a tax levied by OPEC on the consumers of oil-importing countries. To the extent that the OPEC governments do not spend right away the increase of their earnings, aggregate demand in the world will be reduced by a multiple of the increase of the OPEC surplus. As this will be of the order of 1 percent of world GNP next year, it is apparent that this mechanism will make a powerful contribution to the recession which economists expect.

The Keynesian multiplier, like the inflation multiplier, can be moderated by government policy. In 1974–75, quite a few countries followed a policy of restoring by tax cuts and increases in public expenditures the purchasing power taken away from consumers by OPEC. To a large extent, the necessary funds were borrowed on the Eurodollar market—indirectly from OPEC, therefore. The countries which followed such a course did quite poorly in subsequent years: The most successful one was probably the United States. I do not think, therefore, that governments will again offset a significant part of the Keynesian multiplier impact of higher oil prices.

What we seem to be witnessing today is policies which reinforce rather than offset this impact. To stop inflation, central banks have raised interest rates to the highest levels in modern history; a number of countries (the United Kingdom, France, perhaps the United States) have adopted more restrictive budgetary policies.

Another deflationary element of the present situation is that, as pointed out by several speakers at this conference, the OPEC surplus may prove to be more lasting than after 1974. The reasons for this have been set forth by other speakers. The governments of oil-exporting countries may be more prudent in spending surpluses because events in Iran have sharpened their awareness of the social and political tensions which are caused by headlong development. Perhaps (?) they have grown wiser.

Which brings me to the next of the adjustment mechanisms: the recycling of OPEC funds. Here we have a case of déjà vu. In 1974–75 there was already widespread concern about the ability of the world banking system to recycle to countries in need, in particular developing countries, the surpluses deposited by OPEC countries. It should be said that the latter countries have little choice about what to do with their money; the U.S. money market and the

Eurodollar market alone have the strength and flexibility to absorb such large sums. What is at issue, therefore, is whether private banks will be willing to direct funds to the most needy countries, or, as the banks see it, whether the needy countries will remain sufficiently creditworthy to be attractive lending prospects.

Last time around there were optimists and pessimists, and the optimists proved to be right. Governments hardly had to intervene; the banking system recycled funds to countries in need and found this to be quite profitable. Most of the countries which were cited as being on the verge of bankruptcy, like Italy and Korea, achieved a spectacular balance-of-payments recovery. There were difficult cases like Zaire, but these did not turn into outright bankruptcy. There is no reason to think that banks found a larger proportion of poor risks among their international than among their domestic business borrowers.

I was optimistic then, and I am optimistic this time also. The 1980 OPEC surplus will be significantly smaller in real terms than the surplus of 1974; the world banking system has grown stronger, and more experienced in dealing with developing country clients, and developing countries have acquired a much better understanding of the international money market. Debt service ratios are, it is true, higher on the average than they were at the time of the first crisis, but on balance I feel that an optimistic appraisal is warranted. I do wish to mention two unfavorable aspects of the situation. Last time, OPEC's accumulated surpluses, and the developing countries' debts, were conveniently wiped out by an inflation of unexpected magnitude. We now seem to be reentering a period in which interest rates exceed the rate of inflation, so that the real debt service burden of developing countries may be higher than in the past. Also, as was explained, OPEC's surplus may prove to be more durable than last time, and more difficult to recycle in the long run.

I now come to the adjustment mechanism which people think of instinctively with respect to adjustment of developing countries to higher energy prices. According to what in the economist's jargon is called two gaps analysis, the exports and imports of developing countries are regarded as insensitive to prices. What these countries export is primary products, in quantities dictated by available capacities, and at prices dictated by world markets. Imports are essential goods which cannot be produced domestically, and their level depends on the level of activity and not on prices. Such a country is unable to improve its balance of payments by a devaluation. Exports and foreign aid receipts determine foreign exchange receipts, and the only way to tailor foreign exchange expenditures to receipts is to reduce GNP and hence the corresponding imports.

There can be lucky countries of course, like oil producers, who have more foreign exchange than they can use. Their GNP is limited by domestic production capacity: They are in a resource gap situation. Most oil-importing countries have no surplus of foreign exchange. If their balance of payments worsens because oil prices increase, they will have to reduce GNP to bring their balance of payments back into equilibrium. These countries are helpless to improve their situation in the short run. They are in a state of total dependency on OPEC and the developed countries, which set world prices and provide available aid.

It must be realized that the shock which higher oil prices impose on the economies of developing countries is indeed a severe one. Helen Hughes will present to you more accurate figures than mine, but as rough orders of magnitude, we can estimate that oil-importing developing countries spend some $65 billion on oil imports. Brazil alone spends $10 billion, Korea $5 billion. Eighteen months ago this oil cost half as much. The extra cost is as large as all economic aid to developing countries.

I do not want to minimize the pressures to which developing countries will be subjected this year and the following ones. But the record shows that the picture of developing countries which is implied by the two gaps theory is incorrect. If the theory had been true, the 1974 recession and increase in oil prices should have had a disastrous impact on the growth of the developing world. Yet developing countries have weathered the storm far better than we have. It is impossible to discuss in detail in this paper the many ways in which they have adjusted to the difficulties which have confronted them, but I would sum up the experience of this period by saying that the only countries to which the two gaps dependency model is applicable have been the few developing countries which have responded to outside pressures by shutting themselves off from the world, by refusing to allow their economies to adapt to the changed economic realities.

There is yet one more economic mechanism to discuss, which has frequently been invoked, though I am not convinced that it is important. High energy prices, it is said, destroy the effectiveness of a part of the capital goods which were built at a time when energy was cheap. The example usually given is production lines for gas-guzzlers. So far as I know, these lines represent a negligible fraction of the total capital stock, and can be retooled fairly easily. Part of the capital tied up in tankers and refineries has surely been lost since 1974. But apart from these examples, I can see no clear illustration of this theory.

And yet . . . something does seem to have happened to the efficiency of our economic system, as indicated by the break in productivity growth which has occurred in so many developed countries. Nothing in my understanding of economics would tell me that an increase of energy prices can have such an impact on productivity. On the other hand, I do not know of another good explanation.

HIGH ENERGY PRICES AND WORLD TRADE IN THE LONG RUN

In giving us oil, Nature invented a source of energy which is cheap to produce, easy to transport and to handle in production processes, and from which it is easy to fractionate parts which are clean and convenient enough to be used in automobiles and homes. Then it gave us relatively little of it, and apparently decided to put a large fraction of the available supply in a small fraction of the world's area. During the postwar era, this fuel became in a very short period the main source of energy for the world. The problem which confronts us in the long run is to reverse this substitution and thus reduce the dependence of the world's economy on this fuel.

In a way I have witnessed the triumph of oil. I remember directing a study of the evolution of technical coefficients of the Belgian input-output matrix. The study was done in the early 1960s for the Belgian Planning Bureau. And I remember having been amazed at the speed at which Belgian industry ran away from coal and surprised when I calculated how profitable was the switch. About this time my wife and I decided to install an oil burner in our house, and stopped having to worry about what to do with the coke slag.

At $30 a barrel, King Oil does not yet have serious rivals. There would be less talk about conservation and policy intervention if the price of oil were high enough to stimulate alternative sources of supply. Only in electricity production do other types of energy seem to have a moderate edge. But we should remember that part of the oil used for electricity production is in fact a by-product of the distillation of crude and has few other uses. Its price will adjust to prevent its being driven out by the competition of other sources of energy in producing electricity.

The data I have seen on costs of alternative forms of energy lead me to accept the view often expressed that it would take a price of $45–50 to bring about a marked switch to other forms of energy. In fact, an even higher price would be required to bring about the kind of very rapid substitution of one fuel for another which took place in the 1960s, and to provide an adequate stimulus to the production of other forms of energy. If this prediction comes true, developing countries will be spending $150 billion a year, a staggering sum. On the other hand, OPEC will have become a $500–600 billion market, which developing countries can capture in part.

The oil-import figures are entirely hypothetical of course. In recent years more and more developing countries have become oil producers, as illustrated by E. Ehrlich's excellent paper. Drilling activity is increasing rather quickly. Major discoveries in Brazil or in offshore areas near Korea would change these figures significantly.

But what I would like to stress most in closing this paper is that energy prices will not all move up together. The structure of these prices is bound to change markedly, and this will open up research and investment possibilities of which we should be aware.

I have indicated my agreement with the view that in the long run, the price of oil is headed toward $45–50 a barrel, perhaps even higher. But that does not mean that other energy prices will increase equally rapidly. The price of coal, for example, has risen far less than that of oil, and this gap may open even wider in coming years. More generally, oil's competitors are more difficult to transport, and also quite abundant; they should remain rather cheap at the place of production.

There should thus be islands of cheap energy. There is the hydroelectric potential of Nepal and of Inga in Zaire; the coal in Australia, South Africa, the western United States, and Sumatra; the natural gas in the Middle East, Indonesia, and Malaysia; and perhaps the geothermal potential of areas in the Philippines or Central America. These should draw energy-intensive activities, just as cheap labor countries have drawn labor-intensive industries. The moral is, therefore, that in planning for the long run, it is best to forego illusions that in some way—through an OPEC two-tier system or through a rollback of oil prices—the considerable cost to developing countries of increases of the world price of oil will be recouped. The correct strategy is to look at the new markets which will grow up and consolidate in the OPEC countries, to spot the changes of comparative advantage which will result from the changed structure of energy prices, and to design commercial policies which will enable developing countries to make the best possible use of these opportunities.

Chapter 19
Changing Relative Energy Prices, the Balance of Payments, and Growth in Developing Countries

Helen Hughes

The impact of the 1970s petroleum prices has become a highly charged political issue. The industrial countries have failed to deal with the implications of either the price increases or the shift of the higher petroleum rents from industrial-country consumers to petroleum-producing developing countries. Instead, they have used the petroleum price increases to veil their inability to control inflation (which in fact began in the late 1960s, well before the major increases in petroleum prices). Badly managed petroleum-importing developing countries, too, have found the petroleum prices a convenient excuse for slow growth.

The impact of changing petroleum prices on the developing countries has become difficult to evaluate in such a politicized atmosphere. But the increase in the price of petroleum, together with the shift in petroleum rents, has been only one of the many stochastic shocks with which developing countries have had to deal as they attempt to catch up with the industrial countries. In the 1950s and 1960s, the real price of petroleum had been in gradual decline. In contrast, in the 1970s it went up, principally in two sharp jumps. It was the sharpness of the increases, rather than the increase per se (taken as an annual average over the decade), that led to accelerated inflation and balance-of-payments difficulties.

I am indebted to my colleagues in the Economic Analysis and Projections Department of the World Bank for many insights and for the statistical base for this paper. Mildred Weiss helped to edit it, and Jo Saxe's comments in particular were very helpful. The views are my own, however, and should not be attributed to the World Bank.

An analysis of the impact of increased petroleum prices must recognize that developing countries represent a spectrum of development, from very undeveloped countries with low energy use to relatively technologically advanced, high-income countries with energy price adjustment problems similar to those of the industrial countries. The degree of success of an individual country's adjustment to the new prices has varied with its economic policies, just as it has in industrial countries. National policies have determined whether the short-term impact on a country's balance of payments and price structure has turned into a long-term problem. The well-managed developing countries' growth rates were thus not greatly affected. The 1970s' difficulties helped to precipitate major policy changes for some of the badly managed developing countries so that their prospects for the 1980s are better than they were for the 1960s or 1970s. Some countries, however, have not yet changed their policies. These are the ones for which debt and balance-of-payments problems signal critical domestic policy deficiencies.

The developing countries as a group are better endowed with petroleum than the industrial market economies. They also have many other unexploited and as-yet unexplored sources of energy (gas, hydro, thermal, coal, and nonconventional petroleum). They have gained from the resource rents associated with their indigenous energy resources, particularly if they were low-cost, and from the monopoly rents that the petroleum producers have been able to extort. Precisely how much they gained depended on their natural endowment and their domestic policies. Those that were poorly endowed by nature and have had inappropriate policies have lost heavily, but these losers were by no means necessarily low-income countries.[1] Moreover, insofar as developing countries have less commercial energy-intensive economies than the industrial countries, and are investing at a higher rate, they have more room to maneuver. Adjustment is therefore easier.

Their higher priced petroleum exports enabled a number of developing countries to increase their imports of goods and services in the early 1970s. It was easier for the more rapidly growing developing countries to enter these new markets than to struggle to sell their goods in the markets of the slowly growing industrial countries, particularly when the latters' adjustment capacity was weakened by their inept policy responses to the petroleum price increases. For some products exported by developing countries, protectionism grew in the 1970s although most industrial country protectionist actions were taken against other developed countries.[2] Some of the large, poor countries of the Middle East and South Asia found considerable demand for their labor in the petroleum-rich countries, leading to large inflows of workers' remittances and rapidly rising living standards for the emigrant workers and their families.

The low population, low absorption petroleum-exporting countries greatly increased the global supply of savings and made it much easier for rapidly growing developing countries to borrow to supplement their savings. Nineteenth-century conditons, when countries of new settlement borrowed heavily for development, were thus replicated. Inflation gave the developing countries a windfall gain in borrowing abroad, by transferring income from lenders to borrowers.[3]

An overview of the development process and of the impact of the 1973–74 petroleum price increases on it, forms the first part of this paper. The second part focuses on current balance-of-payments concerns and concludes with some thoughts about the likely evolution of developing-country growth during the 1980s.

LONG-TERM TRENDS IN INTERNATIONAL ECONOMIC GROWTH

The industrial revolution now sweeping the developing countries began in eighteenth-century Britain. The first wave of growth in British manufacturing (with accompanying changes in other sectors) did not take place in continental Europe and the new countries of European settlement until the nineteenth and early twentieth centuries. The first socialist country, the USSR, began to rebuild its economy in the 1920s. Some countries—principally in southern Europe and southern Latin America—were left behind until the 1950s.

Technological innovation (in the broad sense, including industrial organization) has played a key role in each phase, determining the leading countries with which the other countries aspired to catch up in their continuing pursuit of increased productivity and higher living standards. But technological change is not the only driving force of industrial progress. It has become increasingly evident that economic policy (broadly defined, to include such areas as education and industrial relations) both within nation states and internationally, is at least as important.[4]

Compared to an estimated 1–2 percent growth in the nineteenth and early twentieth centuries, the post–World War II growth looks spectacular indeed.[5] It must be stressed, moreover, that the use of official rather than purchasing-power exchange rates, Laspeyres indexes (uncorrected for quality and preference changes) to measure prices, and indirectly derived Paasche indexes to measure output, not only distorts comparisons of income levels among countries at a given time, but also understates true growth rates. This is particularly true for the developing countries which were growing rapidly in the 1950s and 1960s.[6]

Many factors—including vastly improved technology, improvements in mass education and hence in human capital, accelerated capital accumulation and investment, structural shifts that moved a significant proportion of the work force from low-productivity agriculture to high-productivity manufacturing and service activities, the movement of women from unpaid household to remunerated work, and flows of temporary or permanent immigrants—have variously been identified as causes of post–World War II growth. The remarkable freeing of capital and trade flows following the dismantling of exchange and other controls on capital movements in the market economy industrial countries, the Dillon and Kennedy Rounds of trade negotiations, and the additional filip given to intra-European trade and capital flows by the formation of the European Economic Communities (EEC) have also played a role in that growth. However, all these explanations are not unlike the blind man's description of an elephant.

The underlying force which made these trends so productive was the focus on social progress (albeit in varying degrees in different countries), backed up by a greatly improved policy formulation and administrative capacity. The 1930s' economic crisis had given rise to changed social perceptions; these were further strengthened by the war. The social waste that had prevailed for centuries became unacceptable, and the wartime experience demonstrated that national economic management was compatible with an essentially market-oriented system. The fear of renewed depression was still strong in the 1950s, and that fear strengthened the political pressures for rapid growth and full employment in the industrial countries, and for social welfare policies in many countries.[7] The protectionism of the 1930s was seen as one of the important causes of a miserable, stagnating world economy, and so the liberalization of the international movement of goods and factors of production was thought essential to growth. Growth became so rapid, and the benefits of liberalization so exceeded its costs, that further liberalization was encouraged. For most of the industrial countries, moreover, catching up to United States productivity and living standards meant following an easily defined growth path. In the developing countries, the internal pressures for catching up were even stronger, and they were endorsed internationally with the spread of social welfare ideals. Once poverty became unacceptable in a national context, inequality among nations also became a concern, leading to a substantial lobby for economic assistance to the developing countries.

Why did these sources of growth dry up in the industrial countries in the 1970s? Why did the increase in the price of petroleum become a serious adjustment problem, rather than lead to policies that would foster a new technological wave? A number of social, political, and economic factors came together in the industrial countries in the late 1960s and early 1970s. Their high general standards of living made further growth seem less necessary to the higher income groups. Indeed, to the dismay of the upper and even the middle classes, growth was spreading the very privileges of wealth which they alone had previously enjoyed.[8] The Club of Rome translated the concerns of the rich into a finite view of the earth's resources, and although this was quickly shown to be erroneous, new variants continued to appear.[9]

Growth thereby became identified with the destruction of the natural environment.

When trade union, farmer, business, and other groups have a short-term horizon, they tend to emphasize increasing their share of the cake rather than raising productivity to expand the cake itself. The slower the growth, the sharper the conflict over the distribution of the shares. The increased cost of petroleum has intensified that conflict. Such social conflict is the principal cause of inflationary pressure, and it frequently leads to a monetary irresponsibility which fuels inflation further. Subsequent monetary restraints and other deflationary policies result in even less willingness to accept more slowly growing incomes.[10] Social services have been greatly improved in most countries. There often is a welfare cushion of some sort for the very poor and the unemployed. Nevertheless, the poor and the unemployed have paid much of the cost of this social tug-of-war. The national characteristics of such pressure groups as trade unions, employers' associations, and farmers' cooperatives have a major influence on the outcome of these socioeconomic struggles. Growth policies have been easier to implement in countries with strong, centrally managed, productivity-oriented industrial trade unions, such as the Federal Republic of Germany, than in countries like the United Kingdom with its highly fragmented craft unions or in the United States where the trade union movement represents only a small proportion of the work force.[11]

In some countries the pendulum has swung too far from the 1930s. Governments which set out to achieve efficiency in production and equity in distribution have started to confuse the two objectives by using production policies for welfare ends. This has led to the nationalization of bankrupt firms and to public guarantees, subsidies, and similar measures that have undermined the competitiveness of the productive sectors. In some countries taxation appears to have reached levels at which it inhibits productivity and entrepreneurship, often undermining national probity as well. In addition, in some countries the educational system, particularly in the technical streams, seems to have deteriorated, with disastrous effects on productivity and innovation.

But to acknowledge something of an overkill in government activities is not to argue that government does not have an important role to play in economic growth. With growing investment lead times in major industries, notably energy, and with growing economies of scale in such industries as computer and aeroplane manufacturing which lead to more monopolistic structures of production, governments can no more abdicate their role in long-run economic planning and arbitration among conflicting economic interests than they can afford to abandon market mechanisms. The art is to balance market forces with government intervention so as to arrive at the best—not the worst—of these two worlds.

Governments can act only if they quickly reach a working consensus, and if they are staffed by capable and honest civil servants. These conditions are most readily achieved in small, relatively homogeneous countries. In a liberal international trading environment, these same small countries can simultaneously benefit from the economies of scale that come with participating in the international economy. It is not surprising that small countries rank among those that have grown most rapidly and become the wealthiest.

In contrast, the two large actors in the world economy, the United States and the EEC, have found policy determination and implementation increasingly difficult. Until the 1950s the United States was the only industrialized market economy with its own large market, accordingly reaping unique benefits of scale and competitiveness. This no doubt contributed to its technological and organizational leadership from the 1920s to the 1960s. But the countries that participated in the liberalization of international trade in the 1950s and 1960s acquired the same advantages of access to large markets. They did not, however, have to create an economic consensus among constituencies as diverse as the northeastern, southern, midwestern, and western states of the United States. In turn the United States, moreover, the division between the executive and the legislature exacerbates geographic heterogeneity, and the politicization of the higher echelons of the public service has reduced its professionalism, further increasing the difficulties of policy formulation and implementation. The EEC is now running into similar and even greater difficulties. Attempts by its 10 sovereign members to arrive at common policies are leading to surrenders to minority group pressures. The Common Agricultural Policy, with its high costs for the EEC as a whole and for its lowest-income countries in particular, is a prime example.

The United States' inflationary pressures were initiated by its political difficulties in absorbing the economic costs of the war in Vietnam, and since then it has had continuing problems in absorbing both its defense and its, albeit limited, social expenditures. To be sure, the United States has not been the only country with inflationary policies, nor have its policies been the world's worst ones. And it is true that the weight of the United States in the industrial countries' national income has dropped (from 60 percent in the 1950s to 40 percent in the 1970s), but it is still important enough to have serious inflationary spillover effects on the international economy, making it difficult for countries that have been able to contain inflation, such as Germany, to pursue reflationary policies.

Some industrial countries' failure to deal constructively with the petroleum price changes must be seen against this background, rather than as an isolated balance-of-payments problem. The well-managed industrial countries have taken several measures to deal with the new petroleum price situation, seeking to bring domestic prices into line with rising world petroleum prices, mainly by allowing prices to rise, by the use of taxes, and by some direct controls. Others, including the United States, have moved very slowly, attempting to stop price rises to cushion the impact on domestic consumers.

Although the detailed objectives and policy approaches with respect to development vary considerably, the growth of the developing countries has always been part of the process of catching up to the countries with higher living standards. Definitions of countries as developing, particularly those with relatively high productivity and living standards midway in a spectrum that reaches from, for example, Chad to Switzerland depend at best on the analytical purpose being pursued, but more often on political convenience. Development is a continuum along which the differences between those countries barely emerging from primitive agricultural production and relatively high-income, industrialized countries such as Spain or Taiwan are much greater than those between the latter and the established industrial countries. Moreover, such countries as India and the People's Republic of China, though still poor and predominantly agricultural, have large, sophisticated, and potentially internationally competitive industrial sectors.

The developing countries have compressed into 30 years much of the development that had taken the industrial countries nearly two hundred years to accomplish. It is impossible to illustrate this process quantitatively because of data limitations. But growth rates as conventionally calculated give some rough indications of relative magnitudes. Developing countries as a group grew somewhat less rapidly in the 1970s than in the 1960s (see table 19.1). Such aggregated figures are somewhat misleading because of the wide variation among countries. (table 19.2).[12]

The very poor countries south of the Sahara, lacking in physical and social infrastructure, entrepreneurs, and administrators, and politically fragmented as well, have had the hardest struggle. These economies are still predominantly agricultural, and harvests tend to dominate their performance. The drought in the Sahel, for example, was substantially responsible for keeping the 1970s growth rates at 1960s levels. Mineral and particularly petroleum development have helped some of these countries, offsetting the disastrous effects of the early 1970s' drought. This region includes some of the world's poorest developing countries, several of them with very rapidly growing populations. However, there have been some success stories in terms of GNP growth. The Ivory Coast, a petroleum-importing country, is the leading example.

The large poor countries of South Asia also grew relatively slowly, but for different reasons. Pakistan and Bangladesh suffered from severe political problems, with their growth rates falling from 7.3 percent and 4.1 percent in the 1960s to 4.1 percent and 2.1 percent, respectively, in the 1970s. India had come to independence with a relatively good infrastructure and well-developed human resources, but its size and heterogeneity made it difficult to create a meaningful political consensus for rapid growth. The development

Table 19.1. GNP and GNP per Capita Growth Trends (average annual percentage growth rates in 1977 US$ at official exchange rates)

	GNP			GNP per Capita		
	1950–60	1960–70	1970–78	1950–60	1960–70	1970–78
Industrial market economies	4.2	5.0	3.4	3.0	3.9	2.6
European centrally planned economies	6.1	4.8	5.5	4.3	3.6	4.6
Capital surplus petroleum exporting economies[a]	—	11.7	9.0	—	7.8	5.3
Developing market economies[b]	5.0	5.7	5.5	2.7	3.1	3.1
People's Republic of China	9.0	5.0	6.0	6.7	3.0	4.6

[a] Iran, Iraq, Kuwait, Libya, Oman, Qatar, Saudi Arabia, and the United Arab Emirates. Population growth includes immigrant inflow.
[b] Spain, Portugal, Greece, Yugoslavia, and other nonindustrial countries of the Mediterranean; Asia and Oceania, except Japan, Australia, New Zealand, and the People's Republic of China; the Middle East and Africa, except countries noted in[a], and Central and South America and the Caribbean.
Source: World Bank data, March 1980.

of agriculture to the point of self-sufficiency has been long and painful. Poor monsoons in the early 1970s (also in part responsible for the low Pakistan and Bangladesh growth rates) led to a fall in India's growth rates from 4.0 percent in the 1960s to 1.4 percent in 1970–74. By the mid–1970s self-sufficiency in food was achieved, and from 1974–78 growth rates rose to an average of 5.6 percent despite fluctuating climatic conditions. However, infrastructure development, particularly in transport and energy, has been slow. Manufacturing production has grown at only 4.5 percent a year, and the manufacturing sector is still very inefficient despite the advantages of a

Table 19.2. GNP Growth of Developing Countries by Principal Regions (average annual percentage growth rates in 1977 US$ at Official Exchange Rates)

	GNP		GNP per Capita	
	1960–70	1970–78	1960–70	1970–78
Africa south of the Sahara	4.4	4.4	1.9	1.6
The Middle East and North Africa	4.2	7.1	1.5	4.4
South Asia	4.3	3.5	1.8	1.3
East and Southeast Asia and Oceania	6.8	8.0	4.2	5.7
Latin America and the Caribbean	5.9	5.6	3.2	2.9
Southern Europe	7.1	5.1	5.6	3.5

Source: World Bank data, March 1980.

large domestic market, because of the early emphasis on capital-intensive basic industries, the fear of excessive private enterprise influence, and an overconcentration on the domestic market. The balance-of-payments situation has been greatly improved with self-sufficiency in food, 85 percent self-sufficiency in energy (total production as a share of total consumption—see table 19.3), a high inflow of workers' remittances, and some growth in nonfactor service exports to the petroleum-exporting countries.

Petroleum and other mineral exports have been important in North Africa and the Middle East, where performance has otherwise been disappointing in relation to the level of economic development at independence. Industrial growth has been rather slow, and industrial output by and large is not competitive with the old industrial countries. However, the direct and indirect effects of the petroleum price increases led to substantial growth in the 1970s. The capital surplus petroleum exporting countries grew more slowly in the 1970s, when petroleum prices increased sharply, than in the 1960s, when the rate of increase of the volume of petroleum exports peaked (table 19.1).

Latin America has had a longer history of political independence than most developing regions. Yet as late as the 1950s, it was still relatively backward economically. Argentina and Chile, together with Australia and New Zealand, had been among the world's highest-income countries in 1900, and were then already taking the first steps toward industrialization, yet by the 1950s they had become developing countries. Latin America is richly endowed with natural resources, including petroleum, and like the islands of the Caribbean, it has a favorable location for exploiting North American markets. But, despite a heavy emphasis on industrialization after World War II and considerable industrial growth, its overall growth has been slow. Latin American development has been hampered by political instability, and the region has been the victim of an inward-looking politicoeconomic ideology which long precluded effective participation in international trade. There were some exceptions: Brazil, a country heavily dependent on petroleum imports, nevertheless averaged a 9.2 percent growth in the 1970s, compared to 6.1 percent in the 1960s.

The East Asian countries have been the outstanding performers, with Korea, Taiwan, Hong Kong, and Singapore—the remarkable gang of four which are all heavy petroleum importers with negligible domestic energy resources—the fastest-growing group among developing countries. Two are city states with little alternative to export-oriented manufacturing development. The other two, Taiwan and Korea, have demonstrated the value of paying attention to all sectors and of maintaining a balance between domestic and externally oriented policies, although Korea may have even overbalanced its economy in the direction of exports.

The growth of the Southeast Asian countries has also been relatively rapid. They all have rich resource bases, although only Malaysia is a petroleum exporter. For the most part they have pursued relatively balanced economic policies, so they now have well-developed infrastructures and agricultural and manufacturing sectors, as well as healthy trade balances. Indonesia, because of its relatively poor policies, has been one of the slower growers of the group, despite its petroleum exports. In fact, the Southeast and East Asian countries grew faster in the 1970s than in the 1960s, with the petroleum-importing countries in the lead.

For several of the Southern European countries, the developing label is more political than analytical. It is not surprising that they have been in the forefront of catching up, with per capita incomes coming close to and, in the case of Spain, bypassing those of lagging industrial countries. Their economies have always been closely linked with Western Europe, and several are soon expected to join the EEC. This close connection, and their approach to the upper end of the catching up curve, has meant that their growth rate was lower in the 1970s than in the 1960s. Turkey, a relatively large country at the bottom end of the Southern European spectrum, performed badly because of inappropriate domestic policies and poor balance-of-payments management. Southern Europe's weight in total developing country income (about 20 percent) depressed the average growth rate in the 1970s for the developing countries. If Southern Europe is excluded, GNP growth for the developing countries was about the same as in the 1960s, although relatively high population growth rates have meant that per capita income grew significantly only in East and Southeast Asia, and in the Middle East and North Africa.

Typically, the developing countries' catching up process has a slow start, while minimal human and physical resources are accumulated and political cohesion is achieved. This is frequently followed by a Rostovian takeoff period when the country makes a rapid transition (which is impossible to measure) to being industrial. It then follows whatever short-term (Juglar) cycles and long-term (Kondratieff) waves dominate the high income, industrial countries' growth paths at that particular time. The growth process is not linear but "S" shaped, and the curve is neither smooth nor uniform for all countries. Some quickly graduate from being low- to middle- and then to high-income countries, while others grow very slowly, and some do not grow at all or even decline.

Political stability is very important to the development process, as are human resources. If the human resources are available, the lack of natural resources need not be a serious barrier to growth. In fact, bountiful natural resources can create problems if human resources are inadequate. Some of the countries with the greatest development difficulties and the slowest growth, including Zaire, Zambia, and Peru, are resource-rich countries. Iran has run into difficulties, and in the future other petroleum-exporting countries may also. Large size and heterogeneity, too, can be a handicap. India, Indonesia, Nigeria, Pakistan, and Bangladesh seem to have been held back by their large, heterogeneous political bases. In contrast, given the liberal trading environment of the last 30 years, a small domestic market need not lead to serious problems. A good location is a plus but it does not guarantee success. Whereas some small Mediterranean countries have exploited their proximity to Europe, most of those in Central America and the Caribbean have failed to take advantage of their proximity to the United States market. Some island economies, such as Mauritius, are growing strongly; others are not. Being landlocked is a handicap, particularly if surrounding countries are poor, yet the world's highest per capita income country (Switzerland) is landlocked. Large injections of aid in per capita terms (for example in Taiwan and Korea) and access to international capital markets have helped many countries, but others (like Turkey) have failed to grow in spite of large aid and commercial capital inflows.

Breaking through balance-of-payments constraints is important, but the claims made for outward-oriented industrialization strategies and export-led growth have tended to exaggerate the importance of exports to development. Rapidly growing countries have had well-balanced overall macroeconomic policies and an appropriate mix of sectoral policies. The principal impact of outward orientation tends to be to improve efficiency. Thus to look at developing country growth from a balance-of-payments point of view is to imagine that the tail wags the dog. In a liberal trading and monetary world, balance-of-payments constraints in the main reflect inappropriate domestic policies. In the early 1970s, analyses focusing on balance-of-payments problems thus failed to understand how the industrial countries' underlying internal economic problems would be exacerbated by increased petroleum prices; they greatly underestimated the petroleum-exporting countries' absorptive capacity; they predicted that the poorest developing countries would be most affected by petroleum price increases, and they were unable to imagine the degree to which the more rapidly growing petroleum importing countries would be able to take advantage of the petroleum exporters' growing markets for goods and services and of the increased availability of foreign capital.[13]

Countries that have introduced social infrastructure policies to overcome human constraints, agricultural policies to break rural production bottlenecks, industrialization policies based on internationally competitive costs of production, and appropriate physical infrastructure investment have enjoyed rapid growth. Energy conservation is merely another component of the efficiency that makes for rapid economic growth. The objective is to make optimal use of all natural resources and factors of production, not to follow now one fad—say, agricultural development— and then another—say, energy conservation. Savings and investment rates and ratios are as limited as indication of good policies as are balance-of-payments surpluses or deficits. The policies of the rapidly growing countries (and their implementation) have not been perfect, but they have been effective enough to release the pent-up energy inherent in the catching up process. A few countries, including Singapore and Malaysia, have been successful in attaining growth with equity by accentuating trickle down effects through paying careful attention to access to public goods, but most countries have paid little attention to welfare. Some, like Sri Lanka and Tanzania, have traded off growth against equity. Others have achieved neither growth nor equity. All of the rapidly growing countries have used market mechanisms as well as government intervention. For example, Hong Kong and Singapore, often regarded as laissez faire economies, have had considerable public works and housing programs which interacted with private sector growth. Indicative planning played a particularly important role in Korea's development. In short, the rapidly growing countries have avoided the shibboleths of both the right and the left. There are many paths to rapid economic growth. Countries must choose the one that suits their particular economic, social, and political conditions. If they choose correctly, they can handle not only the changes in petroleum prices but also the other changes in the economic environment.

THE IMPACT OF CHANGING PETROLEUM PRICES ON DEVELOPING COUNTRIES IN THE 1980s

It is tempting to postulate that radical economic changes and sharp restructuring will be as unlikely in the 1980s as in the 1970s. The countries with appropriate policies will do well, those with inappropriate policies will do badly. Unless Pakistan and Bangladesh resolve their political problems and India improves its policies so that South Asia accelerates its growth substantially, and unless more of the larger Latin American countries catch up to Brazil's performance (and the latter is maintained), the developing countries' average GNP growth rate is likely to continue at about 5.5 percent per annum (as always, with a considerable range in performance among countries). But while history continually repeats itself, it never repeats itself exactly, so actual events in the 1980s will inevitably be very different from the projected ones.

It must be stressed that, just as it makes no sense to talk about developing countries, it makes no sense to speak in terms of petroleum-importing and petroleum-exporting countries, or energy importers and energy exporters. There is a spectrum of developing countries, from major net importers of energy to major net exporters of energy (table 19.3). The ratios of indigenous energy production to domestic consumption range from about 1 percent for some importers to many times 100 percent for some energy exporters. The

Table 19.3. Energy Production as a Percentage of Energy Consumption, 1977

Country	Percentage
Africa south of the Sahara	
Angola	857.5
Botswana	100.0
Burundi	11.1
Cameroon	42.2
Central African Republic	23.4
Congo, People's Republic of	1,062.4
Equatorial Guinea	2.5
Ethiopia	21.1
Gabon	896.5
Ghana	56.5
Guinea	6.9
Ivory Coast	4.4
Kenya	11.3
Liberia	5.1
Madagascar	8.7
Malawi	32.1
Mali	8.1
Mauritius	5.9
Mozambique	145.0
Nigeria	1,962.4
Reunion	11.8
Rhodesia	68.5
Rwanda	48.3
Sao Tome & Principe	9.2
South Africa	89.2
Sudan	5.8
Swaziland	100.0
Tanzania	15.8
Togo	0.6
Uganda	32.8
Zaire	114.1
Zambia	96.3
South Asia	
Afghanistan	381.2
Bangladesh	38.5
Burma	92.2
India	84.2
Nepal	28.5
Pakistan	63.9
Sri Lanka	21.0
East & Southeast Asia & Oceania	
Brunei	1,310.5
China, Republic of (Taiwan)	23.3
Indonesia	279.3
Korea, Republic of	35.6
Laos, Peoples Dem. Republic of	38.2
Malavsia	143.3
New Caledonia	7.5
Papua New Guinea	4.2
Philippines	12.4
Samoa	7.3
Thailand	10.8
North Africa & the Middle East	
Algeria	703.4
Bahrain	123.1
Egypt	167.5
Iran	754.6
Iraq	2,051.6
Kuwait	1,252.2
Lebanon	11.0
Libya	2,607.0
Morocco	22.0
Oman	1,620.0
Qatar	1,194.5
Saudi Arabia	3,717.3
Syria	150.2
Tunisia	212.8
United Arab Emirates	3,495.7
Latin America and the Caribbean	
Argentina	91.5
Barbados	12.9
Bolivia	209.1
Brazil	33.1
Chile	63.3
Colombia	97.5
Costa Rica	31.5
Dominica	26.5
Dominican Republic	1.3
Ecuador	394.5
El Salvador	24.0
Guatemala	8.4
Haiti	19.9
Honduras	16.3
Jamaica	1.0
Nicaragua	4.4
Panama	2.4
Paraguay	32.1
Peru	75.5
Puerto Rico	1.9
St. Vincent	18.5
Surinam	15.3
Trinidad & Tobago	402.5
Uruguay	6.6
Venezuela	515.0

(Continued on p. 288)

Table 19.3. (Continued)

Southern Europe	
Greece	35.5
Israel	1.2
Portugal	29.0
Spain	32.0
Turkey	39.3
Yugoslavia	67.2
Industrial Countries	
Australia	131.8
Austria	47.0
Belgium	17.2
Canada	114.0
Denmark	2.8
Finland	19.0
France	25.9
Germany, Federal Republic of	46.9
Iceland	54.6
Ireland	23.3
Italy	20.5
Japan	8.5
Luxembourg	1.7
Netherlands	115.8
New Zealand	59.2
Norway	125.3
Sweden	40.6
Switzerland	51.7
United Kingdom	77.6
United States	82.3
European Centrally Planned Economies	
Albania	190.5
Bulgaria	33.0
Czechoslovakia	76.6
Germany, Democratic Republic of	70.2
Hungary	57.5
Poland	109.6
Romania	93.6
USSR	126.6
Asian Centrally Planned Economies	
China, People's Republic of	101.9
Korea, Democratic People's Republic of	95.7
Mongolia	65.1

Source: United Nations data.

major importers and exporters are well known, but some of the countries lying between them in the spectrum (such as India, Pakistan, and Argentina) already are or could become self-sufficient in energy (i.e., indigenous production equal to or exceeding consumption) and some others (such as Chad, the Philippines, and Thailand) have good energy production prospects.

Table 19.4 indicates the net volume of petroleum imports for the principal developing country petroleum importers for 1978 and 1979, with estimates for 1980.

The bulk of developing country petroleum imports is concentrated in a handful of countries. Spain, which might be excluded because its per capita income is already higher than one or two of the established industrial countries, is the second largest importer. Brazil is by far the largest, Korea is third, and India, despite its domestic energy resources, is fourth. With the exception of India, the low-income countries, particularly the African countries south of the Sahara, have relatively low petroleum imports.

The countries with the most serious balance-of-payments problems are not among the major petroleum importers. A country's balance of payments reflects much more than the level of imports. It indicates the country's export growth, the growth of such earnings as workers' remittances, its propensity to import, and its general balance-of-payments policies. As al-

Table 19.4. Net Petroleum Imports of Developing Countries

	1978	1979	1980[a]
		(million b/d)	
Middle-Income Countries	5.220	5.150	5.100
Brazil	.900	.960	.950
Chile	.085	.080	.080
Colombia	.007	.025	.025
Greece	.205	.220	.215
Korea	.455	.495	.520
Philippines	.255	.190	.180
Portugal	.125	.140	.145
Spain	.940	.890	.860
Thailand	.190	.230	.230
Turkey	.105	.105	.100
Yugoslavia	.210	.225	.225
Others	1.773	1.590	1.570
Low-Income Countries	.575	.565	.558
South Asia	.465	.455	.450
Bangladesh	.050	.050	.050
India	.315	.300	.295
Pakistan	.065	.065	.064
Sri Lanka	.020	.025	.026
Others	.015	.015	.015
Sub-Sahara Africa	.110	.110	.108
Total Developing Countries	5.795	5.715	5.658

[a] Estimates.
Source: United Nations data.

ready indicated, balance-of-payments surpluses or deficits are the result of external factors only in the short run. In the long run, they flow from national policy decisions. The countries with chronic balance-of-payments difficulties are the ones with low rates of export growth. In some cases, inappropriate monetary policies discourage the inflow of workers' remittances. Some countries have a high propensity to import, often in spite of high protection. For example, on the average, Latin American countries have higher import/GNP ratios than East Asian countries, although the latter's relatively high exports are import- and (particularly petroleum-) intensive. In any case, a balance-of-payments deficit is not necessarily a sign of inappropriate policies. Capital is, by definition, scarce in the developing countries. Even some petroleum-exporting countries borrow against future earnings to accelerate development. The question is whether the developing countries use their borrowed funds productively in the context of sound overall development to ensure their long-run solvency whether they can manage their balance of payments and debt from year to year to ensure adequate cash flows or liquidity, and whether capital markets will be sufficiently liquid and sufficiently competent to continue to make capital available to sound borrowers.

The following assumptions have been made to analyze possible future scenarios. It is assumed that petroleum prices will continue to rise at about 3 percent a year in real terms in the 1980s and 1990s until it becomes clear to investors that a price has been reached at which ecologically satisfactory alternative energy supplies are available in sufficient volume. Such a price is estimated to be about $45–50 a barrel (in 1980 dollars); at an increase of 3 percent a year, it would take until about the year 2000 to reach that level. Within that period prices may have to rise to even higher levels to bring forth sufficient investment, given the high initial costs of new sources of energy. Actual price movements are more likely to continue to take place in jumps and to fluctuate about the trend than to increase smoothly, for a number of reasons—the political difficulties of raising prices in consuming countries, the stickiness of consumer habits, the importance of existing equipment, the possibilities of appropriating some of the rents in consuming countries by taxation, the capital surplus petroleum exporters' awareness of the short-term benefits of exploiting a backward sloping marginal revenue curve (particularly in a period of high inflation), and international and domestic political interests of various kinds. This complicates but does not fundamentally change the policy implications. It is assumed that other energy prices will follow the price of petroleum but that the price relations within the petroleum group and the relations of other energy prices to petroleum prices will continue to vary considerably over time from country to country, according to national policies. This will, of course, be less true for such internationally tradeable sources of energy as gas, coal, and uranium, for which "world prices" in broad terms exist, than for hydro, geothermal, and nuclear electricity.

Table 19.5. Average Annual Increase in Selected Global Indicators (percent)

	1977–80	1980–85	1985–90
GNP in industrial countries (at constant prices)	2.5	3.3	4.0
World trade (at constant prices)	4.8	5.2	5.7
Exports of developing countries[a] (at constant prices)	5.2	5.5	6.4
Price index of traded goods and services (at constant prices)	15.7	8.2	6.4
Supply of external capital[b] to developing countries (at current prices)	24.3	5.9	5.8

[a] Excludes capital surplus petroleum exporters.
[b] Official and private capital flows.

Given the above assumptions and the current levels of debt, financing the prospective balance-of-payments deficits is feasible not only for the developing countries but also for the centrally planned and industrial countries which are expected to run current account deficits during the next few years.

Table 19.6. Global Distribution of Current Account Balances,[a] 1975–1990 (current billion US$)

	1975	1979	1980	1990
Capital surplus petroleum-exporting countries	37	81	109	70
Industrial countries	22	−5	−26	100
Centrally planned economies[b]	−7	1	—	−10
Developing Countries	−49	−53	−74	−140
Petroleum-importing developing countries				
Low-income	−5	−7	−9	30
Middle-income	−34	−37	−55	70
Total	−39	−43	−64	100
Capital deficit petroleum-exporting countries	−10	−10	−9	40
Statistical discrepancy	4	23	9	20

[a] 1980 is estimated and 1990 is projected.
[b] Includes China.

It is assumed that aid will remain at about 0.3 percent of the GNP of the industrial countries and at 3.0 percent or more of the GNP of the capital surplus petroleum exporters. These were the ratios for the 1970s, and they seem unlikely to increase, given the relatively slow growth in the industrial countries and the political reactions against aid in several among them. Such aid flows would be sufficient to take care of the balance-of-payments problems of the very poor countries in Africa south of the Sahara and of Bangladesh, which are not sufficiently creditworthy for commercial capital.

Capital market flows to the other developing countries are postulated to increase less rapidly than prices; that is, they would not increase in constant prices. Given the capital surpluses likely to be generated by the capital surplus petroleum exporters and the industrial countries, the assumptions made here on capital flows to the other developing and centrally planned economies are probably too conservative. As the discussion of the likely trajectory of petroleum prices indicates, there could be additional transfers from the industrial to the petroleum-exporting countries in the 1980s; this would almost certainly lead to a greater demand for, and supply of, capital going to developing countries. However, given current debt levels and the conservative assumptions incorporated in the balance-of-payments projections, plus an assumption that real interest rates will be about 2 percent for the 1980s as a whole, the debt projections shown in table 19.7 emerge.

Table 19.7 Developing Countries' Debt Situation—1970, 1980, and 1990[a] (Current billion US$ and percent)

	1970	1980	1990
External debt outstanding, disbursed only (current billion US$)			
Low-income countries	17	60	240
Middle-income countries	47	340	1,000
Total	64	400	1,240
Debt service (current billion US$)			
Low-income countries	1	6	20
Middle-income countries	7	60	210
Total	8	70	230
Ratio of debt service to exports of total goods and services (%)			
Low-income countries	13	14	15
Middle-income countries	10	14	12
Total	11	14	13
Ratio of interest to exports of total goods and services (%)			
Low-income countries	5	5	6
Middle-income countries	3	5	4
Total	3	5	4
Ratio of international reserves to external debt outstanding (%)			
Low-income countries	16	31	26
Middle-income countries	38	39	41
Total	32	37	38

[a]1980 is estimated and 1990 is projected. Figures exclude the capital surplus developing countries.

The modest increases projected for debt cover great disparities. They include petroleum-exporting countries which now account for more than 30 percent of debt[14] and other mineral-rich countries for whom debt service is not likely to be a heavy burden. However, a few countries would have relatively high debt service without natural resource wealth. The well-managed among them should be able to refinance their debt so that their actual debt service would consist mainly of interest payments. The combination of petroleum imports and interest service would remain manageable even for most of the high borrowers, given appropriate national economic policies. Moreover, although the overall export growth assumptions used for this paper are modest, some countries will have difficulty reaching their share of even that growth. Trade between the developing countries, which depends on the developing countries' reduction of protection, will have to grow, as will trade with the industrial countries. For some countries, workers' remittances (not included in the debt service and interest service ratios) will continue to be an important factor in meeting debt service obligations.

Given the preceding assumptions, by 1990 interest payments would still be only 4 percent of total export earnings. This does not mean that a number of developing countries, including some petroleum exporters, will not find themselves in serious debt-management difficulties. Some countries which did not tighten up their economic management in the 1970s will have solvency problems in addition to cash-flow liquidity problems. That is, their borrowing has gone for high-income groups' consumption, or has been invested wastefully. In addition some countries may become so destabilized politically that their economic and debt management will also deteriorate. Such countries will cease to be creditworthy. Past experience suggests that it is impossible to foretell such cases; in most periods some new troubled countries emerge while others improve their policies. This is also likely to be true in the 1980s, unless confrontation over petroleum supplies leads to destabilization in the Middle East on a large scale.

Data on international flows are still inadequate. Flows to developing countries were estimated at the end of the 1970s to account for much less than 20 percent of total funds available in the principal industrial countries. While some banks, like some countries, may run into trouble, particularly as a result of excessive lending to badly managed countries, the system is not likely to collapse. The capital market is liquid, and relatively high interest rates seem to reflect, in addition to the industrial countries' positive interest rate policies, tough negotiation postures by lenders toward the major borrowers. Maturities vary, influenced by borrowers' preferences for the management of debt through refinancing as well as liquidity. The banks are still active in what seem doubtful borrowing countries, though the bulk of funds is still flowing to established borrowers.

Yet it is frequently being stated that financing the developing country deficits will be difficult. It is argued (with varying doomsday emphases) that the developing countries are too borrowed up to borrow further, that several will be unable to service their debt and therefore will have to repudiate it and so destroy the creditworthiness of all developing countries, or that the major banks in the international lending business are too exposed in certain developing countries and will stop lending either on their own initiative or be forced to do so by their governments. It is even being suggested that some banks may go bankrupt and thereby destroy the structure of the international markets. Some of these views assume that the banks currently engaged in these markets will not show the flexibility they demonstrated in the 1970s and that no new banks and financial institutions will come forward to begin lending, that is, that the international capital markets will stop evolving. Most of these hypotheses have been encouraged by banks seeking publicly financed safety nets for risk-free lending and by other lobbyists. The talk of crisis has lost credibility because major difficulties have not materialized, but it is still dangerous because it could become a self-fulfilling prophecy. In any case, by making the financial markets nervous it leads to unnecessarily high borrowing costs for the developing countries.

Some Alternative Scenarios

The projections discussed so far in this paper are intended to give a reasonable, balanced scenario, but unlikely up and down alternatives should also be considered.

On the optimistic side, the industrial countries might improve their macroeconomic and sectoral policies, including their energy policies. Their lack of investment in the 1970s and the resulting emerging high-capacity utilization in efficient and competitive industries make the situation ripe for more growth-oriented policies based on a macroeconomic policy package that would contain inflation, conserve energy, and stimulate investment in alternatives to petroleum. Firm policies in this direction could lead to high investment and overall recovery in the early 1980s. Energy prices would continue to rise but there would be a reduction in inflation, an increase in GDP growth (and aid flows) and growth in north-south trade. Developing countries would be helped mainly by the more rapid growth of the industrial countries and trading opportunities. (It is assumed throughout this paper that terms of trade trends for primary products are predominantly cyclical, although changes in technology and taste from time to time lead to terms of trade changes for particular products, and climatic changes lead to short-term shifts in agricultural prices.) There might be some short-run crowding out of developing countries in capital markets because of major industrial country investment in the energy sector, but this would soon be offset by increased savings. For developing countries, and particularly for small countries for which international trade and borrowing can be very important, this could mean an acceleration of growth if they are able to take advantage of a more buoyant world economy. They are not dependent on such an improvement for accelerated growth, however. What they do domestically, including the handling of their energy policies and how they expand their mutual trade,[15] will be the main factors.

The prospects on the downside are more problematic. Low growth for the industrial countries in the 1980s and 1990s would be likely to slow the increase in petroleum prices and stretch out the adjustment process.[16] A move toward a regionalization of trade around a protectionist EEC, North America, and Japan would reduce world trade and thus further hurt growth. However, the impact is not likely to be considerable, except in disaster scenarios.

The most likely (though by no means inevitable) disaster scenario would be further political breakdown in the Middle East. If it involved Saudi Arabia, with a concomitant sharp reduction in world petroleum supplies, those developing countries heavily dependent on petroleum imports would be seriously affected even if the industrial countries, which are the principal consumers, and the producers outside the Middle East were to combine in an effective international rationing system. The resultant impact on the balance of payments could take so many forms that it does not make much sense to speculate about it. Nevertheless, it is clear that however unlikely a financial crisis may seem, a Middle East breakdown would most likely be accompanied by one. In this case a war economy that extends beyond the minor wars of the 1950s and 1960s has to be postulated.

A more sensible approach would be to consider conditions that might prevent such a situation. The rapid expansion of alternate energy resources, together with energy conservation, is obviously necessary. Apart from the Middle East and such energy-rich countries as Mexico and Venezuela, only the industrial countries can make a major difference in the demand and supply situations. Even if they take conservation seriously, the developing countries cannot contribute much positively because they are expanding economies. The most important country on both the supply and the demand side is the United States. This is where policy change is critical and urgent. The price of petroleum is not yet high enough to make improved United States policies politically acceptable. Perhaps the most important political economy conclusion to be drawn is that petroleum prices—instead of rising slowly and steadily until they reach the level at which substitutes become ample, as rational economics would suggest—need to be pushed up rapidly, even if this means some additional (and in rational terms unnecessary) costs for both developing countries and industrial countries.

NOTES

1. Although the term "low income" is used from time to time in this paper, it should be noted that it has a descriptive rather than an analytical meaning. Well-managed low-income countries such as Korea or the Ivory Coast have during the last decade become middle-income countries. See also footnote 5.
2. James Riedel and Linda M. Gard, "Recent Changes in Industrial Protectionism: An Assessment of the Implications for Developing Countries," World Bank, October 1979 (mimeo).
3. A phenomenon observed by C. K. Hobson in *The Export of Capital* (London: Constable and Company, Ltd., 1914), when interest rates were positive and inflation was less than 2 percent.
4. It was, of course, Schumpeter's linking of the importance of technological innovations with Kondratieff's statistical observations that brought the Kondratieff cycle into economic analysis. A more up-to-date version might add that an evolution of economic policies is required to catch up to the technological leaders in each successive wave.
5. See D. Morawetz, *Twenty-five Years of Economic Development, 1950–75* (Washington: World Bank Type II Report Service 1977) for a summary of the developing countries' rapid growth since World War II. Despite the large literature of poverty, and whatever

may have happened to income distribution, growth has meant considerable increases in living standards.

6. For purchasing power parity distortions, see Irving B. Kravis, Zoltan Kenessey, Alan Heston, and Robert Summers, *A System of International Comparisons of Gross Product and Purchasing Power* (Baltimore and London: The World Bank, The Johns Hopkins University Press, 1975); and Irving B. Kravis, Alan Heston, and Robert Summers, *International Comparisons of Real Product and Purchasing Power* (Baltimore and London: the World Bank, The Johns Hopkins University Press, 1978). Such distortions are not removed by the use of domestic currency growth rates to estimate past growth (see Irving B. Kravis, Alan W. Heston, and Robert Summers, "Real GDP per Capita for More than One Hundred Countries," *The Economic Journal* 88 (June 1978): 215–242), for these understate the growth of countries moving from simple, nonmonetized to industrial economies. Ongoing work by Kravis, Heston, and Summers, and by Robin Marris, is seeking to identify the character and magnitude of such underestimation of growth. The catching-up process which is understated by growth rate measurements, whether in local or international currencies, is fully captured only when countries revalue their currencies against those of the countries with which they are catching up.
7. H. W. Arndt, *The Rise and Fall of Economic Growth: A Study in Contemporary Thought* (Melbourne: Longman, 1978).
8. See Fred Hirsch, *Social Limits to Growth* (Cambridge: Harvard University Press, 1976) for such a middle class view.
9. Donella H. Meadows, Dennis L. Meadows, Jorgen Randers, and William W. Behrans II, *The Limits to Growth: A Report for the Club of Rome on the Predicament of Mankind* (New York: Universe Books, 1972) gave rise to a debate which quickly negated, on the basis of careful research, this report's findings. Interfutures, *Facing the Future: Mastering the Probable and Managing the Unpredictable* (Paris; OECD, June 1979) sums up the case against the "finite world" concept.
10. See the growing literature on the political economy of protectionism, particularly R. E. Caves, "Economic Models of Political Choice: Canada's Tariff Structure," *The Canadian Journal of Economics* 9 (May 1976): 278–300. A study of 11 industrialized countries is currently being undertaken by the World Bank in conjunction with a number of institutions in those countries, and particularly with the Centre for Political Economy in the Free University of Brussels, on this subject in relation to the penetration of industrial country markets by manufactured imports from developing countries.
11. See Mancur Olson, "*The Political Economy of Comparative Growth Rates*," University of Maryland, November/December 1978 (mimeo). It is interesting, however, that the shop stewards' hold on trades unions seems to be weakening as at least some rank-and-file workers perceive their sectional interests to be bound up with national ones.
12. See World Bank, *World Atlas*, *passim*, for details of country growth that indicate the range of experience.
13. Hollis B. Chenery, "Restructuring the World Economy," *Foreign Affairs* (January 1975): 53, pp. 242-263 for example, a took a pessimistic approach.
14. Including Iran, which is classified as a petroleum-exporting capital surplus country elsewhere in this paper.
15. See Helen Hughes, "Inter Developing Country Trade and Employment," (Paper presented at the IEA Sixth World Congress, Mexico, August 1980).
16. See Interfutures, *Facing the Future: Mastering the Probable and Managing the Unpredictable* (Paris: OECD, June 1979), for low growth scenarios for the industrial countries.

Chapter 20
Energy, International Trade and Economic Growth

Alan S. Manne
with the assistance of Sehun Kim

This paper complements work that is now ongoing in connection with the IBRD's Fourth World Development Report (hereafter abbreviated WDR IV). Results are reported from a small-scale international trade model that focuses on issues related to energy and economic growth. The analysis provides some broad orders of magnitude as a cross-check upon detailed country-by-country projections, but is not to be viewed as a substitute for these "bottom-up" analyses. Our model is compact enough so that all numercial results (and virtually all input assumptions) can fit onto a single sheet of paper. This also facilitates a rapid turnaround in response to changes in input assumptions. The numerical results here are based upon the WDR IV guidelines available in November, 1980.

In the following pages, we review the economic assumptions that underlie the model, and present nine alternative cases—varying the assumptions with respect to energy supplies, demands, economic growth and capital flows. Half of these cases imply much the same price scenario as specified in the current set of WDR IV guidelines—a 3 percent annual real oil price increase between 1980 and 1990. It will be shown, however, that relatively small differences in supplies—and relatively small differences in the ease of de-

---

This work was supported by the International Bank for Reconstruction and Development.

Thanks go to the following individuals for helpful comments and for input data: Robert Cassen, Hollis Chenery, B. J. Choe, Enzo Grilli, Helen Hughes, Peter Pollak and Christine Wallich. On an earlier version of this paper, helpful comments were received from Bert Hickman, Lawrence Lau and Jean Waelbroeck. The author, however, is solely responsible for the views expressed here—and for any remaining errors. This paper is an expanded version of the presentation given by the author at the ERPI Workshop on Energy and the Developing Nations.

mand adjustments to rising prices—can lead to 100 percent differences in the energy prices projected for 1990.

Despite the volatility of energy prices, the GDP growth rates of the industrialized countries do *not* vary widely. Imported energy still remains a relatively small fraction of their GDP, and growth appears largely determined by the underlying assumptions with respect to the productivity of capital and labor. Even a 100 percent difference in energy prices leads to a difference of only 0.3 percent in the annual GDP growth rates of these countries between 1978 and 1990. At first glance, this appears a counter-intuitive result—and quite different from the conventional wisdom expressed in newspaper headlines and editorial pages. Our analysis does not, however, refer to short-term political and military disruptions, but to medium-term economic trends. With a decade for energy demands and trade flows to adjust to higher prices, there is room for guarded optimism that energy constraints will not be the principal determinant of the industrialized countries' growth during the 1980s.

For the oil importing developing countries, the outlook is more problematic. During the 1970s, they succeeded in financing their oil deficits through an unprecedented inflow of capital. If there are even larger capital inflows to compensate for the likely rise in oil prices during the 1980s, the developing countries may be able to achieve the growth targets outlined in WDR IV. Without such flows, the outlook is considerably bleaker. A 100 percent rise in energy prices would then have a major impact upon their terms of trade, and could reduce their GDP growth rates by as much as 1.1.% per year.

General Background

Although the developing countries now consume a relatively small fraction of the world's energy, this fraction could grow rapidly over the coming decades. By comparison with the industrialized nations, the developing countries have high population growth rates and high income elasticities of demand for commercial energy. Moreover, if today's North-South income disparities are to be reduced, the developing countries' per capita incomes will have to grow more rapidly. Taken together, these factors imply a substantial increase in their demand for energy. For example, the WAES study (1977, p. 269) projected that the developing countries' share of the world's commercial energy consumption would grow from 15 percent in 1972 to 25 percent in the year 2000. Meanwhile, however, the principal consumers of energy are the industrialized countries. Their imports—and OPEC's readiness to expand production—are the principal determinants of international oil prices.

In view of the price increases experienced since 1973, it is quite likely that the oil exporting developing countries will continue to enjoy a rapid increase in GDP, and will consume ever-increasing amounts of their own energy production. The oil importing developing countries, however, constitute a far more populous group, and their energy demand projections appear quite uncertain. With sluggish growth in traditional export markets among the industrialized nations, the oil importers are likely to encounter chronic balance-of-payments difficulties. Their prospective trade deficits cannot easily be offset by official development assistance or by private capital flows.

In qualitative terms, it is easy enough to arrive at these generalizations. But what about their *quantitative* impacts? To what extent is energy likely to impose constraints upon economic growth? For this purpose, it appears essential to construct a computable model of international trade. One region's export prospects cannot be assessed without an understanding of the import propensities of its trading partners. International trade may relieve energy constraints upon the growth of individual countries—provided that they can expand their exports of non-energy products to other nations. The ease or difficulty of this adjustment process will be governed largely by the elasticities of substitution in energy and in international trade.

The Basic Simplifications

These are the general considerations that have led to the specific approach taken here—a three-region general equilibrium model. This is not intended for general-purpose analysis of international trade and macroeconomic fluctuations. It is not designed to deal with individual commodities outside the energy sector, hence it is inappropriate for policy issues such as comparative advantage, tariff and non-tariff barriers to trade. For these purposes, one would need far more disaggregation than that adopted here (see Ginsburgh and Waelbroeck, 1975, and Whalley, 1980). Like these more detailed trade analyses, we shall construct a snapshot view (comparative statics) for a single point of time, 1990. There will be no attempt to analyze the year-by-year details of the transition from 1978 (the data benchmark year) to 1990. This is a "surprise-free" scenario.

Aggregating the 19 individual regions distinguished within WDR IV, we arrive at the following three-region classification of the market economies:

1. industrialized countries (region 17; the major OECD nations)
2. oil exporting developing countries (regions 10-16; primarily OPEC + Mexico)
3. oil importing developing countries (regions 1-9).

This model is not addressed to East-West trade issues. In effect, we assume zero net trade in energy with the centrally planned economies (regions 18, 19). This is closely consistent with other WDR IV projections.

Following in the tradition of Armington (1969), Barten (1971), and Hickman *et al.* (1973 and 1979), *non*-energy tradeables are aggregated in dollar terms but distinguished by their region of origin. Energy forms are expressed in terms of their oil-equivalents. Thus, there are only four internationally traded goods whose prices are to be determined: energy and one composite non-energy export product for each of the three regions. All prices are to be viewed as c.i.f. (cost, insurance, and freight). Because of base-year data limitations, we shall project merchandise trade flows only.

Both an optimistic and a pessimistic view will be explored with respect to energy *supplies*. In both cases, these will be taken as exogenous data for each region in 1990. This "bean counting" approach appears more practical than attempting to estimate price-responsive supply curves. Over the next decade, the industrialized nations' domestic supplies appear far less dependent upon future prices than upon institutional and public acceptability factors, e.g. petroleum leasing policies, environmental constraints upon air quality, and nuclear safety regulations. Institutional and political factors also have a major impact upon the supplies likely to be available in the developing nations.

In order to project each region's *demand* for energy and for nonenergy imports, we employ nested CES (constant-elasticity-of-substitution) production functions. With a CES production function, there are diminishing marginal returns from imports into each region. To this extent, the model automatically allows for "absorptive capacity" constraints within the rapidly growing economies of the oil exporters (see Ezzati, 1978).

Each region's future endowments of capital and labor enter into its individual production function. These endowments are estimated in accordance with the region's *potential* GDP growth rates under constant prices of energy and non-energy items. *Each region views these international prices as a datum.* To the extent that OPEC exerts monopoly power, this is already incorporated in the assumptions with respect to the quantities of energy supplied by region 2 (OPEC + Mexico).

Prices are projected so as to equilibrate 1990 supplies and demands. Given the prices of the internationally tradeable goods—and the production functions together with each region's capital, labor and energy resources—it is supposed that each region will choose a mix of energy and of non-energy imports so as to maximize its GDP. The maximization is subject to region-by-region constraints upon the balance of merchandise trade. The outcome of this process is termed the *realized* GDP. This may exceed or fall below the potential GDP, depending upon whether there is an improvement or a deterioriation in the international terms of trade.

For technical details on the three-region model, see Appendix A and the forthcoming dissertation by S. Kim.

Macroeconomic Assumptions and Energy Supplies

Potential GDP growth rates are a key input to the three-region model. These growth rates may be interpreted as an index number of capital and labor inputs—provided that the labor force is expressed in "efficiency units" to allow for productivity growth. One approximation should be noted. In effect, we are assuming that the growth of domestic capital stocks will *not* be significantly affected by oil price increases or by other changes in the terms of trade. For 1990, side-calculations suggest that this feedback effect is minor. Over a longer time horizon, however, we would need an intertemporal model to allow properly for this savings and investment process.

For the "high GDP" cases reported here, the 1978-90 potential growth rates are identical to those being employed elsewhere in WDR IV: 3.5 percent for the industrialzied countries, 6.4 percent for the oil exporters and 5.3 percent for the oil importing developing countries. In general, the *realized* growth rates turn out to be lower for the oil importers and higher for the exporters. This is an immediate consequence of the rise in energy prices projected between 1978 and 1990.

Two alternatives will be considered with respect to the commercial energy supplies available to the market economies in 1990: 122 and 114 mbd. From Table 20.1, note that 122 mbd is broadly consistent with the WDR IV guidelines, but that 114 mbd represents a more pessimistic assessment of the supply prospects within the developing countries. With a continuation of military and political disturbances in the Middle East, it is quite possible that there could be a 5 mbd shortfall in region 2. Similarly, there could be a 3 mbd shortfall in region 3 if energy development is not pursued aggressively there. Note that the 122 and 114 mbd cases bracket the 116 mbd "mid-price" scenario released in October 1980 by the US Energy Information

Table 20.1. Alternative Energy Supply Projections for Three-region Model, 1990 (mbd: million barrels daily oil equivalent)

	Region 1 Industrialized countries	Region 2 Oil exporting DCs	Region 3 Oil importing DCs	Total regions 1-3
IBRD, EPDCE, October 29, 1980	64.3	47.0	14.7	126.0
Cases 1,3,5,7,9	62.3	45.0	14.7	122.0
Cases 2,4,6,8: 5 mbd shortfall in region 2 and 3 mbd shortfall in region 3	62.3	40.0	11.7	114.0
1978 actual supplies	46.9	35.9	6.8	89.6

Administration (1980, p. 4). All these estimates lie well below the earlier ones made by the EIA, IBRD and other organizations. The overthrow of the Shah, and the anti-nuclear fallout from Three Mile Island, have led to drastic downward revisions since 1978.

Energy Demands—the Adjustment Problem

Energy demands depend both upon price and income elasticities. These runs have been based on the following GDP elasticities of demand for commercial energy: 1.0 in region 1; and 1.3 in regions 2 and 3 for 1978, but declining thereafter. That is, at *constant* energy prices and other terms of trade, each one percent increase in GDP will lead to an increase of 1.0 percent in energy demands in region 1; and 1.3 percent in regions 2 and 3. These GDP elasticities seem consistent with time series and also multi-country econometric estimates—made both before and after the oil price increases of 1973 (see Blitzer, Choe and Lambertini, 1979, p. 55).

But energy prices have not been constant during the past decade, and they are likely to rise still further during the 1980s. What clues on price elasticities can we gather from the post-1973 experience with higher energy prices? At the time this report went to press, satisfactory data were available only for the industrialized countries. Between 1970 and 1978, c.i.f. oil prices had tripled (in dollars of constant purchasing power). Over this same period, real GDP increased by 3.5 percent per year, and energy consumption increased at a 2.0 percent annual rate. To some extent, we have already observed a decoupling between energy consumption and economic growth.

For the industrialized countries between 1970 and 1978, the conventionally measured "income elasticity of demand for energy" was 2.0/3.5 = .57. But this calculation tells us little or nothing about long-term price elasticities of demand or the prospects for further adjustments of energy demands to higher prices. If we assume that all of the adjustments to the tripling in oil prices had already occurred by 1978, the implied long-term elasticites are quite low. Conversely, if we assume that only 25 percent of the long-term adjustment had occurred by 1978, the implied long-term price elasticity of demand for primary commercial energy is .27 (in absolute terms). This view of adjustment, plus multi-country cross-section econometric studies, has led the Energy Modeling Forum to specify .40 as the long-term price elasticity of demand for primary energy. (This .40 value is specified for the EMF's reference case and for most of the other scenarios currently being analyzed in the World Oil Study.) If this view is correct, the after-effects of the 1973-78 price increase will continue to help decouple energy consumption from GDP growth during the 1980s. Further assistance will be provided by the 1978-80 price shock and the cumulative impact of the real price increases projected post-1980.

For the developing countries, there have been only a handful of systematic studies of demand growth since the 1973 price rise. Some clues may be obtained from the following comparison between two time periods: 1960-70 and 1970-78. Dunkerley and Matsuba (1980) have calculated that the conventionally measured "energy/GNP elasticity" dropped (from 1.29 to .94) for the oil importers, and that it *rose* (from 1.04 to 1.46) for the oil exporters between the 1960s and the 1970s. The decline for the importers would be roughly consistent with the price elasticities estimated by Blitzer, Choe and Lambertini (1979). The rise for the exporters is counter-intuitive, but may be related to the fact that many of them held their domestic energy prices well below international levels during the 1970s. This stimulated domestic demands and also helped attract energy-intensive industries away from traditional locations.

What follows from all this? On the basis of the actual experience between 1970 and 1978, we cannot be at all certain whether energy demand adjustments will be easy or difficult. The key uncertainty appears to be the diversity in opinions on how long it takes for energy demands to respond to rising prices.

With the CES production functions employed in the three-region model, energy demand adjustments are specified quantitatively through "elasticities of substitution." These are medium-term elasticities for 1990, and therefore lie below the long-term price elasticities of demand for primary energy. In order to achieve comparability with the current set of WDR IV energy demand projections, we have had to adopt rather low values for the 1990 elasticity of substitution between energy and other productive inputs. This will be termed a "difficult" view of the adjustment process. By contrast, when we employ the elasticities corresponding to an "easy" view, the results for the industrialized countries check quite closely with those in the Energy Modeling Forum's reference case. It is assumed that the 1990 energy substitution elasticities will vary by region as follows:

	difficult adjustment	easy adjustment
region 1. industrialized countries	.15	.20
region 2. oil exporting DCs	.10	.10
region 3. oil importing DCs	.10	.15

The elasticity of substitution refers to each individual region's production function. It is defined as the optimal reduction in total demand for energy relative to that of capital and labor in response to a one percent increase in the relative price of energy, total output remaining constant.

International Trade and Capital Flows

For each of the three regions' production functions, there is an elasticity of substitution between *imports* and domestic capital and labor. On the basis of the literature review cited by Whalley (1980, pp. 31-33), we have adopted import trade elasticities of the order of unity: 1.25 for the industrialized nations and .80 for the developing countries. Since each region's imports constitute another's exports, there is no need to specify export elasticities of demand in this model.

Side calculations suggest that the trade elasticities do not have a major impact upon energy price projections. They do, however, affect the relative prices of *non*-energy tradeables. In region 2, for example, case 1 implies an 83 percent real increase of export prices (e.g. through domestic inflation or through currency appreciation). With these changes in the terms of trade, there are strong incentives to import non-energy tradeables. (Between 1978 and 1990, this generates a 14 percent annual increase in imports from region 3.) Moreover, these changes in the terms of trade lead to strong *dis*incentives to export non-energy items from region 2. If the import elasticities were higher, the non-energy price adjustments might be lower, but trade volumes would still have to move in much the same direction so as to restore balance-of-payments equilibrium.

To some extent, international capital flows can cushion the impact of rising oil prices. In order to span a wide range of views, two polar extremes will be examined. Neither is designed to be altogether realistic, but together they cover a realistic range of possibilities. The optimistic view will be described as that of "compensatory capital flows." In these cases, the oil exporters (region 2) recycle 10% of their energy revenues, and the net effect of international capital markets is that this entire amount is available to finance the merchandise trade deficit of region 3. Thus, the higher that oil prices rise, the greater becomes the value of these compensatory capital flows.

By contrast, the pessimistic view is described as "zero trade deficits." In these cases, region 3 succeeds in borrowing just enough new capital to offset its non-merchandise current account deficit (including interest payments on past debts). With zero trade deficits, export volumes must be expanded, and there is a deterioration in the terms of trade. Regions 3's GDP growth is affected by amounts ranging up to 1.2% per year. Under the conditions associated with cases 2 and 6 (table 20.2 below), capital flows of $78 billions would permit a GDP gain of $167 billions in 1990.

Numercial Results

For "high GDP growth" in region 1, eight alternative scenarios have been calculated. These differ with respect to energy supplies, demands and capital

Table 20.2. Alternative Projections of Oil Prices, GDP Growth Rates, and Merchandise Trade

Case number	Capital flows	Adjustment of energy demands	Available energy supplies, regions 1-3, 1990	Results for 1990							
				Oil price (1980 $/barrel)	Realized annual GDP growth rates, 1978–90			Value of merchandise trade, region 3, (billions of 1978 dollars, adjusted for changes in terms of trade)			
					Region 1	Region 2	Region 3				
					Industrialized countries	Oil exporting DCs	Oil importing DCs	Energy imports imports	Non-energy imports	Exports	Merchandise trade deficit
High GDP growth (region 1):											
1.	compensatory	difficult	122 mbd	$45	3.2%	7.1%	4.5%	73	143	156	60
2.	compensatory	difficult	114 mbd	$67	3.0	7.4	4.0	123	106	151	78
3.	compensatory	easy	122 mbd	$31	3.3	6.5	4.6	46	150	155	41
4.	compensatory	easy	114 mbd	$43	3.2	6.8	4.3	79	125	153	51
5.	zero deficits	difficult	122 mbd	$44	3.2	7.4	3.6	51	114	165	0
6.	zero deficits	difficult	114 mbd	$64	3.1	7.7	2.8	86	77	163	0
7.	zero deficits	easy	122 mbd	$31	3.3	6.8	3.9	35	129	164	0
8.	zero deficits	easy	114 mbd	$42	3.3	7.0	3.5	60	102	162	0
Potential annual GDP growth at constant energy prices and other terms of trade, 1978–90					3.5%	6.4%	5.3%				
Low GDP growth (region 1:											
9.	zero deficits		116 mbd	$43	2.8%	7.0%	3.5%	46	107	153	0
					1978 trade flows, region 3			16[a]	122	95	43

[a] This $16 billions refers to net merchandise imports recorded in SITC category 3. At energy import rates of 6.3 mbd and a c.i.f. price of $13.70 per barrel, the 1978 energy import bill would have been $31.5 billions. Since these two methods of estimation are inconsistent, one must be cautious in making comparisons between our 1990 projections based on physical import rates and the 1978 statistics referring to SITC category 3.

flows. For a summary, see figure 20.1 and table 20.2. Details on each case are provided in Appendix B. Merchandise trade flows are expressed in *value* terms (adjusted for changes in terms of trade) in table 20.2 and in *volume* terms in Appendix B.

According to WDR IV guidelines, the international price of oil will rise at 3 percent per year, net of inflation, between 1980 and 1990. This implies a 1990 price of about $45 per barrel, expressed in dollars of 1980 purchasing power. Cases 1,4,5, and 8 have been constructed so that they all lead to this level of oil prices. That is, real prices could increase by 3 percent annually during the 1980s, either with the 122 mbd level of supplies and "difficult" demand adjustments or with 114 mbd and "easy" adjustments. Note, however, that plausible changes in the input assumptions could lead to energy prices that are either much higher or lower than $45. If the "difficult" adjustment view is correct—and if 1990 energy supplies are limited to 114 mbd—there could be a sharp runup. (The price of a barrel of oil rises to about $65 in cases 2 and 6.) Conversely, with "easy" demand adjustments and 122 mbd of supplies, prices could remain close to their 1980 level of $33. Note that the GDP growth rate of the oil exporters is highly correlated with the level of oil prices.

Oil prices are *not* significantly affected by the presence of compensatory capital flows (hatched bars in figure 20.1) or by their absence (no hatching). It appears, however, that capital flows have a substantial impact upon the GDP growth rates of region 3, the oil importing developing countries. If these flows are available to cushion the impact, a 100 percent rise in oil prices leads to a drop of 0.6 percent in their GDP growth rates. (Compare cases 2 and 3.) Without these flows, the drop is projected at 1.1 percent. (Compare cases 6 and 7.) In the most pessimistic scenario examined here (case 6), the GDP growth rate would be only 2.8 percent—barely enough to keep pace with the growth of population.

Despite the wide range of possible energy prices, the GDP growth rates of the industrialized countries do *not* vary widely. Imported energy still remains a relatively small fraction of their GDP, and growth appears largely determined by the underlying assumptions with respect to the productivity of capital and labor. Even a 100 percent difference in energy prices leads to a difference of at most 0.3 percent in the annual GDP growth rates of these countries between 1978 and 1990. At first glance, this appears a counterintuitive result and quite different from the conventional wisdom expressed in newspaper headlines and editorial pages. Our analysis does not, however, refer to short-term political and military disruptions, but to medium-term economic trends. With a decade for energy demands and trade flows to adjust to higher prices, there is room for guarded optimism that energy constraints will not be the principal determinant of the industrialized countries' growth during the 1980s.

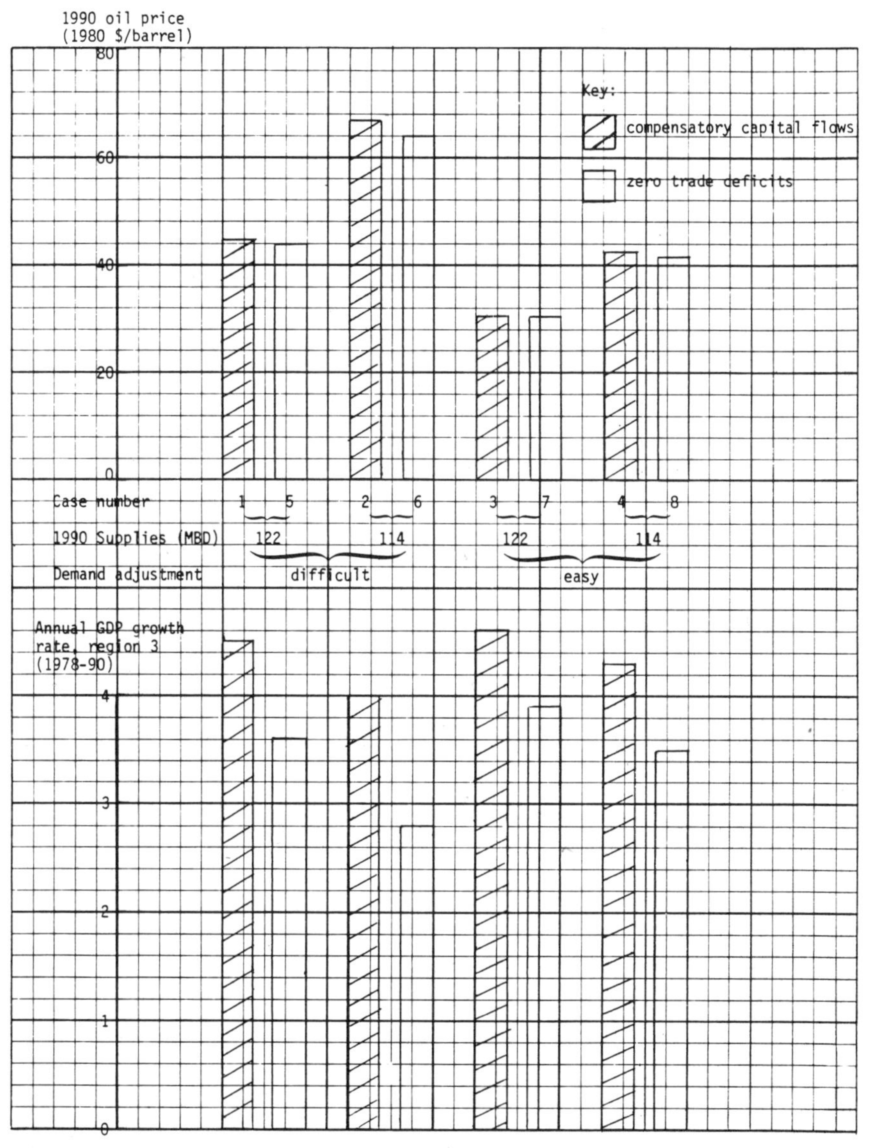

Fig. 20.1. Oil prices and GDP growth rates, region 3 (high GDP growth in region 1)

On the medium-term prospects for oil prices, some insights may be obtained from the Energy Modeling Forum's World Oil Study. See figure 20.2, and note that the WDR IV guideline value for 1990 lies in the middle of the range obtained for the EMF's reference case. There is a wide spread in the price projections from the 10 different models participating in this study,

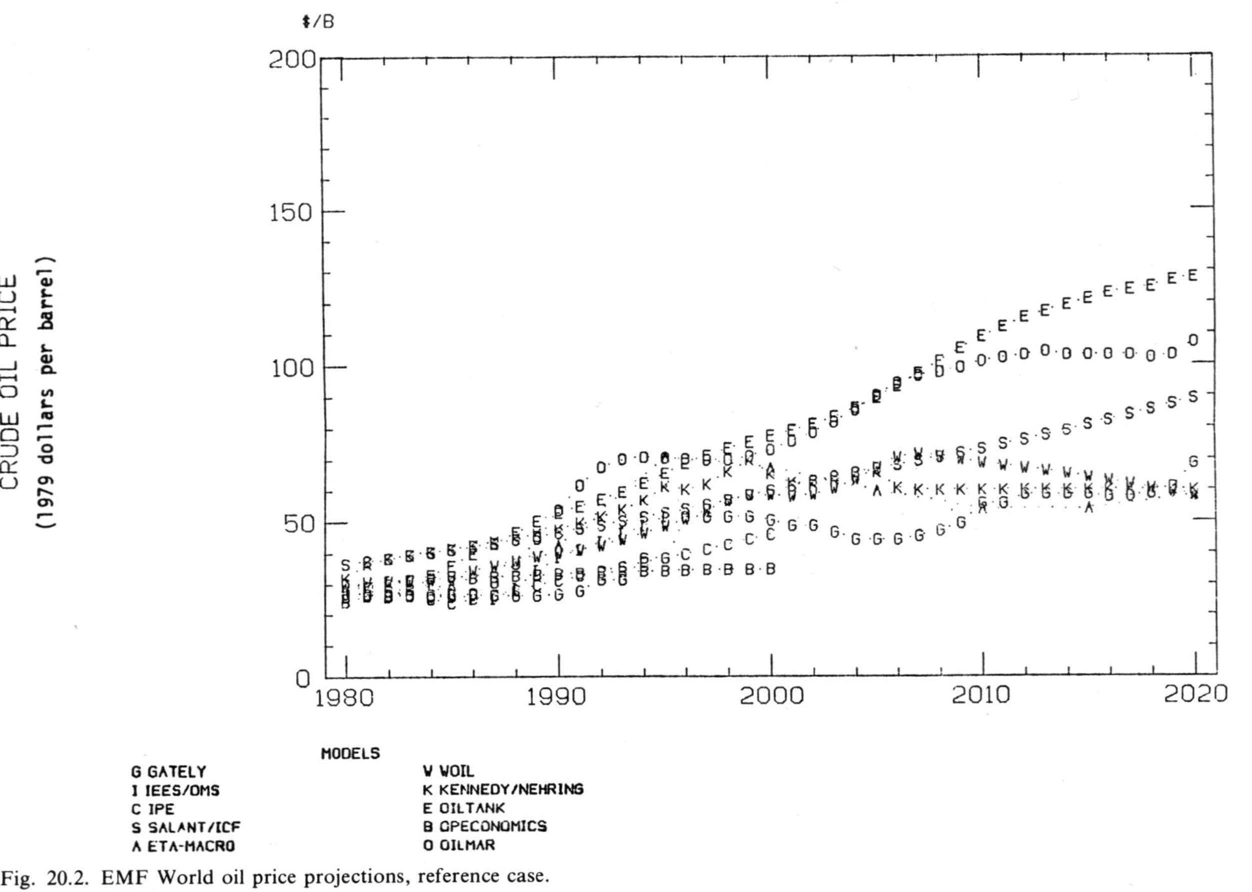

Fig. 20.2. EMF World oil price projections, reference case.

even though it was intended that all would be run on similar assumptions with respect to GDP growth rates, price elasticities of demand, OPEC capacity, oil resources, etc.

ETA-MACRO was one of the models included in the EMF exercise. Its 1990 price projection lies in the central cluster, and cannot easily be distinguished from its neighbors. Among these models, ETA-MACRO was the only one to allow for dynamic energy-economy interactions—via the accumulation of capital and a "putty-clay" model of energy conservation. It is of particular interest, therefore, that the EMF reference case results check so closely with case 4 of the three-region model. Under this view of demand adjustments, it is relatively easy to decouple energy consumption from GDP growth in the industrialized countries:

	Three-region model, case 4 (114 mbd and "easy" demand adjustments)	ETA-MACRO, reference case for EMF World Oil Study
Annual growth rates, industrialized countries, 1978-90:		
Primary energy consumption	1.2%	1.29%
GDP	3.2%	3.05%
Ratio ("energy/GDP elasticity")	.38	.42

Despite extensive econometric research, there is still no general agreement whether the "easy" or the "difficult" adjustment view is more nearly correct for the industrialized countries. See the diversity of results reported in the demand elasticity study by the Energy Modeling Forum (1980). For the developing countries, there is an even wider range of debate—and only a handful of empirical studies. Surely it ought to be a high-priority research topic to study the post-1973 energy price and consumption trends in these countries, and to draw whatever inferences are possible on the basis of this evidence.

A Postcript on "Low GDP Growth"

As of December 1980, WDR IV's "low" case had not been fully defined. It was generally understood that this would refer to a scenario in which the industrialized countries' GDP growth rate would be lowered by at least half a percentage point annually during the 1980s. This decline might be the result of high energy prices, or it might be attributed to declining growth in the labor force and productivity or to other factors. In any case, it was specified that international oil prices would continue to increase by 3 percent annually in real terms.

Case 9 represents an initial attempt to describe the WDR IV "low" case in terms of the three-region model. It is identical to case 5 (zero trade deficits and difficult demand adjustments) with just two exceptions: (a) region 1's *potential* GDP growth rate is reduced from 3.5 percent to 3.0 percent for the years 1978–90; and (b) region 2's energy production is cut back by 6 mbd to avoid the erosion of prices in the face of sluggish demand growth. (See table 20.2 and Appendix B.)

Under these conditions, region 1's *realized* GDP growth rate drops from 3.2 percent (case 5) to 2.8 percent (case 9). Perhaps the principal surprise is that there are no major indirect effects on region 3. Their export markets grow more slowly, but this leads to a minor decline in their GDP growth rate, from 3.6 percent (case 5) to 3.5 percent (case 9). Note, however, that the indirect effects upon region 3 might be far more serious if low domestic growth were to stimulate political pressures for protectionism in the industrialized countries. This possibility has not been incorporated in our "low GDP growth" case.

APPENDIX A. MODEL FORMULATION

The following definitions are adopted here:

Input data (excluding production function parameters)

d_j = domestic energy supplies in 1990, region j (adjusted for non-unitary GDP elasticities and capital transfers, if any)

x_{ii} = index number of capital and labor inputs in 1990, region i; can also be interpreted as potential GDP at constant 1978 prices of energy and non-energy imports

Quantity variables

y_i = realized GDP, region i (in 1978 dollars, adjusted for terms of trade)

x_{ij} = non-energy products of region i imported into region j; ($i \neq j$; i, j = 1,2,3)

x_{4j} = energy consumed by region j

Price variables

π_j = price of non-energy products, region j (j = 1,2,3)

π_4 = price of energy

It will be convenient to report all prices as π_j/π_1. That is, the numéraire is defined as region 1 products, 1978 dollars. For 1980 dollars, add 25 percent.

The nested CES production functions appear within the first three material balances below. There will be considerable debate about the magnitudes of the *exponents* in these CES functions. The ease or difficulty of substituting between domestic and imported non-energy inputs will be determined by the parameter α. The "elasticity of trade substitution" is defined as $1/(1 - \alpha)$. Similarly, the "elasticity of energy substitution" is $1/(1 - \beta)$.

Given the values of the exponents α and β, we employ 1978 data to estimate the a_{ij} constants that appear in the nested CES functions. (The 1978 benchmark estimates are reproduced in Appendix B.) Assuming that the inputs were *optimally* adjusted to the international prices prevailing in 1978, it is straightforward to determine the a_{ij} coefficients from the first-order optimality conditions. That is, the marginal productivity of each input must be equal to its 1978 price. For details, see the forthcoming dissertation by S. Kim. This will also describe our solution algorithm, an extension to nonlinear economies of the procedure described by Manne *et al.* (1980).

In connection with this benchmarking, two observations should be noted: (1) To allow for the incomplete adjustment (approximately 50 percent) that had taken place by 1978 in response to the energy price increases of 1973-74, we assumed that the 1978 "reference price" of oil was only $10 per barrel. This is the price inserted into the first-order optimality equations—not the actual 1978 price of $13.70. (2) The production function elasticities incorporate institutional and behavioral as well as technological constraints. In effect, our benchmarking procedure ignores the possibility of changes in tariff and non-tariff barriers to international trade between 1978 and 1990. Other approaches are required to investigate these barriers to adjustment.

The model is based upon four material balances for the four tradeable commodities:

GNP	+	exports of non-energy products	$\leq$	domestic production, non-energy: non-energy inputs	energy inputs

$$y_i + \sum_{j \neq i} x_{ij} \leq \left[\left(\sum_{j=1}^{3} a_{ij} x_{ji}^{\alpha}\right)^{\beta/\alpha} + a_{i4} x_{4i}^{\beta}\right]^{1/\beta} \qquad (i = 1,2,3)$$

where subscripts for region i are omitted on exponents α and β.

total energy consumption	$\leq$	total energy supplies

$$\sum_{j=1}^{3} x_{4j} \leq \sum_{j=1}^{3} d_j$$

There is also one balance-of-trade constraint for each region:

$$\begin{pmatrix}\text{value of}\\ \text{non-energy}\\ \text{exports}\end{pmatrix} \geq \begin{pmatrix}\text{value of}\\ \text{non-energy}\\ \text{imports}\end{pmatrix} + \begin{pmatrix}\text{value of energy}\\ \text{imports}\\ \text{(exports)}\end{pmatrix}$$

$$\pi_i \sum_{j \neq 1} x_{ij} \geq \sum_{\substack{j \neq i \\ j \neq 4}} \pi_j x_{ji} + \pi_4 (x_{4i} - d_i) \qquad (i = 1,2,3)$$

As written here, the three balance-of-trade constraints imply that net capital transfers will be zero in 1990. To allow for such transfers, we have adjusted the ownership of energy supplies—transferring 10 percent of region 2's resources and placing them at the disposal of region 3.

To reflect 1.3 GDP elasticities of energy demand (at constant prices) in the developing countries, another adjustment is made to the domestic energy supplies, d_j. That is, 30 percent is added to the base-year energy supplies *and* demands. The identical amount is added to the 1990 domestic energy supplies. In this way, we allow for a gradual long-term decline in the GDP elasticity of demands. This 30 percent adjustment can also be interpreted as an allowance for non-commercial to be replaced by commercial energy.

One piece of algebra is straightforward, but is not shown explicitly: In cases 1,3,5,7 and 9, region 2's energy supplies include 3 mbd available at a domestic cost of $12 per barrel (1978 price level). These supplies are *not* available in the shortfall scenarios (cases 2,4,6, and 8).

One final detail: The 1978 base-year energy consumption was adjusted so that it would be identical to the supplies available within the market economies. Consumption was increased by 0.6 percent in each region to account for the net effect of exports from the centrally planned economies, bunkers and other items.

APPENDIX B. NINE ALTERNATIVE PROJECTIONS

High GDP growth:

1.	Compensatory capital flows;	difficult demand adjustments;	122 mbd
2.	” ;	” ;	114 mbd
3.	” ;	easy demand adjustments ;	122 mbd
4.	” ;	” ;	114 mbd
5.	Zero trade deficits ;	difficult demand adjustments;	122 mbd
6.	” ;	” ;	114 mbd
7.	” ;	easy demand adjustments ;	122 mbd
8.	” ;	” ;	114 mbd

Low GDP growth:

9.	Zero trade deficits ;	difficult demand adjustments;	116 mbd

CASE 1.

HIGH GDP GROWTH
COMPENSATING CAPITAL FLOWS
DIFFICULT ADJUSTMENT OF ENERGY DEMANDS
122 MBD ENERGY SUPPLIES IN 1990

(UNIT FOR GDP AND EXPORTS: BILLIONS OF 1978 U.S. DOLLARS)
(UNIT FOR ENERGY: MILLION BARRELS DAILY; OIL EQUIVALENT)

		GDP		EXPORTS TO REGION 1, NON-ENERGY	EXPORTS TO REGION 2, NON-ENERGY	EXPORTS TO REGION 3, NON-ENERGY	ENERGY CONSUMPTION	DOMESTIC ENERGY PRODUCTION AVAILABLE	1990 ELASTICITY	
		REALIZED	POTENTIAL						TRADE	ENERGY
REGION 1	1978	5547.	5547.	-----	85.4	120.4	70.7	46.9		
INDUSTRIALIZED	1990	8075.	8382.		314.6	140.3	86.8	62.3	1.25	0.15
COUNTRIES	ANNUAL GROWTH	3.2%	3.5%		11.5%	1.3%	1.7%	2.4%		
REGION 2	1978	457.	457.	8.4	-----	2.2	5.8	35.9		
OIL EXPORTING	1990	1045.	962.	5.2		1.4	15.0	45.0	0.80	0.10
DCS	ANNUAL GROWTH	7.1%	6.4%	-3.9%		-3.5%	8.2%	1.9%		
REGION 3	1978	819.	819.	85.1	10.1	-----	13.1	6.8		
OIL IMPORTING	1990	1390.	1522.	166.2	48.2		20.2	14.7	0.80	0.10
DCS	ANNUAL GROWTH	4.5%	5.3%	5.7%	13.9%		3.7%	6.6%		
TOTAL	1978	6823.	6823.	93.5	95.5	122.6	89.6	89.6		
REGIONS, 1-3	1990	10510.	10866.	171.4	362.8	141.8	122.0	122.0		
	ANNUAL GROWTH	3.7%	4.0%	5.2%	11.8%	1.2%	2.6%	2.6%		
1990 EQUILIBRIUM PRICES				1.000	1.833 (A)	0.727 (A)	45.39 (B)			

NOTES:
(A) TERMS OF TRADE INDICES. RATIOS TO 1978 DOLLARS FOR REGION 1.

(B) ENERGY PRICES EXPRESSED IN 1980 $/BARREL; COMPARE WITH THE 1978 ACTUAL PRICE OF $13.7 AND THE 1978 "REFERENCE PRICE" OF $10/BARREL EMPLOYED FOR BENCHMARKING PURPOSES.

(C) ENERGY-GDP ELASTICITIES OF 1.3 AT CONSTANT PRICES IN REGIONS 2 AND 3. DECLINING AFTER 1978.

CASE 2

HIGH GDP GROWTH
COMPENSATING CAPITAL FLOWS
DIFFICULT ADJUSTMENT OF ENERGY DEMANDS
114 MBD ENERGY SUPPLIES IN 1990

(UNIT FOR GDP AND EXPORTS: BILLIONS OF 1978 U.S. DOLLARS)
(UNIT FOR ENERGY: MILLION BARRELS DAILY; OIL EQUIVALENT)

		GDP		EXPORTS TO REGION 1, NON-ENERGY	EXPORTS TO REGION 2, NON-ENERGY	EXPORTS TO REGION 3, NON-ENERGY	ENERGY CONSUMPTION	DOMESTIC ENERGY PRODUCTION AVAILABLE	1990 ELASTICITY	
		REALIZED	POTENTIAL						TRADE	ENERGY
REGION 1	1978	5547.	5547.	-----	85.4	120.4	70.7	46.9		
INDUSTRIALIZED COUNTRIES	1990	7940.	8382.		375.1	103.6	80.7	62.3	1.25	0.15
	ANNUAL GROWTH	3.0%	3.5%		13.1%	-1.2%	1.1%	2.4%		
REGION 2	1978	457.	457.	8.4	-----	2.2	5.8	35.9		
OIL EXPORTING DCS	1990	1082.	962.	3.6		1.0	15.2	40.0	0.80	0.10
	ANNUAL GROWTH	7.4%	6.4%	-6.7%		-6.4%	8.4%	0.9%		
REGION 3	1978	819.	819.	85.1	10.1	-----	13.1	6.8		
OIL IMPORTING DCS	1990	1310.	1522.	153.3	58.0		18.0	11.7	0.80	0.10
	ANNUAL GROWTH	4.0%	5.3%	5.0%	15.7%		2.7%	4.6%		
TOTAL	1978	6823.	6823.	93.5	95.5	122.6	89.6	89.6		
REGIONS, 1-3	1990	10332.	10866.	157.0	433.1	104.6	114.0	114.0		
	ANNUAL GROWTH	3.5%	4.0%	4.4%	13.4%	-1.3%	2.0%	2.0%		
1990 EQUILIBRIUM PRICES				1.000	2.238 (A)	0.716 (A)	67.00 (B)			

NOTES:

(A) TERMS OF TRADE INDICES. RATIOS TO 1978 DOLLARS FOR REGION 1.

(B) ENERGY PRICES EXPRESSED IN 1980 $/BARREL; COMPARE WITH THE 1978 ACTUAL PRICE OF $13.7 AND THE 1978 "REFERENCE PRICE" OF $10/BARREL EMPLOYED FOR BENCHMARKING PURPOSES.

(C) ENERGY-GDP ELASTICITIES OF 1.3 AT CONSTANT PRICES IN REGIONS 2 AND 3. DECLINING AFTER 1978.

CASE 3.

HIGH GDP GROWTH
COMPENSATING CAPITAL FLOWS
EASY ADJUSTMENT OF ENERGY DEMANDS
122 MBD ENERGY SUPPLIES IN 1990

(UNIT FOR GDP AND EXPORTS: BILLIONS OF 1978 U.S. DOLLARS)
(UNIT FOR ENERGY: MILLION BARRELS DAILY; OIL EQUIVALENT)

		GDP		EXPORTS TO REGION 1, NON-ENERGY	EXPORTS TO REGION 2, NON-ENERGY	EXPORTS TO REGION 3, NON-ENERGY	ENERGY CONSUMPTION	DOMESTIC ENERGY PRODUCTION AVAILABLE	1990 ELASTICITY	
		REALIZED	POTENTIAL						TRADE	ENERGY
REGION 1	1978	5547.	5547.	-----	85.4	120.4	70.7	46.9		
INDUSTRIALIZED COUNTRIES	1990	8211.	8382.		230.2	147.9	88.2	62.3	1.25	0.20
	ANNUAL GROWTH	3.3%	3.5%		8.6%	1.7%	1.9%	2.4%		
REGION 2	1978	457.	457.	8.4	-----	2.2	5.8	35.9		
OIL EXPORTING DCS	1990	976.	962.	8.2		2.0	14.0	45.0	0.80	0.10
	ANNUAL GROWTH	6.5%	6.4%	-0.2%		-0.9%	7.6%	1.9%		
REGION 3	1978	819.	819.	85.1	10.1	-----	13.1	6.8		
OIL IMPORTING DCS	1990	1399.	1522.	188.3	36.5		19.7	14.7	0.80	0.15
	ANNUAL GROWTH	4.6%	5.3%	6.8%	11.3%		3.5%	6.6%		
TOTAL	1978	6823.	6823.	93.5	95.5	122.6	89.6	89.6		
REGIONS, 1-3	1990	10585.	10866.	196.6	266.7	149.9	122.0	122.0		
	ANNUAL GROWTH	3.7%	4.0%	6.4%	8.9%	1.7%	2.6%	2.6%		
1990 EQUILIBRIUM PRICES				1.000	1.337 (A)	0.691 (A)	31.29 (B)			

NOTES:

(A) TERMS OF TRADE INDICES. RATIOS TO 1978 DOLLARS FOR REGION 1.

(B) ENERGY PRICES EXPRESSED IN 1980 $/BARREL; COMPARE WITH THE 1978 ACTUAL PRICE OF $13.7 AND THE 1978 "REFERENCE PRICE" OF $10/BARREL EMPLOYED FOR BENCHMARKING PURPOSES.

(C) ENERGY-GDP ELASTICITIES OF 1.3 AT CONSTANT PRICES IN REGIONS 2 AND 3. DECLINING AFTER 1978.

CASE 4.

HIGH GDP GROWTH
COMPENSATING CAPITAL FLOWS
EASY ADJUSTMENT OF ENERGY DEMANDS
114 MBD ENERGY SUPPLIES IN 1990

(UNIT FOR GDP AND EXPORTS: BILLIONS OF 1978 U.S. DOLLARS)
(UNIT FOR ENERGY: MILLION BARRELS DAILY; OIL EQUIVALENT)

		GDP		EXPORTS TO REGION 1, NON-ENERGY	EXPORTS TO REGION 2, NON-ENERGY	EXPORTS TO REGION 3, NON-ENERGY	ENERGY CONSUMP-TION	DOMESTIC ENERGY PRODUCTION AVAILABLE	1990 ELASTICITY	
		REAL-IZED	POTEN-TIAL						TRADE	ENERGY
REGION 1	1978	5547.	5547.	-----	85.4	120.4	70.7	46.9		
INDUSTRI-ALIZED COUNTRIES	1990	8130.	8382.		260.7	122.7	81.9	62.3	1.25	0.20
	ANNUAL GROWTH	3.2%	3.5%		9.7%	0.2%	1.2%	2.4%		
REGION 2	1978	457.	457.	8.4	-----	2.2	5.8	35.9		
OIL EX-PORTING DCS	1990	1002.	962.	6.5		1.6	14.1	40.0	0.80	0.10
	ANNUAL GROWTH	6.8%	6.4%	-2.2%		-2.7%	7.7%	0.9%		
REGION 3	1978	819.	819.	85.1	10.1	-----	13.1	6.8		
OIL IM-PORTING DCS	1990	1354.	1522.	181.2	41.6		17.9	11.7	0.80	0.15
	ANNUAL GROWTH	4.3%	5.3%	6.5%	12.5%		2.6%	4.6%		
TOTAL	1978	6823.	6823.	93.5	95.5	122.6	89.6	89.6		
REGIONS, 1-3	1990	10487.	10866.	187.6	302.3	124.3	114.0	114.0		
	ANNUAL GROWTH	3.6%	4.0%	6.0%	10.1%	0.1%	2.0%	2.0%		
1990 EQUILIBRIUM PRICES				1.000	1.547 (A)	0.688 (A)	43.37 (B)			

NOTES:
(A) TERMS OF TRADE INDICES. RATIOS TO 1978 DOLLARS FOR REGION 1.

(B) ENERGY PRICES EXPRESSED IN 1980 $/BARREL; COMPARE WITH THE 1978 ACTUAL PRICE OF $13.7 AND THE 1978 "REFERENCE PRICE" OF $10/BARREL EMPLOYED FOR BENCHMARKING PURPOSES.

(C) ENERGY-GDP ELASTICITIES OF 1.3 AT CONSTANT PRICES IN REGIONS 2 AND 3. DECLINING AFTER 1978.

CASE 5.

HIGH GDP GROWTH
ZERO TRADE DEFICITS
DIFFICULT ADJUSTMENT OF ENERGY DEMANDS
122 MBD ENERGY SUPPLIES IN 1990

(UNIT FOR GDP AND EXPORTS: BILLIONS OF 1978 U.S. DOLLARS)
(UNIT FOR ENERGY: MILLION BARRELS DAILY; OIL EQUIVALENT)

		GDP		EXPORTS TO REGION 1, NON-ENERGY	EXPORTS TO REGION 2, NON-ENERGY	EXPORTS TO REGION 3, NON-ENERGY	ENERGY CONSUMPTION	DOMESTIC ENERGY PRODUCTION AVAILABLE	1990 ELASTICITY	
		REALIZED	POTENTIAL						TRADE	ENERGY
REGION 1	1978	5547.	5547.	-----	85.4	120.4	70.7	46.9		
INDUSTRIALIZED	1990	8108.	8382.		351.8	112.0	87.6	62.3	1.25	0.15
COUNTRIES	ANNUAL GROWTH	3.2%	3.5%		12.5%	-0.6%	1.8%	2.4%		
REGION 2	1978	457.	457.	8.4	-----	2.2	5.8	35.9		
OIL EXPORTING	1990	1075.	962.	4.7		1.0	15.7	45.0	0.80	0.10
DCS	ANNUAL GROWTH	7.4%	6.4%	-4.7%		-6.1%	8.6%	1.9%		
REGION 3	1978	819.	819.	85.1	10.1	-----	13.1	6.8		
OIL IMPORTING	1990	1253.	1522.	214.6	62.4		18.7	14.7	0.80	0.10
DCS	ANNUAL GROWTH	3.6%	5.3%	8.0%	16.4%		3.0%	6.6%		
TOTAL	1978	6823.	6823.	93.5	95.5	122.6	89.6	89.6		
REGIONS, 1-3	1990	10436.	10866.	219.3	414.2	113.1	122.0	122.0		
	ANNUAL GROWTH	3.6%	4.0%	7.4%	13.0%	-0.7%	2.6%	2.6%		
1990 EQUILIBRIUM PRICES				1.000	2.022 (A)	0.598 (A)	44.10 (B)			

NOTES:
(A) TERMS OF TRADE INDICES. RATIOS TO 1978 DOLLARS FOR REGION 1.

(B) ENERGY PRICES EXPRESSED IN 1980 $/BARREL; COMPARE WITH THE 1978 ACTUAL PRICE OF $13.7 AND THE 1978 "REFERENCE PRICE" OF $10/BARREL EMPLOYED FOR BENCHMARKING PURPOSES.

(C) ENERGY-GDP ELASTICITIES OF 1.3 AT CONSTANT PRICES IN REGIONS 2 AND 3. DECLINING AFTER 1978.

CASE 6.

HIGH GDP GROWTH
ZERO TRADE DEFICITS
DIFFICULT ADJUSTMENT OF ENERGY DEMANDS
114 MBD ENERGY SUPPLIES

(UNIT FOR GDP AND EXPORTS: BILLIONS OF 1978 U.S. DOLLARS)
(UNIT FOR ENERGY: MILLION BARRELS DAILY; OIL EQUIVALENT)

		GDP		EXPORTS TO REGION 1, NON-ENERGY	EXPORTS TO REGION 2, NON-ENERGY	EXPORTS TO REGION 3, NON-ENERGY	ENERGY CONSUMPTION	DOMESTIC ENERGY PRODUCTION AVAILABLE	1990 ELASTICITY	
		REALIZED	POTENTIAL						TRADE	ENERGY
REGION 1	1978	5547.	5547.	-----	85.4	128.4	70.7	46.9		
INDUSTRIALIZED COUNTRIES	1990	7980.	8382.		417.1	75.1	81.8	62.3	1.25	0.15
	ANNUAL GROWTH	3.1%	3.5%		14.1%	-3.9%	1.2%	2.4%		
REGION 2	1978	457.	457.	8.4	-----	2.2	5.8	35.9		
OIL EXPORTING DCS	1990	1111.	962.	3.4		0.7	15.9	40.0	0.80	0.10
	ANNUAL GROWTH	7.7%	6.4%	-7.4%		-9.4%	8.8%	0.9%		
REGION 3	1978	819.	819.	85.1	10.1	-----	13.1	6.8		
OIL IMPORTING DCS	1990	1143.	1522.	204.2	76.3		16.3	11.7	0.80	0.10
	ANNUAL GROWTH	2.8%	5.3%	7.6%	18.4%		1.8%	4.6%		
TOTAL	1978	6823.	6823.	93.5	95.5	122.6	89.6	89.6		
REGIONS, 1-3	1990	10234.	10866.	207.6	493.4	75.7	114.0	114.0		
	ANNUAL GROWTH	3.4%	4.0%	6.9%	14.7%	-3.9%	2.0%	2.0%		
1990 EQUILIBRIUM PRICES				1.000	2.434 (A)	0.580 (A)	64.11 (B)			

NOTES:
(A) TERMS OF TRADE INDICES. RATIOS TO 1978 DOLLARS FOR REGION 1.

(B) ENERGY PRICES EXPRESSED IN 1980 $/BARREL; COMPARE WITH THE 1978 ACTUAL PRICE OF $13.7 AND THE 1978 "REFERENCE PRICE" OF $10/BARREL EMPLOYED FOR BENCHMARKING PURPOSES.

(C) ENERGY-GDP ELASTICITIES OF 1.3 AT CONSTANT PRICES IN REGIONS 2 AND 3. DECLINING AFTER 1978.

CASE 7.

HIGH GDP GROWTH
ZERO TRADE DEFICITS
EASY ADJUSTMENT OF ENERGY DEMANDS
122 MBD ENERGY SUPPLIES

(UNIT FOR GDP AND EXPORTS: BILLIONS OF 1978 U.S. DOLLARS)
(UNIT FOR ENERGY: MILLION BARRELS DAILY; OIL EQUIVALENT)

		GDP REAL-IZED	GDP POTEN-TIAL	EXPORTS TO REGION 1, NON-ENERGY	EXPORTS TO REGION 2, NON-ENERGY	EXPORTS TO REGION 3, NON-ENERGY	ENERGY CONSUMP-TION	DOMESTIC ENERGY PRODUCTION AVAILABLE	1990 ELASTICITY TRADE	1990 ELASTICITY ENERGY
REGION 1	1978	5547.	5547.	-----	85.4	120.4	70.7	46.9		
INDUSTRI-ALIZED	1990	8232.	8382.		259.0	126.9	88.8	62.3	1.25	0.20
COUNTRIES	ANNUAL GROWTH	3.3%	3.5%		9.7%	0.4%	1.9%	2.4%		
REGION 2	1978	457.	457.	8.4	-----	2.2	5.8	35.9		
OIL EX-PORTING	1990	1006.	962.	7.3		1.5	14.7	45.0	0.80	0.10
DCS	ANNUAL GROWTH	6.8%	6.4%	-1.2%		-3.0%	8.0%	1.9%		
REGION 3	1978	819.	819.	85.1	10.1	-----	13.1	6.8		
OIL IM-PORTING	1990	1300.	1522.	226.8	46.1		18.5	14.7	0.80	0.15
DCS	ANNUAL GROWTH	3.9%	5.3%	8.5%	13.5%		2.9%	6.6%		
TOTAL	1978	6823.	6823.	93.5	95.5	122.6	89.6	89.6		
REGIONS, 1-3	1990	10538.	10866.	234.1	305.1	128.4	122.0	122.0		
	ANNUAL GROWTH	3.7%	4.0%	7.9%	10.2%	0.4%	2.6%	2.6%		
1990 EQUILIBRIUM PRICES				1.000	1.470 (A)	0.600 (A)	30.90 (B)			

NOTES:
(A) TERMS OF TRADE INDICES. RATIOS TO 1978 DOLLARS FOR REGION 1.

(B) ENERGY PRICES EXPRESSED IN 1980 $/BARREL; COMPARE WITH THE 1978 ACTUAL PRICE OF $13.7 AND THE 1978 "REFERENCE PRICE" OF $10/BARREL EMPLOYED FOR BENCHMARKING PURPOSES.

(C) ENERGY-GDP ELASTICITIES OF 1.3 AT CONSTANT PRICES IN REGIONS 2 AND 3. DECLINING AFTER 1978.

CASE 8.

HIGH GDP GROWTH
ZERO TRADE DEFICITS
EASY ADJUSTMENT OF ENERGY DEMANDS
114 MBD ENERGY SUPPLIES

(UNIT FOR GDP AND EXPORTS: BILLIONS OF 1978 U.S. DOLLARS)
(UNIT FOR ENERGY: MILLION BARRELS DAILY; OIL EQUIVALENT)

		GDP		EXPORTS TO REGION 1, NON-ENERGY	EXPORTS TO REGION 2, NON-ENERGY	EXPORTS TO REGION 3, NON-ENERGY	ENERGY CONSUMPTION	DOMESTIC ENERGY PRODUCTION AVAILABLE	1990 ELASTICITY	
		REALIZED	POTENTIAL						TRADE	ENERGY
REGION 1	1978	5547.	5547.	-----	85.4	120.4	70.7	46.9		
INDUSTRIALIZED	1990	8156.	8382.		292.9	99.7	82.6	62.3	1.25	0.20
COUNTRIES	ANNUAL GROWTH	3.3%	3.5%		10.8%	-1.6%	1.3%	2.4%		
REGION 2	1978	457.	457.	8.4	-----	2.2	5.8	35.9		
OIL EXPORTING	1990	1033.	962.	5.8		1.2	14.8	40.0	0.80	0.10
DCS	ANNUAL GROWTH	7.0%	6.4%	-3.0%		-5.0%	8.1%	0.9%		
REGION 3	1978	819.	819.	85.1	10.1	-----	13.1	6.8		
OIL IMPORTING	1990	1236.	1522.	224.4	53.4		16.6	11.7	0.80	0.15
DCS	ANNUAL GROWTH	3.5%	5.3%	8.4%	14.9%		2.0%	4.6%		
TOTAL	1978	6823.	6823.	93.5	95.5	122.6	89.6	89.6		
REGIONS, 1-3	1990	10424.	10866.	230.2	346.3	100.9	114.0	114.0		
	ANNUAL GROWTH	3.6%	4.0%	7.8%	11.3%	-1.6%	2.0%	2.0%		
1990 EQUILIBRIUM PRICES				1.000	1.703 (A)	0.583 (A)	42.42 (B)			

NOTES:

(A) TERMS OF TRADE INDICES. RATIOS TO 1978 DOLLARS FOR REGION 1.

(B) ENERGY PRICES EXPRESSED IN 1980 $/BARREL; COMPARE WITH THE 1978 ACTUAL PRICE OF $13.7 AND THE 1978 "REFERENCE PRICE" OF $10/BARREL EMPLOYED FOR BENCHMARKING PURPOSES.

(C) ENERGY-GDP ELASTICITIES OF 1.3 AT CONSTANT PRICES IN REGIONS 2 AND 3. DECLINING AFTER 1978.

CASE 9.

LOW GDP GROWTH
ZERO TRADE DEFICITS
DIFFICULT ADJUSTMENT OF ENERGY DEMANDS
116 MBD ENERGY SUPPLIES

(UNIT FOR GDP AND EXPORTS: BILLIONS OF 1978 U.S. DOLLARS)
(UNIT FOR ENERGY: MILLION BARRELS DAILY; OIL EQUIVALENT)

		GDP		EXPORTS TO REGION 1, NON-ENERGY	EXPORTS TO REGION 2, NON-ENERGY	EXPORTS TO REGION 3, NON-ENERGY	ENERGY CONSUMPTION	DOMESTIC ENERGY PRODUCTION AVAILABLE	1990 ELASTICITY	
		REALIZED	POTENTIAL						TRADE	ENERGY
REGION 1	1978	5547.	5547.	-----	85.4	120.4	70.7	46.9		
INDUSTRIALIZED COUNTRIES	1990	7708.	7909.		288.4	105.6	83.0	62.3	1.25	0.15
	ANNUAL GROWTH	2.8%	3.0%		10.7%	-1.1%	1.3%	2.4%		
REGION 2	1978	457.	457.	8.4	-----	2.2	5.8	35.9		
OIL EXPORTING DCS	1990	1030.	962.	5.5		1.1	14.7	39.0	0.80	0.10
	ANNUAL GROWTH	7.0%	6.4%	-3.5%		-5.6%	8.0%	0.7%		
REGION 3	1978	819.	819.	85.1	10.1	-----	13.1	6.8		
OIL IMPORTING DCS	1990	1241.	1522.	217.1	53.5		18.3	14.7	0.80	0.10
	ANNUAL GROWTH	3.5%	5.3%	8.1%	14.9%		2.8%	6.6%		
TOTAL	1978	6823.	6823.	93.5	95.5	122.6	89.6	89.6		
REGIONS, 1-3	1990	9979.	10393.	222.5	341.9	106.7	116.0	116.0		
	ANNUAL GROWTH	3.2%	3.6%	7.5%	11.2%	-1.1%	2.2%	2.2%		
1990 EQUILIBRIUM PRICES				1.000	1.682 (A)	0.567 (A)	43.30 (B)			

NOTES:
(A) TERMS OF TRADE INDICES. RATIOS TO 1978 DOLLARS FOR REGION 1.

(B) ENERGY PRICES EXPRESSED IN 1980 $/BARREL; COMPARE WITH THE 1978 ACTUAL PRICE OF $13.7 AND THE 1978 "REFERENCE PRICE" OF $10/BARREL EMPLOYED FOR BENCHMARKING PURPOSES.

(C) ENERGY-GDP ELASTICITIES OF 1.3 AT CONSTANT PRICES IN REGIONS 2 AND 3. DECLINING AFTER 1978.

REFERENCES

Armington, P. S. "A Theory of Demand for Products Distinguished by Place of Production" and "The Geographic Pattern of Trade and the Effects of Price Changes," *International Monetary Fund Staff Papers*, XVI, March and July, 1969.

Barten, A. P. "An Import Allocation Model for the Common Market," *Cahiers Economiques de Bruzelles*, Université Libre de Bruxelles, No. 50, 2e. trimestre 1971.

Blitzer, C. R., B. J. Choe and A. Lambertini. "Energy in the 1980s: Global Supply and Demand Analysis," International Bank for Reconstruction and Development, Washington, D. C., January 1979.

Dunkerley, J. and S. Matsuba. "Energy Consumption in the Developing Countries," Resources for the Future, Washington, D.C., November 1980.

Energy Information Administration, *Annual Report to Congress, 1979. Volume Three: Synopsis*, U.S. Department of Energy, Washington, D.C., October 1980.

Energy Modeling Forum, "Energy Modeling Forum 4: Aggregate Elasticity of Energy Demand," Stanford University, January 1980.

Ezzati, A. *World Energy Markets and OPEC Stability*, Lexington Books, Lexington, Massachusetts, 1978.

Ginsburgh, V., and J. Waelbroeck. "A General Equilibrium Model of World Trade, Part I, Full Format Computation of Economic Equilibria," Cowles Foundation Discussion Paper No. 412, Yale University, 1975.

Hickman, B. G., Y. Kuroda and L. J. Lau. "The Pacific Basin in World Trade: An Analysis of Changing Trade Patterns, 1955-1975," *Empirical Economics* 4, Issue I, 1979.

Hickman. B. G., and L. J. Lau. "Elasticities of Substitution and Export Demands in a World Trade Model", *European Economic Review* 4, 1973.

Manne, A. S., H. Chao and R. Wilson. "Computation of Competitive Equilibria by a Sequence of Linear Programs," *Econometrica*, November 1980.

Whalley, J. "An Evaluation of the Recent Tokyo Round Trade Agreement Through a General Equilibrium Model of World Trade Involving Major Trading Areas"; University of Western Ontario, April 1980.

Chapter 21

Oil, OPEC, and LDC Balance of Payments: Impacts and Opportunities

Bijan Mossavar-Rahmani
and C. Anthony Pryor

The 1970s were a difficult decade for many of the world's developing countries. The 40 or so whose incomes were lowest continued to suffer extreme poverty, characterized by low productivity, high unemployment, malnutrition, high infant mortality, illiteracy, and low life expectancy. And the gap between rich and poor countries widened.

Undoubtedly the rapid rises in oil prices, first in 1973–74 and then again in 1978–79—and the equally rapid rise in energy prices generally which subsequently followed—contributed to these difficulties. But the precise impact of the oil price increases on the economic performance of the LDCs as a group—or even on individual countries—is controversial, for there is tremendous uncertainty about the relationship between oil prices and inflation, structure of production, and even structure of merchandise trade.

Clearly, many LDC oil importers have experienced large balance-of-payments deficits during the initial phases of rapid oil price increases. Given the limited availability of development assistance on concessional terms, these LDCs increasingly turned, in the 1970s, to private capital markets for short-term, high-interest loans to pay for oil. Those countries who could not raise the necessary cash were forced either to limit their oil imports, or to limit their imports of capital and intermediate goods, thus reducing their existing or future export capabilities.

But high oil and energy prices were only partially responsible for the continuing plight of the LDCs; Egypt and Mexico, for example, while impor-

The authors, Visiting Research Fellows at the Rockefeller Foundation, New York, are solely responsible for the views expressed herein.

tant net exporters of oil, have balance-of-payments deficits far exceeding those of some oil importers. Perhaps the most important culprit has been the sluggish growth in the world economy, which was only partly triggered by the higher oil prices. It both restricted the availability of development assistance and affected demand for the manufactured goods and raw materials exported by the LDCs. Faced with eroding trade balances, the industrial countries launched new protectionist measures, further limiting markets for LDC exports. Finally, prices of raw materials and other primary commodities, on which many LDCs depend for an important portion of their foreign exchange earnings, continued to move erratically, and generally unfavorably, leading to a decline in the LDCs' real purchasing power and a worsening of their terms of trade.

Thus any attempt to explain the poor performance of the LDCs over the last decade in terms of higher oil prices alone is an oversimplification that obscures a far more complex process affecting individual countries in different ways and to varying degrees.

First, it should be noted that the total amount of oil imported by all developing countries—excluding the thirteen members of the Organization of the Petroleum Exporting Countries (OPEC)—is relatively small. In 1977 total net non-OPEC LDC imports of oil amounted to 4 million barrels a day (mbd), or 6.4 percent of total world production that year, and just under one-half of average daily imports into one industrial country—the United States—alone. The total gross imports of oil into the non-OPEC LDCs were somewhat higher, but some 14 non-OPEC LDCs (Oman, Brunei, Mexico, Trinidad and Tobago, Angola, Egypt, Syria, Tunisia, Malaysia, Bahrain, Congo, Zaire, Bolivia, and Burma, in that order) were net exporters of oil; their aggregate oil exports in 1977 stood at close to 1.47 mbd. Significantly, this represented a near doubling of their 1973 net oil exports (see table 21.1).

Second, the bulk of the LDCs' net oil imports are destined for a small number of middle- and higher-income industrializing countries. In fact, in 1977 only nine countries (Brazil, South Korea, India, Taiwan, Turkey, the Philippines, Thailand, Cuba, and Singapore, in descending order of net imports) accounted for a whopping 72 percent of all net imports by non-OPEC LDCs. Brazil alone accounted for over one-fifth of all net oil imports into the LDCs. If these nine countries are excluded, the net oil imports of the rest of the LDCs are very small, and are again concentrated among relatively few countries. Five countries (Pakistan, Morocco, Jamaica, Panama, and Chile) imported between 50,000 and 100,000 b/d each, eight others between 25,000 and 50,000 b/d each, 25 countries between 10,000 and 25,000 b/d, and 33 countries imported less than 10,000 b/d each (see table 21.2).

The non-OPEC LDCs can thus be separated into three distinct groupings for the purpose of this discussion—the 14 net exporters of oil, the nine largest net importers of oil (defined as importing a net volume of over

Table 21.1. Net Oil Exports of Non-OPEC LDCs (in 1,000 b/d)

Country	1973	1977
Oman	283.5	312.0
Brunei	213.8	207.0
Mexico	−93.4	197.0
Trinidad and Tobago	113.8	194.4
Angola	142.6	158.1
Egypt	−27.2	128.0
Syrian Arab Rep.	−78.5	82.7
Tunisia	51.0	47.5
Malaysia	−12.6	58.2
Bahrain	−10.1	42.5
Congo	30.9	24.8
Zaire	−18.2	4.7
Bolivia	32.1	12.0
Burma	−4.7	0.2
Total	738.2	1,469.1

Source: *OPEC Review* 3, no. 3 (Autumn 1979).

100,000 b/d each), and the remaining 72 countries with net imports of less than 100,000 b/d each.

The status of the first group is more or less clear. These countries exported some 1.47 mbd in 1977, which, at the then market crude oil price of $12.70, were valued at over $6.8 billion. Four years earlier, before the first major oil price boost of the 1970s, only eight of these countries were net oil exporters, exporting a total of 738,000 b/d, which at the then market price were valued at nearly $1.4 billion. Needless to say, these LDCs have gained—and will continue to benefit—from oil price increases, not only because of the rise in value of their existing exports, but also because it has become commercially feasible to produce and export reserves that had previously been relatively inaccessible and thus expensive.[1] This year, the value of the oil exports from these 14 countries should top $20 billion, and may be as high as $25 billion. This upward trend is expected to continue into the 1980s. The mean per capita gross national product (GNP) of these countries in 1977 was noticeably higher than that of the other two groups; this gap will, in all likelihood, continue to increase with higher oil prices—and increasing levels of exports.

The second group of countries consists of nine middle- and higher-income industrializing countries, including India and Cuba, which together import as much as three-quarters of the net oil imports of all LDCs. These countries have been able to continue importing large volumes of oil (in fact, the combined net oil imports of the nine rose from 2.40 mbd to 2.87 mbd between 1973 and 1977) largely by borrowing from commercial sources whose liquidity has been enhanced by the OPEC countries' large surpluses. Commercial

Table 21.2. Ranking of Oil-Importing Developing Countries*

	No. of Countries	Percentage Share, 1977				Regional Percentage**			
		Pop.	GNP	Oil Consumption	Oil Imports	AF	ME	LA	A
Countries importing more than 100,000 bbl./day	9	59.9	63.9	66.7	71.8	—	11	22	67
Countries importing 50,000–100,000 bbl./day	5	6.8	6.5	4.8	8.2	17	17	50	17
Countries importing 25,000–50,000 bbl./day	8	10.0	7.9	6.9	7.1	17	17	42	25
Countries importing 10,000–25,000 bbl./day	25	15.1	17.1	19.6	9.9	40	15	35	15
Countries importing less than 10,000 bbl./day	33	8.2	4.3	2.0	3.0	69	7	7	14
(Total number of countries)						(35)	(9)	(19)	(17)

* Excludes Centrally Planned Economies, Belize, Guadaloupe, Malta, and Mauritius.

** Af-Africa; ME-Middle East; LA-Latin America; A-Asia.

Source: *OPEC Review* 3, no. 3 (Autumn 1979), and *World Bank Atlas*, 1978.

lending has readily been made available to these rapidly growing, relatively stable countries, and they are generally considered to be in a good position, in the long term, to turn to commercial sources to maintain the growth in their oil and nonoil imports.

The odds on discovering major oil deposits in these countries as a group do not appear very high, although India and Brazil responded to the 1973–74 oil price increases by stepping up domestic exploration and development and were producing 208,000 b/d and 161,000 b/d respectively in 1977. With additional development of new fields and application of enhanced recovery, these two countries' oil output is expected almost to double over the next few years. All nine countries have also initiated or accelerated programs to develop other domestic sources of energy as well; with few exceptions, these countries have nonoil energy resources that could be tapped further.

In addition to taking up certain supply options, many of these countries could significantly restrain further growth in oil imports by slowing down the present push into energy-intensive industries like steel and petrochemicals, while pursuing, instead, alternative economic development strategies. Yet Brazil, which has accounted for an unproportionately high percentage of all LDC net oil imports, is expected to add more steel-making capacity than all the industrial countries combined between now and 1985. Two other large oil importers, Taiwan and South Korea, are also rapidly building up their steel and petrochemicals capacities, which will result in a substantial increase in their oil import requirements. These countries are well aware of the trade-offs between development of such industries and the high costs of oil imports, and are presumably willing—and able—to meet those costs. The rapid development of these countries was in part based on energy-intensive export policies in the 1960s and 1970s. As energy prices have rapidly increased, there has been much talk about changing development strategies in these countries; however, it appears that several of them may find it undesirable—for a host of internal and external reasons—to do so. This is unfortunate, for even small changes in the oil import requirements of these countries could dramatically alter the import picture for all LDCs; for example, a 5 percent decrease in Brazil's 1977 oil imports would have more than covered the net oil import requirements of all of non-OPEC Africa that year.

How to characterize the remaining 72 LDCs, whose combined net imports stood at 2.57 mbd in 1977, is much more complex. Net oil imports, per capita GNP, potential for indigenous energy production, and ability to finance oil imports vary tremendously. For example, six of these countries imported less than 1,000 b/d in 1977, while another five imported between 50,000 and 100,000 b/d. Most of the rest fell into the 1,000–25,000 b/d category in terms of net oil imports. As for per capita GNP, two countries fell below the $100 mark, while six exceeded $1,000. Most of the rest had per capita GNP levels between $100 and $400 in 1977. One net importing country, Argentina, is itself a major oil producer, whose 1977 production of 431,000 b/d is expected to rise by 50 percent by 1982. Turkey and Colombia are also important producers.

By the mid-1980s, a number of minor and not-so-minor LDC net oil importers are expected to become net oil exporters. However, most of the projected increases in production and exports by some LDCs in the category of current net importers below 100,000 b/d will be offset by increased consumption and imports by others in this same category.

Finally, the ability of the countries in this category to finance oil imports, either through stepped-up exports or through additional official flows or even through private commercial borrowing, differs widely. For example, Afghanistan and Guinea imported roughly the same amount of oil in 1977 and reported similar per capita GNPs. However, Afghanistan's accumulated debt at the end of 1977 was $1 billion, compared to Guinea's $31 million. Elsewhere, North Korea and Jordan also had similar net oil imports and per capita GNPs; North Korea's debt at the end of 1977, however, was $9 billion compared to Jordan's $864 million. Moreover, in 1977 alone, Jordan had received some $338 million in OPEC aid, compared to North Korea's receipts of $48 million.

Thus, the LDCs are singular; each has a different level of economic development and infrastructure, different natural resource endowments, and different strategic connections to the industrial world and to the oil-exporting nations. These differences, however, should not mask the plight of those countries most seriously affected by higher oil prices, increased protectionism, worsening terms of trade, and disappointing assistance from the Development Assistance Committee (DAC) of the Organization for Economic Cooperation and Development (OECD). It is difficult to envisage how anything, short of the drastic steps called for by the LDCs in the so-called North-South dialogue, will prevent the gap between the world's rich and poor countries from widening even further. However, there exist some interim solutions, both internal and external, that can help the "poorest of the poor" to make some gains.

The LDCs cannot easily disengage themselves from the continued—and even growing—use of oil, at least in their modern sectors. Given its convenience and versatility, and until recently, its relatively low cost, oil has come to play an important role, not only in basic support of industrialization within these countries, but in the transport sector and even in household uses. Any discussion of the amelioration of balance-of-payments difficulties must focus on the sectors in which oil is currently consumed. Do they allow for reasonable substitution with domestically available fuels? What is the lead time necessary for the development of these domestic fuels? and Will there be sufficient capital available to develop them?

While it may be technically possible to substitute one type of fuel for another, the economic and social implications of such substitutions are not often known. There are indications that certain economic sectors in the LDCs have reacted surprisingly rapidly to energy price increases because of their relative simplicity. Kenya is one particularly useful example; there, higher electricity rates have resulted in significant conservation in the industrial sector. However, the transportation sector represents the severest constraint to substitution, and, in many countries, transportation accounts for a significant percentage of all imported oil. In addition, the structure of LDC transportation is much more directly related to productive enterprises than it is in the industrial countries. In the United States, private automobiles account for approximately 70 percent of all transport, while in most LDCs, freight accounts for nearly 70 percent. This linkage between oil and the transport of freight suggests that any curtailment of oil use directly affects economic activity, thereby exacting a toll that many LDCs would be unprepared to pay.

Decisions about curtailment of oil imports based on balance-of-payments considerations can have significant—and perhaps again unacceptable—consequences in other sectors of the economy as well. In many rural areas, for example, liquid fuels, though consumed in relatively small quantities, play a critical role in the transport of agricultural products from villages to market towns, and in electric power generation for irrigation and other uses. While it is certainly possible to substitute for these uses over time through the more efficient use of solar power or through the production of liquid fuels from biomass, reaction by governments to a short-term balance-of-payments problem could drastically affect the rural sector's ability to respond to change.

Rather than constrain demand, the LDCs can, of course, seek to increase the supply of indigenous oil through exploratory or development activity. Some have already launched such programs, as noted earlier, but the lead times involved can be quite long. Thus even if commercial deposits are discovered, the new oil can only augment the countries' energy supply at some future time without providing much relief from existing balance-of-payments difficulties.

There are thus several possible solutions to the long-term balance-of-payments problems discussed in this paper. The major approach actually has tended, in many instances, to be an increase in indebtedness. Some countries have also sought to constrain their demand for imported oil, which, as noted earlier, is much more difficult than is often thought, because of the structural, political, and economic concomitants of oil use in the economy. A third, longer-term solution is to increase the supply of oil through exploration, production of nonconventional liquid fuels, or, if possible, substitution by other renewable energy sources such as hydropower. However, if development of these sources, too, requires imported capital, then the problem may not be solved in the short term; in fact, more difficulties may be created, given the capital-intensive nature of the investments that would be required.

This problem, while serious, must be seen in perspective. According to a recent World Bank study, the increase in outstanding LDC debt from 1973 through 1977, though large, was considerably lower in constant dollars than it had been in 1969–73. The reasons for indebtedness may have changed, however, as more and more funds have been required for short term balance of payments support.

Bilateral agreements between developing countries and the major oil exporters have become common, and will probably become more so in the future. One of the important developments in the structure of resource flows to the LDCs in the 1970s has been the increasing volume of concessional aid committed and disbursed by the OPEC countries.

Since 1975, the OPEC countries have sought to redress—at least partially—some of the balance of payments and more general development difficulties created for the LDCs as a result of higher oil prices, both on humanitarian grounds as well as to blunt repeated attempts by the OECD countries to rally LDC support against OPEC pricing policies. Several OPEC countries have, from time to time, simply set up a two-tiered price system, charging a much lower price for oil sold to some LDCs. However, this kind of solution has had limited success. It is difficult to administer, given the problems of assuring that the lower-tier oil does not somehow leak into the world market at the higher prevailing prices. Moreover, a significant number of smaller LDCs, particularly those lacking domestic refining facilities, do not deal with OPEC directly; rather, these countries purchase petroleum products from other private or governmental refineries or from the international oil companies.

So the OPEC countries have chosen to launch an ambitious program of bilateral and multilateral assistance to the LDCs. The LDCs, themselves, seem better disposed to accept aid, soft loans, and other balance of payments assistance for financing their current account deficits than to press the oil exporters either to institutionalize a two-tiered price system or to stabilize oil prices. This reflects both their pressing and immediate needs and also the fact that the LDCs are, by and large, uneasy about any attempt to put a ceiling on, or to limit, what is seen as a sovereign right by exporting countries to set prices for their own commodities.

By DAC's or any other standards, the OPEC countries have been extremely generous in their assistance to the LDCs over the last five years. In 1977, for example, OPEC's official development assistance, consisting of disbursements of grants and loans at concessional financial terms to developing countries, totalled approximately $5.9 billion.[2] Moreover, where loans were involved, these generally carried long-term maturities and low interest rates, if any.

OPEC aid has represented over 2 percent of the donors' combined gross national product, while asssistance from the Western industrialized countries in the DAC during this period totalled only 0.3 percent of their GNP. Four DAC countries (Sweden, Norway, the Netherlands, and Denmark) have had high and rising ratios of official development assistance (ODA) to GNP, but the average ratios for all DAC countries in the 1970s were held down by the disappointing records of the three largest OECD countries, the United States, Japan, and West Germany. While most of the OECD countries had earlier endorsed an aid target of 0.7 percent of their GNP, DAC donors, multilaterally and bilaterally, have provided just half of this amount. Moreover, only half of these countries' ODA goes to the poorest countries. Thus in practice, the actual concessional transfers to the poor LDCs is only one-quarter of what the DAC donors' own collective decision had targeted.

OPEC, in turn, now provides an estimated one-quarter of the total world aid to the LDCs. OPEC countries supply the top six ranks among all donor countries as regards the proportion of aid to GNP. Two of them ranked among the six largest bilateral donors in absolute terms. It should also be noted that while until 1974, OPEC aid was largely concentrated in Arab countries and still is, it now reaches all parts of the world. This trend toward greater geographical diversification is shown by the rapid growth in the number of non-Arab recipients in recent years. According to DAC statistics, of the total 1977 official flow of resources to the LDCs from OPEC countries and Arab/OPEC multilateral institutions, close to one-half went to non-Arab nations. These flows have been channelled in a variety of ways: bilaterally through such state institutions as the Saudi, Kuwait, and Abu Dhabi funds; through regional organizations such as the Arab Fund for Economic and Social Development, the Islamic Bank, and the OPEC Special Fund; and finally, through such international institutions as the United Nations, the World Bank, and the International Monetary Fund. Some of this aid has consisted of outright grants, particularly in emergency situations requiring quick balance-of-payments support, as well as "soft" loans. In 1977, the ODA component of the total bilateral flows from OPEC countries to the LDCs was 81.2 percent, consisting of both grants (63 percent of the total) and loans (37 percent of the total). Moreover, the grant element of ODA loans was 43.5 percent, and of total ODA, 79 percent, according to DAC statistics.

Another fact that should be noted here is that while OPEC has not sought to tie its aid to the recipient countries' oil import bills, the total 1977 official flow of resources from OPEC countries and Arab/OPEC multilateral institutions to the net oil-importing LDCs with per capita GNP less than $400 (excluding India), amounted to nearly one-half the value of the recipients' combined net oil imports.

OPEC concessional aid rebounded in 1979 and will in all likelihood reach record levels in 1980, as the oil exporters move once again to alleviate the LDC balance-of-payments difficulties created by the most recent round of oil price increase. Indeed, the final report of the important OPEC Special Ministerial Commission on Long-Term Strategies recommends a major new commitment to assistance to other LDCs through the following steps:[3]

- The OPEC Special Fund should be converted into a development agency with an initial authorized capital of $20 billion, subject to increase from time to time, by January 1, 1981. The agency would undertake the following types of activities: financing balance of payments deficits; financing, under favorable terms, economic and social develpment projects recognized by the recipient countries as priority projects, including the development of their domestic renewable and nonrenewable sources of energy; financing projects intended to reinforce integration between developing countries in order to promote their collective self-reliance; financing projects intended to upgrade the value of the raw materials produced by developing countries; underwriting export credit between developing countries, especially in the area of energy supply; financing commercial operations of developing countries under market conditions; and underwriting loans floated by developing countries on financial markets. It is also recommended that the agency give appropriate priority to the least developed countries and to the most seriously affected countries, each of which should enjoy special conditions adapted to their respective situation; and promote the procurement of goods and services available in member countries in the context of its loan policy.
- Developing countries should be given assurances about the security of their oil supplies as a matter of priority over supplies to industrialized countries. In times of substantial shortage, an internal agreement among member countries in order to ensure such security of supply may be necessary.
- OPEC should assist developing countries to meet the cost of their oil imports through a series of loans and grants. The loans, to be extended to the better-off among the developing countries, would be on a commercial basis and would account for about three-quarters of the total. The remainder would take the form of long-term loans on concessionary terms plus a relatively small amount which would be direct grants for the poorest countries.
- Substantial assistance should be made available to the developing countries in order to assist them in the development of their indigenous energy resources. Such assistance should be primarily oriented towards financing

exploration for hydrocarbons, but should not be limited to this phase. A separate organization should be established for this purpose, to operate in cooperation with the industrialized countries, which possess the necessary technical expertise.

Thus in the 1980s, OPEC aid may well become the single most important source of external concessional financing for the LDCs. How far this aid will go toward meeting what will almost certainly be substantially higher costs for imported oil remains to be seen. Moreover, it is uncertain whether OPEC will in fact be able to carry out its pledge of assured supplies of oil for the LDCs. As oil supplies become scarcer and prices higher, buyers will be forced to scramble to obtain oil. In the process, the industrial countries and the medium- and high-income LDCs will surely squeeze out the poorer countries.

One area of potential assistance not yet considered by OPEC is aid-in-kind, in the form of secure supplies and even concessional financial terms for liquefied petroleum gas (LPG) sales to the developing countries. LPG is a mixture principally of butane and propane found dissolved in crude oil or in associated and nonassociated natural gas. It is recoverable in a refinery or in special separation facilities. It has a high calorific content, is a clean-burning premium fuel, and has applications in chemical processing as well. Historically, LPG has been produced as a by-product of refining in the petroleum industry, and its markets have been small. The international oil companies made no effort to expand its market or its international trade since LPG, too, was costly to produce, store, transport, terminal, and distribute in an era of cheap and plentiful oil. LPG produced with associated gas was usually flared along with it. Today, close to one-half of all natural gas produced in the OPEC countries is still flared and wasted along with substantial amounts of LPG, which, if gathered and utilized, could go a long way toward meeting incremental energy demand in the smaller LDCs.

LPG is especially appropriate for the smaller developing countries because it does not require a large and costly infrastructure, and because it can be readily substituted for petroleum products in a very wide range of uses, including transport, with only minimal alterations to existing equipment and hence little extra cost.

Although the OPEC countries and several industrial countries, notably Japan and the United States, are all now looking more seriously to LPG to meet their fuel needs, its availability has grown at a rapid pace, as an increasing number of separation and processing facilities continue to come on-stream, particularly in the Persian Gulf region. Availability wil be especially great in Kuwait and Saudi Arabia. In the latter country, for example, expansion of the Ras Tanura Refinery and completion of the massive associated natural gas gathering system could double total Saudi output to some 15 million tons per year by the mid-1980s, or about one-and-one-half times the total world trade in LPG in 1978.

OPEC LPG offers substantial promise as a secure source of fuel for the LDCs. The industrial countries are already starting to recognize its potential as oil becomes more and more scarce. Rather than sell LPG on world markets, the OPEC countries should earmark the bulk of their output for delivery to the LDCs.

A continuing aid package involving assurances of fuel supplies, stepped-up concessional flows, and other development assistance can go a long way toward eliminating important bottlenecks to economic growth in many LDCs. But as assistance from OPEC or from the industrial countries is fashioned, the vast differences among the LDCs should be kept in mind. The common tendency to bunch the LDCs together in discussions of energy and development has obscured the range of problems—and opportunities—facing these countries, and has made them even more difficult to address (see appendix).

APPENDIX

Basic Data on Oil Importing Developing Countries, 1977

	Oil Imports (1,000 b/d)	GNP Per Capita (Current U.S.$)	Terms of Trade (1970=100)	Total Debt (million US$)	Debt Service (million US$)	OPEC Aid (million US$)
Western Samoa	0.3	(na)	(na)	37	3	1.6
Guinea-Bissau	0.4	220	(na)	3	(na)	1.7
Comoros	0.3	(na)	(na)	39	1	3.4
Burundi	0.5	130	(na)	40	3	1.5
Gambia	0.6	(na)	(na)	28	1	3.6
Rwanda	0.7	130	169	71	5	5.4
Central African Empire	1.0	250	(na)	137	7	1.8
Chad	1.3	130	133	115	12	1.2
Nepal	1.5	110	(na)	72	2	8.6
Belize	1.5	(na)	(na)	9	2	0.0
Upper Volta	1.6	130	95	133	7	3.3
Benin	1.7	200	89	137	12	4.2
Laos	1.9	90	(na)	48	3	2.2
Mauritania	2.0	270	79	457	42	116.1
Mali	2.1	110	101	459	12	15.7
Niger	2.3	160	78	209	17	5.8
Malawi	2.6	140	127	357	18	0.0
Togo	2.6	300	136	331	27	0.0
Haiti	3.0	230	(na)	122	10	3.2
Guadaloupe	3.5	(na)	(na)	172	21	18.5
Sierre Leone	4.2	190	83	207	28	1.1
Afghanistan	5.0	190	135	1,059	38	25.2
Somalia	5.2	110	75	422	12	187.4
Guinea	5.6	220	(na)	31	(na)	5.1
Mauritius	5.7	(na)	(na)	75	15	0.0
Barbados	6.3	(na)	(na)	60	11	0.0
Uganda	6.4	270	159	247	28	8.0
Fiji	6.8	(na)	(na)	80	10	0.0
Yemen, A. R.	7.0	430	(na)	315	8	196.9
Cameroon	8.5	340	126	825	79	12.9
Mongolia	9.3	830	(na)	(na)	(na)	(na)
Madagascar	9.4	240	112	217	23	4.1
Malta	9.4	(na)	(na)	51	3	0.2
Liberia	10.2	420	93	514	86	0.0
Ethiopia	10.4	110	177	472	32	2.4
Paraguay	10.7	730	101	356	36	0.0
Guyana	11.0	(na)	(na)	428	46	1.6
Surinam	11.0	(na)	(na)	117	(na)	(na)
Papua N.G.	11.0	490	(na)	355	51	0.0
Honduras	11.3	410	91	475	53	15.3
Senegal	11.4	430	95	479	66	23.7

Appendix (Continued)

	Oil Imports (1,000 b/d)	GNP Per Capita (Current U.S.$)	Terms of Trade (1970=100)	Total Debt (million US$)	Debt Service (million US$)	OPEC Aid (million US$)
Rhodesia	12.1	500	(na)	1,173	42	12.9
Colombia	12.1	720	154	2,956	399	0.0
Mozambique	12.4	150	102	87	12	3.3
El Salvador	14.1	550	151	303	71	18.5
Sudan	14.2	290	97	2,035	136	149.3
Tanzania	14.3	190	127	1,173	42	12.9
Nicaragua	15.2	830	110	868	101	11.9
Zambia	15.8	450	59	1,481	205	0.3
Cyprus	16.1	(na)	(na)	222	34	0.0
Costa Rica	17.4	1,240	114	752	106	24.4
Vietnam	18.2	160	(na)	535	20	0.0
Yemen, D.R.	19.0	340	76	307	6	77.7
Ghana	20.5	380	93	791	38	4.6
Jordan	23.2	710	85	864	51	338.4
Guatemala	23.6	790	142	376	47	23.4
Argentina	23.9	1,730	87	6,160	1,386	0.0
Korea D.R.	25.0	670	76	9,066	1,254	48.4
Peru	29.0	840	(na)	(na)	(na)	(na)
Sri Lanka	32.3	200	141	799	137	15.8
Lebanon	37.4	(na)	83	144	49	14.8
Dominican Republic	38.7	840	79	864	99	60.0
Uruguay	39.1	1,430	74	748	249	0.0
Bangladesh	40.8	90	68	2,305	83	178.9
Kenya	43.1	270	132	1,142	99	5.1
Jamaica	50.8	1,150	87	932	150	32.5
Panama	53.2	1,120	81	1,453	173	18.7
Morocco	72.3	550	90	3,608	307	124.6
Chile	72.4	1,160	50	3,773	857	0.0
Pakistan	78.7	190	80	6,850	391	73.8
Singapore	149.4	2,880	(na)	1,087	135	0.0
Cuba	158.6	115	71	2,906	304	0.0
Thailand	185.8	420	75	1,815	401	0.3
Philippines	213.6	450	68	4,711	533	0.0
Turkey	288.3	1,140	80	5,300	448	28.7
Taiwan	307.5	1,170	80	2,955	567	33.8
India	355.0	150	83	14,928	935	185.6
Korea, Republic of	413.6	820	76	9,066	1,254	48.4
Brazil	807.8	1,360	118	32,100	6,330	0.0

Sources: *OPEC Review* 3, no. 3 (Autumn, 1979). *World Development Review*, World Bank, 1979.

NOTES

1. Egypt, Mexico, and Malaysia, for example, are expected to increase their oil production substantially over the next few years. Smaller production increases are also expected in the Sudan.
2. Total OPEC net disbursements of concessional aid have averaged between $5 and $6 billion a year since 1975 but dropped back in 1978 to $3.7 billion, according to preliminary DAC statistics, although this drop reflected, in part, decreased flows to Egypt by other Arab donors following the signing of the Camp David accords and Egypt's subsequent political isolation. But it should be noted that the DAC statistics of OPEC aid are probably too low; DAC readily admits that its data are incomplete, as not all OPEC bilateral flows are necessarily reported.
3. *Middle East Economic Survey* 23, no. 19 (25 February, 1980).

Chapter 22

The Iranian Syndrome: Political Symptoms and Economic Causes of Too Rapid Modernization

Robert Copaken

The traditional method of forecasting future free-world oil supply-and-demand balance has tended to minimize non-OPEC LDC energy production, treating it as marginal and its contribution to world oil consumption as relatively small, compared with that of either the United States or Western Europe, for example. There are three reasons why this traditional method of projecting the world oil market into the future must be modified if we are to gauge future free-world oil supply, demand, and use patterns more accurately.

First, on the demand side, the developing countries (both OPEC and non-OPEC) account for an already large and rapidly growing share of total commercial energy consumption and of oil in particular. It is true that energy consumption, as a percent of GNP, is less in LDCs than in developed countries and that the bulk of oil consumption is concentrated in 11 non-OPEC LDCs (Mexico, Egypt, India, Brazil, Taiwan, Singapore, South Korea, Malaysia, Thailand, the Philippines, and Pakistan). Yet most energy projections for the period between now and 1990 estimate that consumption growth rates for commercial fuels in developing countries will exceed those of the industrial countries by 2–4 percent. This reflects the developing countries' higher expected economic growth rates and rising levels of industrialization and urbanization, although they generally start from a low base. If those

The views expressed are the author's own and do not necessarily represent those of the Department of Energy.

projections prove correct, the developing country share (including OPEC) of world commercial energy consumption will rise from approximately 19–20 percent in 1980 to about 25 percent by 1990. Those shares are sufficiently large to have a major impact on market conditions. According to most projections, the OECD share of the total will correspondingly decline from 80 percent to 75 percent over that same period.

Second, on the supply side, we expect that non-OPEC oil production will increase by about 1 mbd during 1980. Further non-OPEC oil production increases are projected for 1981. In 1980 most of the increase is seen as coming from the United Kingdom's North Sea production and Mexico. Over the longer term, prospects for increases in non-OPEC oil sources are more problematical because the increases in non-OPEC LDC oil consumption offset production increases, and because of the long lead-times and major uncertainties in new exploration and development. Mexico and Egypt could together account for an additional 2–3 mbd of non-OPEC oil production by 1985. But in Mexico there has been a wave of opposition from those who fear that the revenues resulting from increased oil production would harm the economy. A conservative approach probably will be maintained because the sharp rise in oil prices has already strained the government's ability to handle the surge of resources. Last year, with foreign oil revenues of only $4 billion, inflation exceeded 20 percent, imports rose by 47 percent, and the trade deficit widened to $3–4 billion. President Lopez Portillo seems determined, in his words, "to transform nonrenewable resources into renewable wealth" through sweeping oil-financed industrialization. But the government also recognizes that it may take two decades for the capital-intensive investments of today to be translated into jobs for the millions of unemployed. It seems possible that by 1985, total free-world oil production may increase by as much as 3 mbd from its present level of 52.5 mbd.

Third, while Iran's current policy of limiting production to 2–3 mbd holds that nation's output some 4 mbd below the prerevolution level, other OPEC members have announced production cutbacks in an apparent effort to assure a chronic precarious balance between supply and demand. For example, Kuwait has set a national ceiling for 1980 at 1.5 mbd, (excluding its share of the Neutral Zone) on conservation grounds and because it feels the need to be compensated for higher priced imports from the industrial nations. The ability of certain policy-constrained capital surplus oil-exporting OPEC countries (Saudi Arabia, the UAE, Kuwait, Iraq, and Iran) to expand their production and, in most cases, their capacity, is constrained by their increasing reluctance to do so for a complex and varied set of economic, social, and political reasons. This new concern with OPEC production ceilings represents a fundamental shift in OPEC's outlook since 1973–74. The consequences for oil-importing developing nations are ominous to contemplate.

This complex set of reasons for preferred production profiles that stretch reserve depletion over longer periods and result in a conservative desire to avoid excessive domestic spending because of inflationary and disruptive social effects is what is meant by the "Iranian syndrome," although its origins precede the recent Iranian upheaval. This phenomenon may be applicable to a relatively small number of capital-surplus oil-exporting developing countries, OPEC and non-OPEC alike. It is caused primarily by the effects of sudden and dramatic affluence, generated in large part by massive streams of oil revenues, which result in balance-of-payments surpluses of petrodollars held by these oil-exporting states. These financial surpluses, estimated at $116 billion in 1980, could rise to over $500 billion by the year 2000 and cannot be spent, invested, or even given as aid as rapidly as they are being accumulated. This is partly because of the small population, the underdeveloped infrastructure, insufficient managerial talent, and other preindustrial characteristics that oil-exporting developing countries share with oil-importing developing countries. It is also caused by the fact that, given the relative price and inelastic demand for oil by the world's industrial economies, the producers can maintain constant revenue flows in real terms by raising prices over and above the rate of inflation while reducing production levels correspondingly. Indeed, it has been said that with double digit inflation reducing the value of petrodollar surpluses, the producers ostensibly have little incentive to trade a real asset (oil in the ground) for a paper one of declining value. When these capital surplus countries reach or exceed the limits of their absorptive capacity, the inevitable results are waste, inflation, and misallocation of resources. This is disastrous in economies where petroleum may be the primary if not the exclusive source of foreign exchange earnings. With the exception of Algeria, Ecuador, and Venezuela, all OPEC countries are running substantial balance-of-payments surpluses and can reduce oil exports without impairing internal economic development plans. The situations in other low-absorber OPEC countries vary considerably.

This Iranian syndrome has both economic causes and political manifestations. The revolution and the politically volatile situation which has characterized Iran since the exile of the Shah, had, as one of its roots, the forced pace of economic development, without equivalent political development. This forced draft march toward modern industrial nationhood, overburdened with nuclear power, a steel industry, and military arms, was made possible largely through continued high levels of oil production and gradually increasing oil prices in nominal terms. The members of OPEC have seized control in three vital areas since 1973: control over pricing, control over production levels, and control over the physical oil distribution system. The results of these seizures of control have been, in succession, a dramatic increase in oil prices in 1973–74, followed by a sharp drop in real terms in

1974–78, followed, in 1979, by another sharp rise. This time, an abrupt fall is unlikely in the near future, because OPEC countries seem determined to cut production to keep prices up. These factors, by themselves, did not guarantee that inflationary pressures within the Iranian economy would lead to political unrest, corruption, repression, and opposition to the Shah's authoritarian rule by all sectors of Iranian society. But there is no doubt that these combined pressures significantly contributed to the revolution which resulted in the loss of 2–3 mbd of Iranian oil production from the world market in 1978–79, perhaps forever.

Most of us have tended to divide OPEC countries into two or three major groups—(1) those with relatively modest oil reserves and a high capacity to absorb foreign exchange (the majority), (2) those few countries with relatively large oil reserves and limited absorptive capacity which are capital-surplus oil exporters, (the minority—Saudi Arabia, Kuwait, the United Arab Emirates, Iraq, and Libya), and perhaps (3) a radical fringe group. However, the dividing lines between these groups fluctuate over time as producers reach higher levels of capacity utilization and pricing oil becomes more a matter of national policy objectives than of strict revenue needs per se.

Likewise, the seizure of the Great Mosque in Mecca by an armed group of conservative religious insurgents will have undoubted long-term effects on Saudi Arabia's future political stability. One consequence may be a slowdown of the modernization process as a result of the shifting balance of forces, which was already becoming evident within Saudi ruling circles. Ultimately, unless governmental and societal reforms are forthcoming, the increasingly widespread discontent among the technocratic elites over the corrupting influences of new and unprecedented wealth on the Saudi Bedouins' traditional way of life could spread. The events in Mecca have been minimized publicly by the royal family, but privately they acknowledge that it caught them unawares and unprepared. Traumatic as it may have been, it might be considered beneficial if it acts as a catalyst for growing liberalization within the country. More than half the Saudi labor force is foreign, made up of Palestinians, Yemenis, South Asians, South Koreans, and others. The scale of the economic development started by Saudi Arabia is such that it cannot be accomplished without foreign labor. With a native population of 4–5 million, few skilled workers of their own, and even fewer managers, the Saudis have tied themselves for at least the next two decades to a large contingent of foreign workers. The position of the various communities of migrants differs considerably, depending on the workers' nationality, level of skill, income, education, and religious or ethnic background, as well as on the sector of the economy in which the workers are employed. One inescapable conclusion seems painfully clear: In Saudi Arabia, as elsewhere in the peninsula, policies designed to raise aspirations and to distribute income more equitably have a negative effect on the increasing numbers of non-nationals to whom the government is looking most urgently to participate in the development of the modern Saudi economy. Whether these migrant workers' claims can be satisfied by higher wages alone remains an open question. If these social and political strains cannot be isolated and contained, they could eventually overflow into the all-too-precarious political realm of the kingdom. Particularly in Saudi Arabia, the religious and political realms are inextricably interwoven.

What has happened in the wake of Iran is that among OPEC's key Persian Gulf countries, where oil production and capacity were previously regarded as capable of gradual expansion over time to accommodate increasing levels of oil demand, they now appear to be relatively fixed quantities which the key producers are reluctant to expand significantly. This means that the relative sizes of the OECD, LDC and U.S. slices of that fixed Persian Gulf oil export pie will of necessity be shrinking or expanding to adjust to this new reality.

The issue of OPEC production is, of course, directly related to the energy and economic prospects of the developing nations through the price of internationally traded oil. If the official price of oil increases sharply again as it did this past year (more than doubling its 1978 level in nominal terms), the strain on the balance of payments of oil-importing countries will be increased. This may induce major industrialized nations to adopt unduly deflationary and protectionist policies, slowing their own growth and that of the developing nations which rely on them as major markets for exports. Oil price increases (in real terms), slowed economic growth, and increasing efficiency of energy utilization will all create pressures to help us make this gradual accommodation and may ease the often painful transition to scarcer world oil. But the unpleasant fact remains that the world energy problem reflects both the depletable nature of world oil supplies and the unpredictable course of Middle East politics. The real unanswered question is whether we can expect a return to some sort of world oil market stability or worse price shocks, supply disruptions, and abrupt surprises in the future. Both the developing and the developed nations have an interest in the outcome and both will share responsibility for shaping its outlines.

Chapter 23
Energy Strategies for Oil-Importing Developing Countries

Marcelo Alonso

The effects of energy cost increases on economic growth and overall development are seen by developing countries from many different perspectives. Probably the best criterion to use to identify common positions is whether the country is and will remain in the foreseeable future a net oil exporter or a net oil importer. Here there is a clear line demarcating quite different issues and strategies. While no country, either developed or developing, is totally devoid of energy resources, for most of them the price and availability of oil are more critical than the price and availability of any other sources of energy, given the present energy supply-and-demand patterns and the constraints to change these patterns.

This paper will concentrate on the energy strategies of oil-importing developing countries, most of which are characterized by (1) large population growth rates, (2) profound differences in the quality of life among urban and rural segments of the population, (3) an increasing demand for energy to satisfy minimum standards of living both in urban and rural areas, and (4) chronic scarcity of investment capital and foreign exchange. To place the issue in the proper context, we begin with an overview of energy consumption and development strategies in developing countries, taking as an example the situation in Latin America.

ENERGY CONSUMPTION

The present rate of world energy consumption is of the order of 8 twadells (TW), amounting to about 2.2 kW/cap.[1] This consumption is very unevenly distributed among more and less developed countries and among rural and urban sectors in developing countries. North America, with about 6 percent of the world's population, uses about one-third of the world's energy, about 2.7 TW at the rate of 10 kW/cap, while Latin America, with a slightly larger population, uses eight times less energy, about 0.34 TW, at the rate of about 1 kW/cap. About 72 percent of the world's population uses less than 2 kW/cap of energy, and most of that population is located in LDCs, mostly in rural areas with an average energy consumption of about 0.2 kW/cap.

For LDCs to approach the current world average of 2 kW/cap, an additional rate of energy production of 2 TW would be required. This would imply a 25 percent increase in world energy production, and would require considerable capital investment. As an example in Latin America, and assuming a conservative estimate of $1,000/kWe installed, the direct investment in energy generation would be in the order of $\$3 \times 10^{12}$, and the total mobilization of resources would amount to about $30,000 per capita, that is, about 50 times the average per capita income.

As for the future, energy demand estimates for the world in the year 2030 range between 30 TW, as the maximum sustainable, and 20 TW, as the minimum at which the world economy would essentially be dragging rather than growing.

The figure of 30 TW for world energy consumption assumes, for example, that in North America energy would be used at the rate of 6 TW, or a little more than double the present rate, while in Latin America energy would be consumed at the rate of 3 TW or about nine times the present rate. Assuming that by 2030 the population of North America will increase to about 315×10^6 persons, while in Latin America it will grow to 790×10^6 persons, in North America the consumption would be about 19 kW/cap, while in Latin America it would be 3.8 kW/cap. Therefore, in spite of the fact that Latin America would be using energy at nine times the present rate requiring an investment conservatively estimated as of the order of 30×10^{12} or about $\$6 \times 10^{11}$ per year, the people of Latin America will still be using five times less energy than the people of the developed countries, or less than half what they use now. A similar pricture can be drawn for other regions of the world.

One may wonder whether the world can sustain these fantastic increases in total energy consumption, which also imply severe demand increases in other materials and a corresponding increase in pollution, and whether the LDCs have at their disposal the capital and other resources required even if they limit their goals to the minimal energy consumption increase needed to cope with population increase and the demand for a better quality of life. Another way of stating this question is to ask whether it is possible to improve the quality of life of the majority of the population of the world without such an enormous increase in energy consumption, which in most cases will go to the most favored sectors of the world's population. A third way of putting it is to ask whether the LDCs can achieve an acceptable, less energy-intensive, but equitable development pattern.

Another important aspect is that the energy situation varies considerably from one country to another and therefore has a strong geopolitical aspect which makes world statistics almost meaningless. For example, it is interesting to note that if we look at Latin America as a single geopolitical entity, throughout which energy can flow freely, we could affirm that Latin America would be more than self-sufficient in energy well beyond the year 2030.

Latin America produces oil at the rate of 9.3 Q/y, which is more than enough to meet all its needs for quite a number of years. If to this we add hydroelectric power, which might contribute another 10 Q/y, we might conclude that Latin America could easily meet all its energy requirements for a few decades, and continue developing until other energy sources become fully competitive and technically feasible. If in addition we consider the potential of nuclear, biomass, and gas energy sources, which can contribute substantial amounts in a relatively short time without requiring extensive research and development, the energy picture is even brighter. Thus we may say that there is plenty of energy now in Latin America, taken as a geopolitical unit, and that if energy could flow freely throughout the region, we might not be talking about an energy crisis in the region. Crisis would take other forms such as insufficient capital, large population growth, and extreme poverty for large segments of the population.

DEVELOPMENT STRATEGIES OF LDCs

In energy, as well as in many other sectors of the economy, most developing countries have attempted to duplicate the production and consumption patterns of industrialized countries, for the apparent lack of other viable alternatives. Industrial production was geared to goods and services associated with high end-use energy consumption through manufacturing processes that were also energy intensive. Agricultural development followed a similar path with intensive use of artificial fertilizers, chemical pesticides, high mechanization, and other energy-intensive techniques. As latecomers to economic growth, developing countries favored oil for energy supply, generally neglecting other energy alternatives, which were no longer considered desirable in developed countries either because of higher fixed capital intensity, longer maturity periods, or higher production costs in the short run.

The degree of this dependency on oil for energy production can be illustrated taking again the example of Latin America. From figure 23.1 it is possible to derive the following conclusions:

1. In all except two countries (Haiti and Bolivia), oil contributes more than 50 percent of the primary energy source, in spite of the fact that only four countries are net exporters of oil.

2. With the exception of two countries (Trinidad & Tobago and Venezuela) gas is a relatively minor component of the primary energy production, although gas production is increasing rapidly, particularly in Bolivia and Argentina.

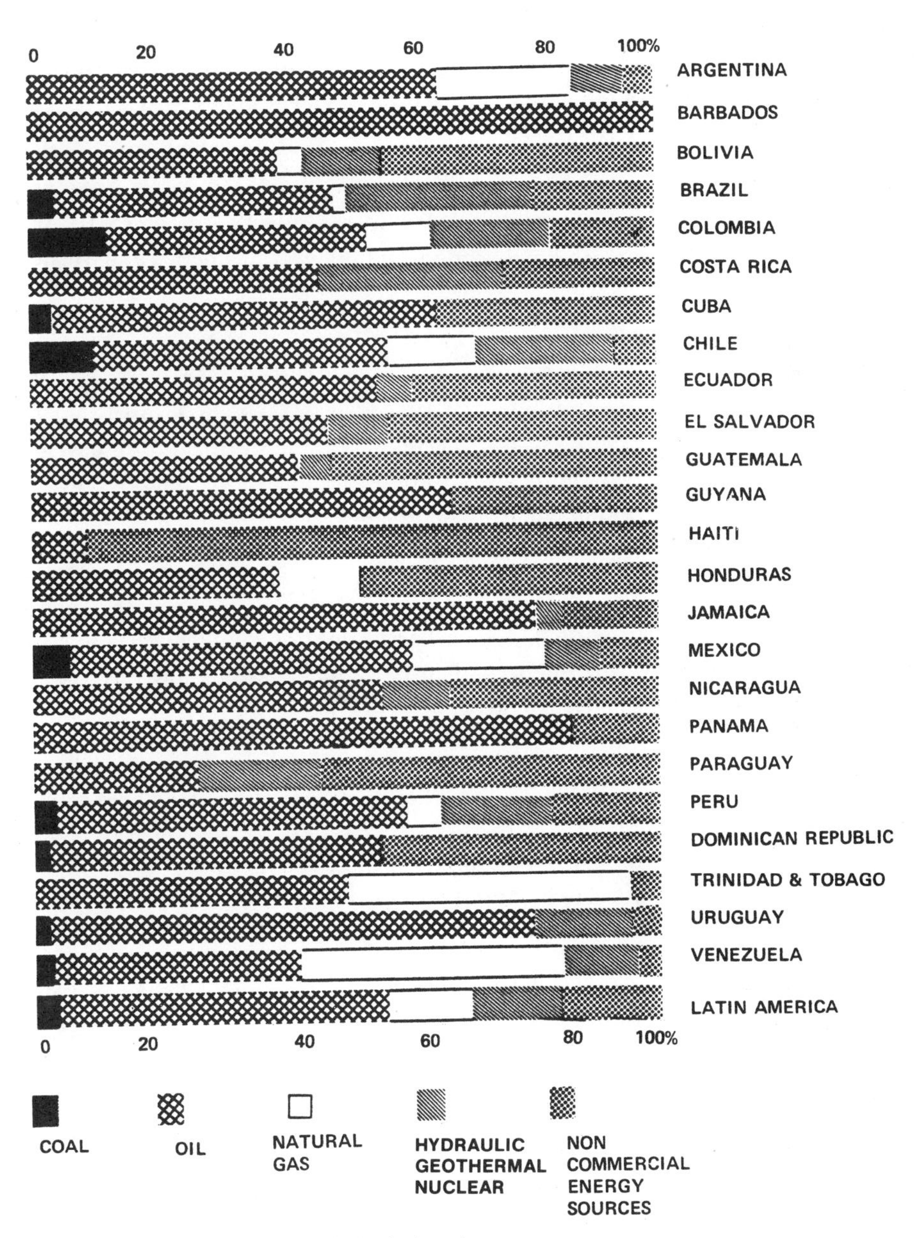

Fig. 23.1. Primary energy sources in Latin America

3. Coal's contribution to primary energy is practically negligible, with the exception of Colombia, although about five countries could become important coal producers.
4. Noncommercial energy production is important in a substantial number of countries, with serious undesirable effects because of the difficulties in properly managing this energy resource.
5. Hydroelectric power, although relatively important, is not yet fully exploited in spite of important new projects, such as the development of the Paraná-Paraguay river basins.
6. Nuclear, solar, and geothermal are still minor components of the energy production, with the exception of nuclear energy in Argentina (and eventually Mexico and Brazil) and geothermal in El Salvador, although the potential for geothermal is important in Central America and Mexico.

For comparison, table 6.1 illustrates the general structure of energy production in Latin America and North America, and clearly shows the larger dependence of Latin America on oil. This heavy reliance on oil, coupled with the various degrees to which Latin American countries can satisfy oil demand with local resources, helps to explain both how critical for a developing region are the availability and cost of oil and the short-run differences in the energy and development strategies of the various developing countries within it.

For example, referring again to Latin America, we see that Mexico, Venezuela, and Ecuador are big oil exporters and that their short-term energy strategies might be characterized by the adoption of a conservative oil-exporting policy compatible with their economic growth plans while simultaneously stretching their resources as much as possible.

Paraguay is a net oil importer but in a few years will have access to huge amounts of hydroelectric power, and in the short run is bound to be a net exporter of electric power to neighboring countries. This should help pay her oil bills.

Countries such as Haiti, the Dominican Republic, and Barbados show a total reliance on oil. Their main problem in the short run is how to get foreign exchange to import oil, since they have no immediate alternate energy sources of significance.

Other countries might be in a less critical situation but still have no simple solutions for a combination of higher oil import bills, reduced foreign exchange availability, and limited possibilities to develop alternative local conventional or nonconventional energy resources. This situation is not yet easy to alleviate by raising capital via export increases, given the high import component of production in developing countries, the worldwide recession, and the balance-of-payments problems of industrialized countries.

It is clear, though, that whatever the energy and development strategies adopted by each developing country, their implementation will require a significant technological change throughout the economy, both to change production patterns, reduce oil imports, and increase export of local products and goods in the short run, and to adjust to higher energy costs in the long run. That is, the energy, economic, and technological policies of developing countries are directly related and cannot be considered separately, even if their respective scopes and time horizons are different. In other terms, in any planning effort energy must be considered as an explicit production factor together with capital, labor, materials, and technology.

This interrelation among the production factors poses another big challenge to LDCs, because these countries face major economic and technological constraints. In most cases they lack the technological decision-making and implementation capabilities required for a successful technological change to different energy usage patterns and production methods. Most developing countries have relied on industrialized countries not only for the supply of products and process technology, but also for the technological capabilities required to select, transfer, and apply technology in their productive systems. As a result technological decision-making capabilities, both in government and throughout the technoeconomic system, are either in an embryonic stage or lack the economic and political support for effective mobilization of both resources and know-how.

As we have indicated, the major constraints arising from this situation are technical and economic. On the one hand, successful technological change requires an in-depth knowledge of local conditions, for the most part found only in the local country's technical personnel and institutions. In many cases these are insufficient or inadequate. On the other hand, extreme reliance on industrialized countries' technologies and processes, although it may solve pressing short-term requirements, adds an important component to production in terms of technical services, capital goods, and intermediate consumer goods. It also limits export possibilities because of competitive considerations and inefficient technological change leading to high-cost production. Thus a hard fact of life is that for most developing countries, the conditions for quick changes in production methods and energy use do not now exist. In most cases they will take a long time to develop unless effective international cooperation is available.

In considering the above, one has to keep in mind that not only the energy resources but also the technological capabilities of developing countries cover a broad spectrum. In the case of Latin America, the situation ranges from significantly developed technological decision-making and implementation capabilities in countries like Argentina, Brazil, and Mexico, to very limited capabilities in most of the smaller countries of the region. Thus we may conclude that unless the countries in a region make a concerted, cooperative effort to improve their energy pictures through the assessment and development of their common resources, and to induce and carry out the necessary technological changes in their production patterns, the smaller countries, with limited energy resources and technological capabilities, face a critical situation both in the short and in the medium terms.

We shall turn now to some general considerations of the common characteristics of energy and technology strategies in oil-importing developing countries, keeping in mind that specific strategies will vary significantly from country to country.

ELEMENTS OF ENERGY AND TECHNOLOGY STRATEGIES FOR OIL-IMPORTING DEVELOPING COUNTRIES

From the preceding presentation it is clear that, in the case of oil-importing developing countries, the adjustment to increasing energy costs adds heavy demands to the development efforts in terms of current consumption, investment capital, foreign exchange, and technology and policy planning, for which the options are very limited.

Preserving and improving current consumption of manufactured and processed goods, clearly low for vast sectors of the population, is threatened by price increases throughout the economy as a result of the higher costs of energy, materials, and labor.

Additional investment capital is required to (1) improve productivity in energy supply and efficiency in energy use, particularly in those countries which are heavily dependent on oil, (2) develop new oil, gas, and coal sources, (3) expand other conventional energy supply sources that for the most part are even more capital-intensive than those based on oil and gas, and (4) develop unconventional sources such as biomass, microhydroelectric, and solar, including those that could contribute to improve energy supply to rural areas. In particular oil-importing developing nations need increasing cash flows to meet the increasing cost of oil and the generally high import component of energy investment. But the energy-induced recession that limits export possibilities, combined with limited flexibility in the use of the technological variables for increasing energy productivity, poses seemingly insurmountable constraints to meeting this pressing requirement of capital for energy development.

In this context, managing a reasonably smooth adjustment to higher costs and less availability of energy places exceedingly heavy demands on policy planning capabilities, particularly on technology policy and planning. Whatever the eventual new structure of production in developing countries, higher energy costs will require a significant technological change, not only in the energy sector but throughout the economy.

The global energy outlook and a review of the main constraints that oil-importing LDCs face suggest some of the elements of energy and technology strategies for development in these countries, which essentially should try to (1) substitute for the use of oil whenever possible, (2) improve the efficiency of energy use, and (3) increase exports to obtain the required foreign exchange.

In designing an energy and technology strategy, it is very important to keep in mind the deep differences between the modern and traditional sectors of the economy, since most of the nonconventional energy sources are more likely to be relevant in the traditional than in the modern sectors, while the modern sector will continue to be the largest user of conventional energy. In addition, in LDCs the modern sector demands much larger increases in energy consumption to achieve desirable improvement in the standards of living across all levels of the population than does the traditional sector.

To carry out this seemingly elementary strategy requires the following immediate steps.

Assessment of Existing and Probable Future Energy and Technology Demand

The first logical step in establishing short- and long-run energy and technology strategies is to know as precisely as possible, how much energy and technology, in what modes, where, etc., is actually being demanded and how this demand is likely to evolve.

This means that for long-run strategies, one should know not only the present energy consumption patterns, but also why they are so. Unless one can relate existing energy demand to economic and social variables through the technology that links them, one will not really be able to obtain a reliable picture of the future. Forecasting is not just extrapolating historical trends, but essentially being able to assess the probable effects that both desired and inevitable changes in each one of the variables, as well as the constraining parameters, will have on the whole technological, economic, and social structure.

In other words, energy-demand assessments, present and future, are not just an exercise in statistics, data processing, and mathematical modeling, but an exercise in technological, economic, and social analysis in which all these tools can be profitably employed, provided their use is guided by common sense and they do not become an end in themselves.

Some of the questions that should be asked for long-run strategy formulation are:

1. How has energy consumption in urban centers been affected by the exodus from rural areas?
2. How can energy consumption be broken down by economic and social strata, so that specific trends and possible corrective measures can be devised?

3. How has energy consumption been affected by technological requirements of the production processes?
4. How much energy intensiveness of production processes has been correlated with economic efficiency in key sectors?

Unless information related to questions 1) and 2) is obtained and conveniently processed, it will be impossible to come up with meaningful, socially relevant proposals for energy development for urban areas, such as investments for mass transportation, energy distribution systems, air and sound pollution control, etc., which would rather address the existing imbalance in social equity and quality of life among urban population groups.

Similarly, unless information related to 3) and 4) is obtained, there will be no clue as to the best dimension and location of key energy development investments to optimize the use of limited capital in technologically well-adjusted plans and equipments.

These examples are given only to underline the fact that the quality and relevance of the answer depends critically on the question asked. It is not just the problem of carrying out energy consumption surveys. To really illuminate the decision-making process involved in energy development, it is essential to have a good understanding of the related technological, economic, and social phenomena.

Assessment and Development of Energy Supply Alternatives

This element of the energy development strategy is the simplest one to justify as a necessary component, although not necessarily the simplest or the easiest to perform or accomplish, not only because of the technical complexities involved, but because it demands, besides extensive technological information, very good judgment on the economic and social issues, as well as the political conditions and implications of any particular choices.

A comprehensive long-term energy supply strategy should provide an assured supply of the amount and kind of energy necessary to maintain development.

In assessing and developing energy supply alternatives, one should use selectivity, in the sense that any commitment is to be made only if it favors the attainment of specific results under well-defined conditions and circumstances. On the other hand, one should make sure that all energy production modes are ideed considered and not discarded offhand, in order to compare all possibilities and take into account their respective merits, in both the short and medium terms, so that concrete and timely solutions can be devised. In other words, diversity in energy supply is as important as selectivity, for while the former provides resilience to the energy supply system by offering multiple and simultaneous back-ups, the latter assures the feasibility of each particular effort.

But it is very difficult to meet the medium- and long-term energy requirements without proper attention to short-term measures. Short-term energy and technology strategies for energy supply are likely to include guidelines for

1. diversification and improvements in the supply of imported oil both in terms of geographical origin and trade arrangements
2. fostering short-run substitution of locally available conventional energy sources, such as gas or coal, for imported oil, for centrally generated energy
3. productivity increases throughout the energy supply sector, through incremental technological improvements and adaptations
4. exploration, both extensive and intensive, and development of indigenous oil and other conventional energy sources
5. assessment of nonconventional indigenous energy resources and development of their exploitation capabilities
6. short-run substitution, of locally available nonconventional energy sources, for imported oil whenever possible

The exploitation of conventional energy resources will probably involve conventional technologies already available, but there exist ample opportunities for technological development even in this area.

The effort to develop indigenous capabilities for exploiting nonconventional energy resources is critical. At the same time, a country must absorb and adapt whatever applicable technologies may exist. In many instances nonconventional resources are site-specific and have received little R & D attention from the developed industrialized countries. A major challenge is to develop the technology associated with these sources to make them effective substitutes for oil. In this respect international cooperation is essential.

Some of the more promising areas are biomass and microhydroelectric resources. These require particular studies of the associated water, soil, climate, and general environment, to be carried out by the individual countries in which these resources occur. It is possible that the related industrial conversion processes could be imported and locally adapted and implemented advantageously.

Matching Demand Reorientation and Supply Development

A third essential element of an energy strategy is to assure a matching of demand and supply for which demand reorientation is a key element. Oil-importing LDCs must carry out an intensive effort geared to a demand reorientation. They cannot rely indefinitely on energy resources that they don't have, with prices that are rapidly increasing, or on sources of dubious availability or which are nonrenewable. An energy demand reorientation, although an important part of the energy development strategy, is not a purely technical problem. It also has political and social components, requiring the establishment of social and economic policies which will eventually produce the desired changes in demand patterns.

For example, directly controlling the price of gasoline affects its demand, but similar effects can be obtained by tax and credit policies applicable to automotive vehicles, as well as regulations on fuel consumption performance, etc.

In summary, formulating energy and technology strategies is not simply attempting to know what the consumption is now or is likely to be in the future, and being able to match the energy supply capabilities to the demand. An energy development strategy is also, and perhaps essentially, an effort to condition and guide energy consumption patterns so that a suitable supply can be obtained by conditioning the technical, economic, and social variables that will influence energy demand. This aspect of an energy strategy is not only much more difficult to accomplish than the others but frequently overlooked, and the required instruments are not adopted or enforced.

As we mentioned earlier, policies can be changed instantly, but they will not produce results unless the instruments for their implementation are created, adapted, and given sufficient time to operate.

In other words, while there are demand analysis and supply assessment phases or steps, there must also be a demand design phase, closely related to the other phases or steps of the energy development process.

COOPERATION FROM INDUSTRIALIZED COUNTRIES

The current situation of the oil-importing developing countries helps to underscore the dangers involved in an excessive reliance on the external supply of the key economic inputs, technology, and technological services without appropriate regional cooperation arrangements.

Industrialized nations, as well as oil-exporting developing countries, have already begun to adopt economic and financial measures to alleviate, at least temporarily, the additional foreign exchange and investment capital burden on oil-importing developing nations.

The World Bank, the Inter-American Development Bank, and other financial institutions have already expanded existing credit levels for conventional energy generation and are in the process of establishing new soft financing and insurance mechanisms for risk-prone petroleum and gas development, UNDP, OAS, OLADE and other international institutions are also expanding their technical cooperation programs addressed to the transfer and adaptation of nonconventional energy technologies.

These initiatives should make a positive contribution to the problem at hand. But there is widespread awareness that the success of these efforts rests to a high degree on strengthening the developing countries' ability to formulate and implement their own energy and technology strategies. Each country has its own specific set of problems and specific endowment of resources. No universal solution can be valid for all of them, at least under present geopolitical circumstances. According to many observers, developing countries themselves should be relied upon to identify specific problems and assess the possible solutions; they should be involved to a higher degree in the development and application of the appropriate solutions, for the effective use of the investment capital, foreign exchange, and technology that could be made available by the industrialized nations.

Industrialized nations can strengthen the decison-making and implementation capabilities of developing countries in a number of ways. One is through technical cooperation to develop human and institutional resources, to search for appropriate problem-solving approaches, and to improve access to technical, economic, and commercial information. Financial cooperation to support the involvement of local capabilities in the development and application of appropriate solutions to the energy requirements may often be even more significant. This kind of economic support is particularly relevant to those developing countries where previous technical and financial cooperation has already helped to develop human and institutional resources that can effectively find and apply solutions to the problems at hand, given the economic means and opportunity to do so. For these countries the utilization of their own technological and economic capabilities may well prove to be not only a way to improve the search for solutions but also a way to make the application of these solutions feasible, given foreign exchange and investment capital constraints. In all probability the utilization of these capabilities is also the best way to develop the kind of self-supporting problem-solving capabilities that developing countries need for self-reliant and fully interdependent development.

Keeping in mind differences in context that will change emphasis in each developing country as well as specific problems of common concern, one may venture some suggestions about areas of action where the contibution of industrialized nations may be most significant.

Energy Policy Planning

Technical cooperation for

1. assessment of current demand and supply patterns
2. assessment of alternative demand and supply patterns
3. assessment of strategies for the development of new supply and demand patterns as a function of the countries' technical, economical, social, physical, and political contexts

4. design of policy instruments to foster productivity increases in energy supply and use
5. design of policy instruments to stimulate conservation
6. design of policy instruments to induce short-run substitution of locally available sources of energy for imported oil
7. technological assessment as well as feasibility studies for investment projects in new energy supply

Technology Policy Planning

Technical cooperation for

1. improving the transfer and diffusion of incremental innovations that can contribute to reduce oil consumption in energy producers' and users' facilities.
2. developing technical standards for increasing efficiency in production, transmission, distribution, and utilization of energy
3. forming strategies for the transfer, local generation, and application of conventional and nonconventional energy supply technology
4. identifying alternative technologies for conventional and nonconventional energy supplies
5. identifying alternative technologies for industrial and agricultural production
6. transferring and adapting site-specific nonconventional energy supplies with focus on rural development

Industrial Cooperation for

1. using local technical and economic capabilities for the development and application to production of incremental improvements and adaptations in existing energy supply and demand facilities
2. fostering the use of local technical capabilities for the development and application of site-specific and conventional technology in rural areas

Although in this paper we have emphasized the energy strategies for oil-importing developing countries, since for them the energy problem is particularly critical, we must keep in mind that the energy problem is a global issue that cannot be solved by a single country independently of the others. As it has been said in a different context, we live in a world of interdependence, and energy has made the world more interdependent than ever.

Table 23.1. Energy Consumption Trend

	Present	2030
World	8 TW	30 TW
	250 Q/y	900 Q/y
	2.2 kW/cap	~4 kW/cap
Western Hemisphere	3 TW	9 TW
	90 Q/y	270 Q/y
	5.4 kW/cap	8.2 kW/cap
North America	2.7 TW	6 TW
	80 Q/y	180 Q/y
	10 kW/cap	19 kW/cap
Latin America	0.34 TW	3 TW
	10 Q/y	90 Q/y
	1 kQ/cap	3.8 kW/cap
Investment Required	$\$3 \times 10^{12}$	$\$3 \times 10^{13}$

Table 23.2. Energy Situation in Western Hemisphere

Gas	500Q
Oil	3,000–5,000 Q
Coal	15,000–30,000 Q
Shale	10,000–25,000 Q
Uranium	800–1600 Q (no converters)
	100,000–200,000 Q (with converters)
Hydro	15–20 Q/y
Biomass	20–40 Q/y
Geothermal	2–4 Q/y
Total	37–64 Q/y

Table 23.3. Oil Production in Latin America (1978)

	10^3 bbl/d	Q/y
Venezuela	2,072	4.560
Mexico	1,226	2.697
Trinidad	230	0.506
Brazil	160	0.352
Ecuador	190	0.418
Colombia	130	0.286
Peru	152	0.334
Argentina	46	0.101
Bolivia	28	0.062
Total	4,234	9.316

Table 23.4. Production of Gas in Western Hemisphere (1977)

	10^9 ft^3/d	Q/y
U.S.	54.85	20.02
Mexico	1.70	0.62
Venezuela	1.32	0.48
Argentina	0.73	0.27

Table 23.5. Export of Oil to the United States (1978)

	10^3 bbl/d	Q/y
Mexico (4.23%)	273	0.600
Venezuela (3.93%)	254	0.559
Trinidad (2.28%)	147	0.323
Ecuador (0.72%)	47	0.103
Colombia (0.08%)	5	0.011
Total	726	1.596

Table 23.6. Sources of Energy (%)

	Western Hemisphere	North America	Latin America
Oil	46	44	66
Gas	29	31	18
Coal	15.8	17	5
Hydro	7	7	11
Nuclear	1.5	2.7	0.1

NOTE

1. For the purpose of this paper the following equivalences have been used:

$1\ Q = 10^{15}\ Btu = 10^{18}\ J = 1\ EJ$
$1\ TW = 1\ TWy/y = 31.6\ Q/y$
$1\ kW = 1\ kWy/y = 3.2 \times 10^{10}\ J/y$
10^6 bbl oil/day $= 2.2\ Q/y$
10^9 ft^3 of gas/day $= 0.365\ Q/y$
1 kg coal eq. $= 2.5 \times 10^7\ J = 0.78 \times 10^{-3}\ kWy$

Chapter 24
The Energy Strategy Planning Process for Developing Countries

Chauncey Starr

INTRODUCTION

Energy is essential to the well-being of any society, and energy consumption is closely related to the production of a society's goods and services. Thus every society seeks the energy supply system which is most appropriate for its current and future needs. This paper addresses the process by which the energy strategy for developing countries may be most usefully planned by them.

The energy strategy most suitable for any country is intimately dependent on its production needs. What, then, are the questions that should be addressed? First and foremost, establish clearly the relevant objectives and their priorities. Second, delineate as fully as possible the social, cultural, and political conditions which limit the scope of options that may be considered. And third, analyze the economic and technical characteristics of the feasible alternatives available. Of course, choosing a mix of these feasible alternatives constitutes the strategy for energy planning.

Let us first consider the range of objectives common to most countries, but whose priority ordering varies depending upon their state of development. They may be listed as follows:

1. *food for sustenance*, ranging from starvation diets through malnutrition to a healthful diet
2. *physical comfort*, ranging from shelter, heat, and food preparation to bathing and sanitation
3. *physical security*, including emergency health services, police services, and on a national scale, military protection

4. *economic growth*, including the accumulation of wealth to permit better health and happiness generally, and the accumulation of an economic surplus to be used as insurance against unforeseen vicissitudes
5. *amenities*, including better health through preventive and curative medicine, education, improved living conditions, and leisure with opportunities for creative development and recreation

These objectives may vary in importance according to the level of development of the country and its trading partners. The problem of establishing priorities among these or similar objectives is common to all nations, and certainly is not uniform within the substructure of any country. It is also necessary to recognize that energy strategies may change as the priorities of these objectives change with the level of development.

The second issue—social, cultural, and political conditions—may be as crucial in determining the feasibility of a strategy as are the technical alternatives. It takes many generations for the customs and traditions of a country and its people to change, and any energy strategy chosen for implementation must certainly be able to be accommodated within existing sociological structures. The demographic configurations of a nation may also be a key to the implementation of a strategy, because the balance between rural and urban populations also takes decades for alteration.

The third broad category of questions which need to be examined is, of course, the economic and technical conditions which will limit the choice of alternatives. The primary economic factor is the development of capital for investments in new technologies and in the industrial infrastructure required to make them function. The ability of a country to develop its own capital resources or to invite foreign resources is clearly a key element.

In addition to all these factors, so much time is required for substantial implementation of strategy that it extends into an unpredictable future. It takes about a decade to demonstrate a new technical strategy, and about two decades to begin to see a substantial output. During such an interval, sociological objectives may change and technical innovations may occur which require altering the strategy. For example, in the early 1920s most agricultural experts predicted that food production in the United States would not be able to support the population which was then predicted for the end of the century; in fact, the population growth predicted for the end of the century has already been exceeded, and food production has grown at such a rate (because of fertilizers and mechanization), that the United States is now one of the world's largest food exporters.

In considering energy options, it is also important to consider the adaptability of the social and political institutions to changes which might result from incorporating these technical options into the total societal structure. For example, the enormous changes in demographic and social structure caused by the automobile have occurred principally within the course of fifty years—the first half of this century. The time frame in which full integration of new energy systems occurs is also roughly fifty years. Therefore, planning horizons of this duration should be the norm in assessments of new energy options.

It is precisely the slow adaptability of societies to change that makes future energy predictions difficult. Although it is tempting to focus on global shortages of fuels and raw materials as setting limits in an eventually steady-state world, the more pressing problem in developing countries is the transient dynamics of maintaining long-term survival and growth.

Because of the length of time required to implement any energy strategy, it is essential that each country thoroughly understand its own dynamics of accommodation to continuing change in the technologies, economics, and societal factors which are keys to the strategy. It is particularly important that there be a recognition that a too rapid or too radical change in any of these factors can produce large instabilities in the social structure and can inhibit, rather than help, the objectives of growth. We have learned from experience with agricultural technology that these potential instabilities should be considered early. For example, the spread of rice-milling machinery in Java and of tractors in Pakistan have both resulted in the displacement of farm labor and reduced the income of the poorest families. Thus the foundations of village life may be impaired even though an aggregate national benefit may accrue.

One of the major factors which will affect the choice of energy system options is the basic problem of population growth. Its influence on the objectives of an energy strategy are well known. From the point of view of technological choices, however, a key element is the growth of skills within the population. If technological and industrial skills are in very short supply, the natural tendency within each country is to centralize these skills so as to use them more effectively. This results in urbanization as well as industrial concentrations and energy centers. If, in parallel with the energy strategy, there is also a strategy for developing supporting technological and industrial skills, there may indeed be greater flexibility in the choice of options. In any event, the nature of the population resources and their change with time is as important as many of the other conditions limiting the choice of energy strategies.

The process of strategy planning, therefore, involves the establishment of a hierarchy of objectives, a hierarchy of technical options, a recognition and clarification of the societal and economic conditions that bound the field of choice, and, finally, the combination of these factors to develop the series of alternatives which are feasible. The choice of these alternatives defines the appropriate technologies for the particular country and its conditions. It is important to recognize that appropriate technology is not necessarily equivalent to the popular image of a cottage industry focused around the individual home and built and operated with the local renewable resources. Such a technology may be appropriate for the poorest societies, but not for others. On the other hand, it need not mean that every region should have a large central power station. Appropriate technologies should provide a system of technical means tailored to the existing and foreseeable circumstances particular to each country.

TRADITIONAL ENERGY SYSTEM ASSESSMENT

The energy system characteristics traditionally analyzed in the industrialized countries provide a basis for a regional assessment. First among these is cost. The capital costs per unit of energy supply capacity will most likely be the single most important factor, particularly in countries with very limited capital. Operating costs, involving fuel and system reliability, do trade off with capital costs, depending on capital availability. The long-term social costs arising from the environmental impact of energy systems should be considered in terms of the present value of future risks and benefits, but these are more difficult to evaluate and assess.

The initial factor in all these issues of energy system planning for underdeveloped areas is capital availability. Besides the normal competition for economic output between consumer goods and capital formation, the competing industrial and public uses for available capital will limit the capital available for energy system expansion.

A rough estimate of the potential rate of installation of an energy system can be inferred from a simple capital availability calculation, except for those rare countries in which capital is not a serious constraint to development (e.g., the OPEC countries). The fraction of economic output used for gross capital formation tends to increase slowly with increasing per capita gross domestic product (GDP). Typical capital formation rates range from 5–15 percent of GDP in the poorest countries (those with GDP per capita of less than U.S. $100) to 15–30 percent for the industrialized nations. A sizable portion of this capital represents depreciation reserves to cover the ultimate replacement of existing stock, and as a result represents capital resources that are not available for expanded facilities. Assuming purely exponential growth of population and GNP, growth curves are shown in figure 24.1 An assumption implicit in the model used for this figure is that economic output will rise in direct proportion to capital stock (hence the purely exponential growth of both GDP and capital stock). For the industrialized countries, this assumption is well supported by the data. Capital-output ratios (the ratio of capital stock to GDP) for the developed countries are remarkably constant.

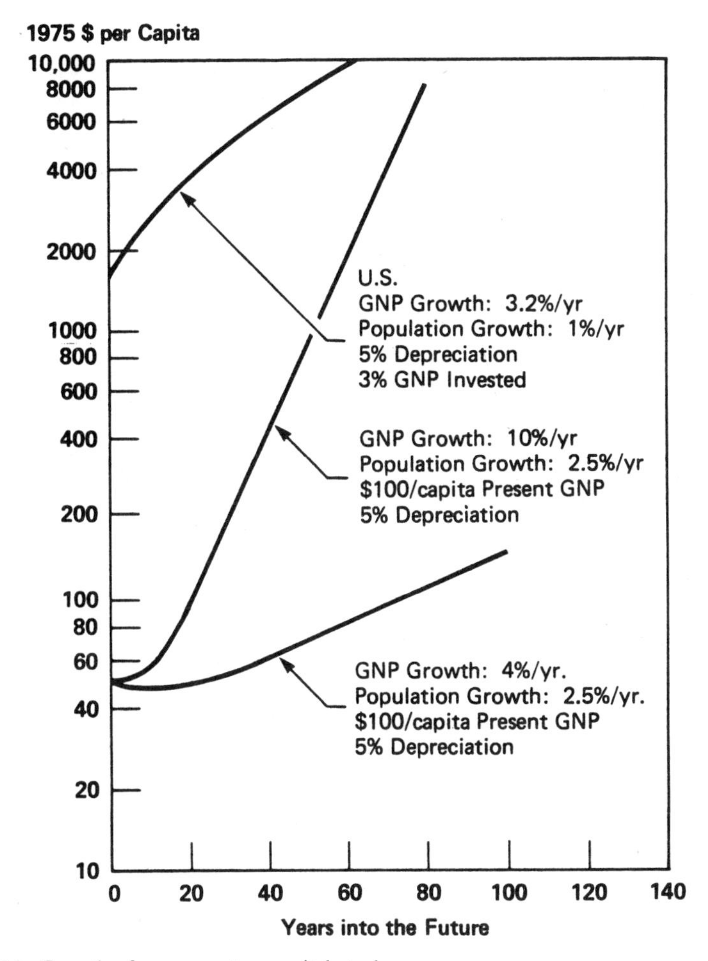

Fig. 24.1. Growth of energy system capital stock

For the underdeveloped areas, generalizations cannot be made; the capital-output ratio can be either higher or lower than that of industrialized nations. The capital-output ratios of the economies of underdeveloped regions are subject to two effects not applicable to developed economies. First, the initial units of capital investment, if well chosen, can yield greater returns in underdeveloped than in developed countries, largely because the initial units of capital invested in any economy are inherently more productive than subsequent investments. The second factor in underdeveloped areas is the need for social overhead capital, such as roads, communications systems, and energy systems. These investments are referred to in the economic literature as indivisible or lumpy, in that there is a certain threshold cost for these systems that is independent of their use, and thus their expansion and use are discontinuous. Yet these systems are needed if other sectors of the economy are to develop. These two effects tend to counteract each other: the initial high marginal return on investment tends to push the capital-output ratio down; the need for social overhead capital tends to raise the ratio.

Determining optimal methods of resource allocation is extremely complex, yet the issue of energy system capital allocations can only be considered within the context of the total problem of development. Assessment of energy options for the underdeveloped countries is more closely coupled with total resource development decisions in the underdeveloped regions than in the developed areas.

ENERGY OPTION ASSESSMENT IN AN ECONOMIC DEVELOPMENT FRAMEWORK

A distinction can be drawn between the energy used to meet direct consumer demand and that used indirectly as an input into a production process. While the direct consumer energy demand produces substantial amenity benefits to the user—comfort (as from space heating) or savings in domestic labor—the focus in underdeveloped countries is likely to be on the other indirect uses—energy for production.

Much has been written about simple small-scale energy systems for underdeveloped areas. For rural areas these small-scale systems appear quite attractive for a number of reasons: Much smaller initial investments are required, less specialized training is needed for construction and maintenance, and the cost of energy transportation is reduced or eliminated. However, while these small systems will improve rural amenities, they make little contribution to GDP. The economic incentive to seek development first in urban areas, and to concentrate on energy for production, leads to a type of energy demand that is much different from that of the rural areas.

The factors which typically contribute to increased production activities—capital stock, division of labor, distance to the markets for the goods produced—all tend to focus such activities in the cities. This has the advantage of best utilizing the existing social overhead capital of roads and energy systems.

In the urban areas, the centralized systems have two substantial advantages. First, the energy cost is lower. Based upon the costs of energy systems in place, energy delivered by wire or pipeline from a central facility is considerably cheaper than energy from small systems, for urban and industrial centers. A recent editorial in *Science*, which recommends rapid development and utilization of small solar systems for rural areas, indicates what some of the typical costs are:

> The cost of utility power in the United States averages 3 to 10 cents per kilowatt hour. It runs as high as 45 cents/kWh in urban areas of developing countries. In rural areas, however, power is available only from diesel generator sets at $1/kWh or more, or from primary batteries at about $12/kWh. Complete solar-thermal power systems costing about $4 per peak watt and capable of providing electricity at less than $4/kWh are already available. Photovoltaic systems costing $1 to $2 per peak watt are expected by 1980. . . .

The cost of urban electricity can, with sufficient load density, be significantly cheaper than 45 cents/kWh. For any rural production process, an energy cost of $1/kWh or more is an enormous competitive disadvantage. (And these cost estimates did not include energy storage costs.) Technical details of typical village-type and irrigation solar systems are given in the Appendix.

At $1/kWh or more, a simple light bulb burning for ten hours a day uses over $300 worth of electricity a year. In many underdeveloped and developing countries, a GNP per capita of less than $300 is common. It would take all of the economic output of the typical laborer just to pay for this energy. This must be viewed as an investment with a very low return relative to its cost. Much more worthwhile investments can be made in urban energy systems with the same amount of capital.

The second advantage of centralized energy systems is due to the form of energy produced. For most production processes, the demand is for mechanical motion (as from motors), high temperature heat, and light. These requirements can be satisfied by electricity, but not by low-temperature energy sources such as solar space heating. For space heating in urban and industrial areas, utilization of waste heat and cogeneration appear to be more attractive than small systems. The one significant exception is that of energy for irrigation, which represents a highly disaggregated load. Improvements in agricultural yields can, in addition to the primary benefit of reducing hunger, contribute favorably to national economics by reducing food import requirements or generating income through export, providing means are available for the transport of agricultural products. For irrigation, small diesel or wind-driven pumps may represent favorable investments.

Option Assessment—Infrastructure

A key factor in determining energy system capital requirements is the cost of developing the infrastructure required to utilize a particular energy technology. It is this factor that contributes substantially to the popularity of oil: An oil-based system requires a very low investment in systems to transport energy. If the oil is imported, a refinery is not even needed, as various oil derivatives can be imported directly. The fuel is easily stored, and the level of technical sophistication required to operate on an oil-fueled economy is much lower than that required for a nuclear or coal-based economy.

As an example of the infrastructure required for a coal energy system, it has been estimated that, in order to double U.S. coal production (600 million tons per year), the following actions would be needed:

- Develop 140 new 2-MTPY eastern underground mines
- Develop 30 new 2-MTPY eastern surface mines
- Develop 100 new 5-MTPY western surface mines
- Recruit and train 80,000 new eastern coal miners
- Recruit and train 45,000 new western coal miners
- Manufacture 140 new 100-cubic-yard shovels and draglines
- Manufacture 2,400 continuous mining machines

These items are in addition to the tremendous capital requirements for coal-fired electric power plants and the transmission and distribution system needed to deliver the energy, or, alternatively, the coal gasification plants with a pipeline network. It is probable that for many countries the number of engineers could be a limitation, so the cost of creating engineering schools or of sending students to other countries must also be included.

For a nuclear system, the material flow is significantly lower than for coal, but the number of areas in which technical specialization is required is greater. A number of developing countries have recognized this problem and established local training facilities to provide the necessary manpower. India has already trained more than 3,000 nuclear engineers, and ambitious plans are under way in Brazil, Iran, and Pakistan.

There can be capital costs that go beyond these issues of industrial infrastructure. When the Tennessee Valley Authority (TVA) began operations in the United States, it was found that many people in the TVA service area could not afford many of the basic energy-using devices. As a result, TVA issued guarantees on extremely low-interest loans to stimulate the purchase of such items as irrigation pumps and refrigerators. Thus capital was needed for the end-use devices as well, increasing the overall requirements.

Effects of Energy System Size

From the preceding discussion, it is clear that the capital required for a complete energy system greatly exceeds the traditional power plant costs. The cost of these systems will also depend on scale factors both in energy distribution and, in the case of electric power plants, unit size. These factors are illustrated in figures 24.2, 24.3, and 24.4.

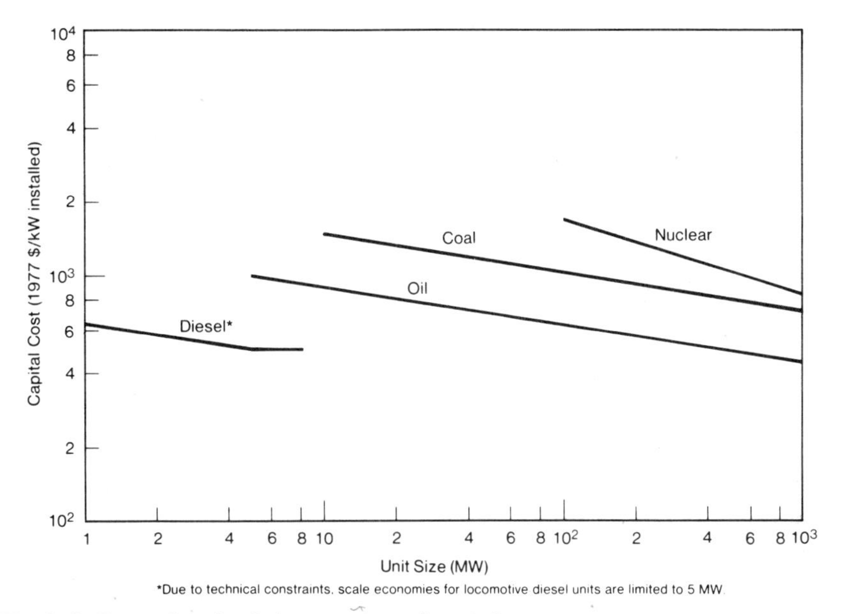

Fig. 24.2. Economies of scale in generating unit capital costs.
Sources: EPRI 1978 Technical Assessment Guide; ERDA WASH-1345; *Fortune*, December 31, 1978.

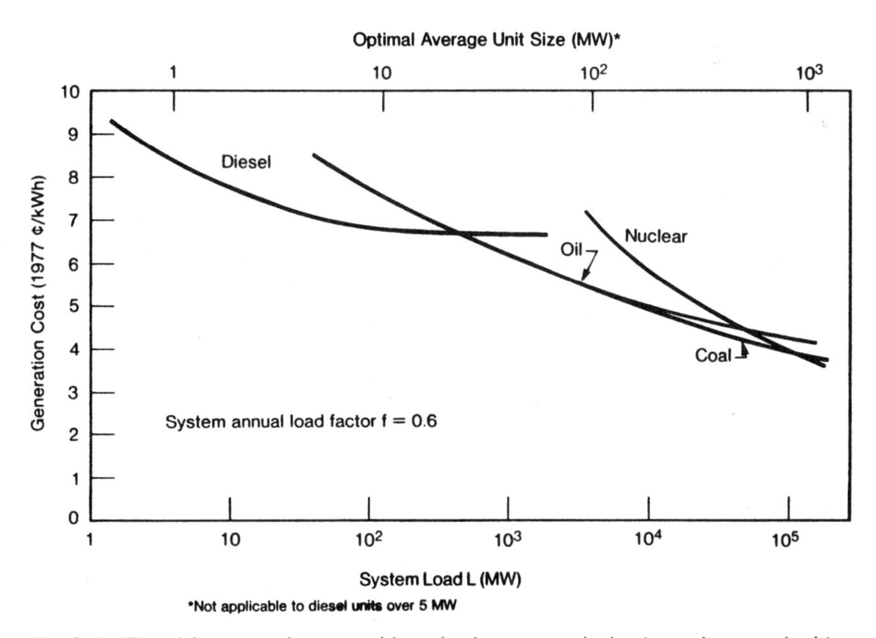

Fig. 24.3. Electricity generation cost with optimal average unit size (new plant cost basis).

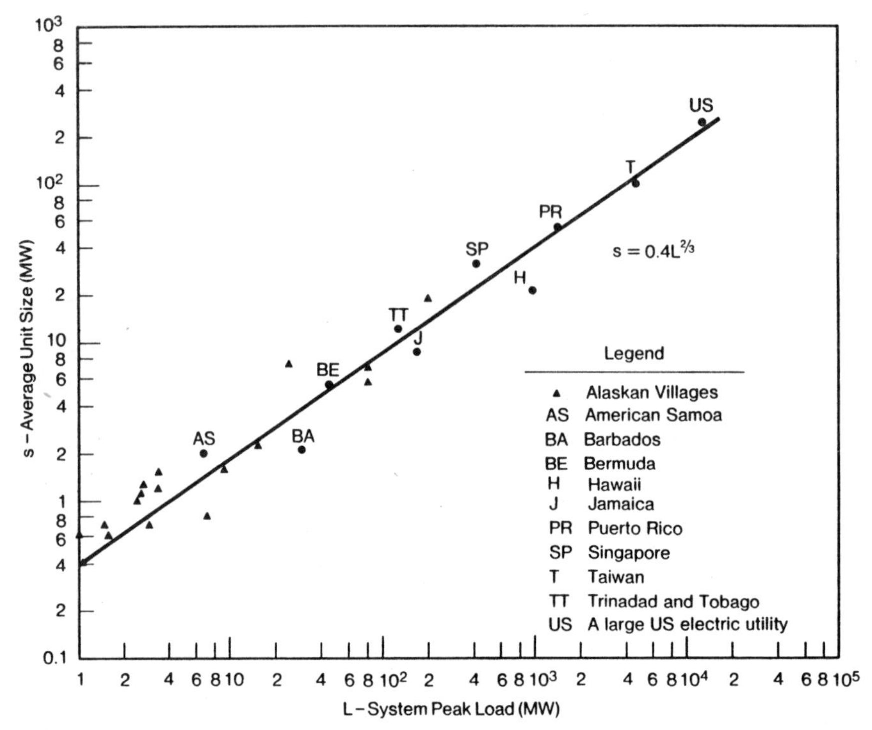

Fig. 24.4. Average generating unit size as a function of system peak load.

Sources: Electrical World, *Directory of Electrical Utilities*, 1977–78; *International Directory of Electrical Suppliers*, 1972–73.

Transportation of energy can be divided into large-scale bulk fuel shipments and local distribution. The more important issue for developing countries is the local distribution of energy. Some minimum capital investment is needed to install a system for energy distribution to each household. For fixed systems, such as those associated with electrical systems or gas pipeline systems, the capital costs are relatively insensitive to the quantity of energy delivered, but, rather, depend on the density of energy consumers. For the rural areas of the underdeveloped and developing nations, this capital cost will be very high. For example, rural electrification in the United States was a very expensive public works project in the 1930s; it was heavily subsidized by the rest of the U.S. economy to meet the sociological objective of economic growth in the rural areas. These rural areas show distribution system capital costs of about $1,000 per customer. On purely economic grounds, at the time it was built, rural electrification could not be justified.

The implication is clear: For those areas in which capital or demand limits construction to small plants, the energy costs will be higher. Combining the distribution and production costs of energy, it seems likely that energy will be most expensive in the least developed areas of the world and in their rural areas. As a result, economic production activities which require a significant energy input will be at a competitive disadvantage in these areas.

Small-Scale Energy Systems

The small-scale energy systems, of which solar heating is an example, can, for reasons relating to energy distribution, be cheaper in areas with low density of energy use. Yet the capital requirements for these systems are high, and the energy provided is more likely to contribute to comfort than productivity growth. In these rural areas, the use of farm waste is traditional, and there are technologies available at low cost which permit more efficient use of such waste (biogas digesters, for example). It should be recognized that large-scale biomass conversion to energy does not generally result in a net energy output, due to all the energy requirements for growth and harvesting. Some exceptions, such as sugar cane, can be found to be net energy producers. However, where biomass waste is already collected and available, it certainly can be used locally. Large-scale biomass projects are likely to compound the problems of global food supply. Energy plantations would compete directly for agricultural land.

Diesel power generators can have special applicability. For high-energy density areas, the extremely high fuel costs and maintenance costs make diesel noncompetitive with centralized electric technologies. In areas where diesels are the primary source of electricity, as in a number of Alaskan villages, the costs range from 14¢ to 48¢ per kWh, much higher than typical U.S. power costs of 4–5¢ per kWh. Yet for these villages, this is the cheapest electricity available. The capital costs of such a system are quite reasonable: Diesel generators are available in the United States for about $100/kW (in 1977 dollars) for 100 kWe output. Little additional capital is needed for ancillary equipment; fuel tanks and fuel delivery trucks are not particularly expensive. While such a system can provide amenities, because of its very high fuel and maintenance costs this type of power system will not lead to light industry that can compete with the urban industries using much cheaper central station power, and capital outlays of this type are made at the cost of more productive investments.

The most important exceptions to the high energy cost in rural areas arise from indigenous resources. Large hydroelectric facilities coupled with energy-intensive industry, such as aluminum production, can provide substantial economic benefits. Dry steam geothermal can also be used as an inexpensive regional power source. For these cases of a specific indigenous resource, the entire schedule of energy production and transportation costs is altered, and regional development plans centered around those resources can be extremely successful.

Implications of Scale Effects

Of all the social overhead items that require large initial investments, energy systems are, in the economist's jargon, the lumpiest. A large power plant, with the related transmission and distribution network, frequently represents an investment in excess of a billion dollars. Most underdeveloped countries cannot allocate capital in this amount to any single project. Two alternatives present themselves: The first is to create a flow of capital to these countries from the industrial and resource-rich areas of the world. In many underdeveloped countries, aid of this sort is the largest source of capital. While there are political issues raised by this method of accumulating capital, it represents an important avenue that should not be overlooked.

If economic growth is the overriding goal, the second strategy suggested by this analysis would be to develop one area at a time. This would best utilize the efficiency of scale, and also serve to produce the infrastructure needed for productivity increases. Several important issues are raised by such a program. Equitable distribution throughout a nation of the costs and benefits of the capital-intensive system might make such a policy politically difficult.

Industrialized and High Capital Availability Regions

In view of the background presented on the relationship between energy systems and regional development, both the industrialized countries and the capital-rich nations can take actions which help the underdeveloped regions achieve their economic goals, while satisfying their own requirements. The spectrum of strategies is shown in table 24.1.

The industrialized countries with capital are, and will continue to be, the technological leaders of the world. They are likely, for economic reasons, to reduce their resource consumption through the use of the high-technology, capital-intensive substitutes which they can best apply. High-technology development is a strategy that the industrial nations short on capital cannot afford. These countries will use coal if they have the source, nuclear converter reactors, and oil and gas. The developing and underdeveloped countries with ample capital (OPEC countries, for example) have the greatest flexibility and widest range of opportunities. An approach to development like that taken in the Tennessee Valley of the United States in the 1930s is a likely choice for many of these countries. Under these conditions, energy system development, coupled with programs for increasing labor productivity, can produce very favorable long-term prospects.

Table 24.1. Energy Strategies

Economic Type	Region Type	Capital Availability: High	Capital Availability: Low
INDUSTRIALIZED	All Areas	Responsible for technical leadership, has capital and infrastructure to utilize high technology sources, seeks to minimize resource competition with underdeveloped	Use lowest cost system that takes advantage of existing capital stock Deemphasize long range energy R&D
UNDERDEVELOPED & DEVELOPED	Urban	Total city planning needed Advanced technologies	Energy density may justify central stations Development of industrial infrastructure is primary goal
	Rural	TVA type approach, smaller capital intensive units, regional industrial centers established concurrently with agricultural capital expenditures	Improve quality of life with small systems–hydro, farm waste, biomass, wind, solar

SUMMARY

- Backyard energy sources for rural areas can improve agriculture, comfort, and lifestyle, but are unlikely to produce substantive economic growth.
- Liquid fuel-based energy systems (diesel and internal combustion) provide the least costly and most flexible intermediate-sized energy sources.
- Electrification of rural areas is a luxury that industrialized countries have achieved only recently; it is economically counterproductive for developing and underdeveloped countries.
- The cities use energy effectively due to economies of scale in generation and delivery. This, coupled with other favorable production factors, makes urbanization a strategy for economic growth, and determines the design of an appropriate energy strategy.

APPENDIX

Small power system requirements for rural applications can be classified in two ways: interruptible and uninterruptible. An example of the former would be power for irrigation, while the latter is more characteristic of a village power system. The energy system determined as most appropriate for a given load will depend upon such a classification, as well as its distance from a commercial supply center.

Below we examine two examples of rural power application, and determine the factors that are more important for the selection of an appropriate technology. The first is a water pumping system for irrigation, which can be considered an interruptible load, i.e., a load which does not require a continuous supply of power. In this case, intermittent sources of power have no inherent disadvantage in providing the power. A village system, on the other hand, may require an uninterruptible and continuous supply of power (power on demand) for serving small, light industrial loads as well as domestic lighting and refrigeration loads. Obviously, intermittent power sources by themselves are not compatible for such needs.

For the irrigation system, we consider three power supply technologies: solar photovoltaic (PV) wind, and diesel generators. We assume a 2-hectare plot of land to be irrigated continuously year-round with a daily average of 50 m^3 of water per hectare. Without considering seasonal variations in wind and solar insolation (which would tend to increase the size of the wind and PV system in order to meet worst day criteria), system requirements have been calculated and are shown in table 24.2.

For the village system, we consider as a reference system the Papago Village PV system recently installed in Schuchuli, Arizona. This system was designed to serve an annual load of 6,255 kWh (17.1 kWh/day) without a backup system. This system and comparable wind and diesel systems are described in table 24.3.

Table 24.2. Irrigation Systems

	Power Output	Annual Fuel Consumption	System Size
PV	1.0 kW peak	—	7,313 lbs.[1]
Wind	1.5 kW rated	—	787 lbs.[2]
Diesel	0.5 kW	180 gallons (1,300 lbs.)	60 lbs.[3]

[1] 15 m^2 array and supports.
[2] 12′ rotor machine, 60′ tower.
[3] generator set only.

Table 24.3. Village Systems

	Power Output	Storage	Annual Fuel Consumption	System Size
PV	3.5 kW	413 kWh	—	38,884 lbs.[1]
Wind	4.0 kW	413 kWh	—	5,684 lbs.[2]
Diesel	0.7 kW	—	550 gallons (3,920 lbs.)	60 lbs.[3]

[1] Includes 71.4-m^2 array, supports, batteries; does not include control equipment and equipment housing.
[2] Includes 20 ft diameter rotor, tower, batteries; does not include control equipment and equipment housing.
[3] Includes generator set only.

The exploitation of a diffuse and intermittent energy source requires large and bulky collector and storage facilities. The PV irrigation system weighs well over 100 times the diesel system, and the village system widens this difference four-fold. Since power can be drawn from the diesel generator on demand and the energy is stored in the diesel fuel, a much smaller system can be employed. This minimizes both system size and initial cost.

One input into the system decision-making process is the transportation requirements of delivering diesel fuel to remote areas. The extent of this concern will, of course, depend upon the distance of the site from a commercial supplier. The fuel requirements of the irrigation and village systems are 4.3 and 13.0 barrels of diesel per year, respectively. In the case of the village, it is not clear that the cost of delivering this amount of fuel for ten years would be greater than the initial cost of delivering the whole PV system (see table 24.4).

Table 24.4. System Costs

	Cost
Irrigation	
PV	$15,000
Wind	6,320
Diesel	500[1]
Village	
PV	$108,483[2]
Wind	44,040[2]
Diesel	500[3]

[1] Without installation, annual diesel fuel requirement of 180 gallons.
[2] Does not include cost of replacing batteries.
[3] Without installation, annual diesel fuel requirement of 550 gallons.

Chapter 25
National Energy Planning in an Energy-Constrained World

Philip F. Palmedo

INTRODUCTION

The last several years have witnessed fundamental changes in the energy conditions affecting nations and in the future energy prospects of the world community. These changes have created severe immediate economic difficulties for many developing countries, particularly those dependent on imported oil for their energy supply. The immediacy and severity of these problems have monopolized attention on short-term remedial measures. The consideration of such measures necessarily takes as a framework the current patterns and technologies of energy supply and current relationships of energy and development.

Changes in world energy conditions, however, have even more fundamental implications for longer-term economic and social development, implications which have received much less attention, and which are much more difficult to comprehend than are current energy problems and possible solutions. The comprehension of those longer-term problems at both national and international levels will require an intellectual liberation from conventional planning approaches and from the energy-development relationships upon which they depend. Attention to the long-term in national energy planning is required by the long-term character of energy problems as well as by the current major shifts in world energy conditions.

In this brief paper I suggest some of the issues that arise when one considers the problems of long-term energy planning in an energy-constrained world.

CHALLENGES TO ENERGY PLANNING

The Implicit Energy-Development Model

Historically, the relationship between growth in energy consumption and economic development—or growth in national income—has been so close that it has suffered the misfortune of being characterized by an elasticity. I term it a misfortune because the pursuit of elasticities in mazes of energy consumption-price-income data so often diverts attention from the pursuit of an understanding of the underlying phenomena. The justification of an income elasticity of energy demand is clear, however, from an examination of either cross sectional data showing energy consumption as a function of per capita GNP or time-series data showing the growth of energy consumption and per capita income with time.

The traditional relationship between energy consumption and economic development is one aspect of what might be called the implicit energy-development model. National development planning requires a view of the future, a view of the character of the society and economy to be attained over the long term. In many cases the explicit or implicit model of development is provided by the fully industrialized countries of Europe, and by the United States and Japan. For many of the poorer developing countries, an intermediate model which has a more direct appeal is provided by the newly industrialized countries, such as Korea, Taiwan, Brazil, and Mexico. To the degree that these countries themselves take the more advanced industrialized countries as their model, the effect is the same. History has been taken as theory.

A country's energy demand pattern and, to a certain degree, its supply pattern, is determined by the structure of its economic and social activities. Thus the acceptance of a development model usually includes, or at least implies, a model of energy use and supply.

The implicit model of future energy consumption is not only conceptual. It is built into the economic and social structure of countries at every stage of development. Although a country may have a low rate of per capita energy consumption, the modern sectors and higher income groups often consume energy at a rate much higher than the national average. For example, figure 25.1 shows an estimate of energy consumption by households of different income levels in Mexico City. As the figure indicates, the higher income groups consume energy at a rate higher than the average in Sweden. The significance of these data lies in the fact that the pattern of energy consumption as a function of income reflects the society's economic and social aspirations. The very large increases in the use of transportation energy with income reflect a shift from public transport to private automobiles as income increases. That trend also reflects the desires and aspirations of those at

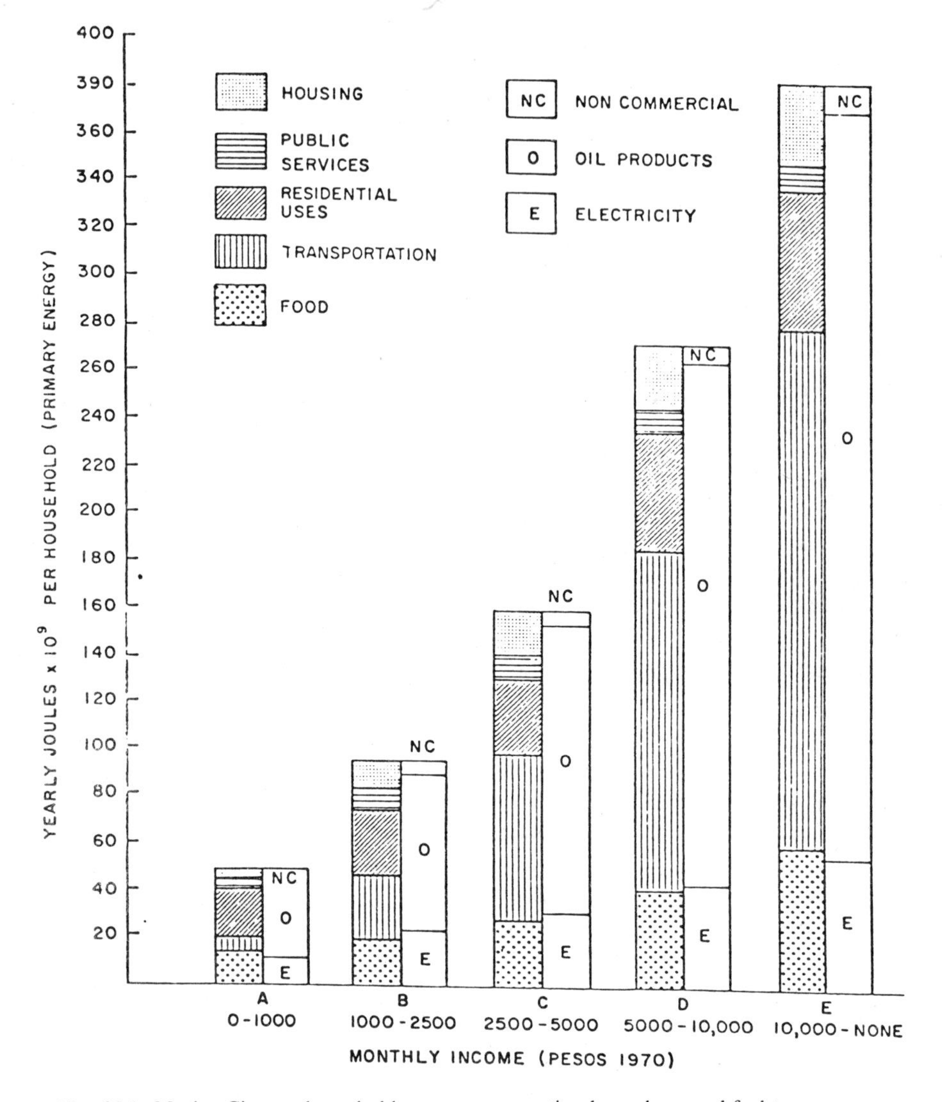

Fig. 25.1. Mexico City per household energy consumption by end-use and fuel type

Source: Gordon McGranahan and Manuel Taylor, "Urban Energy Use Patterns in Developing Countries: A Preliminary Study of Mexico City," Institute for Energy Research, Stony Brook, 1977.

lower income levels. In this sense a more energy-intensive future is already embedded in the structure of many LDC societies.

The energy world that confronts the developing countries is very different, however, from the world in which the OECD countries industrialized. The implicit energy-development model may be valid for a future world in which oil is plentiful and cheap, but it is not necessarily valid for a world in which liquid fuels are rare and expensive. The LDCs, at earlier stages of economic development, are even more dependent on oil than the industrialized countries. The developing countries now depend on oil for 61 percent of their energy supplies in contrast to the U.S. figure of 44 percent (and the world average of 50 percent). Furthermore, the Western industrialization process was powered by indigenous fuels and produced manufactured goods for export, whereas many developing countries use imported fuels to produce goods for internal consumption.

Thus in a number of ways, the implicit model of energy and development provided by history no longer provides a valid guide for energy-economic planning. History can no longer be taken as theory. Developing nations are challenged to think afresh about the relationship between their economic and energy futures. The next section of the paper identifies some of the major issues that should be addressed in this process.

Implications for Energy Planning

The Global Context. As suggested above, a number of factors external to the developing countries will have a strong influence on the energy strategies that those countries should pursue in the future. Those factors include familiar elements such as the competitiveness of products that are possible exports and the cost of imported resources and manufactured goods. A new major determinant will be the future supply and price of petroleum, given the very large dependence of many third world countries on oil as their dominant energy source. Many of the solutions to the energy problems of LDCs and the possibilities of reducing future oil consumption depend on technologies being developed in the industrialized world. To substitute imported technologies that provide solar energy at the equivalent of $6 per million Btu for imported oil costing $5 per million Btu offers no immediate advantage to a developing country. Under certain conditions of future oil prices, however, such a solar strategy would be of real benefit.

Thus it is clear that any energy planning activity in a developing country should take explicit account of future world energy conditions, including the cost and availability of new energy technologies. A wide variety of complex issues enters into this sort of analysis, including trade-offs between current capital expenditures (in a normally highly capital-constrained environment) and continuing payments for fuels at uncertain future prices. Other factors are the implications of dependency on various political entities for the supply of fuels or capital equipment and technological expertise and the willingness of international financial institutions to support the implementation of certain technologies.

Uncertainties. Mention of the future availability and price of world oil immediately raises the issue of the uncertainties which must be attached to critical factors in energy policy analysis. One of the most interesting challenges to energy systems and policy analysis in developing countries is to deal adequately with these uncertainties. A reasonable approach to this problem is the use of simulation models and the construction of alternative scenarios dealing both with the state of the external world, including energy prices, and with future national conditions, including, for example, national economic growth rates and the sectoral composition of that growth. This sort of simulation approach can be used to demonstrate both the implications of alternative futures which may be under the control of national governments and, at the same time, the implications of uncertainties in external conditions which are not under the control of national governments.

The Energy-Economic Relationship. The failure of the implicit energy-economic development model requires that energy strategies be analyzed as an integral part of the economic development context of a country. Coupled energy-economic models are perhaps even more important in analyzing energy futures for LDCs than they are for industrialized countries, since the economic development structures of the LDCs are often less determined and more open than those of the industrialized West. Although a number of approaches to coupled energy-economic modeling have been developed, most of these are now too complicated and costly to apply in many developing countries. In many cases the challenge is to construct as a first step in national energy planning a simplified energy-economic model which can then be elaborated and developed in subsequent steps. A model of this kind has been developed at the Brookhaven National Laboratory and is currently being applied to the case of Portugal. It is based on an energy engineering simulation model, an input-output (I/O) model with an expanded energy service representation, and the World Bank Minimum Standard planning model. A more loosely coupled energy optimization-I/O macro model is being developed in the context of a Tunisian energy assessment being carried out by Gordian Consulting Services, Ltd., and Energy/Development International.

An important point to be stressed in performing energy-economic analyses at the country level is the proper reflection of national fiscal and financial concerns in the analysis of energy strategies. To a certain degree, one can say that the dire impact of increased oil prices on developing countries are not so much the resulting changes in costs to individual industries or consuming sectors but rather the impact on the national balance of payments. Thus one important role of the energy-economic framework is to properly reflect the shadow costs of imported materials and fuels on feasible national energy strategies.

The Geographic Dimension. Most energy assessments carried out in developing countries to date do not give adequate attention to geographical scales other than the nation. The geographical dimension is important in two ways. First, in many countries there are wide geographical differences in the availability of resources and in typical energy consumption patterns. These differences will be even more important as distributed renewable energy sources become more viable. National oil pipelines and national electricity grids are manifestations of the dominance of the national scale in energy thinking to date.

A second way in which the geographical dimension is important is the effect of spatial configuration on future energy demand levels. As the energy costs of transportation become more important, for example, so will the location of industry relative to energy resources. The spatial relationships between living areas and jobs will also become more important, as will the spatial configurations of urban areas.

In the future, therefore, increased attention should be given to the incorporation of spatial considerations into national energy planning and assessment.

Energy Service Demands. As developing countries think about alternative ways of satisfying the energy needs of their economic and social development, it is well to keep in mind the importance of analyzing energy from the point of view of the services to be provided rather than the fuels which have traditionally been the focus of attention. By energy services I mean the transport of materials (in terms of ton-kilometers), the maintenance of a comfortable temperature range, or the provision of heat to certain industrial processes. This concept is particularly important as countries now consider a much wider range of possibilities to satisfy future energy requirements. It is not energy that is required by society, after all, but the services that energy provides. Countries now must consider a much wider range of possibilities of supplying those services and must be able to compare investments in increased energy efficiency, for example, with investments in new energy supply. In fact, a framework must be created and implemented to compare investments in increased energy efficiency in industry with investments in new productive capacity or output. As a further example, rural electrification programs, based on extensions of the national grid, have often been based on tacit assumptions about the benefit of electricity in rural areas. A more basic approach to that question would be to analyze the most effective contributors to rural energy development in an economic or social sense and then to determine the most effective ways of providing the energy services required, considering both electric and nonelectric requirements and, for electricity, both centralized and decentralized options.

The emphasis on energy service demands requires that considerably more attention be given in energy planning models to the physical services or needs satisfied by energy and to the processes at the point of energy consumption where fuels or sources such as solar energy are transformed into those services. This is, of course, not a new concept in developed countries and a number of energy models such as MEDEE and the Brookhaven family of models use this approach.

CONCLUSIONS

The great importance of national energy-economic planning in developing countries combined with the difficulties of carrying out that planning in a responsible fashion, constitutes a formidable challenge. The need to develop the local institutional capacity to carry out planning functions, the need to devote attention to long-term issues in a time of immediate crisis, the very limited experience throughout the world in this type of planning, and the severe shortage of adequately trained personnel, both in the developing countries and in the industrialized countries, are some of the elements of that challenge. The importance of national energy-economic planning in developing countries has been recognized for a number of years; its difficulties are just recently becoming apparent. The creation of this kind of planning capability should remain a high priority on the developmental and assistance agenda of national energy assistance agencies, international assistance organizations, and regional and international financing institutions. It should also present a practical challenge to the academic community interested in energy and economic development.

Chapter 26
Cooperative Energy Assessments with Developing and Industrializing Countries

Robert A. Summers
and Richard R. Cirillo

PART 1—DESCRIPTION OF COUNTRY ENERGY ASSESSMENTS

Introduction—The Energy Problems of Developing Countries

Recent events have had drastic effects on the energy supply problems of the international community and on the economies of individual nations—both industrialized and developing. In the past few years, the view that the United States should be doing more to help resolve these international energy problems has gained increasing acceptance by the U.S. government. There is now also general recognition that the international community as a whole must take vigorous action to reduce demand and increase energy supplies.

As understanding of the international implications of energy has grown, so has awareness of the plight of the developing world. It is the developing world which is hurt the most by rising energy costs and supply limitations and which is, at the same time, least able to do anything about it. The impact of the spiralling oil prices on oil-importing developing countries has been staggering. As the proportion of their national income spent to pay oil bills has increased, the fragile economies of these countries have been brought under greater and greater strain.

The current accounts deficit of the oil-importing LDCs has risen from $22 billion in 1978 to $36 billion in 1979. The 1980 deficit could be over $50 billion, with most of the deficit resulting from the higher prices paid for imported oil. The rest is a result of decreased demand for LDC exports in industrialized countries and increased import costs for the other goods imported by the LDCs.

Unfortunately, the problem may well get worse before it gets better. If current consumption patterns continue, the future oil demand of LDCs will increase sharply both in absolute terms and in proportion of worldwide consumption, reaching perhaps 25 percent of world demand by 1990 and 40 percent by the year 2000. Those countries now at the lower end of the industrialization spectrum must increase their energy consumption to support needed development. The close relationship between the use of commercial energy forms and economic and social development in developing countries makes it particularly difficult to moderate growth in energy demand without adversely affecting the rate at which development goals are achieved. Much of this growing consumption will take the form of demand for imported oil unless these countries are able to adopt other suitable energy technologies.

The need for comprehensive economic and energy planning in many developing countries has only recently been recognized.

Unfortunately, for many countries there is limited knowledge about availability of indigenous energy resources and on the most effective means to exploit them. Necessary data on past and current energy consumption are sparse or have not been collected. Hence there is a growing consensus in the United States and throughout the industrialized world that more must be done to help developing countries address and solve their growing energy problems.

Aims of the Country Energy Assessments

The executive branch and the Congress have, over the last few years, articulated the policy that the United States must increase its energy assistance to developing countries and have put forward a number of policy and program initiatives in this direction. One such initiative is the International Energy Development Program (IEDP) or, as it is generally referred to, the Country Energy Assessments program.

Initiated in 1977 by President Carter, the IEDP program is aimed at helping developing and industrializing countries acquire and improve capabilities for comprehensive national energy planning so that their energy-related decision-making would be consistent with the new realities of energy supply and price.

The program addresses key U.S. foreign policy objectives, including a lessened dependence of these countries on high-priced imported oil, encouraging the use of indigenous energy resources, and avoiding premature or excessive commitments to nuclear power systems. This program also constitutes an important part of the administration's effort to encourage oil exploration and production in non–Persian Gulf developing countries.

Cooperative assessments have been conducted to date with Egypt and Peru. In addition, assessments are now under way with Argentina, Portugal, and the Republic of Korea.

Scope and Structure

Briefly, an IEDP energy assessment is a comprehensive, in-depth technical and economic evaluation of a country's energy supply and demand options or strategies corresponding to a set of economic development scenarios. This is conducted on a cooperative basis with the host government, and requires in-country data gathering, extensive analysis of the data, using technical and economic models, and finally evaluation and interpretation of the results. For each energy strategy, estimates are made of the required manpower and financial resources.

The assessments are managed by the Department of Energy (DOE) with the policy guidance of the Department of State (DOS). The countries are selected by DOS. Each assessment is directed by a USDOE management team supported by the U.S. Geological Survey (USGS), other government agencies as appropriate, national laboratories, and contractors. The U.S. direct effort amounts to about 10–12 professional man-years over a 12–16 month period. The cooperating host country provides a counterpart team of experts which works in close concert with the U.S. team in data gathering in the host country and in the integrated analysis done in the United States. The product is a joint report published after approval by both governments. It is the explicit intent of the assessment to transfer the methodology to the cooperating country so that it may carry on its own national energy planning on a continuing basis.

Importance of Host Country Participation

Active host government participation in all elements of the assessment is an essential feature of the assessment process. This participation should include representatives from energy agencies as well as offices of central planning, finance and budget, industry, mining and minerals, agriculture, transportation, economic development and trade, and housing and civil works, among others. Such participation helps assure that the assessment data, methods, and results will reflect the current energy situation, policies, and plans of that host country. In addition, the host country participants obtain skills derived from interaction with the assessment process, thus allowing those methods to be applied to continued development planning by the country after the initial assessment has been completed.

The Assessment Process and Its Major Elements

A description of the assessment process and its major elements is presented below. A more detailed description of the data collection effort and the analytical methodology is contained in Part II of this paper.

Conducting an IEDP energy assessment involves a well-planned and extensive in-country data collection effort, integration of the collected data into a consistent analytical framework, evaluation of alternative energy options, and the development of conclusions and observations.

The analytical methodology includes seven technical and economic models which can be selected and tailored, i.e., expanded or contracted, to fit each host country's data base and characteristics.

In general, the assessment relies on data already available through various sources in the host country and in the United States. No attempt is made to generate new data, but considerable effort is put into consolidating, verifying, and evaluating the data that already exist.

Data collection can be divided into the three broad categories of energy demand, energy supply, and most importantly, indigenous energy resource assessment and application. This last category includes the very critical elements of the potential for exploiting renewable resources.

Demand Analysis. After the data have been collected, an analysis of current and projected energy demand is conducted. An essential imput to the demand analysis is the development by the host country of a set of alternative development/economic growth scenarios. These scenarios present alternative paths of development that the host country might wish to consider or different rates at which these development and economic growth goals could be achieved.

The scenarios should be accompanied by a description of the underlying assumptions, so that an evaluation may be made as to the potential effect these assumptions might have on the particular character or regional distribution of the projected demand for energy. Examples of these underlying assumptions are population growth and migration trends, balance of trade objectives, unemployment goals, and industrial development goals.

Using these scenarios and the in-country data, alternative levels of demand are developed. These demand projections represent the various demand levels implicit in the different development scenarios.

Along with these projections, energy demand options are identified which could materially reduce or modify the forms of energy that are required. Energy demand options can involve such things as

- changes in the mix of activities within a sector, e.g., a change in the mix of freight movement by trucks vs. railroads

- changes in the industrial processes that are employed, e.g., blast furnace vs. electric furnace steelmaking
- changes in efficiency of end-use devices, e.g., flourescent vs. incandescent lighting

Other factors, such as environmental effects, new technology commercialization limits, social or cultural factors, financing, manpower needs, or other institutional considerations, are examined in terms of the limits they might impose on the use of these energy demand options.

Energy Supply Analysis. Preceeding the energy supply analysis is an energy resource assessment. This activity has two principal parts. The first (conducted by USGS for DOE) relates to resources in the ground: oil and gas, coal, uranium, geothermal, and other energy-related material sources. The second part relates to renewable resources: solar, wind, biomass, hydropower, ocean wave, and tidal activity.

The objective of the resource assessment is to estimate the scope of proven and probable energy resources that exist, taking into account the location and quality of these resources to the extent that such factors could limit their economic utility.

The energy supply analysis then focuses on the current, planned, and potential capabilities for the development and use of energy resources identified in the resource assessment, and attempts to define the energy supply systems that would be needed to provide adequate energy in forms suitable for meeting the projected end-use demand. In addition to the use of nonrenewable and renewable energy resources as commercial energy supplies, the analysis also takes into account current and projected use of noncommercial fuels, (wood, crop residues, and animal waste) as sources of end-use energy. It reviews the various measures that are needed for extraction, processing, transportation, and conversion of all commercially exploitable indigenous energy resources. The energy supply analysis also examines the potential for the application of new technologies, assessing the benefits that would be derived from their use and the extent to which they could be effectively employed. The supply analysis addresses not only the development and use of indigenous energy resources but also the importation and use of foreign energy fuels. When dealing with a country potentially rich in energy resources, the supply analysis identifies and estimates the capability to develop indigenous energy resources as potential export commodities.

Supply/Demand Balance. Using the alternative projections of energy demand prepared in the energy demand analysis and the insights gained from the resource assessment and supply analysis, energy supply/demand (S/D) balances are constructed for each energy demand projection. Each S/D balance uses a framework which displays the full spectrum of energy demands made by the total economy and the various energy supply capabilities that could be developed to meet that projected demand from basic resources through the extraction, processing, transportation, and conversion of these basic energy resources to a form suitable for the identified end-use.

To meet the projected demand for energy, alternative energy supply arrangements are prepared considering such factors as

- the mix of fuels and end-use energy forms needed to meet the projected demand for energy
- attainable production levels of indigenous energy resources
- alternative existing and advanced fuel processing, energy conversion, and transportation technologies
- feasible levels of energy imports and/or exports
- the requirements for financing, manpower, materials, and equipment
- national or regional environmental constraints and potential environmental constraints and potential environmental effects of alternative energy supply scenarios

Overall, the S/D balances attempt to evaluate what is technically and economically practicable for meeting projected demand while, at the same time, considering the above factors.

Evaluation of Alternatives. The next step in the assessment process is to examine alternative energy strategies which the host country might wish to consider beyond the initial S/D balances. For each alternative energy strategy, modified S/D balances are constructed. Such alternative energy strategies could, for example,

- undertake major initiatives to reduce projected demand for energy
- place greater reliance on the use of coal or natural gas, either domestic or imported
- accelerate the use of renewable energy resources irrespective of cost considerations
- alter the planned dependence on nuclear energy

The objectives of these alternative energy strategies for a specific cooperative assessment are jointly defined by the host country and technical specialist participants. These alternative strategies are then evaluated on a comparative basis focusing on

- the extent to which the objectives of each strategy can be attained
- the effect each strategy might have on the use of indigenous and imported energy resources

- comparative requirements of the alternative strategies for finance, manpower, etc.
- the relative risks involved versus the benefits obtained

The next step is to examine the results of these analyses and evaluations and derive the policy, planning, and technical counsel that can be provided to the host country. This counsel is provided as observations rather than as conclusions or recommendations. Three levels of observations are prepared: (1) those for national policy- and decison-makers, (2) those for energy system or sectoral planners, and (3) those which economic sector planners or technical program planners and managers might consider. These observations address such topics as energy demand issues, energy resources and energy supply capabilities, financing considerations, manpower and management capabilities, and energy system planning considerations.

The final step—and the most tangible product of the energy assessment—is the preparation of a joint report by the host country and the U.S. teams and its publication after approval by both governments. The published report is made available to multinational and other organizations that are working, or planning to work, with the host country in its continued economic, energy, or development planning. The assessment report sets forth what was done and how it was done, and presents the results of the analyses and evaluations that were conducted, along with the insights and counsel that can be derived from the analytical results.

Product of the Assessment. The assessment furnishes the host country with an improved information and analytical base for continued energy planning by providing

- a preliminary view of the energy required to meet development and economic goals along with ways to reduce that projected demand for energy
- insights into alternative ways of meeting projected demand for energy, considering known and projected energy resources; the probable limits on those resources; the technologies available to develop and utilize them; as well as other factors which could limit their use
- a preliminary view of the ways the various energy demand and energy supply options might be employed to achieve different energy planning objectives or policies, as well as the financial and other resources that would be required for their realization
- technical, planning, and institutional counsel that the host country should consider if it elects to (a) proceed with the energy options or alternative planning strategies that were examined in the assessment and/or (b) build upon the initial energy assessment to enhance the domestic capability to perform comprehensive energy planning and analysis

The collaborative nature of the assessment also helps identify inconsistencies and gaps in the current planning and data of the participating government. In addition, it demonstrates and transfers to the host country the analytical methods and evaluation processes for achieving more comprehensive and integrated energy planning. Lastly, the improved information, insights, and perspectives presented in the assessment report can be the beginning of an improved framework for the host country's priority assignment of its current and future resources.

Summary

The results of the assessment provide the participating government with a comprehensive framework within which it can make more effective decisions. The assessment establishes a foundation of information and analysis on which more detailed energy sector plans can be developed, while at the same time it identifies possible projects and programs that could be of immediate benefit over the short or medium term. For example, the government of Egypt has established an Interministerial Council on Energy to Coordinate Energy Planning and Projects. The council's near-term objective is to extend the analysis in the recently completed assessment in order to formulate comprehensive energy plans over the next two years. Furthermore, some adjustments to existing plans and policies have already been made. As another example, the government of Peru now plans to establish a semigovernmental institute responsible for energy planning that will build on the analysis and information in the Peru assessment. The enhanced awareness by cooperating governments of the means to examine each in the context of overall energy/economic development will lead them to a balanced approach to meeting their energy needs.

Numerous improvements in methodology, procedure, and counterpart participation have been incorporated into the second round of assessments now underway with Argentina, Portugal, and the Republic of Korea. The timing of these energy assessments is particularly propitious in a number of respects. Although Argentina and Korea are already committed to energy development plans, the assessments stand a good chance of revealing new insights for their energy planning. Korea and Portugal are heavily dependent on imported oil and hence are facing major energy planning decisions which could have significant impacts on their economic development plans. Portugal and Korea have asked that key results of the assessments be made available in the fall of 1980 so that they can be taken into account in major development planning processes.

On the basis of the two completed assessments and the three ongoing ones, we believe that the IEDP Country Energy Assessments have proven their worth as an effective means of helping developing countries address their critical energy problems.

PART 2—THE ANALYTICAL PROCESS

The analytical process can be divided into two parts for the purpose of discussion: data base assembly and integrated analysis.

Data Base Assembly

The assembly of information for an assessment is divided into seven specific areas. These are designed to reflect the various areas of expertise that exist in the energy planning community. Table 26.1 shows the task areas, the type of personnel, and the typical host country organizations that might be involved.

Several general comments must be made with respect to the data base assembly. First, the objective is to gather existing information, not to develop primary data. For example, the resource assessment collects existing information on oil reserves as opposed to conducting a drilling program to identify new reserves. This is to keep the assessment within its objective of conducting a comprehensive evaluation of the energy supply and demand using available data. Major deficiencies in information are identified as requiring more detailed study in the future.

Table 26.1. Data Base Assembly Task Areas

	Task Area	Personnel Involved	Government and Private Organizations Involved
Demand analysis	Development analysis	Economists, planners, demographers, financial specialists	Ministries of planning, finance, economic development; national bank, universities
	Sectoral demand analysis	Enginers, planners	Ministries of industry, agriculture, housing, transportation, rural development; industrial trade organizations; private industries, developers
Supply analysis	Resource evaluation	Geologists, geophysicists, hydrologists	Ministries of natural resources, mining, energy; geological survey
	Fossil energy evaluation	Engineers, resource economists	Ministries of energy; coal, oil, gas companies; refinery operators; construction firms
	Renewable resource evaluation	Engineers, researchers	Ministries of energy, water supply; research organizations; universities
	Electric sector evaluation	Power engineers, system planners	Ministries of energy, electricity; electric utilities; construction firms
	Environmental evaluation	Environmental engineers, ecologists	Ministries of environment health

Second, the data base assembly is more than simply collecting information and preparing a report on it. A significant amount of analysis is required to screen the available data and put it into the proper format for the integrated analysis. For example, the renewable resource evaluation does more than collect solar insolation data, it also evaluates the potential applications of solar systems in the country, considering things such as system performance, economics, and other factors.

Third, the involvement of host country personnel in the data base assembly can vary over a wide range. The most desirable situation is to have active participation in collecting data and in doing the analyses required to prepare the information for the integrated analysis. This is the most effective means of exchanging energy planning experience and information between the U.S. team and the host country team.

Demand Analysis. The analysis of energy demand is carried out at two levels, as indicated by the first two task areas on Table 26.1 The development analysis focuses on the overall growth and development plans of the country while the sectoral demand analysis deals with the details of the energy consumption pattern in the major energy-consuming sectors.

Development Analysis. The principal objective of this part of the data base assembly is to prepare a set of scenarios of economic growth and development for the country. These long-term projections, covering a period of 20–25 years, are designed to describe the alternative development paths that the country might take in the course of its growth. Because of the uncertainties of forecasting the future, the projections are interpreted as a set of alternative future scenarios rather than a set of predictions of what will happen. Thus the set of assumptions used to develop the scenarios must be clearly defined. Table 26.2 shows an example of the kinds of alternative scenarios that are developed as part of this activity.

The development of these scenarios must be followed by a conversion of the economic projections to energy demand projections. This can be carried out in a variety of ways, from the use of very simple energy/economic ratios to the use of complex energy/economic mathematical models. The choice of method depends on the level of information and the amount of experience with the sophisticated analytical tools available in the host country. This will be discussed further under integrated analysis. Independently of how the conversion from economic to energy projections is made, the results of this macroeconomic review must be coordinated with the sectoral demand analysis to develop an internally consistent set of projections.

It must be emphasized that the development of these scenarios is crucial to the assessment. Most of the balance of the analysis will be based on these data. It is extremely important that people from the host country be deeply

Table 26.2. Examples of Alternative Development Scenarios—Country A

Scenario	Assumptions	Projections (Value Added—Billions U.S.$)			
			1977	*1985*	*2000*
I	In the period 1977–85: • continued balance of payments deficit • weak growth of private consumption • restructuring of agriculture • development of intermediate goods industries • improvements in infrastructure • slight population shift from rural to urban In the period 1985–2000: • external indebtedness reduced • new markets opened • significant increase in net income • basic industries established • equipment and consumer goods industries developed • significant development of services sector	Agriculture/Fishing	72	109	243
		Mining/Manufacturing	204	354	994
		Energy/Water	15	26	74
		Construction	36	65	178
		Transport/Communications	36	56	154
		Services	200	306	837
		Total	564	916	2479
			1977	*1985*	*2000*
II	As above, but growth rates by sector are based on the best historical rates.	Agriculture/Fishing	72	108	236
		Mining/Manufacturing	204	415	1619
		Energy/Water	15	31	120
		Construction	36	87	410
		Transport/Communications	36	67	270
		Services	200	368	1470
		Total	564	1076	4126

involved in this activity. Only they can adequately define the types of development alternatives appropriate for their country. The U.S. team can assist in the application of various methodologies, but the identification of the assumptions and goals for development must come from the host country.

In addition to preparing the development scenarios, the development analysis must also provide insight into other aspects of the general economic picture. The assembly of data on historical energy use patterns, economic and social factors affecting energy consumption, demographic characteristics, labor availability, and financial resources is also included in this task area.

Sectoral Demand Analysis. This part of the data base assembly is designed to provide more detailed information on the energy use patterns in the major energy-consuming sectors: industry, agriculture, residential/commercial, transportation, and noncommercial fuel use. In each of these sectors, the energy use pattern is evaluated on the basis of the current structure of the sector, current energy use patterns, projected growth in the sector, and alternatives for changing the energy use pattern. Table 26.3 shows an example of the type of information assembled for one portion of the industrial sector.

The assembly of this part of the data base is the most effort-intensive since it covers a wide variety of activities and involves a large number of diverse organizations. This is where the various government and private organizations, both in developed and developing countries, generally have had the most difficulty in developing effective coordination. It is an area where the assessment process can make a significant contribution.

It should be pointed out that the principal objective of this portion of the data base development is to identify opportunities for changing the energy use pattern in each sector. These changes can have significant impact on the energy requirements of the country and include actions such as reducing total energy demand through improved efficiency, fuel switching, substitution of industrial processes to use less energy-intensive materials, and others. These energy-demand–modifying actions can become part of an overall country plan to meet growth and development objectives in spite of the high price and limited availability of energy. In order to carry this analysis out, the energy demand is expressed in terms of end-use energy rather than fuel type. Table 26.4 shows the end-use categories used. By expressing demand in this format, it is possible to identify opportunities for fuel substitution. For example, heat requirements for cement kilns can be met by a variety of fuels. The integrated analysis will evaluate the alternatives on the basis of overall cost competitiveness, among other factors.

In addition to identifying ways of changing the energy use, this task adds more detail to the very aggregate scenarios of the development analysis. The growth projections from the two activities can be made consistent in a

Table 26.3. Typical Industrial Process Data

Industry	Process Used	Output		Fuel Used	
		Capacity (ton/day)	Production (106 ton)	Fuel Oil (tep/ton)	Electricity (kWh/ton)
Cement		Base Year—1978			
Plant A	Wet	1700	0.31	.146	93
Plant B	Wet	3250	0.41	.155	99
Plant C	Dry	2250	0.54	.083	99
Plant D	Dry	3200	0.77	.088	110
Totals	Wet	4950	0.72	.152	97
	Dry	5450	1.31	.086	108
		Projection Years			
1990	Wet	—	0.80	.145	95
	Dry	—	1.90	.080	102
200	Wet	—	0.80	.140	95
	Dry	—	2.30	.075	95

Table 26.4. Typical End-Use Energy Demand Categories

Sector	End-Use Demand Category	Additional Disaggregations
Industry	Direct heat	By industrial classification
	High temperature	
	Medium temperature	
	Low temperature	
	Indirect heat	
	High temperature	
	Medium temperature	
	Low temperature	
	Electro-Mechanical	
	Feedstocks	
Residential/Commercial	Heat	By urban/rural
	Electro-Mechanical	
Transportation	Auto	By technology, by passenger/freight
	Aircraft	
	Rail	
	Truck	
	Bus	
	Water vessel	
Agriculture	Machinery (tractors, combines, etc.)	
	Irrigation	

number of ways, such as using the sectoral information to disaggregate the macroeconomic projections or using the sectoral projections in place of the macroeconomic information.

Supply Analysis. The supply analysis is carried out in five parts, as indicated on table 26.1. This division is designed to reflect the different aspects of the supply system.

Resource Evaluation. This activity is designed to address the availability of energy resources in the country. The location, quantity, and quality of reserves are investigated in cooperation with geological personnel from the host country. The review includes fossil fuels (coal, oil, gas, oil shale, tar sands, etc.), nuclear materials (uranium, thorium), energy-related minerals (copper, nickel, iron, etc.), and water supply.

For the fossil fuels and nuclear materials, the intent is to estimate reserves and identify possible constraints on the exploitation of those reserves. These evaluations will be used to determine what the practical limits are on the use of indigenous resources. For the evaluations of energy-related minerals and water supply, the focus is on whether these materials may present a constraint on development in the energy sector.

Fossil Fuel Evaluation. This activity is designed to focus on all the stages of the fuel cycle, from resource extraction through to delivery for end-use. Table 26.5 shows the types of activities included.

The primary objective of this evaluation is to review the means by which fossil fuels are currently used and to identify alternative ways in which they might be used in the future. This activity will provide the necessary information to determine the most cost-effective way to use fossil fuels to meet end-use energy demands.

Data are collected on the existing fossil fuel supply system (e.g., oil and gas wells, coal mines, refineries, fuel processing plants, pipelines, etc.) in terms of technical performance parameters, capital and operating costs, prices charged, and a number of other items. Plans for the expansion of the fossil fuel systems are reviewed, and the set of alternative fossil fuel supply system configurations that might be considered are identified.

The purpose of this evaluation is not to project fossil fuel demand into the future. This must be done in the integrated analysis when the supply system is matched to the demand requirements. Rather, this review is designed to set the bounds and constraints under which the fossil fuel supply system must operate. For example, it will identify the maximum possible coal production from a given mine or the maximum petroleum product output from the country's refineries. The details of how much of these maximum amounts will actually be required will be determined in the supply/demand balance.

Renewable Resource Evaluation. The evaluation of renewable resources is analogous to the fossil energy analysis in that it focuses on the entire cycle from resource extraction through to delivery for end-use, considers technology operating parameters and economics, and addresses the current level of resource use and projections for future utilization.

The types of technologies considered are shown in table 26.6 The evaluation includes estimating the resource base, potential technology applications, system capital and operating costs, and factors that may hinder or enhance the implementation of the technology.

As with the fossil fuels, the extent to which renewable resources will be used in the future is determined in the integrated analysis, when energy demand is matched with alternative supply systems. The penetration of these renewable technologies into the market is based on a comparison of their costs to those of conventional systems and on policies that may be imposed to accelerate this penetration. This part of the anlaysis is designed to establish the limits to which the technologies can be employed.

Electric Sector Evaluation. The electric sector is treated as a separate entity because many countries have a separate group dealing with electrical system planning. The assembly of information in this area encompasses a review of

Table 26.5. Typical Stages of the Fossil Fuel Cycle

Resource	Extraction	Processing	Transport	Conversion
Oil	Onshore	Refining	Tanker	Combustion
	Offshore	Well-head processing	Pipeline	Feedstock
	Secondary recovery		Rail	Lubricants
	Tertiary recovery		Barge	
Gas	Onshore	Well-head processing	Pipeline	Combustion
	Offshore	Liquefaction	Rail	Feedstock
	Associated		Ship	
	Nonassociated			
Coal	Underground	Cleaning	Rail	Combustion
	Surface	Solvent refining	Barge	Coke
	In-situ combustion	Gasification	Slurry pipeline	
		Liquefaction	Truck	
Oil shale	Mining	Retorting	Tanker	Combustion
	In-situ combustion	Shale refining	Pipeline	Feedstock
			Rail	Lubricants
			Barge	
Tar sands	Mining	Retorting	Tanker	Combustion
	In-situ combustion	Refining	Pipeline	Feedstock
			Rail	Lubricants
			Barge	

Table 26.6. Typical Renewable Resource Evaluation Parameters

Technology	Resource Base Definition	Typical Potential Applications
Direct solar	Solar insolation	Water heating
Thermal		Space conditioning
Photovoltaic		Electricity generation
Wind	Wind velocity profile	Electricity generation Motive power
Biomass	Crop residues Animal wastes Urban wastes	Gas generation and combustion Direct combustion
Ocean systems	Temperature gradients	Electricity generation
Thermal		
Tidal	Tidal head	
Wave	Wave duration	
Geothermal	Temperature profile	Space heating Process heat Electricity generation
Small-scale hydroelectric	Flow rates Hydraulic head	Electricity generation Motive power

the current electrical system (including generation, transmission, and distribution) and an evaluation of future electric system expansion. Since a good deal of long-range planning is done in the electrical sector, there is usually a good base of information to work with.

The matching of supply and demand in the integrated analysis may result in different electrical requirements under various assumptions. The country's electrical system plan is used as the starting point, and modifications to that plan that would be needed to meet the alternative future scenarios are developed.

The electric sector evaluation includes all types of generation equipment (coal, oil, gas, nuclear, hydroelectric) and reviews both grid-connected systems and dispersed systems.

Environmental Evaluation. The environmental evaluation is designed to determine where developments in the energy sector are likely to be hampered by environmental considerations. The goal of the data collection is to establish what the current environmental conditions are and where serious environmental impacts could result from energy system development. It should be emphasized that no attempt is made to conduct a complete environmental assessment. Rather, the effort is made to determine, in a qualitative way, where possible environmental problems may exist.

Integrated Analysis

The discussions of the previous section have focused on the assembly of a data base. This includes the collection of data and the analysis of that information to put it into a format for an integrated analysis. The process of carrying out this integration is one of the significant analytical efforts of the assessment. It puts all of the data collected into a consistent framework and provides a means for evaluating alternative energy plans on a consistent basis. There are two major steps in the integrated analysis: construction of supply/demand balances and evaluation of alternative strategies. Before describing how these two steps are taken in the integration process, a brief review of the analytical tools available is presented.

Computational Methods. A number of computational procedures are available for use in the integration process. There are three basic classifications of computational methods: supply/demand models, which focus on constructing a balance of the energy supply system against anticipated demand; impact assessment models, which determine the resultant effects of a particular supply/demand balance; and energy/economic models, which deal with the interaction of energy and overall economic growth. Table 26.7 lists the major analytical tools used in the cooperative country energy assessments.

Supply/Demand Models. A supply/demand model is a method of matching the demand for energy against available supplies. The term "model" is used in its broadest sense to describe a formal mechanism for carrying out a calculation. It need not refer to a computerized program. The specific method to be used in a country energy assessment is chosen on the basis of available data and the level of detail used in the assessment from a number of methods selected for use in the program (see table 26.7). The manual network energy flow system, the Decision Focus, Inc. (DFI) model, and the Brookhaven Reference Energy System (RES) deal with the entire energy system. The Wein Automatic System Planning (WASP) model and the Reliability Computation (RELCOMP) model focus on the electric sector.

Independent of which of these tools is used, the common procedure is to construct a supply/demand balance that traces the energy flows from raw resources (e.g., coal, oil, gas, solar insolation, etc.) through to end-use demand. This information is displayed in the network format shown in figure 26.1. The manner in which the various energy flows are determined for the future years is what distinguishes one methodology from another. In the manual network energy flow system and in RES, the distribution of energy among the various fuel types is made by considered judgment on the part of the analyst. In the DFI model, the computer package uses an allocation algorithm to distribute the fuel used to meet each end-use demand on the

Table 26.7. Integrated Analysis Computational Methods

Type of Method	Methods Available	Typical Outputs
Supply/Demand	Manual network energy system	Balance of supply and demand, based on exogenously defined demand, efficiencies, market allocations.
	Reference energy system (RES)	As above. Can be run with linear optimizing package to generate market allocations on the basis of least cost.
	Decision Focus, Inc. model	Balance of supply and demand, based on competitive economics of all stages of the supply system.
	Wein Automatic System Planning Model (WASP)	Electric system expansion plan, based on least cost generation system.
	Reliability Computation Model (RELCOMP)	Electric system reliability for exogenously defined system configuration.
Impact assessment models	Energy Supply Planning Model (ESPM)	List of energy facilities required for a given supply/demand configuration and cost, manpower, resource requirements.
	World Bank Minimum Standards Model (WBMS)	Balance-of-trade, balance-of-payments, investment requirements for exogenously determined growth rate.
Energy/Economy models	Brookhaven Energy Economy Assessment Model (BEEAM)	Input/output formulation of final demand in the energy and nonenergy sectors of the economy.

basis of the comparative costs of each fuel type. The WASP model uses a linear optimization procedure to select the cheapest electrical system configuration. RELCOMP uses a simulation routine to determine the reliability of a given electrical system structure.

Impact Assessment Models. Impact assessment models focus on the resultant impacts of a given supply/demand balance. They are applied once the supply/demand balance is constructed, and they give the decision-maker information on the implications of following a particular path to meet given energy requirements. The impacts that are determined for each supply/demand balance configuration considered include the following:

- number and type of new energy facilities required
- capital cost of new facilities
- operating costs
- manpower requirements in the energy sector
- materials requirements in the energy sector
- efficiency of energy use

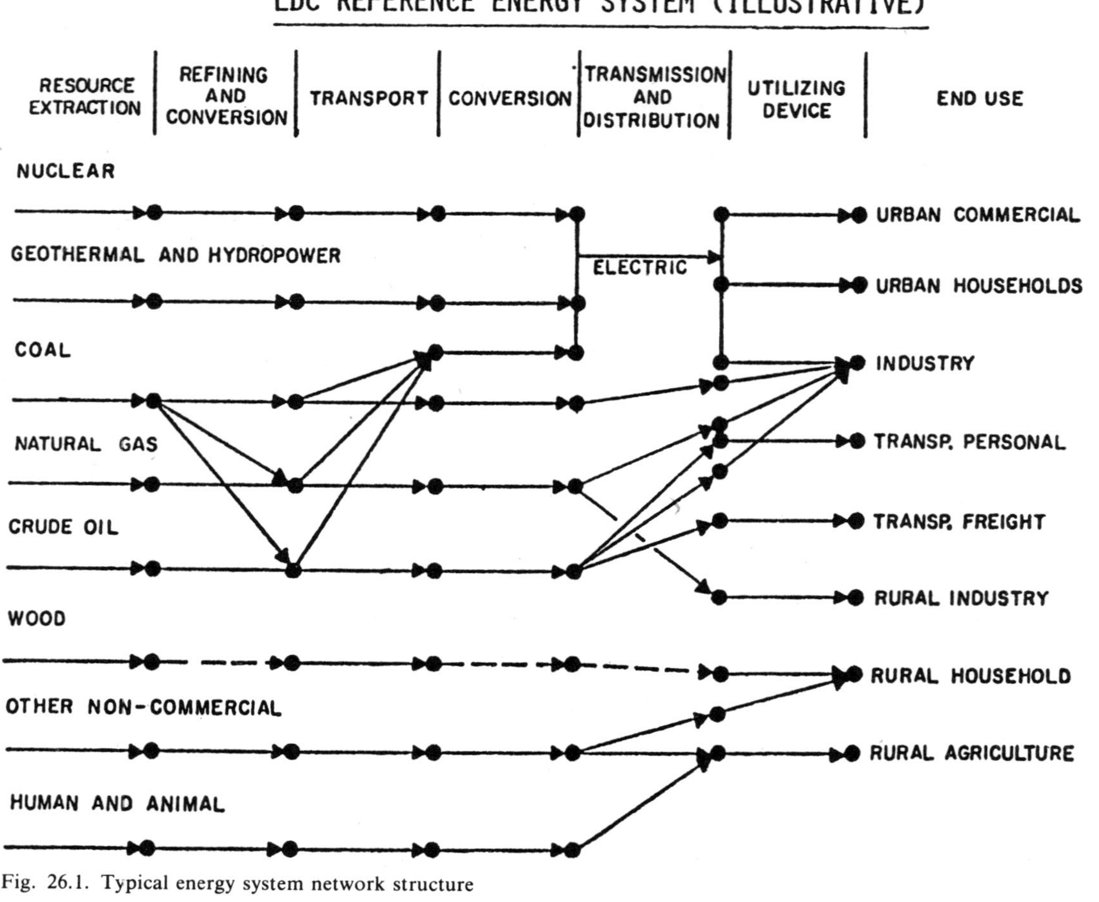

Fig. 26.1. Typical energy system network structure

- imports and exports in the energy sector
- investment requirements
- environmental effects

The term model is again used in a generic sense to describe a computational procedure and not necessarily a specific computer program. The impact assessment models that have been selected for use in the country energy assessment are shown on table 26.7. The use of these tools depends on the data availability and quality.

Energy/Economic Models. These types of computational procedures are designed to relate energy consumption to overall economic activity. They generally deal at a much more aggregated level of detail in the energy sector than do the supply/demand models or the impact assessment models. They simultaneously use more disaggregated data on the rest of the economy.

As shown in table 26.7, BEEAM is the only one currently available in the country energy assessments program. It has an economic input/output formulation to derive a consistent set of economic and energy projections. Its use is aimed at ensuring internal consistency among the macroeconomic development projections, the sectoral demand projections, and the supply/demand balances constructed.

Construction of the Reference Case. The process of carrying out the integrated analysis begins with the construction of the reference case. This is the assembling of the information from the data base into the framework prescribed by the computational procedures chosen. In general, the supply/demand balance is set up first, and the impact assessment is carried out once the balance has been constructed. The energy/economy analysis is conducted parallel to this process, to verify the consistency of the information used in the demand projections. In some cases it may be possible to exercise the energy/economy models prior to carrying out the supply/demand balance in the interest of beginning with an internally consistent set of energy demand projections. This is strictly a function of the level of sophistication of the country's economic planning.

The reference case is designed to be the starting point for the analysis. It is not intended to be the projection of the most likely future conditions.

Construction of the Supply/Demand Balance. The supply/demand balance for the reference case begins with the tabulation of the base year historical information in a network formulation; that is, the set of end-use demand categories is selected and the set of energy supply system components that is used to meet those demands is identified.

For the projection years, the construction of the supply/demand balance requires the use of one of the analytical tools described previously. Although the details of the method vary depending on the particular model used, the process is basically the same. The approach is demand-driven in that a particular demand scenario is selected first and the supply system that is used to meet those demands is then built.

The demand scenarios come from the development analysis and the sectoral demand analysis. The future years used in the analysis depend on the planning horizon chosen. The end year of the planning period and at least one intermediate year are used as calculation points.

Included in the reference case are the set of alternative development scenarios chosen by the host country for analysis. The scenarios reflect the planning activities that have gone on in the development analysis task.

The energy supply system assumptions used in the reference case analyses reflect a business-as-usual approach to meeting the supply requirements; that is, either the current energy supply system is assumed to expand in its current structure to meet the projected demand or, if available, the projected changes to the energy supply system structure anticipated by the host country are assumed to take place.

Figures 26.2 and 26.3 show supply/demand balances for Egypt for 1985 and 2000. These were both constructed using the manual network methods. The allocation of market shares to different energy supply systems was made on the basis of considered judgment.

Evaluation of the Supply/Demand Balance. The construction of the balance for the reference cases implies a set of impacts of the energy system on the economy. This is in the form of the need to invest capital in the construction of energy facilities, the need to expend operating funds in order to import energy resources and/or technologies, the need for manpower to build and operate the energy system, the resultant price of energy to the other sectors of the economy, and other effects. The reference case must be evaluated in the light of these impacts for the purpose of providing decision-makers with the necessary information for informed choices. The evaluation is carried out by exercising one or more of the impact assessment methodologies. The specific impacts that are computed are determined by the availability of information and by the issues that are significant to the country. Figure 26.4 and table 26.8 show some of the impact parameters computed for the Egypt assessment.

Construction of Alternative Strategies. The alternative strategies to be considered in the analysis are perturbations of the reference case conditions. The fundamental assumption with respect to these strategies is that the basic economic development scenarios chosen for the analysis remain the same;

that is, it is assumed that the economic growth rate alternatives and the structural versions of the economy are unchanged. The strategies, therefore, are designed to investigate alternative means of structuring the energy supply system to meet the projected demand or means of changing the end-use energy demand level, for example, through the imposition of more efficient technologies. It is important to recognize this distinction between a development scenario and a strategy in this assessment process. Any and all alternatives that result in different structural features of the overall economy or different levels of output from various sectors of the economy are classified as development scenarios. In contrast, a strategy is defined as a set of alternatives in the energy supply system or a set of changes to the energy demand levels that is imposed while maintaining a constant level of economic acitivity.

Demand-Modifying Strategies. The strategies that result in changes in demand levels while preserving activity levels result from changes in technologies used, changes in system configuration, or policy constraints. In the analysis of sectoral energy demand, a set of options was identified in each sector that would change the energy consumption pattern without changing overall activity level. A demand-modifying strategy is a collection of a set of these options.

To illustrate the construction of this type of strategy, the development of an energy efficiency strategy can be postulated. The set of options included in such a strategy might be the use of more fuel-efficient automobiles, changes in industrial processes to emphasize more energy-conservative technologies, and increased use of central district heating for residential space heat. In these cases the activity levels (i.e., the number of passenger trips, the level of industrial output, and the number of residential dwelling units, respectively) is assumed to remain constant while changes are made in the energy consumption pattern.

From the analysis standpoint, these types of strategies affect the entire chain of the supply/demand balance since they affect the demand levels, which are the starting point of the computation. (See figures 26.5 and 26.6).

Supply System Strategies. Supply system strategies are also composed of a set of technological, system, and policy options, but they focus on the problem of meeting given end-use energy demand. Examples of supply strategies include an emphasis on coal as a replacement for oil, on electrification, on renewable resources, or on fuel switching. In addition to dealing with choices of entire fuel cycles (e.g., coal versus oil), the strategies require choices within each fuel cycle (e.g., use of low-Btu coal gasification versus direct coal combustion).

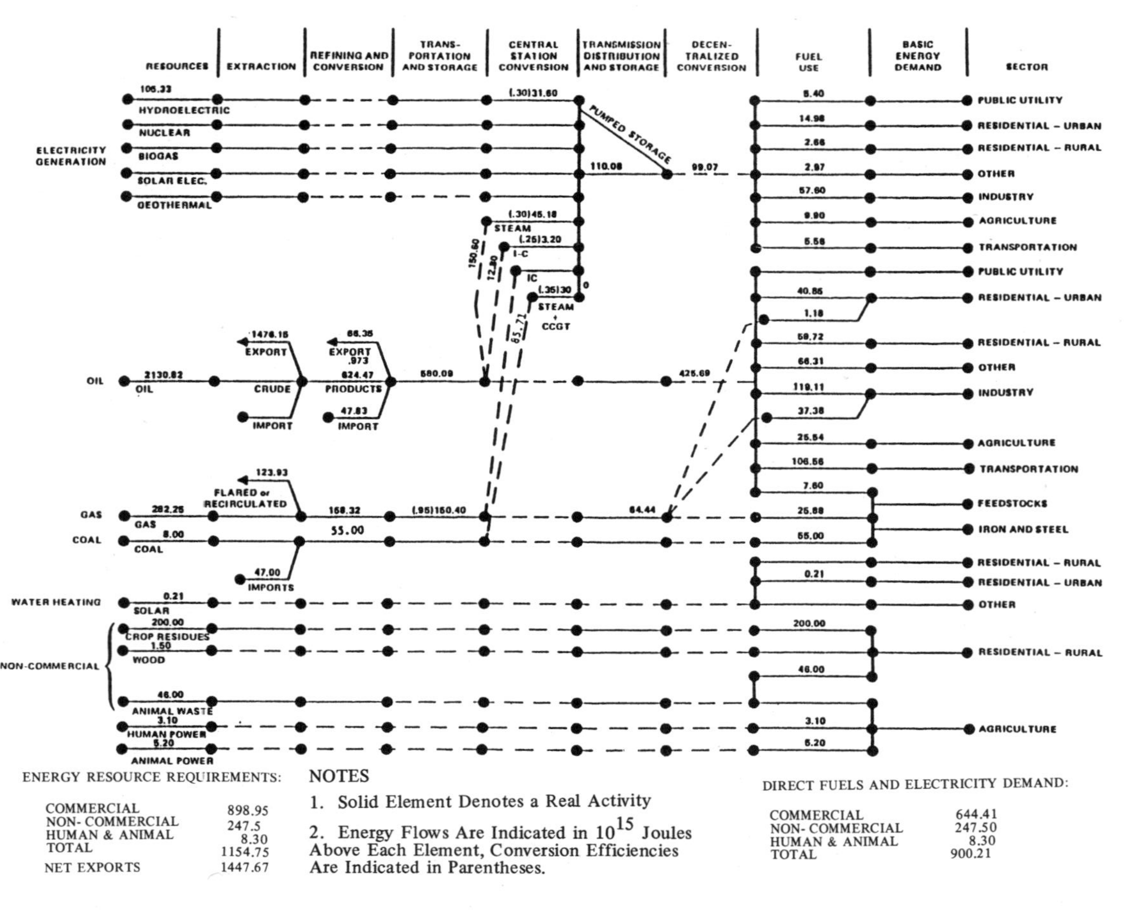

ENERGY RESOURCE REQUIREMENTS:

COMMERCIAL	898.95
NON-COMMERCIAL	247.5
HUMAN & ANIMAL	8.30
TOTAL	1154.75
NET EXPORTS	1447.67

NOTES

1. Solid Element Denotes a Real Activity
2. Energy Flows Are Indicated in 10^{15} Joules Above Each Element, Conversion Efficiencies Are Indicated in Parentheses.

DIRECT FUELS AND ELECTRICITY DEMAND:

COMMERCIAL	644.41
NON-COMMERCIAL	247.50
HUMAN & ANIMAL	8.30
TOTAL	900.21

Fig. 26.2. Reference Energy System - Egypt: 1985 Reference Case

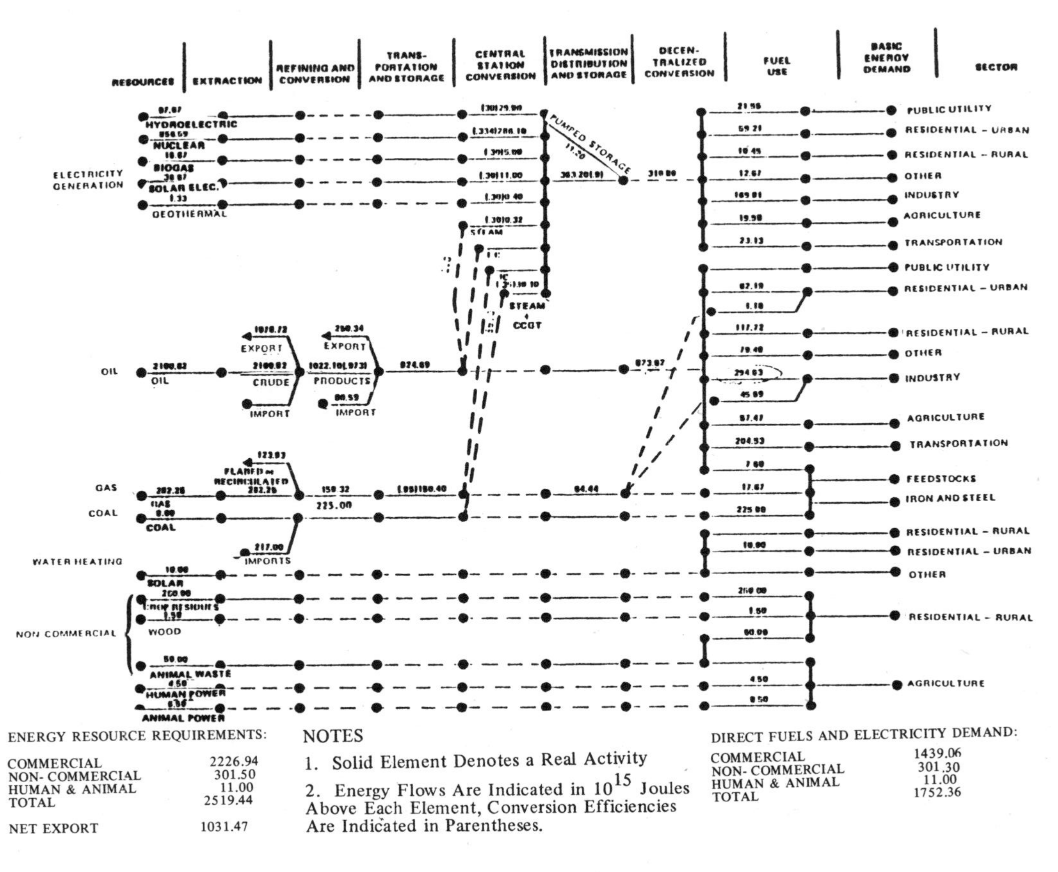

ENERGY RESOURCE REQUIREMENTS:

COMMERCIAL	2226.94
NON- COMMERCIAL	301.50
HUMAN & ANIMAL	11.00
TOTAL	2519.44
NET EXPORT	1031.47

NOTES

1. Solid Element Denotes a Real Activity
2. Energy Flows Are Indicated in 10^{15} Joules Above Each Element, Conversion Efficiencies Are Indicated in Parentheses.

DIRECT FUELS AND ELECTRICITY DEMAND:

COMMERCIAL	1439.06
NON- COMMERCIAL	301.30
HUMAN & ANIMAL	11.00
TOTAL	1752.36

Fig. 26.3. Manual Network Balance - Egypt: 2000 Reference Case

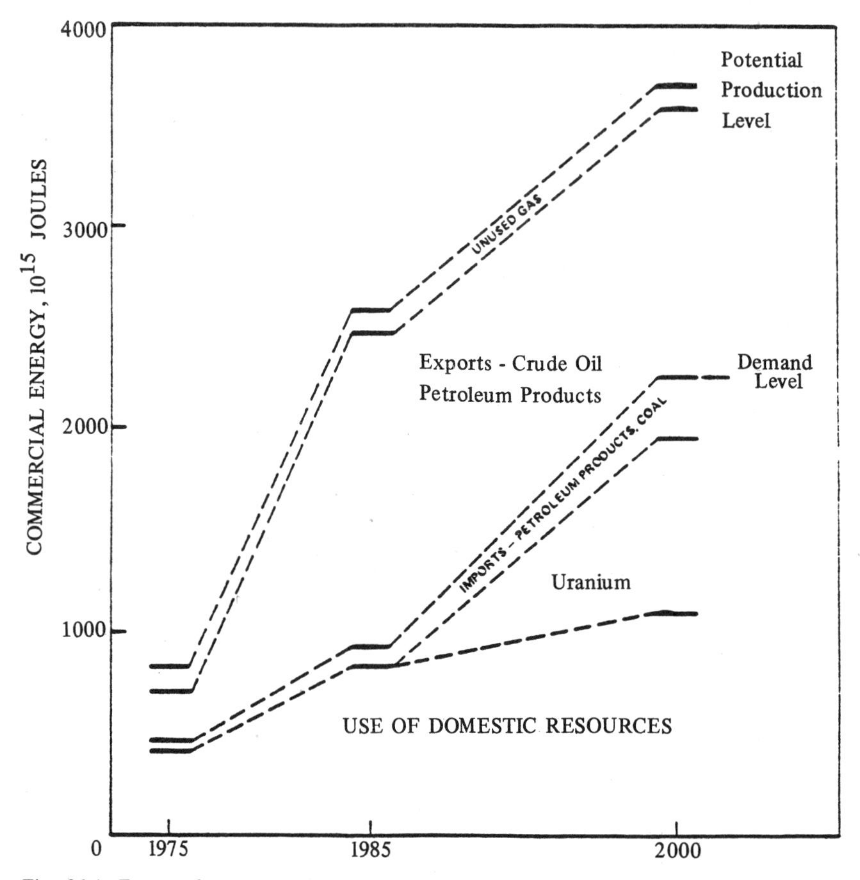

Fig. 26.4. Egypt reference case impacts on energy imports and exports

The construction of supply system strategies is carried out to the level of detail permitted by the data available. This must be done with a measure of concern for the level of detail required by the supply/demand modeling approach. It is important to select a set of alternatives without going through laborious detail that does not add to the insights achieved. It is also important to recognize that the supply strategies should represent true alternative courses of action. It is easy to construct a set of supply strategies that are not mutually exclusive. If this is desired, it might be recognized that the set of conditions considered will not represent alternative paths of action. It is more desirable to construct a set of strategies that are bona fide alternatives that combine the best options for each set of conditions assumed.

Construction of the Supply/Demand Balance. The supply/demand balance is constructed for each of the strategies in the same fashion as for the reference

Table 26.8. Egypt Reference Case Capital Requirements

Facility	1978–2000 Capital Requirements (Egyptian £ 10^6)
Oil	
Exploration, development, and production	1922
Refining and processing	1469
Pipelines	80
Tankers, barges, trucks	385
Bulk stations	25
Other	0
Total Oil	3881
Oil shale	0
Gas	
Exploration, development, and production	23
Geopressured methane	0
Pipelines	0
LNG tankers	46
Distribution	0
Other	0
Total Gas	69
Coal	
Underground mining	0
Surface mining	0
Rail	73
Pipelines	0
Barges, trucks	10
Total Coal	83
Nuclear fuel cycle	108
Solar facilities	1339
Electric generation (excluding solar)	
Coal-fired	0
Sulfur removal facilities	0
Coal-fired (fluid bed)	0
Nuclear	8943
Hydra and pumped storage	968
Oil and gas	452
Geothermal	0
Total generation (excluding solar)	10363
Transmission	670
Distribution	1051
Total utilities (excluding solar)	12084
Total Energy Industry	17564

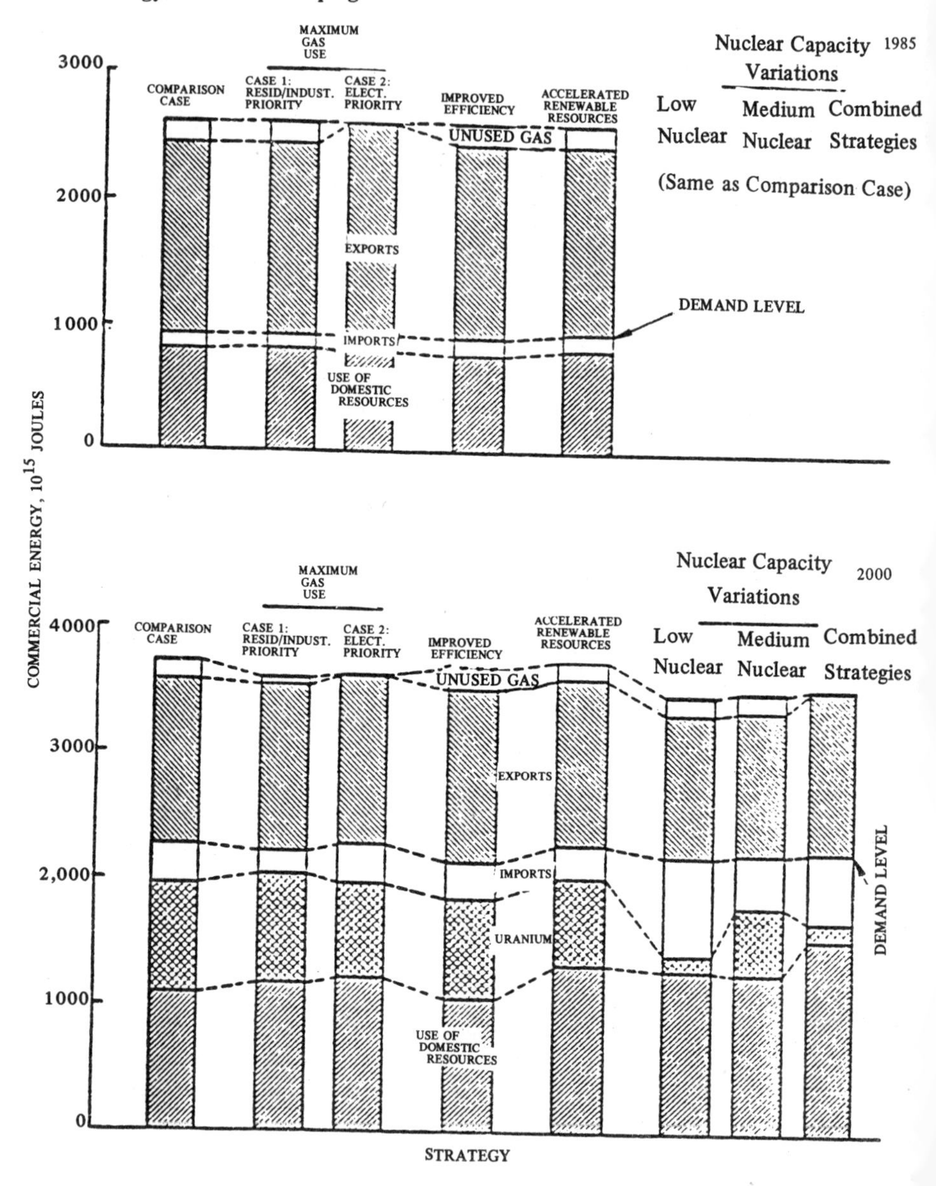

Fig. 26.5. Impact of strategies on imports and exports for Egypt

case. The process is the same, with the exception that the starting demand levels are changed to reflect the effect of the demand-modifying strategy, and the supply system configuration is changed to reflect the supply strategies. The networks are reconstructed to reflect these new conditions.

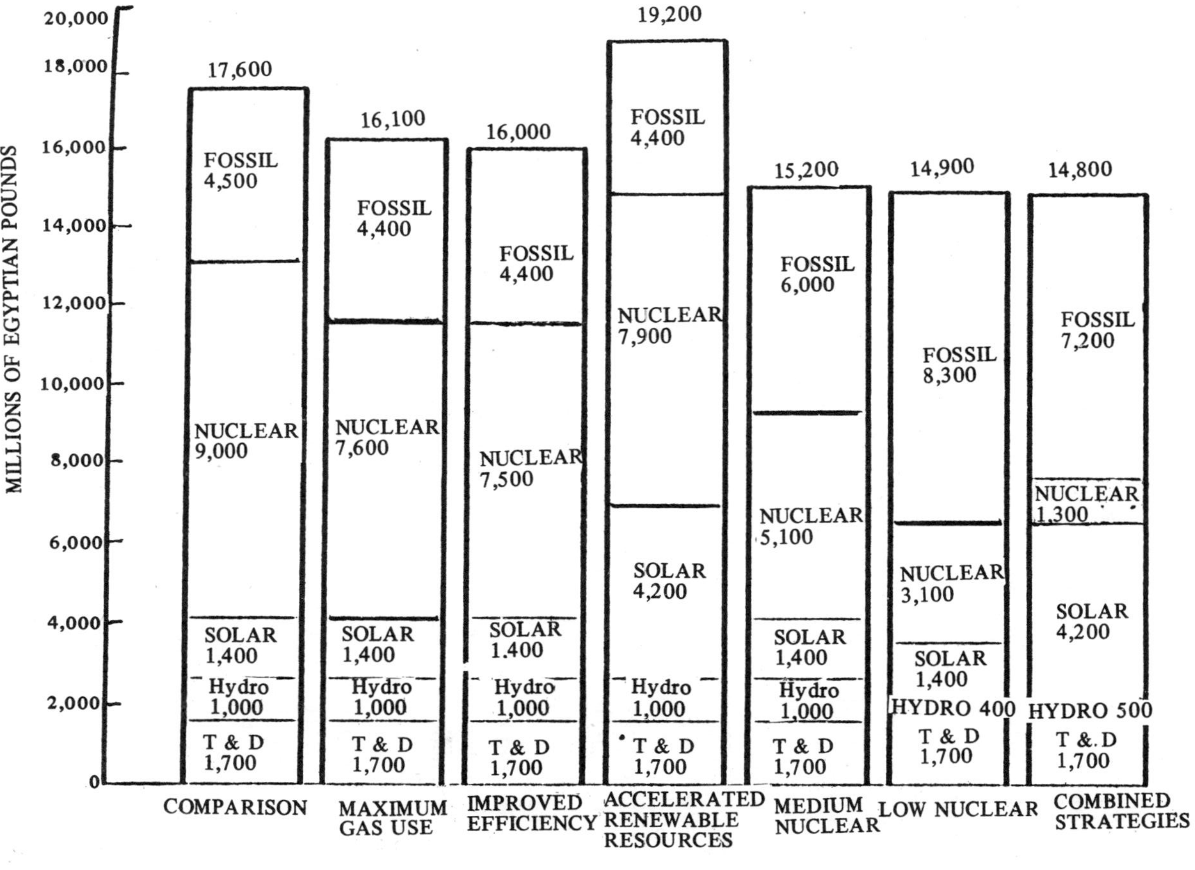

Fig. 26.6 Assignment of cumulative capital costs by supply sector

SUMMARY AND CONCLUSIONS

As a result of the experience with the energy assessments in Egypt and Peru and the initial phases of the data base assembly in Argentina, Portugal, and the Republic of Korea, a number of observations can be made.

Relevance of the Method

The analytical methodology described here provides a systematic structure for conducting energy analyses. As such, it provides a means for organizing the collection of energy information into manageable segments and provides a means of synthesizing those segments into an integrated analysis using the supply/demand balance approach. The methodology as applied in Egypt and Peru was certainly far from perfect. Some portions of the analysis were treated in only a superficial way due to time, resource, and data constraints while others received perhaps too much emphasis. Nevertheless, the analytical framework provides a means for carrying out this process in a systematic fashion that can be improved upon continually. Improvements are already being implemented in the Argentina, Portugal, and Republic of Korea assessments. The application of the methodology did provide both the Egyptian and Peruvian governments with information necessary for decision-making. It displayed the implications of several alternative approaches to energy development and has resulted in some further evaluations of the next steps to be taken in energy planning. It has also identified a number of short-term actions that could be taken and a set of long-range research and development activities that could be carried out.

An indirect benefit of the application of the methodology has been the bringing together of all of the governmental agencies dealing with energy problems to focus on an integrated analysis. The cooperative assessment provided a mechanism for agencies with varying mandates (e.g., energy, electricity, industry, transport, etc.) to pool their capabilities and focus on a unified approach to energy planning. As such interaction is crucial to meaningful energy planning, the assessment has provided a useful start to the systematic exchange of perspectives.

It is important to note that the methodology can be employed at a level of detail appropriate for the country under study. There is no need to employ all of the sophisticated analytical tests if the situation does not warrant it. Nevertheless, the structure of the data base formulation and the integrated analysis still provides the format under which an energy analysis can be conducted. The various stages of each of these activities can easily be modified to reflect an increased or reduced level of detail based on the data available. In essence, the structure of the analysis is generally applicable, and the specific analytical tools can be tailored to the country's situation.

Data Problems

As with all quantitative analyses, the quantity and quality of data are the most constraining factors in the conduct of the assessment. Availability of information, particularly time-series historical data, is a substantial problem in the developed nations of the world and is even more critical in the developing world. As these assessments demonstrated, the information is sometimes available in great quantities but has not been synthesized into an easily usable form. In other cases, the data are simply not available or are of questionable validity due to nonstandardized collection procedures.

The assessments did show, however, that with a concerted effort and the considered judgment of people who have been involved in various areas of the energy supply/demand problem, enough information can be assembled to provide a first-order estimate of energy conditions. Results so obtained must, of course, be interpreted cautiously and should be used to identify areas where data collection needs to be improved. Of particular value would be an inventory of current energy consumers disaggregated by end-use (e.g., industrial process heat, residential space heat, private auto transportation, etc.); historical series of energy use by fuel type; cost and price information in the energy sector; and historical series of economic parameters that can be used in relating energy consumption to the structure of the economy.

Analytical Methodologies

The analytical methodologies applied to the Egypt and Peru assessments were necessarily limited by resource constraints but did, nevertheless, demonstrate the applicability of systematic procedures that can be used in other developing countries. Particularly, the manual energy network analysis and the Bechtel Energy Supply Planning Model were successfully used. In the Argentinian, Portugese, and Korean assessments, the other analytical tools will be applied for the first time.

In the construction of the supply/demand balances, the manner in which decisions on market shares of alternative energy supply systems are determined will be improved through the application of the DFI model. More extensive economic analysis is required to simulate these decisions on the same basis as energy system planners do. That is, more extensive evaluation of capital costs, fuel prices, financing requirements, and other parameters are required. This requires the collection and interpretation of more data, and so the application of these procedures will depend heavily on the availability of information in each country.

In the evaluation of the interaction between energy and the economy, a number of new approaches will be tried. These include the BEEAM model and macroeconomic models developed specifically for each country. These tools will be used to improve the quality of the demand projections and to enable these projections to be more closely related to economic development patterns.

Strategy Formulation

It is evident that in carrying out these types of analyses, the consideration of a set of development scenarios and a set of energy supply strategies for each results in a large amount of information to be handled. While it is highly desirable to consider a wide variety of alternatives, caution must be exercised to prevent the quantity of information from being so large as to obscure the key items for decision-making. Emphasis must be placed on those alternatives that, in fact, represent significant differences in energy patterns and/or those alternatives that have the highest priority for the development planners.

It must also be noted that the separation of development scenarios from energy supply strategies is useful only insofar as it facilitates the collection of information by specialists. The interaction between energy plans and development plans is becoming increasingly complex as the world leaves the era of relatively cheap and abundant fossil fuels. This interaction must not be overlooked, and the assessment process outlined here provides the mechanisms, through the integrated supply/demand balance, of dealing with these issues.

Chapter 27
Energy Analysis in Countries of Latin America

Chester L. Cooper

There is a growing recognition that an assessment of the nature and dimensions of a developing country's energy problems and review of its energy prospects can make an important contribution to the choice and planning of development strategies. And so, during the past few years, teams of energy analysts from United States government agencies, universities, and consultant groups have spent a great deal of effort and a substantial amount of money conducting surveys, collecting data, and writing reports throughout Asia, Africa, and Latin America. Although these enterprises have been undeniably useful, much more of lasting value could have been gained if local analysts and institutions had been more directly involved, so that after the outside experts have returned home to Palo Alto, Cambridge, Stony Brook, and Washington, the country's own specialists could continue the process of analysis and planning on their own.

We at the Institute for Energy Analysis (IEA) believe that any approach along the following lines can be applied by local analysts in any one of a number of less industrialized societies. Obviously, it should be modified to reflect country-specific economic, social, and institutional considerations. The Institute hopes to focus on Latin America for an early trial because of the successful workshop based on this methodology which we conducted last year for energy analysts from a dozen or more Latin American nations.

Two propositions guided us in designing this approach: First, government officials should regard energy analysis not as an end in itself, but rather as a contribution to economic development planning; the approach should be a mix of economic, political, social, and engineering considerations. Second, developing-country analysts and planners should, after a modest amount of orientation and training, be able to understand, undertake, and expeditiously complete an energy analysis for their own country; the methodology should be straightforward (no black boxes) and lend itself readily to explanation to and understanding by national policy-makers.

For both substantive and practical reasons, an approach along these lines should not be launched as a full-blown, region-wide enterprise; it would be more sensible to move forward in stages such as the following:

1. the construction of a general methodology that could be adapted to specific countries (spelled out below)
2. the adaptation of such a methodology to two or three specific countries
3. the training of analysts in the methodology from local institutions in the selected countries
4. the actual undertaking of energy analyses by the local analysts and institutions with whatever assistance they may seek from outside experts
5. a critique of the experience by energy experts from the United States, Latin America, and international institutions.
6. the development of an approach for other countries in Latin America and elsewhere based on the experience in the experimental cases and on the critique.

The approach used by IEA is primarily noneconometric. The methodology is flexible and can be adapted to the special circumstances and characteristics of the particular country under study. The essential components of the model could be computerized to assess the impact of different weights for individual variables toward improving the procedures and the analysis the next time around.

In its most aggregate form, the IEA energy estimate is based on projections of gross national product (GNP) times an efficiency factor (E/GNP ratio). The analysis begins with the study of the expected path of population and labor force growth. It then introduces projections of hours worked and unemployment to estimate the contribution of labor to GNP.

The next step is to construct an output series, measured in constant value national currency units. The two keys to estimating future output increases are: (1) investment trends and (2) total factor productivity-per-worker trends. This gives us GNP.

The next step is to calculate what we identify as the intermediate factors. These factors are intermediate between population and GNP, on the one hand, and calculated energy requirements, on the other. There are four principal categories, all familiar to you: household; commercial (including tourism, and public services); transport, and industry. The forecasted level of activity in each sector should be consistent with overall GNP and population trends, and be based on an analysis of social, political, economic, and technological trends which dominate the sector.

For example, household sector energy demands are driven by the number and size of households and income levels. Similarly, the growth of the commercial sector depends on the number of households it serves, the changing age distribution of the population, the stage of economic development which affects the government's provision of services, and the level of income in the economy. Tourism and public service are separately projected where appropriate.

Transportation of people is projected from income and demographic statistics as well as social trends, such as the participation of women in the commercial labor force. Assessing the stage of economic development is also necessary. For instance, Bolivia is a relatively poor nation with fewer than 5½ cars per 1,000 population. Yet Bolivia is participating in the world automobile population explosion; its fleet has been growing at more than 8 percent per year since 1965. Transportation of goods is primarily a reflection of industrial activity.

The industrial sector is disaggregated into three primary categories: agriculture, mining, and manufacturing. Where agriculture is a major economic activity, as it is in most developing countries, more elaborate analyses of agricultural energy demands are desirable, particularly the identification of fuel use patterns and trends—commercial or noncommercial energy, for example.

Having constructed the size and shape of the economy currently in place and expected in the future, the next step is to calculate, in some detail, the energy needs of each sector. The base for each series is the latest year for which actual data are available. The base series is then modified in two ways to project through the period of study—first, for the expected sector growth and second, for the energy efficiency estimated to be achieved in each sector. The expected improvement in household and commercial energy use, for example, is a consequence of official conservation programs and of economic influences, particularly the sharp escalation of energy prices. In the transport sector we would adjust automotive, truck, and other transport uses for improvements required by law or stemming from voluntary programs.

The largest efficiency improvements are likely to come in the industrial sector. This is especially true in those developing economies where the growth process is already taking hold. The faster an economy grows, the younger is the average age of its capital equipment. Industry has had a history of increased energy efficiency even when energy prices were falling.

As to energy supply, past trends in the consumption of primary fuels, including the conversion of hydro and nuclear fuels into electricity, are more easily constructed than a sectoral breakdown of energy use. However, the availability of needed supplies in the form of imported oil can no longer be assumed, since escalating oil prices have brought many developing countries to a balance-of-payments crisis. Kenya, for example, is now using 25 percent of its export earnings to pay for oil, and its 1980 trade deficit is estimated at between $200 and $300 million. Credit will be a lot harder to come by because debt service is absorbing a large share of GNP in many of these countries.

Future energy supplies, therefore, will be much dependent on the exploitation of indigenous resources than in the past. Governmental or private sector planning should be examined to calculate the likely quantity of required imports and the reasonableness of plans for exploiting domestic resources. Forward planning in Colombia, which imports only 8 percent of its oil, is already resulting in new hydro, gas, and coal supplies. Colombia, as a consequence, is able to maintain a GNP growth of better than 6 percent a year. In contrast, many African nations, including Ghana, Zaire, the Ivory Coast, and Zambia, may not be able to achieve planned economic growth because of energy supply problems.

Very little hard data exist on conventional energy sources which can be commercially exploited in many of the LDCs. One of the benefits of making projections of energy supply/demand balances is to improve national development planning. Supply estimates, particularly in the longer run, need to embrace potential output of energy from new, nonconventional technologies which could replace or supplement oil, gas, and coal. These would include solar, geothermal, small hydro, and ethanol which, from the standpoint of benefits to economic growth and balance of payments, could be promising.

Whether noncommercial sources of energy should be included or not depends upon the scope of the energy survey. If the primary focus is on a national economic development strategy, it may not be necessary to cover rural noncommercial sources. On the other hand, if the rural areas contain extensive supplies of potential energy, or if the analysis is geared primarily to rural development programs, noncommercial fuels should obviously be included.

It is often assumed that data needed to prepare useful energy studies are sufficiently detailed only in advanced industrial nations. But anyone who has worked on national energy demand/supply balances is aware that data problems exist in every country—rich or poor, developed or developing. The energy analyst's task is to get around data obstacles and to make use of the demographic, labor force, household, and fuel consumption data which are available. In many instances, these sources have already been used to make estimates of gross domestic product. Further, yardsticks can be used to make estimates of fuel consumption where detailed statistics are not available. For example, in the Caribbean region, about one-third of a barrel of kerosene will provide light for a one-room house for three hours a day, 52 weeks a year.

We do not believe that the data problem is necessarily a serious obstacle to preparing energy supply/demand balances for Latin American countries, although the information base must obviously be surveyed prior to an energy analysis. Once an initial demand/supply study has been completed, major information gaps will be identified. Fortunately, the reduction of energy demand gaps is neither as costly or as time-consuming as correcting data deficiencies in resource availabilities, particularly where oil and natural gas surveys are needed.

A final note of caution as regards both methodology and data: Two keys to effective energy analysis in the developing countries are simplicity and realism. Thus the contrivance of supply and demand balances should be developed in the light of the situation in the country involved; outside experts should not impose methodologies which the local data cannot support.

Chapter 28
Energy and Development: China's Strategy for the 1980s

Thomas Fingar

INTRODUCTION AND OVERVIEW

While people elsewhere discuss the high costs of modernization, the merits of "soft energy paths," and the virtues of dispersed and humane development, China's leaders and economic theorists extol the importance of economics of scale and the need to build a conventional energy system.[1] Although Chinese leaders seek to avoid the "mistakes" of other countries, their basic developmental strategy is to follow the beaten track rather than experiment with unproven theories and technologies. As a result, they seem unreceptive, sometimes even hostile, to suggestions that the People's Republic of China (PRC) should follow a different strategy of development. They also seem slightly bemused at the way their approach has often been characterized by foreign observers.

The following description of China's policies and overarching strategy contrasts sharply to the common perception of the PRC's approach to development. But the fact is that the supposed "Chinese model" has enjoyed more support abroad than in the People's Republic. Despite past paeans to local self-reliance and small-scale development, few Chinese leaders or ordinary citizens have ever considered methane digesters and tiny hydrogenerators more than temporary expedients on the way to "real" modernization. Modernity, as understood in China, requires much more: extensive indus-

The author wishes to acknowledge the many helpful suggestions of Douglas P. Murray and John W. Lewis.

trial development; vast improvements in housing, transportation, and material well-being; and massive amounts of energy.[2] For a variety of military, economic, ideological, and political reasons, China aspires to great-power status and to "all" the conditions associated therewith.[3] Senior leaders now recognize that these aspirations might not be fully realized for a very long time, but they seem adamantly opposed to lowering national objectives. Merely to become less backward is not enough; China must strive to be "modern."[4]

Although recent commentaries have seldom mentioned specific target dates, current policies are supposed to transform China into a "powerful, modern, socialist state by the end of this century." To achieve this objective, China must realize the "four modernizations," namely, in agriculture, industry, national defense, and science and technology.[5] Speeches and commentaries reveal a lack of clarity and/or agreement on precise goals and detailed policies, but the overall objectives seem quite clear:

Agriculture. Modernization is to be accomplished through increased use of machinery and agricultural chemicals (fertilizers, herbicides, pesticides, etc.), and through specialization based on comparative advantage. Peasant well-being is to improve steadily through the introduction of labor-saving devices and greater availability of such amenities as electricity, gaseous fuels for cooking, consumer goods, and more animal protein. After announcing an all-out push for full-scale mechanization, officials realized that the economic, social, and energy costs were much too high and have since retreated to calls for "Chinese style" agricultural modernization.[6] This appears to mean a type of agriculture that remains labor-intensive but avoids the backbreaking toil typical of peasant societies.

Industry. Current policies stress the importance of expanding both light and heavy industry, but most investment continues to be channeled into the producer goods sector. Officials insist that China must first strengthen the industrial base in order to produce the equipment needed by other sectors (e.g., defense, transport, mining, agriculture), but they also have adopted measures to expand light industrial production.[7] Such expansion is needed both to meet the high and growing domestic demand for consumer goods and to provide exportable commodities.[8] One frequently noted advantage of light industries is that they come on stream faster and have higher profit rates than do heavy industrial plants.[9] Another is that they require less energy both to build and to operate.[10] Extensive development of the coal, petroleum, electric power, transport, communications, and building materials industries has received special attention.[11] Within the industrial sector as a whole, the emphasis is on consolidation, specialization, and concentration of investment.[12] Small-scale collective enterprises are encouraged but seemingly are left to their own devices when it comes to finding money, resources, and energy supplies.[13]

National Defense. Understandably, much less information has been released on China's policies and plans in the military sector, but the ultimate objective is to transform the People's Liberation Army into a modern fighting force. In the short term, emphasis is on improved training and the gradual acquisition of modern weapons systems.[14] As naval, air, and ground forces are modernized and training increases, energy demands will increase sharply. "People's war" has been redefined to look very much like conventional warfare as understood in other countries.[15]

Science and Technology. The announced objectives here include "reaching or surpassing advanced world levels" in most technical fields and training a total of 800,000 professional researchers by 1985.[16] Technological advances in other countries have contributed greatly to the widening gap between the PRC and the advanced industrial states.[17] Committed to closing the gap, Chinese officials have attached high priority to training scientists and engineers, promoting research and development, and acquiring and absorbing advanced technologies.[18] Energy research was included in the list of eight critical areas approved by the National Science Conference held in March 1978.[19] The importance assigned to energy R & D is also evident in the scientific exchange agreements concluded with other countries and by the fact that technology imports are heavily concentrated in the energy sector.[20]

Even though several specific measures have been announced, and officials have cited the need for comprehensive guidelines, individual regulations have not yet been integrated into a formal energy policy. Nevertheless, the policies announced thus far fit together reasonably well, and it is unlikely that a future energy program will differ significantly from that outlined below. Highlights of the program include:

Conservation. Waste is said to be the most important problem on the energy front. According to official spokesmen, waste ensues from inappropriate use of fuels, inefficient equipment, and careless procedures.[21] In contrast to American commentators who urge greater use of home insulation and reduced use of private automobiles, Chinese officials have focused on conservation in the industrial sector, where more than half of all commercial energy is consumed. The key to more efficient use of both fuel and power is said to be better management.[22]

Coal. Current policies call for greatly increased use of coal as an industrial fuel. Boilers are to be converted from oil-burning to coal-burning, virtually no new oil-fired plants are to be built, and new mines are to be opened as quickly as possible.[23] Foreign assistance and imported equipment are to play a major role in both expanding and opening large-scale coal mines.[24]

Concentration. Investment in the energy sector is to be concentrated in relatively few facilities.[25] The state will channel its resources into large coal mines, the petroleum industry, and large-scale thermal and hydroelectric generating facilities. Priority has been assigned to raising production in existing facilities through use of more efficient equipment and techniques; new facilities will be built thereafter.[26] The renovation of existing industrial enterprises and construction of new ones must consider the availability of fuel and power; for the next few years, at least, industrial development will tend to cluster around existing centers.

Conventional Facilities. State resources will be used primarily to modernize and expand the production of fossil fuels and hydroelectric power.[27] To develop reserves as rapidly as possible. China has actively sought the assistance of foreign firms. Energy exports (especially oil) are expected to provide a major source of foreign exchange and thus to pay for a large part of the modernization effort.[28] Even with primacy assigned to coal, domestic demand for other fuels will increase steadily, and China will be hard pressed to satisfy this demand even if exports are limited and new facilities come on stream quickly.[29] Limited research support will be provided for work on unconventional fuels such as biogas and mixtures of coal and oil, but, relatively speaking, this effort will be trivial.[30] By the same token, investment will be concentrated in large facilities; although small coal mines and hydro facilities will be encouraged, such development will be left primarily to local units.[31] China has recently announced renewed interest in nuclear powerplants, but no concrete plans have been revealed.[32]

Coordination. China will henceforth pay greater attention to matching supply and demand for fuel and power. According to official statements, many enterprises operate below capacity because of fuel or power shortages.[33] Energy allocation now is to be monitored and controlled more carefully, and particularly important or efficient enterprises will have sure and adequate supplies of both fuel and power.[34] New enterprises will not be authorized and expansion plans will not be approved unless adequate energy supplies are assured.[35]

Incentives. Since demand still far exceeds supply, officials have decided to use energy as an incentive to improve enterprise and managerial performance. For example, enterprises that turn out high quality products and show a profit will obtain additional energy. Conversely, firms that operate at a loss or produce inferior goods may face reduced allocations.[36] In the latter case, it will be difficult for managers and workers to obtain bonuses, better housing, or other benefits. The goal is clear, but it is not at all clear how this policy will work in practice. To note just one potential difficulty, an enterprise that is punished by having its energy supply reduced will find it even more difficult to satisfy performance criteria since it will have to operate below capacity unless the management performs truly Herculean feats. Another problem concerns the potential for conflict between planned distribution of energy and policies that shift fuel and power to punish or reward performance.

The remainder of this paper is divided into three sections. Section II examines the origins of the present strategy of development; Section III focuses on the developmental strategy *per se*; and Section IV outlines China's emerging energy policy for the 1980s. This organization reflects the order and underlying logic of China's apparent approach to energy policy.

ORIGINS OF CHINA'S STRATEGY OF DEVELOPMENT FOR THE 1980s

China's approach to development or, more accurately, the approach favored by the dominant leadership group, derives from broadly shared objectives and assumptions. The principal objectives are to strengthen security and independence, to raise living standards, and to "create conditions for the transition to communism." Realization of all three goals is said to depend on the pace and degree of modernization.[37] National experience, communist ideology, and perceptions of foreign development underlie both the general objectives and the chain of relationships that leads, ultimately, to a particular set of developmental and energy policies. Each of these elements merits a brief explanation.

The century of "national humilitation" which China suffered at the hands of stronger powers has had a profound impact on the thinking of PRC officials. Indeed, a principal objective of the Revolution that began more than eighty years ago has been to restore China to its proper place in the world and to ensure that it will never again be victimized by other nations. In attempting to explain the past and to frame policies for the future, Chinese leaders have focused on economic and technical factors. According to their analysis, other nations enjoyed military—and therefore political—superiority because they were "more modern." The key to modernity is a strong economy based on advanced technology. The conclusion is that large-scale industrial facilities, energy-intensive development, and continuous technological advances are the *sine qua non* of security and independence.[38]

China's pre-1949 experience was generally consistent with Marxist doctrine. One need not accept Marxist teachings to acknowledge that PRC officials still consider basic economic conditions (the "factors of production") to be of fundamental importance. During the past three years, senior leaders have stated repeatedly that the political, social, and economic goals of the Revolution cannot be attained without rapid modernization of the economy.[39] On this point, ideology reinforces history, and vice versa.

Ideology also is important in another way. Most of China's present leadership received the Marxist gospel from Russian apostles who obscured the difference between doctrine and Stalinist practice. PRC leaders still believe in the essential wisdom of the Marxist (read Stalinist) path. Whereas many of the Khrushchev-Brezhnev reforms have been excoriated by China, there has been remarkably little criticism of Stalin's strategy of development. Hints that the strategy might not have been completely correct are beginning to appear, but PRC commentaries still acknowledge and applaud Stalin's successes.[40] This assessment encourages a doctrine of central planning and a tendency toward gargantuan, energy-intensive development.

Experience and ideological teachings are buttressed by a third element new to policymaking in the PRC: perceptions of how developments have unfolded elsewhere. By many standards, and certainly by those that weigh most heavily in the present Chinese calculus, the technological gap between China and the industrialized world has widened dramatically during the past two decades.[41] After closing the gap slightly during the 1950s, China slipped even further behind in the '60s and early '70s, i.e., at precisely the time it was experimenting with "unique" or "Maoist" strategies of development.[42] Because the gap widened, China became more vulnerable to external pressure and less capable of influencing events beyond its borders. This accounts, in part, for the sense of urgency that permeates official statements on the need to "make up for lost time." The following excerpt from Hua Guofeng's report at the Second Session of the Fifth National People's Congress (June 18, 1979) reflects the dominant mood:

> Whether we succeed or fail in our endeavor to modernize China by the end of the century will decide the future of our country and people. . . . Much time has been lost and we must speedily make up for it. We have finally created the internal and external conditions favorable for rapid, peaceful construction. If we do not make good use of this precious, hard-won opportunity, go all out and do everything possible to speed up socialist modernization, our generation will be unworthy of our country and people.[43]

Confronted with the inadequacies of past experiments (most notably during the Great Leap Forward of 1958–59 and the Cultural Revolution of 1966–68) and the demonstrated effectiveness of strategies employed elsewhere, PRC leaders began to study the experience of successful modernizers. The United States and Japan are described as having been particularly successful, and Japan's success in adopting American models is singled out for special study.[44] However, the way in which Yugoslavia and Romania have utilized certain "capitalist" practices has also been mentioned with favor.[45]

One result of the effort to "learn from" the experiences of others—which, rhetoric to the contrary, often comes very close to "copy"—is a tendency to retrace many steps in research, development, and deployment. This stems, in part, from a rational desire to "master" the process, technology, or strategy being studied on the assumption that this is the only way fully to understand what takes place. But it also derives from a strong desire to avoid the kind of dependency that plagues much of the developing world and that enfeebled China during its alignment with the Soviet Union. To minimize dependence on foreign suppliers and the resultant vulnerability, China seeks to master all aspects of imported equipment and technology. Repeating every step of both micro and macro developmental processes is regarded as essential for avoiding dependency.

China's leaders hope to telescope the modernization process—to enable the PRC to go through "All" the steps taken by successful modernizers in a much shorter time. The example of Japan is considered instructive because it passed through the respective stages much more quickly than did the United States. Whether mere rhetoric or real belief, Chinese officials assert that "the superiority of the socialist system" and the fact that China is a late starter will, in combination, enable their country to outperform even Japan.[46]

There are at least two "mistakes" in foreign experience that China does not want to repeat: environmental damage and "excessive" reliance on oil. Even with benefit of hindsight, however, it is difficult to distinguish "mistakes" from essential components of the modernization process. It is far easier to judge what should be different than to prescribe ways to realize preferred outcomes. In the case of China, the difficulty is compounded by an understandable distrust of "new" or "revolutionary" strategies of development. Such strategies just have not worked very well in the PRC, and the operative assumption is that China should follow almost exactly the same path of development as did others so that informed judgments about what to retain, change, or discard can be made from direct experience.

Historical, ideological, and perceptual incentives to follow a particular strategy of development are reinforced by the personal biases of China's senior policymakers. Many of today's leaders held comparable positions during what are now regarded as the halcyon days of the mid-1950s. In many ways, these were very good years for the PRC. Economic progress was steady and impressive, policymaking was still collegial and pragmatic, and officials had a clear sense of what should be done to overcome deficiencies.[47] For example, in 1955–56 Party leaders devised policies to overcome problems inherent in the Soviet model and to take advantage of new opportunities in the international environment. Overtaken by the Great Leap Forward in 1958, these measures were never really implemented. Nevertheless, those who had devised the new strategy—many of whom have now returned to power—remained convinced that their policies were essentially correct. Following the Great Leap, they attempted, with partial success, to resuscitate the approach and policies devised for the Second Five-Year Plan.[48] Faith in the wisdom and appropriateness of "their" approach was heightened during the Cultural Revolution when many were subjected to personal hardship and banished to the political wilderness. Restored to positions of influence during the mid-1970s, these "veteran cadres" attempted to revive policies that would accelerate the pace of China's modernization and, not at all coincidentally, to "prove" that they had been right all along.

CHINA'S CURRENT STRATEGY OF DEVELOPMENT: SEVEN PRINCIPAL THEMES

Taken as a whole, the domestic and foreign policies of the People's Republic of China constitute a carefully conceived and reasonably well-integrated response to the challenges and opportunities confronting the country. Integration and mutual reinforcement derive, in part, from political compromises and repeated marginal adjustments occurring naturally in both planning and policymaking, but they also reflect a conscious attempt to fit policies into a specific strategy of development.[49]

PRC leaders have felt it necessary to insist that, irrationalities and failures notwithstanding, there is a "scientific" path of development that can be discovered and followed.[50] Following this "scientific" path requires definitions of and adherence to a set of interconnected propositions or "lines." The fundamental determinant of specific policies is supposed to be the "ideological line." Stripped to its essentials, the correct ideological line "for the present stage of development" is that everything depends on the level of economic development. The "political line" derived from this "ideological line" calls for uniting with all who can help to realize the "four modernizations."[51] In other words, rather than pursue "class struggle" for its own sake or exclude all but a small group of purists from the developmental effort, the Party (and the State) should utilize the talents of all who can contribute. More specifically, this means that cadres are not to discriminate against scientists and engineers trained in "bourgeois" or "revisionist" countries;[52] non-Party specialists are to play an important role in formulating and monitoring developmental plans;[53] and "friendly" foreign companies and governments are to be drawn into China's drive for rapid modernization.[54] Adherence to these "lines" has produced a "strategy" of development with seven general themes. These themes are summarized below.

Combining Elements of Planned and Market Economies

Even though PRC officials continue to assert the value of central planning and the need for overall coordination, they now add that it is inappropriate, even impossible, to effect comprehensive planning at the present stage of development.[55] Their arguments echo those advanced in the mid-1950s by Xue Muqiao, Hu Qiaomu, and other economic theorists. Then, as now, leading economists argued that premature or overly ambitious planning was bound to have deleterious results. Rather than attempt the impossible, they argued, central planning organs should concentrate on areas of decisive importance to the overall development of the country.[56] During the 1950s, advocates of this view maintained that all activities not covered by the plan should be left to market forces, albeit with close monitoring by central authorities. After an eight-year effort to restore comprehensive planning modeled on the First Five-Year Plan, PRC officials have now returned to the alternative approach suggested in 1955–56.

The shift to policies associated with the Second Five-Year Plan was not heralded with an official announcement; indeed, it appears to have occurred incrementally as officials began to grapple with specific problems. Efforts to be "comprehensive" had led to paralysis and/or totally unrealistic targets. To escape from this dilemma, political leaders began to stress the positive role of the market.[57] Overall coordination and detailed planning in certain areas of activity continue to be emphasized, but the scope has been greatly reduced. The energy sector is among those singled out for continued, even strengthened, control by central planners.

Since the energy sector is relatively well defined and involves few units (if unconventional fuels and small mines and hydrogenerators are excluded), it seems more susceptible to planning than, say, agriculture or commerce. That assessment may be wrong, but it is clearly what China's leaders believe. Energy also qualifies for comprehensive central planning because it is critical to all other sectors of the economy; it is simply too important to be left to the market. Even though officials appear resigned to the need for continued, even expanded, reliance on "regulation through the market," they do not really trust the "invisible hand," and are reluctant to entrust critical parts of the economy to forces they cannot control. Their reluctance is traceable, in part, to the political judgment, which is probably correct, that they must demonstrate the advantages of the current developmental strategy if the regime is to retain popular support.[58] As a key to success, energy *must* be monitored and controlled by the central government.

Combining "Administrative" and "Economic" Controls

This theme is related to, but distinct from, combining the elements of planned and market economies. Recent PRC commentaries often depict "administrative" methods as a kind of bureaucratic tyranny whereby superior units merely issue orders and impose targets.[59] However, other statements insist that certain bureaucratic controls are essential for rapid modernization. They are henceforth to be combined with "economic" methods such as flexible pricing mechanisms and rewards and penalties based on economic performance. For example, rather than simply ordering enterprises to produce more of a particular commodity, responsible officials are supposed to secure desired ends by providing financial inducements (e.g., higher prices to recover costs or authority to retain a higher percentage of profits earned). Manipulation of tax rates and the "market" and imposition of fines and bonuses are being substituted for cruder—and presumably less effective—means of control.[60] But, rhetoric to the contrary, PRC officials have not abandoned the vertical bureaucratic structure inherited from Chinese tradition and reinforced by adoption of the Soviet model. Until the *structure* is changed, there will be a tendency to rely on administrative methods.

Publicly announced energy policies illustrate how administrative and economic measures are to be used. The allocation of fuel and power to various consumers is supposed to be controlled partially by planned allocation (administrative measures reflecting priorities and political considerations) and partly on the basis of economic performance.[61] Enterprises that produce high-quality goods and profits are to be rewarded with increased allocations. This is described as an economic measure because increased energy will enable the enterprise to earn even higher profits, which, in turn, will translate into higher wages, more money for investment in new plants, equipment, or improved housing for workers, and more high-quality goods for the State. Administrative and economic considerations will conflict whenever the energy demand of profitable enterprises exceeds the unallocated supply. Current policies call for reducing the amount of energy supplied to unprofitable and/or badly managed enterprises. In theory, this will free supplies for better-run facilities, but problems and conflicts seem inevitable. How, for example, should officials choose between meeting the energy requirements of the best/most productive enterprises and satisfying the needs of collectively run service facilities employing many people? Given realities in China and inherently contradictory guidelines, the choice will be neither easy nor uncontested.

Combining Centralization and Decentralization

Centralization and decentralization, like the other dichotomies in this summary, are vague terms providing only imprecise guidance to policymakers and those charged with implementing policy decisions. Officials often use the above formula to make the rather simple point that activities should be neither completely "centralized" (i.e., controlled by authorities in Beijing) nor completely in the hands of "local" units (provinces, municipalities, enterprises, or production teams, among others), but they rarely provide useful guidance on how to strike a proper balance.[62] Nevertheless, operative guidelines are emerging.

Basically, the larger, more complex, and more important an activity or installation, the greater the "appropriate" degree of central control. Conversely, activities or units that have only local impact, or those that cannot be managed effectively from the center, should enjoy a high degree of autonomy. While China's petroleum industry and large hydroelectric facilities are considered proper objects of close central control and supervision, small-scale hydrogenerators and biogas units are to be managed in a highly decentralized fashion.

Current policies call for further decentralization of some activities and for consolidation or greater centralization in others. The dominant trend in agriculture, for example, is clearly toward decentralization.[63] Commentators also have called for greater enterprise authority in most industrial fields; factory managers are henceforth to have more latitude to decide what to produce, how to produce it, and how to distribute profits.[64] At the same time, however, officials have called for the consolidation and amalgamation of individual factories. "Small and complete" enterprises that attempt to do everything themselves are now denounced as a manifestation of the "small producer mentality"; specialization and clear divisions of responsibility are strongly encouraged.[65]

Although the magnitude of centralization/decentralization varies with the activity or sector in question, the common objective is to foster local initiative, to break down bureaucratic rigidity, and generally to minimize obstacles to rapid economic and technical modernization.[66] Each level or unit is supposed to concentrate on relatively few matters, and the degree of centralization is to be no higher than necessary to ensure rapid attainment of developmental goals. Neither centralization nor decentralization is described as inherently good or bad; the goal is to use whatever formula produces the best results. As noted previously, the "best" results, in the present calculus, are those that produce the highest profits, the greatest output, the most rapid introduction and absorption of advanced technologies, and thus the most "modernization." Equality and other social values once touted by PRC officials and their foreign boosters no longer weigh so heavily in the calculation of costs and benefits.

Combining the State and Collective Sectors

Foreign observers have often contrasted the vitality of collective enterprises to the inertia and inefficiency of the State sector.[67] PRC officials apparently have reached a similar judgment, recently affirming the importance of collective units to the modernization process. Administrative controls, planning, and centralization tend to predominate in the State sector, while economic controls, market forces, and decentralization are more important in the collective enterprises. Current policies call for taking full advantage of the flexibility and vitality of collectives.[68] The importance attached to the collectives is striking because (a) virtually all large, capital-intensive, and energy-intensive enterprises fall within the State sector, and (b) the structural and procedural legacy of the "Soviet model" is most pronounced in State enterprises. This is true not only in terms of size, reliance on quotas, and amount of direct investment (as opposed to loans), but also in terms of waste and efficiency. Although this point should not be pushed too far, Chinese commentators have implied that problems of worker motivation, inefficient use of fuel and raw materials, and irrational expansion of facilities have been most serious in the State sector.[69]

Collective enterprises are now being given an unprecedented mandate to explore new procedures, managerial forms, and production possibilities. Recent policy changes have removed the limitations on wages within collective enterprises, enabled collectives to recruit workers without going through the Labor Bureau, assured adequate supplies of fuel and power when the enterprise meets certain criteria, and authorized price adjustments, altered production runs, and access to new markets.[70] These policies are designed to foster entrepreneurial activity by making it easier for able, talented, and ambitious individuals to organize economic activities. But since large-scale energy facilities are State-run, the obstacles to their rapid advance may be more formidable than in the case of small-scale collective projects.

Combining New and Existing Facilities

In 1977–78, PRC plans and statements stressed the importance of expanding capacity to satisfy the large and growing demand for commodities of all types. One hundred and twenty major new projects (including ten new oilfields, five major harbors, and several new steelmills) were ballyhooed as the centerpiece of the modernization strategy, but numerous smaller facilities were also to be constructed "immediately."[71] Thereafter, officials realized that it would be impossible to implement such grandiose construction plans and that even trying could be disastrous.[72] They also concluded that sizable short-term increases could be realized by improving the efficiency of existing facilities, accounting for the current emphasis on modern management.[73] Indeed, renovation and more efficient use of existing facilities have now been accorded higher priority than construction of new installations.

Renovation has the advantage of being faster and cheaper than construction of new facilities; it also attacks energy-related problems. Officials have acknowledged that fuel and power supplies cannot be increased significantly for several years and that short-term gains must be achieved primarily through conservation.[74] Retrofitting existing plants will make them more energy-efficient, thereby enabling China to accomplish more with the same energy inputs and distribution network. The other side of this particular coin is that, if new facilities were built before substantially more energy could be supplied, power would have to be drawn from that now supplied to existing plants. Robbing Peter to pay Paul would not only result in economic costs far in excess of productivity gains but would also entail high political and social costs because existing facilities, many of which already operate inefficiently and/or at a loss, would find themselves in even more difficult straits.[75] This would have a pronounced impact on both the lives of the workers in the affected factories and on perceptions of regime performance.

Making Full Use of Both Domestic and Foreign Resources

In many ways, this simply restates the general proposition that China must do everything possible to speed up modernization, but it also has specific policy implications. For example, the injunction to make full use of indigenous resources requires, i.e., that skilled and experienced people be assigned to positions commensurate with their training. "Veteran" cadres are being restored to positions of responsibility; "bourgeois" teachers, scientists, and engineers are being reassigned to suitable positions; and "former landlords and capitalists" are urged to contribute what they can to the modernization effort. In short, China will make "full use" of its human resources.[76]

Foreign resources include everything from loans and credits to training opportunities and advanced technologies. Narrow definitions of "self-reliance" are condemned for having widened the gap between China and the industrialized world.[77] Technology, scientific knowledge, and other elements that can contribute to the modernization of China are now described as part of the universal heritage of mankind, and as being "without class content."[78] China, the argument holds, should avail itself of anything and everything, whether found at home or abroad, that will contribute to rapid development and increased national independence.

Combining Short-Term and Long-Term Considerations

From the early 1970s until late 1978, PRC policy looked to the long term and generally slighted the existing human and material base. By focusing on the future, officials missed opportunities and frustrated popular expectations. But by the end of 1978, they had begun to concentrate on more immediate problems.[79]

In addition to the need to demonstrate that current policies—and leaders—would succeed where others had failed, the shift to a shorter-term perspective was prompted by a growing appreciation of China's underutilized capacity and of the potential for dramatic advances if certain immediate problems were solved.[80] Improved management, assured supplies of energy and raw materials, and all the measures described above were designed, in part at least, to make better use of existing capacity.

The Chinese have not, of course, abandoned the search for broader long-term solutions. In the energy sector, the balance between long- and short-term considerations is reflected by the effort to enlist foreign assistance to raise production quickly and by the injunction to rely on coal rather than oil.[81] Development of hydroelectric power is a second illustration of the effort to make short-term solutions conform to longer-term plans.

While big dams and turbines are regarded as the wave of the future, much of China must continue to rely on small hydrogenerators. Construction of small units is encouraged, but current policies require greater planning and coordination than in the past, especially for the purpose of adding new units and building local grids. Under the old *laissez faire* approach, expansion of facilities or building additional powerplants raised problems; for example, constructing a higher dam might submerge small upstream generators and reduce the flow of water to those located downstream. Now, new facilities and short-term solutions must also facilitate the transition to better and longer-term solutions.[82]

The desire to make optimal use of existing energy supplies is reflected in the fact that, although officials continue to channel an overwhelming percentage of State investment into the producer goods sector, some funds (and the rhetorical emphasis) have been shifted to industries that can generate high productivity and profits in a very short time, e.g., textiles and consumer goods. Light industry as a whole has received special attention on the grounds that such facilities can be constructed or renovated more rapidly than can heavy industrial plants, produce higher rates of return on investment, earn badly needed foreign exchange, create more jobs in the short run, and provide goods demanded by the populace. Moreover, they require relatively little energy. To ensure that light industry develops rapidly, this sector is supposed to obtain all the energy it needs.[83]

At least in terms of articulated policy, China is emphasizing investment in fields that will produce results quickly, although it is not ignoring those that will generate benefits only in the long run. In the energy sector, as noted below, the choice is clear: energy resources are to be developed *at once* because they are essential for all other sectors of the economy.

ENERGY POLICIES AND THE STRATEGY OF DEVELOPMENT

Unlike most developing countries, China is richly endowed with energy resources.[84] Because much of the country remains unexplored and the quality of data is uneven, there is great uncertainty about the magnitude of PRC reserves, but unquestionably they are substantial. Coal reserves, recently estimated at some 600 billion metric tons (bmt), are of the same general magnitude as those of the United States and the Soviet Union.[85] Estimates of total and/or recoverable oil reserves vary widely, but, on the basis of current information, they probably fall within the range of 3–10 bmt.[86] In 1979, China produced 106.1 million metric trons of crude.[87]

After 1965, the petroleum industry grew dramatically for several years, but technical and economic constraints have begun to take hold and annual production increases have become quite modest. Oil output increased by only 1.9 percent in 1979.[88] Since oil production has failed to reach projected levels, China has been unable to satisfy growing domestic demand or to provide even the modest quantities pledged to Japan and Thailand.[89]

Little is known about the magnitude of natural gas deposits, but they too are sizable. Optimistic estimates place China's reserves on a par with those of the United States, and even the most conservative projections indicate vast reserves.[90] Whatever their extent, the country's gas reserves are still virtually untapped, in part because the largest known fields are in Sichuan province, far from China's major industrial centers.

China's hydroenergy potential of more than 500 million kilowatts (theoretical) is the largest in the world. However, only about 3 percent has yet been tapped, and formidable difficulties must be surmounted to exploit even one-quarter of this potential.[91] Nearly three-quarters of total capacity is located in the southwest, far from industrial and population centers. The extraordinarily heavy silt load carried by the Huang He (Yellow River) and other major rivers makes it difficult and expensive to exploit that portion located closest to (but still quite far from) industrial consumers.

In aggregate terms China's solar potential is very large, but the areas that receive the most sunlight (e.g., Xizang [Tibet] and Xinjiang) are remote and sparsely populated. Without major technological breakthroughs, this potential will remain largely untapped. Advances in solar engineering make it possible, in theory, to install collectors in the more populated regions of the country, but the prerequisites and opportunity costs of doing so are enormously high.

Indirect utilization of solar energy in the form of biomass (wood, crop residue, leaves, etc.) is and will continue to be extremely important. Most is burned directly as fuel for cooking, but a small portion of the biomass is converted to methane in household or community digesters.[92]

China probably has sizable geothermal resources (almost completely untapped) as well as significant deposits of uranium. Although reserves of these energy resources are substantial, the present rate of exploitation cannot be increased significantly without major infusions of capital and technology.

Throughout most of the 1970s, PRC leaders apparently regarded oil as the panacea for China's energy and developmental woes. Annual production was increasing at the rate of 20 percent, OPEC-led price increases opened the possibility of using petroleum exports to finance the drive for modernization, and advances in China's petroleum industry were being made with minimal foreign assistance. Continued growth of the oil industry was expected to cover increased domestic demand, produce sizable surpluses for export, and allow China to follow the same general path of development as countries that had already "modernized."

Unwarranted euphoria about the growth potential of the petroleum industry (fed, in part, by political considerations that sometimes overruled technical judgments)[93] and a strong desire to follow the proven path led PRC leaders to a series of unfortuante energy-related policies. Switching from coal to oil as the preferred industrial fuel is now regarded as a mistake. The "mistake" has been blamed on the "gang of four," but in this case the hapless cohorts of Madam Mao were probably joined by many members of the present leadership. Conversion from coal to oil was regarded as a proper, even necessary, step on the path of development traversed by others. The following excerpt illustrates the current view:

> For some time we lacked an adequate understanding of the strategic importance of oil as one of the vital energy resources. Our lopsided view of the energy structure, with a number of industrially advanced countries having the proportion of 60 or 70 percent in the use of oil and gas and with ours a little over 12 percent, made us think that our country's energy structure was rather backward and required an increased proportion of oil and gas. After discovering new oil and gas fields, we inappropriately encouraged coal consuming enterprises to switch to using oil as fuel. As a result, the amount of oil consumed by the country's industrial enterprises and transport departments rose more than 16 times in the 11 years between 1966 and 1977. . . . Over 65 percent of the total amount of oil consumed was used to heat boilers, while only about 35 percent was used for industrial purposes. This was irrational and caused serious waste.[94]

Slower growth than expected in the oil industry and the increased severity of energy bottlenecks prompted a fundamental rethinking of the developmental strategy that had been pursued, albeit intermittently, since the early 1970s. Commentators began to complain about fuel and electricity shortages in early 1977, but it took two more years of slow energy growth before officials decided to place energy at the top of the national agenda.[95] It is interesting to speculate about the extent to which PRC officials became concerned with the "energy crisis" as a result of developments and alarmist predictions elsewhere; but there also were compelling domestic reasons for focusing on energy policy during 1979.[96] Officials began to talk about the need for a comprehensive energy policy and for concrete measures to solve both short- and long-term problems of supply.[97] Before summarizing the most important aspects of the resultant policy, it is useful to note one additional and rather surprising feature of energy policymaking in China: Until the Spring of 1980, when the State Energy Commission was created, there was no organizational equivalent of the U.S. Department of Energy. The new commission is headed by Yu Qiuli, a Vice Premier who previously served as head of the State Planning Commission.[98]

The long delay in creating the State Energy Commission is striking because PRC officials usually attack highly visible problems by creating new, specialized organizations. Thus, for example, when agricultural mechanization was elevated to paramount importance, they (re)created the Ministry of Agricultural Machinery. The underlying philosophy seems to be that a specialized agency, as opposed to an interagency task force, will assure that the problem receives proper attention and that guidelines are properly implemented. Creation of a special commission signals to all that energy has risen to the top of the policymaking agenda.

Prior to establishment of the State Energy Commission, energy decisions were made in several organizations operating at many different levels of the system. The highest such body was the State Economic Commission (SEC) headed by Vice Premier Kang Shi'en. Although neither he nor the SEC was identified as having special responsibility for energy questions, Kang emerged as a principal spokesman on such matters.[99] Yu presumably will play such a role in the future and the new commission will assume the task of coordinating the efforts of all agencies and activities involving the production, distribution, and consumption of energy. To do so, it will have to take over responsibilities and power previously exercised by many other organizations, including the ministries of coal, petroleum, and power, the State Scientific and Technical Commission, the State Commission on Foreign Investment, and the Ministry of Foreign Trade, as well as by the administrative organs of regional power networks. Regional, provincial, and myriad subprovincial units have played and presumably will continue to play an important role (e.g., in decisions to construct small hydrogenerators or to provide fuel and power for local industries).

Although China only recently created the State Energy Commission and still lacks a formal "energy policy," it did develop a series of relatively well-integrated measures to ensure that energy bottlenecks are overcome and that energy considerations are given proper weight in the drive for modernization. The central ingredients of China's energy policy for the 1980s were outlined in Section II; the pages that follow describe the relationship between those guidelines and China's overall strategy of development.

Coal, petroleum, and electric power have emerged as areas of preeminent importance. One indication of their importance and of the significance of energy more generally is that efforts to acquire foreign technology are focused primarily in the energy sector.[100] Officials seemingly have concluded that the financial and technical obstacles to rapid exploitation of the country's energy resources are so formidable that developmental objectives cannot be realized without large-scale foreign involvement. Whatever the costs of relying on foreign firms and governments, the costs of slow growth in the energy sector are higher, at least according to the view which now holds sway in Beijing.

While seeking foreign assistance in long-term development projects (e.g., coal mines, offshore and onshore oil fields, and hydroelectric facilities) the PRC also is pursuing several policies designed to produce substantial results in the short term.[101] The most important of these measures center on the drive for energy conservation launched in late 1979. Because it will take several years to bring new fields into production, and both political and economic factors preclude significant oil imports, immediate supply difficulties can best be alleviated by making more efficient use of available supplies. The severity of the shortfall is indicated by an article published in May 1979.

China possesses extremely abundant energy sources, but there are serious problems in the field of energy supply. Due to shortages of fuel and electric power, many enterprises are unable to operate at full capacity. This affects the speed of development of the national economy. It is not possible to fully meet the fuel and power needs of the people in urban and rural areas. This affects normal life. The energy problem is therefore a problem which we must take urgent steps to solve.[102]

Speaking at a rally held in October 1979 to hail the start of the "first nationwide 'energy conservation month.' " Kang Shi'en expressed alarm at the extent to which scarce supplies of energy were being wasted. Allowing for the hyperbole and use of extreme examples that are common in PRC statements, the following passages provide useful insights into both the magnitude of the problem and the thinking of senior officials:

Why do we particularly stress energy conservation at present? As everyone knows, due to interference and sabotage by Lin Biao and the gang of four, the management of energy is now in a state of confusion, the effective energy utilization rate is very low, and wasteful losses of coal, petroleum, and electricity are frightening. China now consumes 40–50 million tons more coal, 3–4 million tons more oil, and 20–30 billion kilowatts more electricity than is necessary each year. This waste is even more shocking when compared with the effective energy utilization rate in advanced foreign countries. In China the effective utilization rate of thermal energy derived from fuel is only about 28 percent, while in the developed countries it has reached around 50 percent, almost double ours. . . . We are consuming too much petroleum and not conserving our natural resources in a rational way. Various enterprises on the industrial and communications front are now consuming over 30 million tons of oil. Over ten million tons of this petroleum are consumed directly. Most of the direct consumption of oil may be eliminated through using coal. If we can save over 10 million tons of petroleum for export, we can earn more than 2 billion US dollars in foreign exchange a year. We really cannot afford the consumption of oil in this manner.[103]

At present China's total energy consumption amounts to approximately 600 million tons of standard coal [coal equivalent], almost equalling that of Japan. However, Japan's national output value is three times higher than our country. . . . The State demands that in 1980 all trades and professions must achieve a 10 percent savings in petroleum, 5 percent in coal and 3 percent in electricity.[104]

To implement the call for energy conservation, "All departments and localities" were instructed to prepare concrete plans to reduce wasteful consumption.[105] Enterprises must monitor the amount of fuel and power used and strive to reduce consumption per unit of output at least to the best ratio achieved in the past.[106] Inefficient factories will be penalized, presumably by having their allocations of energy reduced, and those able to use energy most efficiently will be given additional supplies if the addition will enable them to operate at higher capacity. This is in keeping with the shift to "economic" methods described in Section III.

Other conservation measures include the refitting of existing facilities as part of the emphasis on tapping the full potential of existing enterprises, the introduction of "modern" principles of management, and techniques associated with systems engineering and operations research, and the installation of meters so that energy consumption can be monitored.[107]

To conserve oil, oil-burning industrial boilers are to be converted to burn coal. Current policy specifies that "all" oil-fired boilers must be converted and no new ones may be constructed except in extraordinary cases.[108] Reasons for making such a change include freeing more oil for export, other forms of domestic consumption, and for taking full advantage of China's large coal reserves, but there are also costs and problems associated with this policy. For example, increased use of coal will require additional transport capacity, opening and/or expanding mines, and greater attention to pollution problems—to mention the direct costs of conversion. Thus far, officials seem to be paying little attention to the negative consequences.[109]

China must overcome enormous problems on the energy front if it is to "realize the four modernizations." The effort to enlist foreign support to expand existing mines and open new ones, to search for and extract petroleum, and to design and construct new hydroelectric installations suggests that officials are willing to go further to solve these problems than anyone would have predicted a few years ago, but they may not be willing to go far enough. For example, although foreign oil companies have begun seismic exploration in offshore areas, it is not at all clear that they will agree to extract the oil under the terms that China has proposed. Similarly, recent efforts to entice foreign firms to help develop onshore fields located deep in the interior have been largely unsuccessful because of the costs involved, and because the companies want to see what happens with respect to offshore negotiations before committing themselves further to China's modernization effort.[110]

The development of coal reserves has likewise proved to be a more protracted, complex, and costly process than had been envisioned. New mines can certainly be opened, and production eventually will rise, but development will be slow.[111] This imposes significant short-term constraints on the growth and modernization of China's economy. If, as officials assert, they plan to base the economy on coal and severely to limit the role of oil (and, presumably, gas, although here cost and technical considerations appear more significant than any conscious decisions to export or conserve energy resources), then expansion of industrial capacity may occur more slowly than if the country relied on oil. Over the long run, China's coal-based strategy is probably wise; in the short run it will make it difficult for the PRC to realize other developmental objectives.

SUMMARY

While virtually all of the developmental themes described in Section III are reflected in energy policy, the first four are particularly relevant.

Planned/Market Economies. Energy is clearly identified as a critical area, that "must" be controlled and subjected to more comprehensive planning than other less important sectors. This is especially true for the development of large coal mines, the oil and gas industries and large scale hydro facilities. Since energy is scarce, it must be allocated in a planned way. Current policies specify that light industries (and possibly agriculture) are to have priority in the allocation of fuels and power because they require less, produce goods for export, meet particular domestic demands (e.g., for consumer goods that will give value to incentive bonuses now being paid to workers), and produce high rates of profit for reinvestment. Other regulations specify that no new facilities can be constructed and no enterprises can be expanded without assured supplies of fuel and electricity. Supplies must be regulated according to plan in order to ensure that "key" facilities are not deprived of energy. Energy suppliers cannot allocate fuel or electricity simply on the basis of ability to pay.

Administrative/Economic Controls. Since energy is a critical sector and requires centralized planning, administrative methods must play an important role. Economic controls will play a minor role, for example, in the distribution of energy on the basis of performance, but in general bureaucratic regulation will be more important than pricing and similar mechanisms.

Centralization/Decentralization. This is reflected in the energy sector by the fact that large conventional facilities are to be controlled by units at or near the center; but unconventional installations such as small hydrogenerators and biogas digesters remain under local jurisdiction. One reason for placing large energy facilities under direct central control is that foreign investment is supposed to play an important role. Both the delicacy of dealing with foreigners and the need to maintain coordination militate against much decentralization, at least in the Chinese context.

State/Collective. Since large-scale energy installations fall within the State sector, investment will be administered directly by the relevant ministries, and the kind of bureaucratic controls and procedures noted above will apply. At the same time, the collective sector is almost completely free to develop and exploit small-scale facilities. It is not yet clear how much latitude the collectives will have in such matters as the prices charged for locally produced electricity or coal, but it seems possible that prices can be raised to generate more revenue and to encourage conservation. That will require important changes in social as well as economic practice. In this, as in many other areas, policy changes have opened myriad possibilities that may be unpalatable because they are unfamiliar. Despite all the rhetoric about revolution and social change, China is extremely conservative. Changing the nation's energy "system" will be a long, expensive, and politically difficult process.

NOTES

1. Typical discussions of China's developmental and energy needs include Yu Qiuli, "Mobilize the Whole Party and the Nation's Working Class and Strive to Build Daqing-Type Enterprises Throughout the Country," *Renmin Ribao* (hereafter RMRB), May 18, 1977, in *Peking Review* (hereafter PR), No. 22 (May 27), 1977, pp. 5–23; Yu Qiuli, "Report on the Draft of the 1979 National Economic Plan," RMRB, June 29, 1979, in Foreign Broadcast Information Service, *Daily Report: People's Republic of China* (hereafter FBIS), July 2, 1979, pp. L13–28; and Chen Xi, Huang Zhijie, and Xu Junzhang, "The Effective Use of Energy is an Important Issue in Developing the National Economy," RMRB, May 13, 1979, in FBIS, May 29, 1979, pp. L7–10. See also Yang Zhirong, Zhu Bin, Xu Junzhang, and Zhang Zhengmin, "Several Technical and Economic Questions on Increasing Energy Sources," RMRB, February 28, 1980, in FBIS, March 27, 1980, pp. L14–17.
2. Visions of China's future are outlined in Hua Guofeng, "Unite and Strive to Build a Modern, Powerful, Socialist Country!" (Report at the First Session of the Fifth National People's Congress, February 26, 1978), RMRB, March 7, 1978, in PR, No. 10 (March 10, 1978), pp. 7–40; Xue Yunying, "The Accomplishment of the Four Modernizations is a Profound Revolution," RMRB, July 4, 1978, in FBIS, July 14, 1978, pp. E7–10; and Editorial, "Let the Whole Party and the Whole Nation Unite as if with One Heart and Work Together to Achieve Socialist Revolution," *Hong Qi* (hereafter HQ), No. 1 (January), 1979, pp. 22–26, in FBIS, February 2, 1979, pp. E12–16.
3. See, for example, Thomas Fingar, editor, *China's Quest for Independence: Policy Evolution in the 1970s* (Boulder: Westview, 1980).
4. One of the charges levelled at the "gang of four" (a sobriquet for Jiang Qing and others purged after the death of Mao Zedong; current usage refers more to a way of thinking

than to the individuals who were removed from office in October 1976) is that of seeking to keep China weak and backward. See, for example, "'On the Ten Major Relationships' is a Powerful Ideological Weapon to Criticize the Gang of Four" (second in a series of sixteen lectures on Mao's "On the Ten Major Relationships"), Beijing Radio, February 1, 1977, in FBIS, February 10, 1977, pp. E2–7.

5. The goal of transforming China into a powerful, modern, socialist state was first proclaimed in early 1977, but it did not receive wide attention until Hua Guofeng used the phrase in his report at the First Session of the Fifth National People's Congress (NPC) in February 1978. See Hua Guofeng, *supra* note 2; and "The Basic Policy for Building a Powerful Socialist Country is to Mobilize All Positive Factors," Beijing Radio, January 31, 1977, in FBIS, February 9, 1977, pp. E6–10. The "four modernizations" were announced by Zhou Enlai in his report to the Fourth National People's Congress (January 13, 1975). His speech was printed in RMRB, January 21, 1975; and in PR, No. 4 (January 24), 1975, pp. 21–24.
6. The push for full-scale mechanization is described in Yu Qiuli, "Mobilize the Whole Party, Wage a Decisive Battle for Three Years, and Strive for Basic Realization of Agricultural Mechanization" (Summary Report at the Third National Conference on Agricultural Mechanization), RMRB, January 29, 1978, in FBIS, January 31, 1978, pp. E6–25. For discussions of "Chinese style modernization," see Fang Yuan, "Talking Briefly About Studying the Experiences of Agricultural Mechanization Abroad," *Guangming Ribao* (hereafter GMRB), April 29, 1979, excerpted in FBIS, May 23, 1979, pp. L8–9. For a more general discussion of "Chinese style" development, see Sun Yefang, "A Talk on the Need to Reform an Equipment Management System Which 'Reproduces Antiques and Freezes Technical Progress,'" HQ, No. 6 (June), 1979, pp. 24–31; and Zhang Bizhong, "We Must Start from China's Special Characteristics in Carrying Out the Modernization Program," *Gongren Ribao*, May 3, 1979, in FBIS, May 21, 1979, pp. E5–7.
7. See, for example, Zhou Shulian and Wu Jinglian, "Accord Priority Status to the Development of Light Industry," RMRB, August 31, 1979, in FBIS, September 12, 1979, pp. L8–11; "It is Necessary to Handle Properly the Relationship Between Heavy Industry, on the One Hand, and Light Industry, on the Other," Beijing Radio, February 2, 1977, in FBIS, February 10, 1977, pp. E8–12; and Hua Guofeng, "Report on the Work of the Government" (Delivered at the Second Session of the Fifth NPC, June 18, 1979), RMRB, June 26, 1979, in *Beijing Review*, No. 27 (July 6), 1979, pp. 5–31. Also see, Thomas Fingar, Developments in *PRC Science and Technology Policy: January–June 1979*, especially pp. 21–23. (Number 10 in a series of summaries published by the United States-China Relations Program at Stanford University. Future references will cite the author, *Developments*, and the period covered.)
8. See Fang Yan, "Light and Textile Industries Must Be Developed at a Relatively Fast Rate," RMRB, May 25, 1979, in FBIS, June 21, 1979, pp. L18–21.
9. Editorial, "Light Industry Must Develop More Rapidly," RMRB, February 20, 1979, in FBIS, February 23, 1979, pp. E8–11
10. Editorial, "Light Industry Must Grow by a Relatively Big Margin," RMRB, December 3, 1979, in FBIS, December 19, 1979, pp. L10–14.
11. Hua Guofeng, "Report . . . ," *supra* note 7, pp. 15–16; and Fingar, *Developments: January–June 1979*, pp. 23–25.
12. Hua Guofeng, "Report . . . ," *supra* note 7, pp. 11–13; and Zhang Jingfu, "Report on the Final State Accounts of 1978 and the Draft State Budget for 1979," RMRB, June 30, 1979, in FBIS, July 3, 1979, pp. L6–19, especially p. L18. Also see the Beijing *Xinhua* broadcast of July 26, 1979, in FBIS, July 30, 1979, pp. L6–8.
13. Commentator, "Appraise and Treat Urban Collective Ownership Correctly," RMRB,

August 4, 1979, in FBIS, August 23, 1979, pp. L10–16; Liaoning Radio, August 26, 1979, in FBIS, August 30, 1979, p. S6; and Qinghai Radio, December 14, 1979, in FBIS, December 19, 1979, p. T1.

14. See the discussion in John W. Lewis, "China's Military Doctrines and Force Posture," in Fingar, editor, *supra* note 3, pp. 147–198; and Jonathan Pollack, "The Modernization of National Defense," in Richard Baum, editor, *China's Four Modernizations: The New Technological Revolution* (Boulder: Westview, 1980), pp. 241–261.
15. See Lewis, *supra* note 14, pp. 180–181, and sources cited therein.
16. See Fang Yi, "Report at the National Science Conference," RMRB, March 29, 1978, in FBIS, March 29, 1978, pp. E1–22; and Thomas Fingar and Genevieve Dean, *Developments: January–March 1978.*
17. See, for example, Qian Xuesen, "Science and Technology Must Catch Up With and Surpass Advanced World Levels Before the End of This Century," HQ, No. 7 (July, 1977, pp. 14–18, in FBIS, July 8, 1977, pp. E2–5.
18. I have discussed this point in "Recent Policy Trends in Industrial Science and Technology," in Baum, editor, *supra* note 14, pp. 61–101.
19. See Fang Yi, *supra* note 16.
20. For information on science exchange agreements and on PRC technology imports in recent years, see issues of *China Aktuell* (Hamburg); and *China Business Review*. Also see "Western Technology in the People's Republic of China," in U.S. Congress, Office of Technology Assessment, *Technology and East-West Trade* (Washington, D.C.: Government Printing Office, 1979), pp. 245–283.
21. Typical assessments include Kang Shi'en's speech at the Beijing rally for "Energy Conservation Month," reported by Beijing Radio, October 31, 1979, in FBIS, November 1, 1979, pp. L12–18; and Chen Xi, Huang Zhijie, and Xu Junzhang, *supra* note 1.
22. See, for example, *ibid.*, especially p. L9; and Commentator, "Strengthen Centralized Management Over Electric Power, Increase Generation and Save Electricity," RMRB, February 27, 1980, in FBIS, February 28, 1980, pp. L10–12.
23. Kang Shi'en, *supra* note 21; Chen Xi, Huang Zhijie, and Xu Junzhang, *supra* note 1; and former Minister of Coal Xiao Han's comments on the development of coal production, Beijing Radio, January 3, 1978, in FBIS, January 5, 1978, pp. E9–11.
24. See, for example, the *Kyodo* report on Prime Minister Ohira's meeting with Premier Hua Guofeng, December 8, 1979, in FBIS, December 10, 1979, p. D1.
25. The "readjustment" strategy announced at the Second Session of the Fifth NPC in June 1979 called for concentration of investment in all areas, not just in the energy sector. See Hua Guofeng, "Report . . . ," *supra* note 7, pp. 11–13; and Zhang Jingfu, *supra* note 12, p. L18.
26. See Hua Guofeng, "Report . . . ," *supra* note 7, p. 12; Editorial, "Existing Enterprises Are the Base for Modernization," RMRB, May 10, 1979, in FBIS, May 15, 1979, pp. L4–6; and "Assist Older Enterprises to Transform Themeslves," RMRB, May 12, 1979, in FBIS, May 15, 1979, pp. L7–8.
27. See the discussion of energy-related measures in Yu Qiuli, "Report . . . ," *supra* note 1, especially pp. L19–20.
28. This is implied in Kang Shi'en, *supra* note 21, p. L13. Also see Kevin Fountain, "The Development of China's Offshore Oil," *China Business Review* (January–February 1980), pp. 23-36. For a more modest projection of PRC export potential, see Kim Woodard, "China's Energy Prospects," *Problems of Communism* (January–February 1980), pp. 61–67. China exported approximately 12.7 million metric tons of oil in 1979.
29. See generally Vaclav Smil, "China's Energetics: A System Analysis," in U.S. Congress, Joint Economic Committee, *Chinese Economy Post-Mao* (Washington, D.C.: Government Printing Office, 1978), pp. 323–369.

30. See Fang Yi, *supra* note 16, pp. E9–10; Zhu Shida, "Efficiency: Key to China's Energy Policy," Beijing *Xinhua* broadcast, January 12, 1980, in FBIS, January 14, 1980, pp. L1–2; and the recent call for research on nuclear power issued by Jiang Shenjie at the First Congress of the Chinese Nuclear Society, Beijing *Xinhua* broadcast, February 23, 1980, in FBIS, February 25, 1980, pp. L2–4.
31. That is the implication of Hua Guofeng, "Report . . . ," *supra* note 7, p. 12; and Beijing *Xinhua* broadcast, January 20, 1980, in FBIS, January 22, 1980, pp. L2–3.
32. Hua Guofeng announced plans to build nuclear power plants in his address at the First Session of the Fifth NPC (*supra* note 2, p. 27) and orders were placed for two French plants in late 1978. This order was cancelled in July 1979 (see *Asian Wall Street Journal Weekly*, July 23, 1979, p. 6). A new call for nuclear power was issued in early 1980 (*Xinhua*, February 23, 1980, *supra* note 30).
33. Chen Xi, Huang Zhijie, and Xu Junzhang, *supra* note 1, p. L7.
34. Yu Qiuli, "Report . . . ," *supra* note 1, p. L19.
35. Ibid., p. L20.
36. Ibid. Reduced supplies are implied in Kang Shi'en's warning to firms which fail to use resources efficiently. See Kang Shi'en, *supra* note 21, p. L17.
37. I have discussed this point in "Introduction: The Quest for Independence,' in Fingar, editor, *supra* note 3, pp. 3–8; and in "China's Quest for Technology: The Implications for 'Arms Control II,'" in John Barton and Ryukichi Imai, editors, *Arms Control II* (Cambridge: Oelgeschlager, Gunn and Hain, forthcoming).
38. See, for example, Qian Xuesen, *supra* note 17; Commentator, "Get Mobilized and Accelerate the Modernization of Science and Technology," HQ, No. 7 (July), 1977, pp. 3–5, in FBIS, July 5, 1977, p. E5; and Joint Editorial, "Foster Lofty Ideals, Set High Goals and March for the Modernization of Science and Technology," RMRB, March 18, 1978, in FBIS, March 20, 1978, pp. E14–17.
39. See, for example, Hua Guofeng, "Unite and Strive . . . ," *supra* note 2, p. 14.
40. For example, Xin Guangmin, "Reading a Speech by Stalin," GMRB, November 12, 1978, in FBIS, November 20, 1978, pp. E26–28.
41. Qian Xuesen, *supra* note 17.
42. Elements of the "Maoist" strategy are sympathetically described in John G. Gurley, *China's Economy and the Maoist Strategy* (New York: Monthly Review Press, 1976). Also see the special issue entitled "China's Road to Development," *World Development* (July–August 1975).
43. Hua Guofeng, "Report . . . ," *supra* note 7, pp. 8–9. Also see Yu Qiuli, "Mobilize . . . ," *supra* note 1, pp. 16–18.
44. See, for example, Xi Zhengyong, "Introduction to Quality Control in Foreign Countries," GMRB, October 21, 1978, in FBIS, November 2, 1978, pp. E12–14.
45. See, for example, Zhang Qihui, "How Do Romanian Factories Effect Net Output Value?" RMRB, June 25, 1979, in FBIS, July 7, 1979, pp. H1–2.
46. On the "superiority of socialism," see, for example, Commentator, "It is Necessary to Uphold the Four Principles to Achieve the Four Modernizations," HQ, No. 5 (May), 1979, pp. 11–15, in FBIS, May 22, 1979, pp. L1–6.
47. On the 1950s, see, for example, Jan S. Prybyla, *The Political Economy of Communist China* (Scranton: International Textbooks, 1970), chapters 4–8; and Thomas Fingar, *Politics and Policy Making in the People's Republic of China, 1954–1955* (unpublished Ph.D. dissertation, Stanford University, 1977).
48. For a discussion of the Second Five Year Plan, see Zhou Enlai, "Report on the Proposals for the Second Five-Year Plan for Development of the National Economy" (September 16, 1956), in *Communist China, 1955–1959* (Cambridge: Harvard University Press, 1962), pp. 216–241.

49. Although it is true that both explicit and implicit strategies do shape policy, "strategy" often emerges *post hoc* as the summation of actions and decisions taken without regard for overarching guidelines or fundamental principles. Much of what appears to be purposive and coordinated to the outside observer may, in fact, represent little more than the aggregation of sequential events. China does have a "strategy" of development, but it is as much the product as the determinant of sequential policy decisions.
50. Commentator, "It is Necessary . . . ," *supra* note 46.
51. Although doctrinal underpinnings have been refined in recent months, current policies date from early 1977. See, for example, Thomas Fingar and Genevieve Dean, *Developments: January–March 1977*, and references cited therein.
52. See, for example, "Comprehensively and Accurately Understand the Party's Policy on Intellectuals," RMRB, January 4, 1979, in FBIS, January 5, 1979, pp. E7–13.
53. See the discussion and sources cited in Fingar and Dean, *Developments: July–September 1977*, pp. 15–17; *Developments: October–December 1977*, pp. 13–20; and *Developments: July–December 1978*, pp. 22–24.
54. Yao Zhuang, "The Positive Role of the Law on Joint Ventures Using Chinese and Foreign Investments," RMRB, July 24, 1979, in FBIS, August 8, 1979, pp. L8–11.
55. See, for example, Xue Muqiao, "How Can We Effect Planned Management of the National Economy," RMRB, June 15, 1979, in FBIS, June 20, 1979, pp. L10–16; and Yu Guangyu, "Certain Economic Policies and Theories are Worth Exploring," Beijing *Xinhua* broadcast, April 9, 1979, in FBIS, April 12, 1979, pp. L8–10.
56. Xue Muqiao, *supra* note 55; and "Appraise and Treat . . . ," *supra* note 13.
57. Xue Muqiao, *supra* note 55. Recent statements are very similar to those made—often by the same spokesmen—in the mid-1950s. Dorothy Solinger (University of Pittsburgh) is preparing an article on this point and was kind enough to share with me a preliminary draft with the working title "Economic Reforms Via Reformulation: Where Do Rightest Ideas Come From?"
58. On the relationship between performance and legitimacy, see Thomas Fingar, "China's Quest for Technology . . . ," *supra* note 37; and Frederick C. Teiwes, "The Legitimacy of the Leader in China: Mao, Hua, and the Shifting Bases of Authority," paper prepared for the California Regional Seminar, Berkeley, Calfiornia, March 15, 1980.
59. See, for example, Hu Qiaomu, "Act in Accordance with Economic Laws, Step Up the Four Modernizations," RMRB, October 6, 1978, in FBIS, October 11, 1978, pp. E1–22; and Xue Muqiao, *supra* note 55.
60. Ibid.
61. See, for example, the *Xinhua* broadcast on the State Economic Commission's December 2nd Circular to Industrial and Transportation Departments, December 6, 1979, in FBIS, December 13, 1979, pp. L4–7.
62. Examples include "It is Necessary to Bring into Full Play the Initiative of Both Central and Local Authorities," Ninth Beijing Radio Commentary on Mao's "On the Ten Major Relationships," February 8, 1977, in FBIS, February 24, 1977, pp. E1–5; and Hua Guofeng, "Unite and Strive . . . ," *supra* note 2, p. 25.
63. Hua Guofeng, "Report . . . ," *supra* note 7, p. 14; and "Central Committee Decision on Some Questions Concerning the Acceleration of Agricultural Development" (September 28, 1979), RMRB, October 6, 1979, in FBIS, October 25, 1979, Supplement, pp. 1–18.
64. Editorial, "It is Necessary to Give More Authority to Enterprises," RMRB, February 19, 1979, in FBIS, February 26, 1979, pp. E9–12; and Qing Wen and Xu Guang, "How Can the Decision-Making Power of Enterprises be Expanded?" RMRB, June 4, 1979, in FBIS, June 21, 1979, pp. L16–18.
65. See, for example, Li Xiannian, "Speech at the Opening Session of the National Conference on Learning from Daqing and Dazhai in Finance and Trade Work," RMRB, June

27, 1979, in FBIS, June 28, 1978, pp. E1–11; Hu Qiaomu, *supra* note 59; and Ji Di, "Bring About Industrial Modernization," RMRB, May 23, 1978, in FBIS, June 1, 1978, pp. E13–15.

66. This theme was revived in late 1976 with the first official publication of Mao Zedong's 1956 speech entitled "On the Ten Major Relationships." That speech was published in RMRB, December 26, 1976, and in PR, No. 1 (January 1), 1977, pp. 10–25.
67. See, for example, "Three People's China," *The Economist*, December 31, 1977, pp. 13–42.
68. See, for example, the *Xinhua* broadcast on Li Xiannian's remarks to the National Forum of Directors of Commerce Bureaus, August 8, 1979, in FBIS, August 9, 1979, pp. L5–6.
69. Commentator, "Collectively Owned Enterprises Have Real Vitality," RMRB, July 30, 1979, in FBIS, August 7, 1979, pp. L4–7.
70. Ibid.
71. Hua Guofeng, "Unite and Strive . . . ," *supra* note 2; and Editorial, "It is Necessary to Greatly Develop Commune and Brigade-Run Enterprises," RMRB, April 4, 1978, in FBIS, April 6, 1978, pp. E9–13.
72. Cautious injunctions against trying to do everything at once began to appear in early 1978, but the point was not emphasized for another year. See Editorial, "In Capital Construction It is Necessary to Concentrate our Forces and Fight a Battle of Annihilation," RMRB, May 9, 1978, in FBIS, May 10, 1978, pp. E1–5; and Editorial, "Make Adjustments in the Course of Advance, Advance in the Course of Making Adjustments," RMRB, March 24, 1979, in FBIS, March 27, 1979, pp. L10–14.
73. Editorial. "Existing Enterprises . . . ," *supra* note 26; and "Increase Production by Practicing Economy," *Xinhua* broadcast, May 12, 1979, in FBIS, May 17, 1979, pp. L9–11.
74. Editorial, "Firmly Grasp Energy Conservation as a Matter of Great Importance," RMRB, November 1, 1979, in FBIS, November 7, 1979, pp. L8–11; and Yu Qiuli, "Report . . . ," *supra* note 1, p. L19.
75. In his report to the Second Session of the Fifth NPC, Hua Guofeng revealed that twenty-four percent of state industrial enterprises were operating at a loss. See Hua Guofeng, "Report . . . ," *supra* note 7, p. 12.
76. See, for example, Huo Shilian, "Continue to Firmly Grasp and Implement the Party's Policy on Cadres," HQ, No. 5 (May), 1978, pp. 19–23, in FBIS, May 22, 1978, pp. E5–12; "Boldly Promote Professionals Who Are Knowledgeable in Technology to Serve in Leading Groups," RMRB, March 21, 1979, in FBIS, March 22, 1979, pp. L8–12; and "Comprehensively and Accurately . . . ," *supra* note 52.
77. Jin Ying, "Commenting on the Social Autocracy of Lin Biao and the 'Gang of Four,'" GMRB, February 11, 1979, excerpted in FBIS, February 28, 1979, pp. E20–22.
78. See, for example, Commentator, "Combine the Import of Advanced Technology with the Spirit of Self-Reliance," RMRB, June 29, 1979, in FBIS, July 6, 1979, pp. L20–21. I have discussed the view of PRC leaders on questions of technology transfer in "Transferring Technology to China: Patterns, Prospects and Policy Implications," U.S. Congress, House Committee on Science and Technology, Hearings before the Subcommittees on Science, Research and Technology, and on Investigation and Oversight, November 13–15, 1979
79. This point is discussed in Fingar, *Developments: January–June 1979.*
80. See "Existing Enterprises . . . ," *supra* note 26; and Zhang Jingfu, *supra* note 12, especially pp. L15–16.
81. See the *Xinhua* broadcast on an interview with an official of the State Economic Commission, November 18, 1979, in FBIS, November 19, 1979, pp. L14–15; Commentator, "Resolutely Curtail the Consumption of Fuel Oil and Practice Economy in Oil," RMRB,

September 12, 1979, in FBIS, September 25, 1979, pp. L5–7; and the *Xinhua* report of a press conference given by Yang Bo, February 8, 1980, in FBIS, February 12, 1980, p. A4.
82. This seems to be one of the implications—and motivations—of the meeting on small hydropower stations convened by the Ministry of Water Conservancy, the People's Bank of China, and the Agricultural Bank of China (but not the Ministry of Power). See the *Xinhua* broadcast of January 20, 1980, excerpted in FBIS, January 22, 1980, pp. L2–3.
83. Yu Qiuli, "Report . . . ," *supra* note 1, p. L18.
84. See the inventory of resources and list of sources in Smil, *supra* note 29.
85. See the *Xinhua* broadcast report of an address by Gao Yangwen (Minister of Coal), March 3, 1980, in FBIS, March 5, 1980, pp. L4–5.
86. Smil, *supra* note 29, p. 347.
87. See *Xinhua* English release, December 31, 1979, in FBIS, December 31, 1979, p. L13. It should be noted that this is almost exactly the figure (106 mmt) predicted by Yu Qiuli in his report at the Second Session of the Fifth NPC (June 21, 1979). See Yu Qiuli, *supra* note 1, p. L19.
88. *Xinhua*, December 31, 1979, *supra* note 87.
89. See Smil, *supra* note 29, pp. 360–361; and the *Kyodo* report of January 23, 1980, in FBIS, January 23, 1980, p. D3. The amounts pledged for 1979 were 7.5 mmt to Japan and .73 mmt to Thailand.
90. Smil, *supra* note 29, p. 347.
91. See Smil, *supra* note 29, p. 348; and the *Xinhua* English release of November 18, 1979, in FBIS, November 19, 1979, pp. L14–15.
92. Smil, *supra* note 29, pp. 331–332.
93. I have discussed an example in "Pursuit of Political Interest: Scientists, Policy Makers, and Policy Making in the PRC," paper presented at the Workshop on the Pursuit of Political Interest in the People's Republic of China sponsored by the Joint Committee on Contemporary China, Ann Arbor, Michigan, August 10–17, pp. 19–22.
94. Commentator, "Resolutely Curtail . . . ," *supra* note 81 (translation corrected to correspond to Chinese text).
95. For an early commentary on energy bottlenecks, see "After the Masses Were Informed of Difficulties," RMRB, March 11, 1977. Also see the *Xinhua* release of December 18, 1977 (reporting on the National Conference on Power), in FBIS, December 19, 1977, pp. E2–3.
96. PRC commentators have discussed U.S. energy problems. See, for example, Bao Jinhua, "U.S. Energy Problems and Energy Policy," RMRB, February 5, 1980, in FBIS, February 21, 1980, pp. B2–4.
97. See, Xu Shoubo, "A Proposal for Initiating Scientific Research on Energy Policy," RMRB, March 3, 1980, in FBIS, March 14, 1980, pp. L21–23; and Yu Qiuli, "Report . . . ," *supra* note 1, p. L20.
98. See "Energy: A Bottleneck in China's Industrial Drive," *Business Week*, May 19, 1980, p. 59; Lo Bing, "Yu Qiuli Deprived of Part of His Power," *Cheng Ming*, No. 33 (July 1) 1980, pp. 7–8, in FBIS, July 9, 1980, pp. U3–6; and the *Xinhua* release of August 26, 1980, in FBIS, August 26, 1980, pp. L1–2.
99. See, for example, Kang Shi'en *supra* note 21; and the *Xinhua* report of September 5, 1979, in FBIS, September 6, 1979, p. L6.
100. See the *Xinhua* report of Yang Bo's press conference in Geneva, *supra* note 81.
101. Increasing emphasis on short-term problems and short-term solutions is discussed in Fingar, *Developments: January–June 1979.*
102. Chen Xi, Huang Zhijie, and Xu Junzhang, *supra* note 1, p. L7.

103. Kang Shi'en, *supra* note 21, pp. L12–13.
104. Beijing *Xinhua* broadcast, January 25, 1980, in FBIS, January 30, 1980, p. L1.
105. Kang Shi'en, *supra* note 21, pp. L15–16.
106. Ibid.
107. Ibid; and Editorial, "Firmly Grasp . . . ," *supra* note 74.
108. Commentator, "Resolutely Curtail . . . ," *supra* note 81.
109. Some people are at least thinking about the problem; see, Xu Shoubo, *supra* note 97.
110. See Barry Kramer, "China Seeking Help From Foreigners in Effort to Explore for Inland Oil," *Asian Wall Street Journal Weekly*, February 25, 1980, p. 2.
111. For a similar assessment, see Smil, *supra* note 29, pp. 339–341.

SELECTED BIBLIOGRAPHY

Much has been written about energy in China and a comprehensive bibliography would run to several pages. However, many articles present esentially the same information and some of the older pieces were prepared at a time when political rhetoric and fragmentary statistics substituted for hard data. The items listed below provide excellent summary analyses as well as more extensive lists of data compilations and secondary analyses.

Vaclav Smil. "China's Energetics: A System Analysis," in Joint Economic Committee, *Chinese Economy Post-Mao*. Washington: Government Printing Office, 1978.

———. *China's Energy: Achievements, Problems, Prospects*. New York: Praeger, 1976.

Bobby A. Williams. "The Chinese Petroleum Industry: Growth and Prospects," in Joint Economic Committee, *China: A Reassessment of the Economy*. Washington: Government Printing Office, 1975.

Kim Woodard. *The International Energy Relations of China*. Stanford: Stanford University Press, forthcoming.

Chapter 29
Preliminary Energy Sector Assessments of Jamaica

Herbert C. Yim

INTRODUCTION

Jamaica's principal source of energy is imported oil. Foreign oil accounts for approximately 99 percent of the country's commercial energy consumption. There is little likelihood that this dependence will decrease during the early 1980s.

Jamaica's oil imports rose in price from an average of U.S.$61.6 million in 1971–73 to U.S.$195.2 million in 1974 and to U.S.$225.6 million in 1977. Based on end-use forecasts, the 1982 oil import bill is estimated to be U.S.$357 million. Further, Jamaica has no known coal or gas deposits. Test drilling for oil since the 1960s has proved unsuccessful. The only commercially exploitable and proven resource of solid fuel is swamp peat. These reserves are sufficient to generate about 80 MW of electric capacity for 30 years. Renewable energy resources have yet to be examined in sufficient detail to accurately determine future potential.

Jamaica and approximately 80 other oil-importing developing countries are hard hit each time there is an OPEC oil price increase. Unlike developed countries, Jamaica and such oil-dependent developing countries are affected two ways. They must not only pay higher prices for imported oil but they must also pay higher prices for imported goods, services, and technologies.

Among developing countries, Jamaica is one of the highest in per capita energy consumption. As a consequence, in an environment of rapidly escalating oil prices, Jamaica's total economic situation has become increasingly desperate. Directly due to the oil crisis and the deteriorating economic situation, Jamaica faced an exchange rate devaluation which led to the following situation in 1978:

- Average real wage had declined by 30 percent by the end of the year.
- Real per capita personal consumption was 13 percent below the previous year.
- Real GDP continued to decline at 1.7 percent.
- Unemployment climbed steeply from 23 to 26 percent between April and October.
- There was a widening trade gap and the balance of payment deficit grew because the Jamaican economy became burdened with foreign exchange costs for imported energy.
- The average level of consumer prices increased steeply and was approximately 35 percent higher than the previous year.
- The average price of electricity per kilowatt hour increased by 33 percent across the board to U.S. 11.61 cents. The residential users bore the brunt of this because of the regressive electric utility rate structure of the public utility.

In response to the deteriorating economic situation, the government of Jamaica (GOJ) declared a state of economic emergency in January 1977. The Five-Year Development Plan (1978–83) was prepared, adopted, and implemented for all sectors of the economy in an attempt to control and reverse the economic downturn. The separate Energy Sector Plan, a by-product of the energy crisis, formed the basis for the government's strategy for solving the energy problem. Needless to say, continued imported oil price increases since 1973 have made Jamaica's economic recovery and growth objectives exceedingly difficult to achieve.

Despite the difficulties encountered, Jamaica has an energy plan that is being implemented on a country-wide basis. Although in a difficult situation when compared to many of the other oil-importing developing countries, Jamaica appears to be ahead of many such countries in the energy planning phase.

Various developed countries, multilateral assistance agencies, and international financial institutions are assisting Jamaica in various ways to achieve energy independence. To be effective, such assistance, must be clearly focused, properly directed, and well integrated. The U.S. Agency for International Development (USAID) energy assessment of Jamaica is one step towards such an integrated approach.

In conjunction with the government of Jamaica, USAID sponsored this particular study for the following reasons:

- A national energy sector plan must be further developed in order for Jamaica to survive and prosper.
- There is a sense of urgency requiring innovative ways to help shorten

the gap between the study phase and the implementation phase for viable energy projects.
- USAID's Jamaica Assessment represents a potential study model for replication in other developing countries needing U.S. assistance.

Based on the above considerations, USAID, in conjunction with the various energy-related agencies of the government of Jamaica, identified a number of energy areas for technical assessment and potential development. Recognizing that the energy assessments could not cover all possible energy areas due to urgent schedules and budgetary constraints, the assessment was referred to as the "Preliminary Energy Sector Assessments of Jamaica" and included the following eight tasks:

1. Study integration and project management
2. Economic assessment
3. Solar energy—commercial and industrial
4. Solar energy—agricultural
5. Biogas applications
6. Energy conversion from waste
7. Coal prefeasibility study
8. Electric utility rate analysis

A brief description of each of the eight study tasks follows.

Study Integration and Project Management. The project management team was responsible for the overall management of the study. The management team developed a project management and detailed study plan that coordinated and guided individual study development and provided for periodic study reviews. Study integration and the preparation of the final reports were also the responsibility of the project management team.

Economic Assessment. The objective of this study was to integrate the specialized study results into a combined energy program. The study was designed to prioritize energy implementation programs so that the greatest gains could be realized in the shortest time. All recommended programs would be reflected in Jamaica's Five-Year Development Plan and Energy Sector Plan.

Solar Energy—Commercial and Industrial. The objective of this study was to provide a technical and economic assessment of the feasibility of near-term applications of solar water heating in the commercial and industrial sectors. Various specific feasibility studies of commercial solar air-conditioning and solar hot water applications were requested by the Ministry of Mining and Natural Resources.

Solar Energy—Agricultural. The objectives of this study were to assess the possibility of using solar energy in Jamaica for drying agricultural products, and to identify the technical and economic needs for making the application of solar energy a reality. The assessment for drying included corn, peas, beans, peanuts, onions, pepper, coffee, cocoa, turmeric, cassava, timber, and fish.

Biogas Applications. The objectives of this study were to assess the feasibility of biogas generation in Jamaica and to recommend projects that might be undertaken. Included were energy and economic evaluations, as well as a review of low-cost facility designs to determine which would be most suitable for the conditions which prevail in the country. The study also addressed the feasibility of generating biogas as an integral part of the sewage treatment facilities of the Mona Campus of the University of the West Indies (UWI) in Jamaica.

Energy Conversion from Waste. The objective of this study was to assess the feasibility of a prototype demonstration unit for energy recovery from the solid wastes generated on the Mona Campus of UWI.

Coal Prefeasibility Study. The objective of this study was to determine if the use of steam coal for electric power and steam generation appears significantly attractive, both technically and economically, to justify a full-scale feasibility study. The study assessed the technical, economic and environmental feasibility of diversification from oil-fired units to coal. The possibility of retrofitting existing units to burn 3-percent sulfur coal or a coal-oil mixture was also investigated. Since any benefits realized will accrue to both the alumina companies and the Jamaica Public Service Company (JPS), plant sites and harbors were considered during the analysis of coal transportation, coal handling, and storage facilities, and the infrastructure required to support coal operations. Price projections for coal and the potential for obtaining long-term coal supply contracts were also addressed.

Electric Utility Rate Analysis. The objective of this study was to conduct an examination of the JPS electric utility rate structure with a view toward determining whether it (a) promotes an economically efficient use of electricity in the society, (b) is consistent with basic principles of equity and fairness, and (c) provides JPS with sufficient resources to maintain an acceptable quality of service.

Areas Not Included in the Study

Because they either were already being studied by other foreign entities or they were not prospective candidates for immediate study, the following energy areas were not included in the study:

- energy conservation
- wind
- hydropower
- geothermal
- photovoltaics
- peat
- ocean thermal
- bagasse
- charcoal

Contract Award

Energy Systems International (ESI) of McLean, Virginia was awarded the contract to conduct the "Preliminary Energy Sector Assessments of Jamaica." The contract duration was from August 10, 1979, to January 31, 1980, of which fifteen weeks were to be spent in Kingston, Jamaica, by the energy team.

BACKGROUND INFORMATION

Due to the urgency of the study to meet milestone objectives of the GOJ Five-Year Energy Sector Plan, and also due to the fact that this particular study represented USAID's first major overseas energy assessment of a developing country, USAID focused special attention on a novel proposed study approach. It became apparent that due to the diverse technological requirements and high degree of specialization required by each of the studies, the formulation of a United States energy team of experts would best meet the study's stringent demands. It was imperative that prospective team members have the proper balance of academic and hands-on project experience so that the studies would have near-term practical value. Team members were expected to

- be recognized experts in their fields of speciality
- be highly recommended by appropriate agencies
- have appropriate hardware experience
- have applicable international consulting experience
- be available for an extended in-country assignment
- be strong team players
- be willing to establish close working relationships with their host-country counterparts
- recognize the importance of study integration

After extensive screening and interviewing of candidates, primary and back-up teams were formed. All team members fully understood that special emphasis was to be given to employing general systems engineering/technical direction techniques to mold a diverse group of technical experts into a cohesive project team able to work in close cooperation with all Jamaican counterparts.

The team members selected for the assessment are eminently qualified in their fields of specialty and came highly recommended from various government agencies or institutions.

STUDY METHOD

Project Management

To provide program coordination and direction, a project management and detailed study plan were developed early in the study. The plan was prepared in three distinct parts:

1. Project management—specifying team responsibilities, working relationships
2. Team organization—delineating technical requirements
3. Detailed study plan—depicting study schedules, responsibilities, and task completion requirements

Before the arrival of the individual U.S. energy team specialists, the project management staff conducted a series of interviews with key personnel and Jamaican team counterparts associated with each study. During the discussions, the contract's original terms of reference were updated if required; current program activities were reviewed for continuity and applicability to each study task; interface organizations and contacts were identified; and key study issues and areas of concern were highlighted.

Throughout the study, the project management staff continued to develop the study plan and to involve the Jamaican counterparts in all scheduled events. A key event during each of the specialized studies was a training seminar given by each U.S. team member. In addition to technology transfer, the seminars provided the opportunity to stimulate Jamaican private sector/GOJ agency/U.S. energy team interface and interaction, as well as joint review of current Jamaican program plans and policies.

In addition to the seminars, progress reviews were held for each specialized study. Each specialized study's scheduled assessment reviews—preliminary, intermediate, and final—were well attended by all Jamaican study participants and counterparts.

Before the energy team's departure from Jamaica, a final report conference was held at the Jamaica Pegasus Hotel on November 13 and 14, 1979.

The conference encouraged the participation of the Jamaican public and private sectors, USAID, and many international organizations. Its objective was to present a joint forum on the results of the Jamaica Preliminary Energy Sector Assessments, and it was very well attended and highly successful. The results of the seven specialized studies were presented by the U.S. energy team specialists and their Jamaican counterparts on the first day. The second day was reserved for splinter discussion groups in which the team members fielded questions from the conferees.

Specialized Studies

Due to the diverse requirements associated with each of the specialized studies, a completely standard methodology or study approach was not possible. In each case, however, similar activities were undertaken during the early stages of the study. Each of the reports details these activities, but in general the following areas were covered during study development:

1. Personnel interviews—key personnel from all sectors pertaining to the specific study were interviewed.
2. Sites visited—specific site visits were conducted, and others were added as required, to establish the data base.
3. Surveys—two studies, solar energy and biogas, required the use of surveys to determine market potential.
4. Pertinent data—pertinent data for team use were gathered from all participating government agencies and private organizations.
5. Plans, programs and policies—all existing plans, programs, and policies were reviewed for compatibility prior to the development of recommendations.

SUMMARY OF MAJOR FINDINGS AND CONCLUSIONS

Economic Assessment

Almost 4.2 percent of Jamaica's energy imports can be replaced by solar water heating systems (3.5 percent) and biogas systems (.7 percent) if exploited in accordance with the study recommendations. If the net fuel savings of a coal-fired electrical generating plant are included, the potential savings increase to 7.3 percent of energy imports. If the new electrical utility tariff structure that JPS adopts results in even a 1 percent savings in fuel, the potential savings of the program amount to 8.3 percent. Using the 1978 figure of 8,887,200 barrels (net) f.o.e. (nonbauxite energy consumption) for Jamaica's consumption of imported fuel, 737,637 barrels f.o.e. could be saved annually.

A total economic assessment of each of the specialized studies was conducted. The implementation costs were estimated and the effects of the recommended actions on Jamaica's economy measured. Finally, a cost/benefit analysis was completed to test program potential and feasibility.

A final, comprehensive set of integrated recommendations was then chosen for inclusion in a combined energy program. Groups of highest priority, higher priority, and high priority recommendations were identified.

Perhaps the most surprising aspect of the combined energy program is the modest cost of all these improvements in Jamaica's energy situation. The initial costs (capital costs plus initial operating costs) of the entire program are only approximately (1979) U.S.$6 million.

Solar Energy—Commercial and Industrial

1. Jamaica's high level of solar radiation (450 langleys per day mean) and relatively moderate ambient temperatures provide an ideal climate for the implementation of proven solar technologies.

2. Study results show that the immediate potential is primarily water heating—either domestic water heating or preheating of boiler make-up water. These applications are divided into four categories: residential, commercial (hotels), institutional (hospitals), and industrial.

3. Since all residential and institutional and a number of commercial applications in Jamaica utilize electricity for heating water, these are prime areas to study. Realizing that electricity costs on the average are J.$0.17 per kWh (U.S.$0.10), solar water heaters are economically justifiable.

4. On the other hand, solar cooling, even with high-performance collectors and systems, is not yet economically justifiable.

Solar Energy—Agricultural

1. Jamaica's climate and average solar radiation figures make the collection of heat from solar energy attractive because solar collectors operate at higher efficiencies. Therefore, Jamaica offers ideal conditions for the use of practical, proven, low-cost solar systems for drying agricultural products.

2. In Jamaica, the agricultural sector plays a major role in the country's economy. Crop drying by solar technologies can help to minimize food spoilage in most environments, preserve agricultural products for nonharvest seasons, and reduce consumption of fossil fuels.

3. Important areas in which programs can be immediately implemented include:

1. Establishment of climatic stations
2. Development of air-type solar collectors
3. Research and development on drying of agricultural products

4. Sun drying of agricultural products, and research and development on cost-effective drying methods.

Biogas Applications

1. Jamaica has many conditions which could lead to successful application of biogas as an alternate energy source. Much of the economy is agricultural, and ambient temperatures are high enough throughout the year so that auxiliary heating of biogas digesters would not be necessary. This obviates the need for the costly equipment required in colder climates.

2. Biogas energy can be provided directly to the rural areas of Jamaica, where butane and kerosene are becoming increasingly unavailable and require a substantial portion of the family's budget. A biogas energy program can, therefore, play a significant role in contributing to Jamaica's effort to achieve greater energy independence.

3. Jamaica is behind the Asian countries which are developing this alternate energy source in spite of studies of biogas by The Scientific Research Council (SREC) over the past several years. Designs which have been slow in developing have finally emerged to the point where at least one such design has been field-tested and found to be satisfactory. There is sufficient design development now to proceed with at least an initially modest biogas program. Design will be the key to the degree of ultimate success of the program. Design development should, therefore, be accelerated with SRC in a major role and as much support as is required should be provided so that this responsibility can be accomplished in a timely manner. Jamaica can take advantage of the experience in low-cost design of biogas plants in other countries. One of the major costs is for steel reinforcing bars and gasholders. Some designs, however, such as those used in China, require only local construction materials and little, if any, steel. Use of such a design can result in a family-size unit costing even less than the steel gasholder designs which have been used successfully in large-scale biogas programs in other countries.

4. Many institutions in Jamaica, such as schools and centers for the handicapped, are associated with medium-sized to large farms, which could provide biogas for cooking. This biogas could substitute for fuel costing several hundred dollars per month per institution.

5. On the basis of field surveys, data analysis, and meetings with the Biogas Working Group and other associated personnel, it is concluded that a national biogas program would be successful in Jamaica. Indicators used in arriving at this conclusion were climate, precentage of the population living in rural agricultural areas, animal numbers, and Jamaica's dependence on imported oil and gas.

Energy Conversion from Waste

Urban Solid Wastes—Kingston. 1. The only alternative worthy of consideration for energy recovery from urban solid wastes in Kingston is waterwall incineration with steam generation. This application appears to be cost-effective. The major problem is inefficient and unreliable refuse collection, making it impossible to recover energy from the wastes.

2. It may prove cost-effective to apply the biogas approach in small cities where incineration or RDF are not feasible. This would be especially true in small communities that are constructing a new waste-water treatment plant. In such cases, the sludges from the waste-water and the organic solid wastes could be treated in a common digestion system, especially designed to handle the refuse and sludges.

3. For health, environmental, and energy reasons, a careful study of urban refuse collection in Kingston should be conducted.

4. The proposed field-scale biogas demonstration at the UWI Mona Campus would not be a very meaningful or helpful exercise in terms of proving a technology for wider use in Jamaica.

5. The results of this study lead to the conclusion that the UWI Mona Campus should not get involved in a field-scale demonstration of the anaerobic digestion of campus refuse for methane generation.

6. A viable alternative would be for the UWI Mona Campus to develop an alternative energies laboratory and for the university to be assisted financially in the amount of U.S.$182,500 over the next three years.

Coal Prefeasibility Study

1. The installed system capacity of the existing JPS oil-fired electric power plant is approximately 455 MW, which is greatly in excess of the present 210 MW peak. Thus a new power plant unit is not needed to meet the expected system load at the present time. However, unit outages cause frequent rolling blackouts. A major program is needed to improve unit availability, particularly of the newer, more efficient units. If the four newest units, having a total capacity of 250 MW, could be upgraded to have improved availability and operate at a high capacity factor, there would be a large dual saving without any capital expenditure for a new unit.

2. Existing oil-fired boilers for JPS or the alumina companies cannot be retrofitted to burn coal directly.

3. The installation of a coal-fired unit may seem attractive if a conservative differential oil-coal price escalation rate of only 1 percent exists, and if joint use of harbor facilities and sites can be achieved at Port Esquivel, or at the proposed JBM South Manchester site at Cuckold Point.

4. The possibility of obtaining a favorable long-term coal contract is very good at the present time.

5. The schedules of the bauxite and alumina ships are tight, but the transportation of coal from the U.S. Gulf Coast on the backhaul is feasible.

6. Coal can also be transported by non-U.S. ships at a cost approximating that of the bauxite ships.

Electric Utility Rate Analysis

1. The pricing of electricity should always be given considerable attention no matter what type of generating fuel is used, regardless of the utility firm's operating environment. The situation confronting JPS dictates that an even greater than normal attention be given to the pricing structure of electricity. The size of JPS oil expenditures alone, considering the constant OPEC pressure for price increases, would focus attention on pricing structures which are more closely related to marginal costs than the current pricing structure.

2. Various tariff options, based on marginal cost pricing principles, have been presented for each rate category in the JPS system. These options demonstrate immediate or interim steps that can be taken to gradually approach marginal cost pricing so that the implementation rate of tariff changes can be completely controlled. The effects of each step in tariff development and the associated customer bill impacts are indicated.

3. The following marginal cost pricing principles were used to develop the tariffs for each customer category:

- Total revenue requirements will not be changed.
- Total revenue allocations between customer categories will not be altered.
- Declining block pricing will be eliminated.
- Ratchet provisions will be eliminated.
- Expander tariffs will be eliminated.
- Flat (both all-energy and two-part) demand and energy tariffs will be considered.
- When metering and customer acceptance seem reasonable, time-of-use pricing will be considered.
- The very progressive (from an efficiency stand point) fuel adjustment clause will be continued.
- The innovative cost-of-service adjustment will be retained.
- The residential customer flat rate with the first 10 kWh included free will be continued.

4. The overall conclusion is that JPS and public officials should review these options and develop strategy compatible with economic goals and

plans, social considerations, and other factors. From this an implementation plan should be established to sequence rate tariff changes consistent with Jamaican needs and plans.

SUMMARY OF MAJOR RECOMMENDATIONS

General Economic Assessment

The main recommendation of the economic assessment is that the Combined Energy Program is a sensible way for Jamaica to direct its efforts in the energy sector over the next five years and that it should be accepted. The recommendations of the special studies that have been incorporated into this program have been shown to produce benefits well in excess of their costs. The costs themselves have been found to be modest, and the government should be able to find the necessary funding.

It is recommended that immediate action be taken to implement the Combined Energy Program. The government of Jamaica should consider the program without delay and move quickly on the recommendations it accepts. Financing should be sought as soon as possible.

Solar Energy—Commercial and Industrial

It is recommended that a solar water heating program be developed that will be implemented over a three-year period. The program should

1. establish 10 solar radiation stations
2. establish a solar collector certification program
3. eliminate the 37.5 percent water tank tax
4. establish a consumer information office
5. establish training programs for engineers and installers
6. initiate solar hot water feasibility studies in selected building sectors
7. investigate export potential of solar hardware
8. support the JPS residential solar demonstration program
9. establish solar tax incentives
10. initiate solar hot water demonstration projects at hospitals

Solar Energy—Agricultural

It is recommended that a solar agricultural drying program begin immediately and be implemented over a five-year period. The program should

1. develop solar-heated air collectors
2. establish R&D programs for drying agricultural products

3. design and construct solar collector systems
4. provide comprehensive training for the recommended solar agricultural program
5. develop community drying systems

Biogas Applications

It is recommended that a biogas applications program be implemented over a five-year period. This program should

1. implement the Option A program, which consists of 2700 family-size units and 41 medium-size units immediately
2. begin a detailed marketing study to assess commercialization potential
3. examine the feasibility of a large-scale demonstration unit for electric generation
4. explore international funding options for an expanded program
5. begin a comprehensive training and public awareness program.

Energy Conversion from Waste

It is recommended that the following programs be pursued:

1. a detailed study of waste collection problems in Kingston, with an evaluation of other alternatives for energy recovery
2. An Alternative Energies Laboratory at UWI Mona Campus.

Coal Prefeasibility Study

It is recommended that the following actions be taken:

1. The government of Jamaica and JPS should obtain funds to upgrade existing oil-fired electric power plant units.
2. Exploratory negotiations should be made on the use of a joint port facility for a coal-fired power plant at Port Esquivel or Cuckold Point respectively.
3. If adequate cost benefits are realized from recommendation 2, a coal supply and shipping study should be authorized.
4. Coal-oil mixture development progress should be monitored.
5. Coal gasification technology progress should be monitored.

Electric Utility Rate Analysis

As requested by JPS, specific recommendations as to which type of tariff should be adopted are not presented in this report. JPS shall review and analyze the various options and make its own recommendations to the Ministry of Public Utilities.

Chapter 30
Energy Management Training for Developing Countries in the 1980s

T. Owen Carroll

THE ENERGY MANAGEMENT TRAINING PROGRAM FOR DEVELOPING COUNTRIES

During the last two years, the Institute for Energy Research at Stony Brook and the National Center for Analysis of Energy Systems at Brookhaven have been involved in planning and conducting a number of several-month courses dealing with national energy planning and management for developing countries. This Energy Management Training Program (EMTP) was organized early in 1978 as a central part of the ongoing activities of the LDC Energy Program at Brookhaven National Laboratory and the Energy/Development Assessment activities at the Institute for Energy Research at Stony Brook. The work of these groups includes (1) energy technology assessments for renewable resources utilization, (2) the development of methodologies for carrying out national energy assessments, energy assessments, energy planning models, and energy information systems for LDCs, and (3) field studies in LDCs dealing with energy use in urban and rural households.

The course we offer is directed at middle- and upper-level government officials who are now working or expect to work in some aspect of energy planning and management. The course is one of two or three now available to persons from LDCs wishing to receive training in this pivotal area. The others are offered at Saclay in France and the Baraloche Institute in Argentina. All three programs have a very pragmatic bent. In ours we employ lecturers from our own staff, the World Bank, and other U.S. and Third World institutions to train participants to become actively and productively involved in some aspect of national energy planning on their return to their countries. We have also developed problem sets and case studies to reinforce what is learned in the lectures. Topics covered include energy-economic systems analysis techniques, collection and analysis of energy information, assessment of energy resources, energy pricing, and finally, energy project evaluation and design and the preparation of proposals for securing financing. Support for the program comes from the USAID, The aL Dir'iyyah Institute, and several other sources.

The first session of the program was given in October 1978, the second in September 1979, and the third for May 1980. If all goes as planned, we expect to be offering the course twice a year starting in 1981. The response to the course from its inception has been enthusiastic. We received a total of 60 applications for the 30 openings in the first session. Applications for the second session increased to 85, and for the third we have received over 1200. The relevance of what we teach to Third World country concerns is reflected not only in this increasing number of applications, but also in a change in the affiliations of the persons applying to the program. Individuals applying to Session I came mostly from utilities. This shifted to a 50–50 split between utilities and economic planning ministries and central banks in the second session. For Session III, we have received numerous applications from newly set up energy ministries, offices, commissions, etc. All in all we have trained about 60 persons from 35 developing countries.

As part of our training program, our staff has traveled extensively to a large number of LDCs over the past two years. These visits provide us with a basis for recruiting not only the best qualified persons to the training session, but also those who are most likely to occupy key positions in the energy planning hierarchy on their return. At the same time it permits us to develop a hands-on appreciation of the energy concerns of government officials in Third World countries, to learn what new additions should be made to the training program, and finally to make sure that our program is identified as a training program for developing countries that happens to be located in the United States and *not* a U.S. program which has been slanted to the needs of the LDCs. This is an important distinction. Viewed as a program that happens to be located in the United States we can and do take a stand on such controversial points as the need for LDC governments to incorporate the consideration of alternative development strategies in their energy planning process. Being viewed as a U.S. program would require us, for example, to defend U.S. energy use practices against the accusations of energy waste and extravagance.

ENERGY PLANNING ISSUES IN THE DEVELOPING COUNTRIES

If there is one thing we have learned from our prolonged association with the participants and lecturers in the EMTP and our visits to developing countries over the past two years, it is that both the real and the perceived energy problems of the developing countries differ widely throughout the third world. There is no doubt that the crisis in all the oil-importing developing countries arises out of such factors as the severe balance-of-payments deficits caused by the precipitous rises in oil prices, the growing shortages of noncommercial fuels, and the lack of capital for investment in indigenous energy resources. Yet in identifying the context for energy planning training in the developing countries, it is important to distinguish between the lower-income, predominately primary-exporting LDCs like Tanzania, Malaysia, and the Dominican Republic, the small, industrialized countries like Korea, Taiwan, and Thailand, and the large, more industrialized developing countries like Brazil, India, and China.

The energy situation in the Dominican Republic is more or less typical of that to be found in many of the smaller, primary-exporting developing countries. The Dominican Republic relies almost exclusively on imported oil to meet its commercial energy needs, and on sugar and nickel for its export earnings. Recent price increases in oil have outpaced increases in either nickel or sugar prices. At the same time, the steady increase in population combined with increased migration to Santo Domingo have brought about a sharp increase in consumer consumption of commercial fuels. Given the urgency for increasing national economic output, energy use per capita is also increasing as the Dominican Republic government struggles to increase the industrial share of its GNP. At the same time, as in many of these lower income countries, the so-called noncommercial fuels, which of course are no longer actually noncommercial, are disappearing; in fact, in the Dominican Republic, deforestation is so extensive that it is against the law to cut down a tree anywhere. A consequence of this scarcity is that more and more families in urban areas and small industries are turning to the use of commercial fuels such as electricity, out of necessity. However, most of the electricity produced in the Dominican Republic is from oil-fired plants, which further aggravates oil demand.

In the long term, the Dominican Republic, like many other small, primary-exporting developing countries, faces the necessity of making a transition away from oil to, if at all possible, a much greater reliance on indigenous fuels. This transition will require a tremendous investment of capital and human resources. It will also require major changes in the mandate of organizations which have, up to now, dealt with the delivery of energy. Moreover, the sheer size of the undertaking required will reduce the capital resources available for investment in other sectors of the economy. As a result, the traditional patterns of economic and social development in countries like the Dominican Republic, which have aimed at diversifying their export products to make them less sensitive to shifts in commodity prices and putting in place light industries capable of reducing the need to import consumer products, may have to be delayed or at best altered to reduce their energy intensiveness.

In the near term there will be a strong need to allocate oil and other scarce energy fuels, as well as the capital being invested in energy infrastructure, in a more economically efficient manner. This means more government involvement in the pricing of both commercial and noncommercial energy fuels, in controlling the import of devices using energy, and in setting mandatory controls for use of energy-intensive industries and transportation.

While the small industrialized developing countries must address many of the energy planning issues noted above, their combined patterns of rapid industrialization and urbanization over the past decade have already introduced into their economies a number of structural energy-economic linkages that will make it difficult for them to abruptly reduce their relatively high per capita commercial energy consumption without invoking serious economic dislocations, a fact of which most of these governments are already well aware. On the other hand, many of these countries may be able to price their exports to reflect increases in oil imports, and more quickly put into place new conventional and nonconventional energy supplies.

Korea, for example, has moved to install a large number of nuclear reactors and, at the same time, to shift from oil to indigenous coal in industry. At the same time, it is moving rapidly to restrict the use of oil as a consumer good and to mandate increased energy use efficiency in medium heavy industry. Energy planning issues for countries like Korea center around questions such as (1) how changes in the energy price structure will affect their economy, (2) how they can maintain the export market for their manufactured goods and still price products to reflect the increase in the costs of imported fuels, and (3) how they can plan for the long-term global situation of a diminished supply of oil and perhaps uranium.

Both because the economies in these countries are more diversified and the energy supply/demand system itself includes a wider mixture of fuels, these countries have a wider range of options available to them. At the same time, their wider access to capital markets and more extensive supply of skilled manpower allow them to consider a more multifaceted approach to addressing their energy problems.

For the larger newly industrialized countries in the third world, like India, Brazil, and soon China, the primary issue is one of finding the means to quickly exploit their indigenous resources in an attempt to continue their headlong rush toward developing a modern economy. In India and Brazil, the rapid increases in oil import costs have been met by aggressive programs to find the short-term solutions which allow them to maintain growth in the presence of these added costs. India and China are moving to increase the use of coal, and Brazil to produce large quantities of sugar-based alcohol. For the long term, energy supply enhancement issues continue to be the focus of national energy planning activities. All, for example, are investing heavily in exploring for and exploiting oil and natural gas resources. In the near term, energy planning issues are focused on industrial energy conservation and, to a lesser extent, on developing a better basis for a rational energy-pricing policy.

All of these large developing countries possess large rural populations that face shortages of energy because of the declining supply of traditional fuels. This is true even for China, if its population continues to grow and migration to the urban areas continues to be restricted. The problems surrounding the provision of new energy resources to these rural inhabitants are substantially different from those in the urban/industrial sectors. In a sense, they are also more difficult because we know much less about the intervening technologies that will have to be utilized. Nevertheless the planning and management of rural energy delivery systems is bound to become a major preoccupation in these countries over the next decade.

ENERGY PLANNING AND MANAGEMENT TRAINING NEEDS

To deal with the energy-planning and management issues outlined above, people need both scientific and engineering training directed toward conventional and nonconventional resource exploitation and the technologies employed to convert these resources to useful fuel forms, as well as training in the management of national energy systems. The latter includes such subjects as micro- and macroenergy economic analysis, energy demand forecasting, pricing, project evaluation, and monitoring of government actions. While this paper deals primarily with energy management training, the obvious interactions between the two areas in the development of a comprehensive national energy policy require us to comment briefly on developments in the technology areas.

The most pressing need for energy technologists in the developing world, particularly among oil-importing countries, is for persons trained to help find, assess, and exploit conventional fossil resources. This is particularly true for the sizable number of countries that expect to locate usable quantities of oil, but in amounts not sufficient to open up export markets. This makes it very difficult for them to attract foreign companies as partners in exploitation on terms that the governments are likely to find acceptable. Coal technologies will be much in demand. The fact that geologists are predicting that coal of low and medium quality will be found in many areas of the Third World and the fact that there are relatively few experts in this area are both important contributing factors.

The basis for rapidly expanding the supply of fossil fuel technologies is to be found in expanding first the resources of a number of university and technical institutes in the Third World, and then those of the numerous institutions in the advanced and industrial countries. No similar basis exists for expanding the supply of technologists who can deal with the assessment and utilization of nonconventional resources and technologies for increasing energy end-use efficiency, in spite of the fact that both of these areas represent the only options available to many of the oil-importing LDCs if they are to reduce their dependence on oil imports. Moreover, in the case of a number of the nonconventional resources, the training needed does not lend itself to the structures to be found in conventional engineering schools and technical institutes to be found in most developing countries. This is not the place to get into the complexities of the small decentralized versus the large centralized controversy. Nevertheless, it is evident that training large numbers of people in the use of nonconventional energy resources will require the development of a wide range of institutional arrangements.

While it is not our intent in this paper to get into a detailed description of the requirements for training large numbers of energy technologists, it is appropriate to comment on the fact that neither international nor national assistance agencies have undertaken any large programs to support the expansion of training efforts in the Third World. Nor have there been significant efforts to support institutions in the advanced industrial countries who seem willing and able to undertake this kind of training on a large scale. There are a few exceptions, of course. The United Nations is sponsoring a series of short seminars on renewable resource use in the developing world. USAID is supporting a special program at the University of Florida which will offer an intensive fifteen-week course on renewable resource technologies for participants from developing countries. A number of other small efforts are underway at the U.S. Department of Energy and AID. However, we should note that there are serious discussions underway at AID, the UN and several European assistance agencies on funding much larger efforts in this field.

One many argue that the most serious impediment to mobilizing large-scale technology efforts in the developing world to reduce their oil dependence may not be the lack of energy technologists but rather the lack of the tremendous amounts of capital that will be required to support the massive programs that will have to be initiated. If this is true, then the demand for energy technologists in the developing countries may not begin to accelerate until the middle or the end of this decade. But at that time, the demand is likely to far exceed the capacities of the educational institutions in the Third World and the advanced industrial countries to meet it, unless major efforts are made within the next few years to develop centers of excellence at top-rated institutions in the developing world. Institutions from the advanced industrial countries can and should be involved in these undertakings, but only in a supporting role.

Unfortunately, the case we have made above for expecting that a few years may go by before the demand for energy technology training accelerates, does not apply to energy management training. The situation described in the previous section with respect to diminishing energy supplies and expanding demand, in fact, suggests the need for prompt action by large numbers of developing country governments to bring energy demand into balance with the supplies likely to be available without producing severe dislocation in the economies and social structures. This concern for the kinds of energy management and planning decisions will, no doubt, remain at center stage throughout the remaining years of the 1980s and beyond.

If it is clear there is already a strong and urgent need for large numbers of energy management experts in the developing world, it is not so clear how we should go about training the large numbers of people who must quickly be made familiar with appropriate background material, methodologies, and techniques and the specialized knowledge experiences that this training calls for. Table 30.1 classifies the different audiences at whom we see energy management training being directed. Described briefly are the background materials, the methodologies and techniques, and finally a series of special topics we consider appropriate for each group.

There is clearly much to be done and not much time to do it in. In thinking about mechanisms which would allow the kinds of management training we have outlined above to be carried out on a worldwide basis, we have concentrated on trying to maximize the utilization of Third World institutions for at least two very important reasons. First, there is no other practical way of training the large number of people involved, particularly at the lower levels. Second, and perhaps even more important, when it comes to the teaching of energy management for developing countries, it is difficult to separate the political from the analytical dimensions, even if one thought it best to do so. At the same time one must recognize that there are only a handful of institutions in the developing world where engineers, economists, and others have come together to form groups that are working actively in this field. In table 30.2 we describe a number of mechanisms for mobilizing such efforts and the kinds of efforts we believe it will take to train individuals in the subjects noted for the various levels shown in table 30.1

We have repeatedly noted the large audience for these training activities. To estimate roughly how many people are likely to be trained over the next decade, we extrapolate from our work with some 30 countries in recruiting for the EMTP, field work in rural and urban energy system analysis, national energy assessments in selected developing countries, and special seminars and training for individuals from a number of countries.

For a typical small primary exporter like the Dominican Republic, we estimate there are 5–10 officials at the senior level in the various ministries

Table 30.1. Energy Management Training Audiences and Subject Matter

Level I:	Senior government and private sector officials involved directly or indirectly in decisions affecting energy supply and/or demand.
Background:	Global energy situation, national energy situation; recent energy technology developments, energy conservation; constraints attached to changes in national system of energy supply, distribution, and end-use.
Methodologies & Techniques:	Concepts of national energy planning, estimating economic, social, and environmental impacts of alternative energy technologies; revised pricing policies, mandatory controls on energy use; elements of energy infrastructure investment planning.
Special Topics:	Renewable resource utilization; potential for energy savings in industrial and mining operations; policies of development banks and private banks with respect to investments in energy infrastructure; alternatives to conventional rural electrification.
Level II:	Middle- and upper-level staff in energy units, utilities, central-planning ministries responsible for ongoing energy management-related activities.
Background:	Global energy situation and effects on energy management decisions; linkages between energy consumption and growth and development of agricultural, industrial, transportation, government services, and commercial sectors as well as urban and rural development.
Methodologies & Techniques:	Energy systems and economic frameworks, models, information requirements; sources of energy information, surveys, computerized energy information systems; energy audits for estimating conservation potential, costs, and effectiveness of pricing and mandatory controls; quantitative basis for preparing comprehensive energy pricing policies; resource and new technology assessment; design and evaluation of energy projects for securing loans.
Special Topics:	Design of demonstrations for evaluating alternative rural energy delivery systems, energy conservation audits for large commercial and government buildings, maintenance of services, etc. and large-scale agrobusinesses; forecasting electricity demand, monitoring quality of imported refined oil products, increasing production of fuelwood.
Level III:	Junior staff in energy management units seeking graduate degrees in a related area (economics, applied math, statistics, computer sciences, resource assessment, urban planning, etc.).
Background:	Depending on the discipline, students would be required to take several courses dealing with energy management and/or planning topics in some of the other disciplines.
Methodologies & Techniques:	Depending on the discipline, several courses should be available which deal with the application of disciplinary techniques and methods to specific energy issues and problems.
Thesis Topics:	Depending on the discipline, theses should deal with subjects of direct relevance to the energy management issues of concern to the country at present or sometime in the near future.

Table 30.1. (Continued)

Special Comment: It is assumed that these graduate programs will be offered either in a university in the country of the applicant or a nearby regional technical center. If offered in the United States or other advanced industrial countries, it is vital that the faculty involved have had experience in dealing with energy analysis in the developing country context.

Level IV: Professional staff involved in collection data, performing energy audits, serving as extension staff of the energy units in the field to monitor energy usage, offering advice and counsel to users on energy conservation, alternative fuels, and technologies.

Special Comment: These training programs would be offered in-country, if at all possible, or if necessary, at a nearby regional training center set up especially to train energy management professionals. Students enrolling in the courses would be expected to have had a technical education up to the junior college level.

Background: The national energy situation and energy use in major end-use sectors; energy policies; role of the government energy office to which the professional is to be connected; purposes of energy management; energy management and information systems.

Methodologies & Techniques: Depending on the special concentration in which the professional is to work on completion of the course. A major portion of the course would be devoted to on-the-job training at an on-site location.

who would be expected to be participants in the short executive seminars. For many of these countries, the seminars are best organized at the regional level (South America, the Caribbean, Central America, South East Asia, the Middle East, West Africa, and East Africa) where joint and common interests in energy-economic policy and planning are best emphasized. There are, in addition, some six to eight individuals within the Dominican Republic who would be expected to attend courses like the EMTP. These are the individuals who would attend the EMTP, or other programs like it, over the next decade. The number of professionals required to support energy audits, make field surveys in the rural energy system, and monitor price impacts would be considerably larger. In the Dominican Republic, for example, we estimate the staff needs at 150 persons over the decade. The academic graduate-level training, which includes large numbers, lies partly outside the realm of the specialized energy-planning and management activities. We will not include this area in energy-training estimates here. We estimate that a total of 160–170 persons from the Dominican Republic should be coming into all the programs cited over the next decade. At present costs for such programs, which are in the range of $5000 per participant, the training cost

Table 30.2. Mechanisms for Energy Management Training

Level I:	Senior officials
Mechanism A:	Two-week executive seminars are held in the country in question, organized jointly by the appropriate in-country energy unit and an outside group. Lecturers come from the country itself, other Third World countries, international agencies, and advanced industrial countries.
Mechanism B:	Key officials are brought to an international or national energy management center from a number of countries sharing similiar energy situations for a period of several weeks. Center organizes a series of these for interested countries. Lecturers are selected on the same basis as above.
Level II:	Middle and upper level staff
Mechanism A:	The Energy Management Training Program (EMTP) as outlined in the first section of this paper.
Mechanism B:	A regional version of the EMTP, organized by an institution with ongoing activities involving energy management. Were several regional versions of the EMTP to be organized, the institutions involved might be organized into a consortium for the purpose of agreeing on a curriculum and sharing the course and case-study development efforts.
Mechanism C:	Workshops of 6–8 weeks would be organized on some of the special topics listed. The workshops would be held at a central location. Lecturers would be recruited from national and international agencies, universities, consultants, and private firms.
Mechanism D:	On-the-job training of individuals for periods of several months at national and international institutions, private firms, universities. Particular attention would be paid to finding locations for on-the-job training in developing countries.
Level III:	Junior staff
Mechanism A;	Outstanding Third World universities and technical institutes would receive support to develop special graduate programs dealing with some aspect of energy management. A few of these programs are already in operation. If receptive to taking additional students from other countries, they would receive support to diversify their curricula. In terms of needs, these graduate programs should place their emphasis on Master's rather than Ph.D. programs.
Mechanism B:	Universities and other institutions in the advanced industrial countries wishing to develop special programs to meet the needs of this audience would receive support to do so.
Mechanism C:	Existing graduate programs both in the developing countries and in the advanced industrial countries, in the contributing disciplines enumerated in table 30.1, would receive support to develop new courses to meet the needs of this audience. Although they would tend to be less topical, these graduate programs would probably be more rigorous in a disciplinary sense.
Level IV:	Professional staff
Mechanism A:	Outstanding Third World institutions interested in taking on a long-term obligation to train energy management professionals would receive long-term institutional support. This support would be used to develop special faculties

Table 30.2. (Continued)

	for this kind of training, laboratory and field facilities, and a staff to administer the programs.
Mechanism B:	Government units in Third World countries involved in ongoing programs covering one or more of the areas specified in table 30.1 would be supported to take on-the-job trainees from their own and nearby countries.
Mechanism C:	Outside consultant groups would be retained to post staff in interested developing countries for extended periods to undertake the training of these energy management professionals. These consultant groups would be required to work with an in-country university which would be expected to take over responsibility for the training program at the end of the contracted period.
Mechanism D:	Regional centers would be set up specially to train energy management professionals. During their start-up, these centers would be associated with institutions in the advanced developing countries with experience in training these kinds of energy management professionals.

would be about $800,000 for the decade, or $80,000 per year. Dominican Republic oil import costs are presently $150 million per year. If therefore, as a result of attendance at energy planning and management courses and seminars, the individuals were to be instrumental in reducing oil consumption by even less than one percent, they would have paid for their training expenses. The point is simply that the cost of developing a well-trained staff for energy sector investment planning is well within the budget capability of even smaller developing countries, while at the same time the benefits of better energy management, measured by any reasonable standard, are going to be enormous.

In the small industrial states, more extensive planning efforts are underway. Thailand, for example, has a National Energy Administration staff of about 150 persons, many of whom should be expected to participate in some kind of midcareer energy management training program during the decade ahead. Additional staff dealing with energy matters in other ministries raises this to 400. A rough training cost over the decade is $2.5 million, or $250,000 per year. A similar situation exists in Korea. Again, compared to the cost of oil imports, the training costs in both countries are small. Korea and Thailand have in recent years reflected this high payoff by seeking to send large numbers of individuals to programs such as our own and graduate programs at a variety of other institutions in the United States.

The energy training needs in Brazil, India, and in a few years China, are apt to be similar to those of smaller industrial states. Brazil already sends large numbers of governmental agency and energy research institute personnel to the United States and Europe for advanced training in the energy area. In-country professional training is beginning to evolve at some of the university centers. With oil imports at $12 billion, there are clear and strong incen-

Table 30.3. Estimated Training Need

	Level I Senior Officials—Executive Seminars	Level II Middle and Upper Management Staff—Energy Management Training Program	Level IV* Professional Staff—In-country or Regional Training Centers
Large industrial (Brazil, India China)	One seminar per country each year—50 persons/country	200–300 persons/country	several thousand
Small industrial (Thailand, Korea)	One seminar per country each two years—50 persons/country	30–50 persons/country	several thousand
Small primary exporter (Dominican Republic, etc.)	Eight seminars every two years—3 persons/country	6 persons/country	less than 10,000

* About 3 persons per million population in small primary exporter countries; 10 persons per million population in industrializing countries.

tives to establish a governmental capability to perform the broad range of energy management, planning, and monitoring tasks. Moreover, this planning capability is being complemented by growing and specialization in energy resource technologies of special import—oil shale, coal, and alcohol in Brazil.

It is difficult to extrapolate to worldwide non-OPEC developing country demands for energy-related training over the next decade, but a crude estimate puts the total at perhaps 100 executive seminars (level I), 2000 persons entering energy management training programs (level II), and perhaps 10,000 persons entering professional staff training (level IV) as shown in table 30.3. From this, it must be obvious that the industrialized world will not be the fundamental source of training activity. The development of the variety of Third World institutional settings described earlier must be viewed as the basis for long-term energy planning and management programs.

Chapter 31
Energy and Development: An Overview of Selected Issues

Charles R. Blitzer

INTRODUCTION

There is growing recognition of the importance of the energy issues confronting the oil-importing developing countries. There are two fundamental concerns. First, that many developing countries will be unable to adjust successfully to higher energy costs. In the short run, there are uncertainties about recycling OPEC financial surpluses and the impact of global recession on LDC growth rates. Long-run issues include supply uncertainty and the effect of rising energy prices on investment and growth rates and general development prospects.

The second concern is the impact of growing LDC energy demand on the global oil supply/demand balance. Developing countries already use about one-sixth of the world's oil and are likely to use close to one-quarter by 1990.[1] While it is not possible to estimate accurately the effect these additional demands might have on future oil prices, their impact almost certainly will be significant. Moreover, developing countries can help themselves and ease future oil-market pressures by rapidly exploiting their indigenous energy resources.

Only recently have planners, researchers, and policy-makers begun to address these problems seriously. If the world economy is to deal effectively with the energy challenges of the 1980s, a way must be found to ease the

---

I would like to thank Jane Barber of IDCA and Cathy Gleason of AID for their assistance in preparation of this paper. I am also grateful for the many comments offered by the conference participants to the oral version presented them. The views expressed are entirely my own and do not reflect those of IDCA or the U.S. government.

development constraints imposed by rising energy costs, while avoiding upsetting the global energy balance. This is one area in which both OPEC and OECD countries agree with the developing countries on the need for effective action. The last two economic summits, the long-run strategy committee of OPEC, declarations of the G–77, and the Brandt Commission have made pleas for increased assistance to help the oil-importing countries deal with their energy problems.

Public statements aside, relatively little new has emerged to date. There are disagreements between donors as to the scope of the problem, how it should be addressed, and who should pay what shares. The LDCs themselves have not been clear about their priorities, either in terms of macro issues related to finance and world trade, or micro issues involving energy assistance more narrowly defined. Given the recent nature of the problem and its recognition, this is not necessarily a cause for despair. There is still much to learn about the nature of specific problems before agreeing on specific solutions. Better analysis can help build the political will to take new multilateral initiatives.

In this spirit, I will review the dimensions of three important, and underresearched, LDC energy issues—the relationship between energy prices and development, the rate of oil/gas exploration, and the substitution of coal for oil imports. My purpose is not to provide answers, but to focus issues and perhaps stimulate additional needed analysis.

ENERGY AND DEVELOPMENT

There has been much rhetoric about the impact of higher energy prices on LDC growth prospects. There are claims that the LDCs, especially the poorer ones, have been the hardest hit by post–1973 oil prices. There have also been counterclaims. Some of the specifics have been well discussed in other papers at this conference, from which two conclusions seem to emerge. First, the increased oil bill directly, and reduced OECD growth indirectly, have led to slower LDC growth. Second, the LDCs, with only a few notable exceptions, weathered the 1973–74 storm far better than most experts had anticipated.

In this section a simple dual economy model is used to attempt to put some limits on the long-run relationship between rising energy prices and future growth prospects. But before discussing the model, it may be useful to review briefly the wider dimensions of energy/development interactions.

As in the industrialized countries, the cost of imported oil is the most severe immediate energy problem facing the developing countries. It is estimated that the oil-importing developing countries will import about 5 million b/d in 1980 at a cost of $65–70 billion. The increase in the past two years, in excess of $30 billion, outweighs the entire OECD development assistance effort. Oil imports as a share of export revenues are likely to exceed one-third in 1980, in comparison with less than one-fifth in 1978 and less than one-tenth in 1973. With each OPEC price increase, oil-importing LDCs are forced to choose between reducing development expenditures, cutting essential social services, or going deeper into debt.

The rise in petroleum prices is a serious impediment to increased food and agricultural production, particularly where increases in production depend on petroleum-based transportation and irrigation systems and chemical fertilizers. At the same time that commercial energy resources are increasing in price, the world's tropical forests are disappearing rapidly with alarming energy implications for the rural poor of the developing world. The ever more demanding search for firewood diverts villages from agricultural tasks and denudes the landscape, causing widespread soil erosion and additional food production problems.

The indirect impact of higher energy prices on developing countries through economic relations with the industrialized world has been equally severe. Stagflation in the OECD economies has put a damper on the markets for LDC exports, both primary commodities and manufactured exports. Not only has demand been growing more slowly than previously, but protectionism against manufactured exports is on the rise. For developing countries as a whole, terms of trade have seriously worsened, reducing the potential for increased development through international trade.

Despite considerable efforts, net oil imports of the non-OPEC countries will increase substantially in the 1980s as will their need for rapid export expansion as well as continued deficit financing. Most developing countries are going through an energy-intensive phase of growth similar to that experienced by Western countries in the hundred years or so before World War II. Consumption in developing countries is projected to grow faster than in developed countries because of the expected increases in the shares of industrial output in total output and in the urban population share, which generally tend to be more energy-intensive. Even if conservation and energy-efficiency are actively pursued, LDC demand for commercial forms of energy is, therefore, likely to keep pace with, and even exceed, GNP growth rates.

The noncommercial energy problems are also likely to become more acute. Growing populations and rising prices of kerosene will keep demand for firewood and biologic wastes high, further damaging the natural ecology. The obstacles to rapid reversal are formidable, and the problems are not clearly understood. New and decentralized institutions must be created to provide effective land management, but the financial costs are high. The Food and Agriculture Organization (FAO) has estimated the cost of replacing the industrial lumber and firewood cut each year in the developing countries at more than $3 billion. Only a small fraction of that is being invested today.

The principal medium-run commercial energy problem is a projected shortage of petroleum during the second half of the 1980s. Even if the OPEC countries were to produce at their full projected capacities—an unlikely assumption—the imbalance could amount to 10–13 million barrels per day by 1990.

Underlying this possibility are the following important assumptions:

- Oil production in the industrialized countries will peak early in the 1980s and decline afterward.
- The industrialized countries will be able to reduce by about 30 percent the incremental energy-intensity of their economies through a variety of announced conservation measures.
- Increased fuel exports from China will not be able to overcome reduced oil exports of the Soviet Union, at least until after 1990.
- The lack of basic physical and technological knowledge, combined with long lead-times for investment, will limit the quantitative significance of new energy sources, including renewable energy, in developing countries for at least another 10–15 years.
- Demand for commercial forms of energy will continue to grow relatively rapidly in developing countries, as a consequence of industrialization and urbanization.

This picture raises some serious concerns. Instability, in terms of limited supply bottlenecks and rapidly rising oil prices or a combination of both, can easily occur. The high dependence on foreign sources of energy supply leads to concerns about the issue of access to energy in the required volumes at reasonable prices and on a continuous, uninterrupted basis. As the residual supplier of petroleum, which is the critical balancing factor in global energy demand/supply balances, the rate of production of OPEC countries is very important. Changes in their production, either because of political disturbances or a deliberate decision to conserve oil resources, could alter the world demand/supply balances.

All the major groups of countries have their own problems of transition. For the developing countries, the main issues are exploration and development of domestic commercial resources, more efficient use of commercial and traditional energy sources, and the need to adjust to a higher energy price situation. Virtually all developing countries share these concerns and have taken steps to manage their energy sectors more efficiently. These steps range from increasing prices, to expanding investment in domestic energy sources, and organizing energy ministries and planning boards. But rapid development of their energy sectors is limited by

- lack of adequate economic incentives
- lack of basic knowledge of their own resources
- lack of appropriate technology
- lack of financial resources
- lack of adequately trained personnel

There has been little systematic analysis of the quantitative impact of any of these issues on broadly defined development prospects. Not only is there a lack of relevant data, but there are no tested methodologies for incorporating the variety of energy issues into one economy-wide framework. Most econometric work has focused on demand forecasting not on the relation between growth and energy usage. The model discussed below represents a partial attempt to develop a simple comprehensive framework. Stylized facts are freely used to show the magnitude of some aspects of the energy and development problems for a typical LDC.

The model itself focuses on the impact of the price of imported energy on development. Specifically, a dual economy framework is used to examine the effects on rural and urban growth and employment, sectoral income distribution, investment, and GNP. Alternative time paths of the endogenous variables are generated, using different assumptions about future energy prices.

There are two sectors—rural (A) and urban (M)—in which production requires inputs of labor, capital, and energy.

$$X_i = X_i(L_i, K_i, E_i), \qquad i=A,M, \tag{1}$$

where X, L, K, and E represent output and inputs of labor, capital, and energy.[2] In the short run, capital stocks are fixed.

Energy inputs are chosen efficiently, so that the marginal product of energy in each sector equals its exogenous price.

$$X_{i,E} = P_E \qquad , i = A, M \qquad , \tag{2}$$

where $X_{i,E}$ stands for the marginal product of energy in sector i and P_E is the price of energy

Labor is treated in the dual economy tradition—exogenous in rural sectors and hired in urban sectors. This simplification implies that only in the urban sector is labor's marginal product equated with its wage.

(3) $$X_{M,L} = w ,$$

where $X_{M,L}$ is labor's marginal product in the urban sector and w is the wage rate.

These equations, (1)–(3), can be solved for the X's, E's, L's, and GNP once a wage and energy price are specified. The next step is to trace the distribution of this income and its expenditure. In the rural sector, employment is fixed in the short run, but the average income level is endogenous. Letting Y_A stand for average rural income, we have

(4) $$Y_A = \frac{X_A - P_E E_A}{L_A}$$

In the urban sector, the wage is exogenous, but the unemployment rate, U, is endogenous.

(5) $$U = \frac{L_u - L_M}{L_u} ,$$

when L_u represents total urban available work force.

To derive consumption and investment, labor income is divided into savings and consumption. Nonlabor income is divided into private and government components, the former being added to private savings and the latter to public consumption and investment in fixed proportions. These variables determine endogenously migration and sectoral investments, which in turn determine the following year's initial conditions.[3]

Table 31.1 illustrates how this simple model can be used. Three cases are examined: In case I, energy prices remain constant; in Case II, energy prices rise at 2 percent per year; and in Case III, energy prices rise at 7 percent per year. For each of these cases, average growth rates are calculated over a 15-year period. To avoid overoptimism, it is assumed that net foreign borrowing will not adjust to higher energy prices and that the entire adjustment burden must be born domestically.

The parameters were chosen to reflect a reasonable initial division between the two sectors and a potential for healthy growth, 5.5 percent annually, if real prices remained constant (Case I). The implied income elasticity of aggregate energy demand is about 1.2 and reflects the more rapid growth of the urban sector which uses energy more intensively. Increased urban employment affords room for modest per capita growth in rural income levels.[4]

By contrast, when energy costs double within 10 years (Case III), the growth rate is reduced by about 30 percent. That is considerable, but not necessarily catastrophic. But it appears that the impact of this reduction is likely to fall disproportionately on the rural sector, where growth in income

Table 31.1. Energy and Development: Results of Stylized Dual Economy Model (units: average annual growth rates, %)

	Case I	Case II	Case III
Output			
Agriculture	2.9	2.8	2.4
Urban	7.4	7.0	6.0
GNP	5.5	5.1	3.9
Energy			
Volume	6.8	5.4	2.7
Total cost	6.8	7.4	9.7
Employment & Income			
Rural income, per capita	1.4	1.2	0.9
Urban employment	6.1	5.6	4.8
Expenditures			
Consumption	4.4	4.1	3.5
Investment	7.7	7.0	4.4

Case I : constant energy prices.
Case II : energy prices rise at 2% per year.
Case III: energy prices rise at 7% per year

levels falls by more than one-third. This is caused less by the reduced agricultural output than by slower employment growth in the urban sector.

In terms of energy, the growth rate of demand falls by 60 percent, implying a long-run price elasticity of about 20–30 percent. However, despite slower demand growth, the total energy bill rises considerably, approaching 15 percent of GNP, or double the ratio for Case I, after 15 years.

Case II is intermediate, indicating that if properly managed, the macroeconomic impact of modest energy price increases might not be too severe. These results are meant to be illustrative. The numbers would be somewhat different if different parameters or functional forms were chosen. More significantly, a simple model such as this can only be useful as a beginning effort in setting bounds on certain relationships and in pointing out which parameters are most crucial in analyzing the impact of alternative energy scenarios.

HYDROCARBON EXPLORATION IN NON-OPEC DEVELOPING COUNTRIES

There is some evidence that at least until the end of the century, oil and gas resources have the greatest potential for increased energy supplies in the non-OPEC developing countries. In the oil-importing developing countries alone, oil production could go from 1.25 million b/d in 1976 to perhaps 3.80 million b/d by 1990 and 4.75 million b/d or more by 2000. There is also some probability of major new finds in explored regions; these might amount to perhaps an additional 4 million b/d by 2000.

Accelerated development of these potential resources would not only ease the development constraints caused by reliance on imported oil, but would also reduce future market pressures and have an indirect effect on the balance of payments situation of all oil-importing countries.

At present, the non-OPEC developing countries account for a small proportion of total world oil and gas production. Despite occupying large land masses (including offshore continental shelves), these countries produce only 11 percent of global oil. About one-half of this is from China and Mexico. The remainder, about 3.5 million barrels per day is produced in more than 20 countries, about half of which are minor exporters. Other than China and Mexico, the largest producers are Malaysia, Egypt, and Argentina, each of whom produces about .5 million b/d.

The long-run oil supply prospects of the non-OPEC developing countries are murky. Future production levels are entirely dependent on reserves. At present, the non-OPEC developing countries account for about 4 percent of proven global reserves, or about 25 billion barrels (see table 31.2). However, there is considerable uncertainty and disagreement about the outlook for finding new reserves. On the positive side, it is evident that in contrast with other regions, these countries have been less intensively explored. Some geologists estimate that up to 40 percent of future discoveries will be in non-OPEC developing countries. On the negative side, most experts seem to believe that total future discoveries worldwide will be only on the order of 1,000 billion barrels, less than 40 years' supply at current rates of consumption. The majority opinion also suggests that the bulk of future discoveries will be in areas where sizable reserves have already been discovered and in areas of difficult access (arctic and deep offshore).

This debate is not just of academic interest, but has very important social, economic, and public policy implications. Energy planning to a considerable extent is a long-run exercise. The lead times of investments in nuclear power, hydroelectric power, and fossil fuel fields range from 5 to 15 years. In the case of power generation, capital facilities can last many decades. The time perspective for research and development of new energy technologies is considerably longer. Rational planning of long-run energy investments is sensitive to estimates of how long oil will be available, in what amounts, and at what cost. Therefore, reducing the uncertainty about future reserves can have important benefits in improved energy planning.

While it is obvious that future reserves and discoveries will have an important effect on future energy prices, it is becoming clear that current perceptions of future reserves also have an impact on near-term price formation. Current theories of cartel behavior and natural resource economics, along with several specific analyses of world oil markets, indicate that expectations

Table 31.2. World Oil and Gas Reserves and Exploratory Wells

	Oil Reserves		Gas Reserves		Total Wells Drilled in 1978	1978 Wildcat Wells	1978 Discoveries	1978 Discovery Rate (Column 5/ Column 4)
	(Billion Barrels) January 1, 1980	% World	(10^{12}Cu Ft) January 1, 1980	% World				
Industrialized western countries	59.4	9%	452.6	18%	56,505	14,249	4,505	.32
U.S.	26.5	4	194.0	77	48,513	10,677	2,728	.26
Canada	6.8	1	85.5	3	7,170	3,144	1,711	.54
Western Europe	23.8	4	135.4	5	659	297	63	.21
Others	2.3	.4	37.7	2	163	131	33	.25
OPEC countries	436.6	68%	959.9	39%	1,970	362	90	.25
Saudi Arabia	163.3	25	93.2	4	145	25	3	.12
Iran	58.0	9	490.0	19	180	10	0	0
Others	214.2	33	412.7	16	1,645	327	87	.27
Communist countries	90.0	14%	935.0	36%	NA	NA	NA	NA
USSR	67.0	10	900.0	34	NA	NA	NA	NA
China	20.0	3	25.0	1	NA	NA	NA	NA
Others	3.0	.5	10.0	.4	NA	NA	NA	NA
Non-OPEC developing countries	56.7	9%	189.7	7%	2,647	637	160	.25
Exporters	45.8	7	99.2	4	797	211	64	.30
Mexico	31.3	5	59.0	2	306	83	23	.28
Egypt	3.1	.5	3.0	.1	64	24	9	.37
Malaysia/Brunei	4.6	.7	24.7	1	123	23	3	.13
Others	6.9	1	12.5	.5	304	81	29	.36
Importers	10.8	2	90.6	3	1,850	426	96	.22
Argentina	2.4	.4	15.2	.6	865	81	30	.37
Brazil	1.2	.2	1.5	.06	256	89	26	.29
India	2.6	.4	9.3	.4	119	66	0	0
Others	4.6	.7	64.6	2	610	190	40	.21
Total World	641.6	100%	2,573.2	100%	61,122	15,248	4,755	.31

Source: *Oil and Gas Journal*, December 31, 1979.

about reserves, and their location, are important factors in determining the equilibrium rate of increase of oil prices. While finding more oil will generally have a negative impact on price, the impact on prices of reducing the uncertainty is ambiguous in theory and not yet determined empirically.

Oil is not found in random locations, but generally in sedimentary basins or "provinces" that can be identified on the basis of general geologic knowledge. So far, about 600 basins have been identified as having oil potential. In roughly 400 of these basins, exploratory drilling has been done in varying amounts and with varying degrees of success.

About 160 of the basins in this group are now capable of producing commercial quantities of oil and gas. A very large amount of oil is judged to be yet discoverable in these 160 basins where the existence of oil has already been demonstrated. A World Energy Conference poll indicated that about 60 percent of the remaining discoveries were likely to be made in these already productive basins.

The other 240 partly explored basins have been drilled without yielding commercial discoveries to date. Statistically, perhaps 40 of these basins might eventually be found productive.

The remaining 200 basins have had essentially no drilling as yet in most cases, because they are in harsh physical environments, or because market restraints have restricted access to them. Perhaps half of these may eventually prove productive. The potential of these basins is the most speculative part of the assessment, and the furthest in the future. The underlying basis for the pessimistic view of future resources is the observation that most of the oil found so far has been in a few super-giant fields and the assumption that there are few of these which remain to be found. This view has been forcefully argued by Nehring (1978) in a study conducted for the CIA and by Exxon. They base their pessimism on the following points:

- Super-giant fields have always been discovered relatively early in the exploration of a province.
- Giant fields have generally been found in large "traps" which are detectable by surface techniques which are quite advanced. Hence less drilling is needed in the exploration process than in the past.
- Within each basin, it is often necessary to drill only a few wildcat wells to determine the largest oil-bearing structures.
- From past experience, we would be unusually lucky to discover as much oil in the 200 unexplored basins as in the 64 basins with proved giant fields. Nehring offers a complicated geologic theory to explain the apparent bad luck that most oil exploration in the really promising regions has already occurred.

Associated with this position is the view that past oil exploration has been economically efficient in choice of location. Exxon (1978) argues that for cost and technology reasons, the most readily accessible regions were tested first, and as a result, exploration has moved progressively into ever harsher environments—the deeper offshore and more remote continental interiors—where the time to position and drill a well is much longer and the cost much higher.

The optimists rest their case mainly on a comparison of the relative intensity of drilling in prospective areas. In a series of papers, Bernardo Grossling and others have argued that to extrapolate into the longer-term outlook for oil and gas resources, the present geological distribution of known reserves is not valid. Rather, future reserves, probably will be found in closer proportion to the global distribution of potentially oil/gas-bearing basins. Odell (1980) and others explain the present high degree of concentration of proved reserves not as the result of economic efficiency, but as nothing more than a historical accident arising out of the relationship between the Great Powers during the latter part of the nineteenth century and the first half of the twentieth century.

This basic point of argument can be seen most straightforwardly in figure 31.1. The relative sizes of the regional boxes reflect the relative distribution of areas with oil and gas potential worldwide, with each full-sized dot representing 50,000 wells. Those with a very optimistic view of the prospects for non-OPEC developing countries say that if we believe, as a matter of national policy, that more oil is to be found in the United States which has already been so well explored, then surely we should not exclude the possibility of additional large reserves from the underexplored regions, especially Africa and Latin America. The more conventional, conservative viewpoint rests on the observation that most proved resources have been found in the relatively unexplored regions, mainly the Middle East and Latin America, implying a low correlation between discoveries and drilling intensity.

The evidence supports few hard conclusions about the total amount of resources which might be discovered or its geographic distribution. This caveat in mind, the following conclusions seem reasonable:

- Additional drilling in the non-OPEC developing countries almost surely would lead to some discoveries. Experts differ in their estimates of the size of discoveries and their concentration.
- It is likely that future discoveries in these countries would be at least in proportion to their existing proved reserves.
- There is some probability of finding much larger resources if extensive exploration is undertaken. How much larger and their possible locations are impossible to judge.
- The need for additional basic research and data analysis is clear-cut. Neither the U.S. government nor international organizations collect or analyze data on the oil/gas prospects of these countries on any regular or systematic basis.

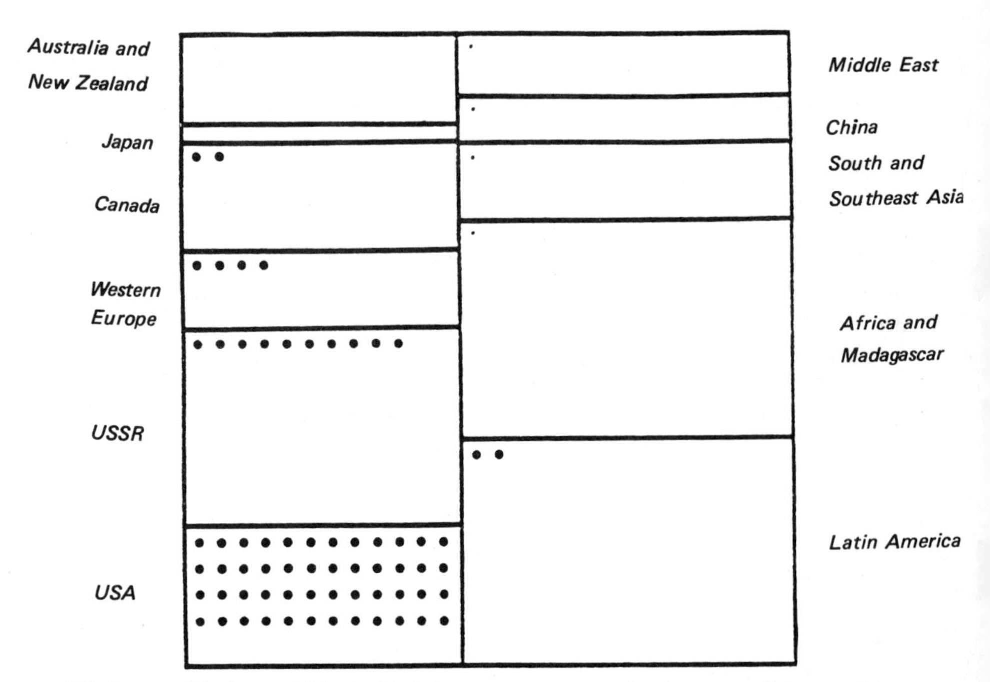

Window on Oil: Areas within the black bars represent, to scale, the extent of the petroleum prospective areas (onshore + offshore to 200 m water depth). Each full black circle represents 50,000 wells (exploratory + development) drilled in each partition of the chart. Numbers of wells smaller than 50,000 are represented approximately by segments of the full black circle.

Fig. 31.1. Relative drilling intensities.
Source: Bernardo Grossling, London Financial Times, 1976.

Only additional exploration can resolve these issues. Despite some recent upturn, only 4 percent of the wildcat wells completed in 1978 were drilled in the non-OPEC developing countries. Of the 637 wildcat wells completed that year in these countries, half were drilled in Argentina, Brazil, Mexico, and India. The situation is most severe in the three dozen or so countries which do not yet produce oil/gas, but have unexplored sedimentary basins. In 1978, a total of only 42 exploratory wells were drilled in 17 of these countries. Interestingly, some prospects were found in at least 7 of these.

Although geophysical activity, the first stage of hydrocarbon exploration, has increased rapidly worldwide, there has been a noticeable fall-off in the developing countries (see table 31.3). This trend, which is continuing, indicates that exploration drilling will not increase very much in these countries in the short term.

A variety of constraints have been identified as limiting the rate of exploration in those countries. Some are general factors, such as cost differentials, which affect many countries in similar ways. Others are country- or region-specific, such as boundary disputes or exclusionary investment regulations. However, broad categories of constraints emerge—technical, economic, institutional, and political.

The principal technical constraints are a lack of knowledge about resource potential and a lack of technical expertise at the national level in most of these countries. Much better use can be made of existing geologic and geophysical data both at the country level and for cross-country analysis. Additional geologic work will be required as a precondition for accelerated exploration.

The lack of technical expertise is perhaps an even more serious problem. In the first place, even when countries choose to rely on foreign companies, complex negotiations are required necessitating specialized skills. It is clear that at least a small cadre of local professionals is required. Many countries, especially those not producing oil or gas, do not yet have any expertise. Second, most developing countries eventually want to assume national control over the exploitation of their natural resources. Especially where national oil companies have been organized, accelerated exploration and development may progress only as fast as the human capital of local institutions.

Relative costs play an important role in determining where exploration occurs. The oil companies are quick to point out the higher costs of exploration in the developing countries as an explanation of their relative lack of drilling. Exxon has calculated that the average cost of a well in North America is less than 20 percent of the cost of a well in the third world. This is

Table 31.3. Geophysical Activity in Developed & Developing Countries, 1970–78 (Seismic party-months)

	1970	1972	1973	1974	1975	1976	1977	1978
Total	6,155	7,230	6,810	7,565	7,475	6,692	7,706	8,354
Canada	663	403	477	439	597	421	664	893
United States	2,340	3,016	2,999	3,661	3,416	3,140	4,036	4,223
Western Europe	667	608	508	546	574	638	634	624
Other developed	184	360	240	187	90	171	122	304
Dev'g countries:								
OPEC	772	1,005	894	926	938	772	757	919
Venezuela	(16)	(35)	(43)	(56)	(64)	(38)	(52)	(65)
Other exporters	526	741	643	780	760	521	529	545
Mexico	(276)	(336)	(288)	(252)	(264)	(230)	(276)	(336)
Peru	(2)	(141)	(237)	(271)	(105)	(30)	(2)	(4)
Oil importers	1,003	1,097	1,049	1,026	1,100	1,029	964	843
Producers	(855)	(835)	(826)	(842)	(879)	(872)	(783)	(709)
Nonproducers	(148)	(262)	(223)	(184)	(221)	(157)	(181)	(134)

largely because the non-OPEC developing countries lack essential infrastructure. Typically, the oil companies must develop the minimum infrastructure themselves. For example, the Congo Basin in Zaire has good potential, but there has been very little exploratory activity, in part because of difficult terrain with little existing infrastructure. Similar examples exist on all continents.

It is likely that most discoveries, especially in the oil-importing developing countries, will be in relatively small fields, with little export potential. Small discoveries, especially located far from ports, may be considered uneconomical to foreign companies. One instance of this phenomenon is Chad, where a small but significant strike was made deep in the interior of the country. Export potential may not justify construction of an expensive transportation system, but the strike almost certainly will be sufficient to provide local self-sufficiency and some regional export. Similarly, recent activity in the Sudan is limited somewhat by uncertainty about whether reserves in the southwest are sufficiently large to justify a pipeline to the Red Sea.

As development of low-cost Middle East reserves expanded after 1960, interest in underexplored regions far from Europe, Japan, or North America waned. Economics dictated that production be from very cheap sources or near to centers of demand. Immediately after the 1973–74 price rise, many expected exploration in non-OPEC LDCs to increase dramatically. The World Bank and others felt that the price incentive would be sufficient to produce an exploration and development boom without any official financial support. Obviously, this has not occurred.

The straightforward explanation is that the investment climate was chilled. Oil companies and host governments, having somewhat different objectives, became more distrustful of each other. Negotiations have become so difficult or uncertain that many potential investors and explorers have shied away. The reasons are both general and country-specific. To get the flavor of the principal general considerations, we look separately at the problems from the perspective of the companies and the host governments.

Compared with most industries, investments in oil exploration are unusually risky. The chances that any one drilling program will be successful and lead to commercial development normally are high only near existing discoveries. While on a basin-wide basis, history indicates that about half will contain commercial reserves, the odds are much longer on successful wildcatting of individual smaller tracts of land in relatively unexplored basins. Generally, oil companies assume the risks of an unsuccessful exploration in return for acquiring some specified rights to whatever oil or gas is later produced. To make such an investment attractive, the profits from the few discoveries must be sufficient to cover the full costs of dry holes elsewhere. In other words, under a production-sharing agreement with a foreign company in return for the company's assuming the direct exploration risk, the country assumes an upside risk of paying high rates of return to its foreign partners, when commercial reserves are discovered.

Over the last decade, major new trends have appeared in the world oil industry. Host governments no longer act as passive partners in the oil development process. Most countries with oil production or resource potential have sought to increase control over the development and disposition of their oil resources. This new approach by host governments is often in conflict with the objectives of the private international oil companies and those of the oil-consuming countries.

The oil companies have, for the most part, lobbied against governmental activity in exploration and development. They have argued that current activity is consistent with prudent profit maximization. The two main themes they most often recite to explain their relative lack of interest in additional exploration in developing countries are the following:

- Geologic prospects are not very great, considering cost factors and the probable size of discoveries.
- Developing countries are difficult business partners with whom the rules of the game constantly change.

Regarding geologic prospects, as discussed earlier, the data can be interpreted in different ways. With the growth of national oil companies, multinational oil companies are less interested in small deposits with minimal export potential. Partnership with a national oil company implies loss of direct control over operations and the added difficulties of training and technical assistance, all of which reduce profit rates. As long as the exploration firms in the LDCs remain companies principally in the development, refining and distribution business, some bias against looking hard for medium-sized deposits is likely to continue.

The oil companies are correct in pointing to increased political risks as a major barrier to be overcome to accelerate exploration. Widescale nationalizations, the oil-price revolution, and calls for a new international economic order began to happen almost simultaneously. With the Western governments no longer able to enforce contracts for them, the companies have protected their interests by seeking higher returns in more geologically certain regions. Level of interest has been proportional to the perceived stability of governments and their friendliness to conventional business arrangements.

Somewhat surprisingly, the LDCs have said very little explicitly about oil exploration. Rather, their concerns have been included in their somewhat wider condemnation of existing economic institutions and demands for a new international economic order. Rhetorical extremes aside, the issues the LDCs seem to raise include:

- The oil companies ask too high a price, implicitly forcing them to assume an unfair upside risk burden.
- They are at a disadvantage when bargaining about production-sharing before exploration occurs.
- The pace of technology transfer in all its dimensions is too slow, forcing them into increased dependency.

The factors have led the oil-importing LDCs, as a group, to adopt a go-slow attitude in making exploration/development deals. Until they feel they are getting an equitable deal from the foreign oil companies, the LDCs may continue pinning their hopes on the development of national oil companies even though it will take many years to develop exploration capabilities. Thus accelerated exploration will require that host countries be offered more attractive terms (financial, training, management, etc.).

Developing countries see the oil companies as part of and not separate from the industrialized countries. This has important implications for their view of risk distribution. The oil companies earn high rates of return on successful explorations in order to cover the costs of failures incurred elsewhere and to provide a "bonus" for risk-taking. But from a single country's viewpoint, the LDC where oil is found sees itself as paying for the dry holes drilled in another country. Since the industrialized countries benefit considerably from oil found and produced anywhere, it is often felt that these countries should assume a portion of the inherent geological risks.

Moreover, it seems considerably easier to strike a deal *after* some drilling has already occurred. The LDCs apparently are more comfortable about negotiating once the geologic uncertainty is partially removed. Presently, countries which wish acreage to be explored must either assume all the risk themselves or make a long-term production-sharing deal with a foreign oil company.

Quite clearly, a vicious circle exists which feeds on itself. LDC reluctance to do business as usual only reinforces the uncertainties of the oil companies, driving up their terms, which in turn leads to more distrust by the host governments. For exploration to accelerate, efforts will have to be made to break the circle in one or more places. More widespread and in-depth seismology, together with expected future oil prices, will determine the attractiveness of various locations. However, new innovative institutional arrangements must emerge for the investment climate to thaw significantly.

There are definite signs of improvement in some countries. China has successfully attracted much of the world's seismologic talent to conduct offshore surveys. Argentina and Brazil have liberalized their policies toward private foreign investment in oil and gas exploration, the result being an upturn in oil company interest in these countries. Even India seems to be considering opening offshore acreage for foreign exploration. Several countries are experimenting with new model contracts which they believe will lay out more clearly what is expected of both sides.

As the energy situation in the oil-importing developing countries continued to worsen after 1973, with little or no signs of large-scale private investment in oil exploration or development being made in these countries, the question arose of what, if any, response should be made by the development assistance agencies. The issue was most focused at the World Bank, the largest multilateral lending institutions. At the urging of executive directors from several donor countries and bank staff, management agreed to look closely at what role the bank could play.

In July of 1977 the World Bank expanded its lending activities to include oil and gas production. In approving this expansion, its board stressed that the bank should serve as a catalyst for mobilizing increased private investments and should emphasize technical assistance for improving national energy planning and knowledge regarding energy resources.

In January 1979, the board reviewed its experience under the program and approved expansion to include financing for exploration. Under this program, the bank has identified approximately 60 LDCs that would benefit from expanded exploration and has budgeted approximately $1 billion (in 1978 dollars) for lending to exploration projects over the fiscal years 1980–83. The bank estimates that 24 oil and gas projects—both exploration and development—totaling $814 million are likely to be presented to the board in the next two years. Of these, about 40 percent should be for exploration. A much larger fraction of the program is aimed at the development of known hydrocarbon resources.

The principal weakness of this program from the point of view of stimulating exploration is that the repayment of the loans is not contingent upon the eventual development of commercial discoveries but rather on the country's general creditworthiness. A borrowing country assumes the downside risks itself. Except for more favorable terms, this is not much better than financing exploration at commercial rates on the government's full faith and credit. While the program is too new to make definite judgments, exploration has so far focused primarily on the most promising regions of oil-importing countries already producing oil.

Some governments offer political risk insurance for oil/gas projects in developing countries. In the United States, for example, the Overseas Private Development Corporation (OPDC) has insured a number of such projects in recent years, but it is not obvious that these insurance policies can play a decisive role in stimulating exploration investment. The insurance provided covers only the value of physical capital, less remitted earnings. For a successful exploration project, the future profits are in the ground and not in physical assets. So in fact, not much protection can be offered. To be effective coverage would probably have to be extended well beyond direct capital investment.

One idea which has won increasing attention is the creation of a new multilateral facility to finance exploration. The new facility could finance exploration through lending and equity sharing. Unlike the loans now available, the fund would only require repayment if commercial discoveries were made. Such an arrangement would be attractive to many developing countries if the repayment terms were not onerous. Host governments would not welcome a scheme which required repayments much in excess of the costs actually incurred in their territory, except perhaps if very large reserves were found. If the repayments from countries with successful exploration programs fully covered the costs of unsuccessful drilling elsewhere, the fund would completely "revolve," and subscribers could expect to have their risk capital returned eventually. To the extent that repayments fell short of this mark, the subscribers would be sharing with the developing countries the cost of exploration drilling in countries where no commercial discoveries were made.

Equity ventures could work much the same way, with the facility entering into partnership with host governments and oil companies. Here, revolving would work through selling off equity interest in case of discoveries.

For all the reasons discussed earlier, a new exploration facility should choose as its principal target of opportunity those countries with promising geology but little production so far. There may be 25–40 such countries. Additional exploration expenditures might be on the order of $1–2 billion per year for a 10- to 15-year period, as part of an objective to double the pace of exploration in the non-OPEC LDCs and triple the rate in the oil-importing countries as rapidly as possible.

It is likely that public expenditures for this purpose by the OECD countries could be justified on narrow economic grounds alone, leaving aside the direct benefits to LDCs. An exploration program of this magnitude is quite likely to discover sufficient reserves to produce at least an extra 1 million b/d by the mid-1990s. Even if this leads to an increase of only 0.5 percent in global supply, the price elasticity is low enough to imply savings of at least $0.20 per barrel in terms of future price levels.

For the United States alone, that should represent annual savings, at 1980 prices, of about $500 million. If the costs were widely shared by OPEC and other OECD countries, the annual net costs to the United States might be only $250 million or less.[5]

This may be one North/South issue on which agreement can be reached which serves the interest of all parties. Exploration and development are very expensive activities, almost necessitating a multilateral approach to concessional financing. Both OPEC (strongly and explicitly) and OECD (Tentatively and obliquely) have been proposing establishing a risk-taking new facility if funds from other donors could be obtained. The upcoming UN-conducted global negotiations offer an opportunity to see if all parties can agree to new initiatives in developing Third World energy resources.

COAL-FOR-OIL SUBSTITUTION

During the next decades, coal can play a key role in alleviating the energy problems of many developing countries. For some of these countries, coal is a significant indigenous resource, and its production should be expanded as rapidly as possible. But the majority of developing countries have no known coal resources, and for them, the relevant issue is whether to substitute imported coal for imported oil in the generation of electricity and industrial heat.

Existing forecasts indicate that these countries are likely to install new thermal generating capacity in the 1980s using about 1 million b/d of oil. At current and projected prices, coal-fired generation offers considerable opportunities for cost savings, even when all environmental protection measures are taken. Moreover, unlike oil, it is possible to negotiate long-run contracts for coal which provide reasonable supply and price security. Very few noncoal producing Third World nations have thought seriously about coal imports. The three main constraints are a lack of technical experience with coal, the high cost of the required initial infrastructure, and —most important—an inability to finance long-term contracts with outside suppliers of coal.

U.S. coal reserves are more than ample to meet any foreseeable increase in demand from the LDCs, perhaps 50–75 million tons annually by 1990. The U.S. government is already committed to supporting a major coal export programim aimed at the IEA countries; it should not be technically difficult to increase the export capacity to take account of new LDC exports.

Exporting coal and industrial equipment associated with coal to developing countries could become a major element in U.S. export promotion efforts. The target of 50–75 million tons in annual exports to developing countries by 1990 represents, at current prices, $2–3 billion in exports. Coal exports to developing countries only amounted to $219 million in 1978. Similar potential exists for export of industrial equipment associated with coal—electricity generating plants, boilers, and pollution control equipment. This market could grow to $5 billion or more annually.

To bring about a rapid expansion of U.S. coal exports to these countries, the United States could consider undertaking a comprehensive initiative bringing together our technical, financial, and development assistance capabilities. A comprehensive package would include elements aimed at each of the major constraints, namely:

- bilateral technical assistance provided through AID and DOE
- multilateral development bank financing for coal-using equipment and related infrastructure
- U.S. government export finance support for coal and coal-related industrial equipment

Such a program would have a number of attractive features. It would ease the balance of payments problems of many developing countries by taking advantage of coal's growing cost advantage over oil and would act to promote the exports of U.S.-manufactured equipment and raw materials. The provision of financing for American equipment and raw materials would add considerably to coal's attractiveness, by easing the foreign exchange bottleneck that many countries now face. More important from the viewpoint of national security and peace, such a program would offer an economically attractive alternative to nuclear power and reduce future pressures on the world oil supply, perhaps by as much as 1 million b/d by 1990.

NOTES

1. This presupposes that the world economy is healthy enough to sustain 4–5 percent growth in the developing countries and that the real price of energy less than doubles by 1990.
2. For the numerical calculations, Cobb-Douglas production functions are employed.
3. The migration function used in the numerical calculations is a linear function of the ratio of the average urban income (including the unemployed) to the average rural income. Sectoral shares of investment are exogenous and not necessarily optimal.
4. It is assumed that the reproductive rate is 3 percent per year in rural areas and 2 percent per year in urban areas.
5. This calculation is based on assumptions that the United States share would be 25 percent and that successful projects would pay for half the costs of the total exploration program.

Author Index

Subject Index

Participants and Observers

Dr. Achilles Adamantiades
Nuclear Power Division
ELECTRIC POWER RESEARCH INSTITUTE
3412 Hillview Avenue
Palo Alto, CA 94304

Dr. Marcelo Alonso
Executive Director
FLORIDA INSTITUTE OF TECHNOLOGY
Melbourne, FL 32901

Dr. John H. Ashworth
Senior Policy Analyst
Policy Analysis Branch
SOLAR ENERGY RESEARCH INSTITUTE
1617 Cole Boulevard
Golden, CO 80401

Dr. Peter Auer
Upson Hall
CORNELL UNIVERSITY
Ithaca, NY 14853

Professor John Barton
Crown Law Quad, Rm. 319
STANFORD UNIVERSITY
Palo Alto, CA 94305

Dr. Paul S. Basile
INTERNATIONAL ENERGY DEVELOPMENT CORPORATION
IEDC S.A.
18, rue le Corbusier
CH-1208 Geneva, Switzerland

Dr. Charles R. Blitzer
Sr. Advisor, Energy Planning Office
INTERNATIONAL DEVELOPMENT COOPERATION AGENCY
Washington, DC 20523

Mr. Robert Blum
1301 20th Street
Washington, DC 20036

Dr. Chaim Braun
Energy Study Center
ELECTRIC POWER RESEARCH INSTITUTE
3412 Hillview Avenue
Palo Alto, CA 94304

Professor Nathan Buras
Dept. of Operations Research
Terman Engineering Center
STANFORD UNIVERSITY
Palo Alto, CA 94305

Dr. T. Owen Carroll
Institute for Energy Research
STATE UNIVERSITY OF NEW YORK
Stony Brook, NY 11794

Dr. Vashek Cervinka
Long Range Planning Staff
Office of the Director
CALIFORNIA STATE DEPARTMENT OF FOOD AND AGRICULTURE
1220 N Street
Sacramento, CA 95814

Dr. KunMo Chung
NATIONAL SCIENCE FOUNDATION
1800 G Street NW
Washington, DC 20550

Dr. Richard Cirillo
ARGONNE NATIONAL LABORATORY
9700 South Cass
Argonne, Illinois 60439

Professor Thomas Connolly
Dept. of Mechanical Engineering
STANFORD UNIVERSITY
Palo Alto, CA 94305

Dr. Chester L. Cooper
INSTITUTE FOR ENERGY ANALYSIS
11 Dupont Circle, Suite 805
Washington, DC 20036

Dr. Robert Copaken
International Energy Affairs
U.S. DEPARTMENT OF ENERGY
7F031 Forrestal Building
Washington, DC 20585

Mr. John Crutcher
NATIONAL FOREIGN ASSESSMENT CENTER
P.O. Box 1925
Washington, DC 20013

Mr. John Day
STRATEGIES UNLIMITED
201 San Antonio Circle, Suite 205
Mountain View, CA 94040

Mr. David Deese
CSI/KSG
HARVARD UNIVERSITY
79 Boylston Street
Cambridge, MA 02138

Mr. John Douglas
360 Maclane
Palo Alto, CA 94306

Dr. Richard J. Eden
Cavendish Laboratory
UNIVERSITY OF CAMBRIDGE
Cambridge, England

Dr. Everett Ehrlich
Natural Resources and Commerce Div.
CONGRESSIONAL BUDGET OFFICE
2nd and D Streets SW
Washington, DC 20515

Dr. Fereidun Fesharaki
Resources Systems Institute
EAST-WEST CENTER
1777 East-West Road
Honolulu, Hawaii 96849

Dr. Thomas Fingar
U.S.-China Relations Program
STANFORD UNIVERSITY
Palo Alto, CA 94305

Dr. John Foster
PETRO-CANADA
306-350 Sparks Street
Ottawa, Ontario K1R 7SB
Canada

Professor Dermot Gately
Terman Engineering Center, 432A
STANFORD UNIVERSITY
Palo Alto, CA 94305

Professor José Goldemberg
Instituto de Fisica
UNIVERSIDADE DE SÃO PAULO
Cidade Universitária, C.P. 20516
CEP 05508, São Paulo
Brasil

Dr. Lincoln Gordon
Senior Review Panel
for National Foreign Assessment
Central Intelligence Agency

Mr. Gerald Graves
Manager, Energy Industry Division
Corporate Planning Office
EXXON CORPORATION
1251 Avenue of the Americas
New York, NY 10020

Dr. Dale Gray
Energy Laboratory, Bldg. E-38, Rm. 566
MASSACHUSETTS INSTITUTE OF TECHNOLOGY
Cambridge, MA 02139

Dr. Martin Greenberger
1850 Willow Road, Apt. 3
Palo Alto, CA 94304

Professor Michael Grenon
INTERNATIONAL INSTITUTE FOR APPLIED SYSTEMS ANALYSIS
2361 Schloss Laxenburg
Austria

Mr. Nathaniel B. Guyol
135 Highland Avenue
San Rafael, CA 94901

Professor Bert Hickman
Department of Economics
STANFORD UNIVERSITY
Palo Alto, CA 94304

Professor Dr. Lutz Hoffman
Lehrstuhl für Volkwirtschaftslehre
UNIVERSITÄT REGENSBURG
84 Regensburg
West Germany

Dr. James W. Howe
SOLAR ENERGY RESEARCH INSTITUTE
1617 Cole Boulevard
Golden, CO 80401

Mrs. Helen Hughes
Director, Economic Analysis and Projections Department
WORLD BANK
1818 H Street NW
Washington, DC 20433

Mr. John Hurley
Deputy Director, BOSTID
NATIONAL ACADEMY OF SCIENCE
2101 Constitution Avenue NW
Washington, DC 20418

Mr. John R. Kiely
Executive Consultant
BECHTEL POWER CORPORATION
50 Beale Street
San Francisco, CA 94105

Mr. Ji Soo Kim
Terman Engineering Center, #593
STANFORD UNIVERSITY
Palo Alto, CA 94305

Sehun Kim
Departmental Operations Research
STANFORD UNIVERSITY
Palo Alto, CA 94308

Mr. Charles Klotz
ARGONNE NATIONAL LABORATORY
PMS, Bldg. 362
9700 South Cass
Argonne, Illinois 60439

Dr. M. V. Krishna Kumar
School of Business Administration
350 Barrows Hall
UNIVERSITY OF CALIFORNIA
Berkeley, CA 94720

Dr. Franklin Long
Henry Luce Professor of Science and Technology, Emeritus
Clark Hall
CORNELL UNIVERSITY
Ithaca, NY 14853

Professor Thomas V. Long, III
Committee on Resources and Public Policy
UNIVERSITY OF CHICAGO
Chicago, IL 60637

Professor Alan S. Manne
Department of Operations Research
STANFORD UNIVERSITY
Palo Alto, CA 94305

Professor Gil Masters
Dept. of Chemical Engineering
STANFORD UNIVERSITY
Palo Alto, CA 94305

Dr. Helio Mattar
Head, Dept. of Industrial Economics
INSTITUTO DE PESQUISAS
TECNOLÓGICAS
Cidade Universitária 05508
São Paulo C.P.7141 (CEP 01000)
Brazil

Ms. Susan Missner
International Energy Program
Energy Information Center
STANFORD UNIVERSITY
Palo Alto, CA 94305

Dr. Matthias Mors
UNIVERSITAT REGENSBURG
84 Regensburg
West Germany

Mr. John Morrison
Energy Study Center
ELECTRIC POWER RESEARCH
INSTITUTE
3412 Hillview Avenue
Palo Alto, CA 94304

Mr. Bijan Mossavar-Rahmani
THE ROCKEFELLER
FOUNDATION
1133 Avenue of the Americas
New York, NY 10036

Mr. James Mulvaney
Policy Planning Division
ELECTRIC POWER RESEARCH
INSTITUTE
3412 Hillview Avenue
Palo Alto, CA 94304

Jaime Navarro
RAND CORPORATION
1700 Main Street
Santa Monica, CA 90406

Ms. Jean Neuendorffer
Policy Analysis Branch
SOLAR ENERGY RESEARCH
INSTITUTE
1617 Cole Boulevard
Golden, CO 80401

Professor Robert Packenham
Department of Political Science
STANFORD UNIVERSITY
Palo Alto, CA 94305

Dr. Philip F. Palmedo, President
ENERGY/DEVELOPMENT
INTERNATIONAL
505 Main Street
Port Jefferson, NY 11777

Dr. Uwe Parpart
Director of Research
FUSION ENERGY FOUNDATION
888 7th Avenue, Suite 2404
New York, NY 10019

Dr. Lewis J. Perelman
JET PROPULSION LABORATORY
4800 Oak Grove Drive, Bldg. 506-316
Pasadena, CA 91103

Mr. Robert Phillips
DECISION FOCUS, INC.
1801 Page Mill Road
Palo Alto, CA 94304

Mr. Jeremy Platt
Energy Analysis and Environment Div.
ELECTRIC POWER RESEARCH
INSTITUTE
3412 Hillview Avenue
Palo Alto, CA 94304

Dr. James Plummer
Director, Energy Analysis Department
Energy Analysis and Environment Div.
ELECTRIC POWER RESEARCH INSTITUTE
3412 Hillview Avenue
Palo Alto, CA 94304

Mr. Steve Powell
Energy Modeling Forum
Terman Building
STANFORD UNIVERSITY
Palo Alto, CA 94305

Dr. Martin Prochnik
Director, Office of Energy Technology Cooperation, Rm. 4327A
U.S. DEPARTMENT OF STATE
21st and C Streets NW
Washington, DC 20520

Mr. C. Anthony Pryor
CENTER FOR INTEGRATED DEVELOPMENT
Suite 9A
777 UN Plaza
New York, NY 10017

Daniel A. Relles
Economics Department
RAND COPORATION
1700 Main Street
Santa Monica, CA 90406

Dr. Roger Revelle
Program in Science, Technology and Public Affairs, B-007
UNIVERSITY OF CALIFORNIA, SAN DIEGO
LaJolla, CA 92093

Mr. Eli B. Roth
Consulting Engineer
6800 Fleetwood Rd., Apt. 820
McLean, VA 22101

Dr. Henry Rowen
School of Business
STANFORD UNIVERSITY
Palo Alto, CA 94305

Mr. Lee Schipper
Lawrence Berkeley Laboratory
UNIVERSITY OF CALIFORNIA
Berkeley, CA 94720

Professor Richard Schuler
313 Hollister Hall
CORNELL UNIVERSITY
Ithaca, NY 14853

Mr. Sam H. Schurr
Deputy Director, Energy Study Center
ELECTRIC POWER RESEARCH INSTITUTE
3412 Hillview Avenue
Palo Alto, CA 94304

Mr. Milton Searl
Energy Study Center
ELECTRIC POWER RESEARCH INSTITUTE
3412 Hillview Avenue
Palo Alto, CA 94304

Dr. Corazon M. Siddayao
Resource Systems Institute
THE EAST-WEST CENTER
1777 East-West Road
Honolulu, Hawaii 96848

Professor Vaclav Smil
Department of Geography
UNIVERSITY OF MANITOBA
Winnipeg R3T 2N2
Canada

Dr. Chauncey Starr
Vice Chairman
ELECTRIC POWER RESEARCH INSTITUTE
3412 Hillview Avenue
Palo Alto, CA 94304

Dr. Robert A. Summers
Office of Country Energy Assessment
U.S. DEPARTMENT OF ENERGY
IA-23 MF/7 F032
1000 Independence Avenue SW
Washington, DC 20585

Dr. James Sweeney
Director, Energy Modeling Forum
Terman Building, Rm. 406
STANFORD UNIVERSITY
Palo Alto, CA 94305

Mr. Blair Swezey
Energy Study Center
ELECTRIC POWER RESEARCH INSTITUTE
3412 Hillview Avenue
Palo Alto, CA 94304

Professor Jean Waelbroeck
Center d'Économie Mathematique et d' Économétrie
UNIVERSITE LIBRE DE BRUXELLES
Avenue F.D. Roosevelt, 50
1050 Brussels
Belgium

Dr. Chris Whipple
Energy Study Center
ELECTRIC POWER RESEARCH INSTITUTE
3412 Hillview Avenue
Palo Alto, CA 94304

Dr. Mason Willrich
Vice President, Corporate Planning
PACIFIC GAS AND ELECTRIC COMPANY
77 Beale Street
San Francisco, CA 94106

Dr. Charles Wolf
Head, Economics Division
RAND CORPORATION
1700 Main Street
Santa Monica, CA 90406

Mr. Lawrence Yamron
Energy Analysis and Environment Div.
ELECTRIC POWER RESEARCH INSTITUTE
3412 Hillview Avenue
Palo Alto, CA 94304

Mr. Herbert Yim, President
ENERGY SYSTEMS INTERNATIONAL
8301 Greensboro Drive, Suite 30
McLean, Virginia, 22102